American Institute of Chemical Engineers

10th International Conference on Microreaction Technology

IMRET 10

Topical Conference at the
2008 AIChE Spring National Meeting

April 1-4, 2008
New Orleans, Louisiana, USA

Printed from e-media with permission by:

Curran Associates, Inc.
57 Morehouse Lane
Red Hook, NY 12571
www.proceedings.com

ISBN: 978-1-60560-211-0

ISBN: 978-1-60560-211-0

American Institute of Chemical Engineers
Proceedings
Three Park Avenue
New York, NY 10016-5991
Phone: 212-591-8100

www.aiche.org

American Institute of Chemical Engineers

10th International Conference on Microreaction Technology
2008

TABLE OF CONTENTS

Investigation into the preparation of synthetically useful α-aminonitriles under continuous flow

Charlotte Wiles

The Department of Chemistry, The University of Hull, Cottingham Road, Hull, HU6 7RX, UK.

Abstract: We report herein the initial evaluation of an integrated micro reactor capable of performing solution phase and heterogeneously catalyzed steps in series (Figure 1). Using this approach, two solid-supported Lewis acids are evaluated, enabling an array of α-aminonitriles to be synthesized with typical yields in excess of 99.6 % and single reactor throughputs of 17.2 to 63.6 mg hr^{-1}.

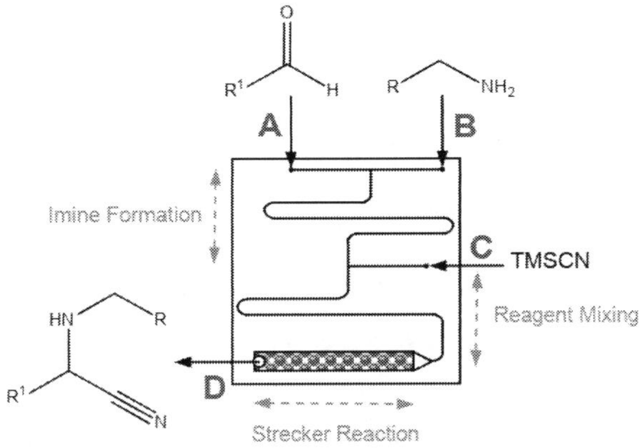

Figure 1. Schematic illustrating the reactor manifold employed for the evaluation of the multi-component Strecker reaction.[1]

Introduction: With interest in non-proteinogenic α-amino acids on the increase, there is a growing demand for novel and efficient techniques that enable the synthesis of such non-natural compounds. Although many synthetic procedures have been reported, few are suitable for the incorporation of alkyl imine substrates, often resulting in low yields and selectivities. One synthetically useful approach is the Strecker reaction,[2] which as Scheme 1 illustrates, is a three-component reaction between a carbonyl containing compound, amine and a cyanide source. The resulting α-aminonitrile can then be hydrolysed, followed by deprotection, to afford an array of non-natural amino acids, or be used as a precursor in the preparation of nitrogen containing heterocycles and 1,2-diamines.[3,4] Despite its synthetic utility, the use of expensive Lewis acid catalysts, harsh reaction conditions and somewhat variable yields preclude the use of the Strecker reaction on a production scale. In addition a major drawback of the one-pot Strecker reaction is the competing cyanohydrin formation, which is particularly problematic with respect to aromatic aldehydes; as the respective aldimine forms slowly, leaving the aldehyde available for competing cyanohydrin formation, resulting in poor reaction selectivity (Scheme 1). It was with these factors in mind that we embarked upon an investigation into the continuous flow synthesis of α-aminonitriles, proposing that sequential reagent addition, coupled with the use of a solid-supported Lewis acid catalyst, would allow for the efficient and selective synthesis of α-aminonitriles.

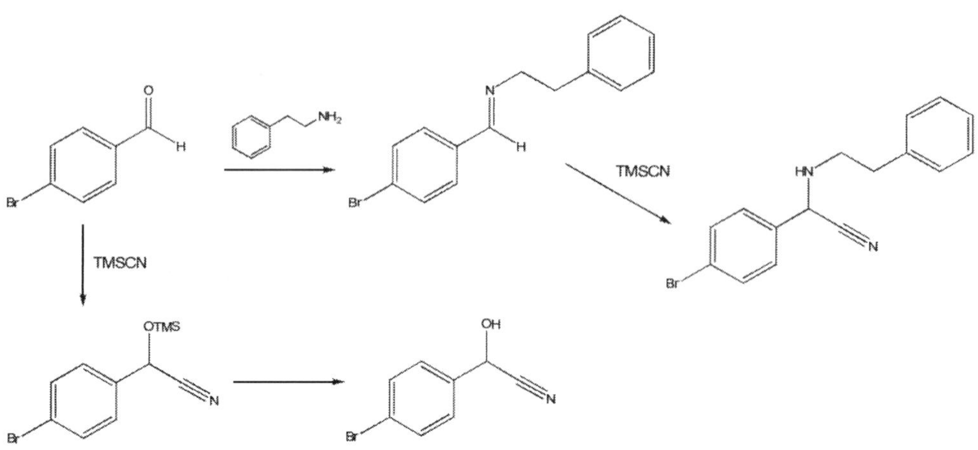

Scheme 1. Schematic illustrating the synthesis of α-aminonitriles and some of the by-products encountered in conventional batch reactions.

Experimental: Owing to the hydrolytic instability of imines, which leads to difficulties in their isolation and purification, it was firstly desirable to generate these reactive intermediates *in-situ* in order to afford the selective synthesis of the desired α-aminonitrile. In the first step towards this goal a borosilicate glass micro reactor was fabricated (footprint = 3.0 cm x 3.0 cm x 0.6 cm), which consisted of a 'T'-intersection where the aldehyde (inlet A) and amine (inlet B) react within a central channel (150 µm (wide) x 50 µm (deep) x 5.6 cm (long)) to afford the respective aldimine; with Figure 2 providing a diagrammatic representation of the interconnect strategy employed herein. Subsequent introduction of the cyanide source, in this case TMSCN, through inlet C enabled the reagents to mix under diffusion before nucleophilic addition of the cyanide anion to the imine occurred within the catalyst bed (0.3 cm (wide) x 200 µm (deep) x 2.1 cm (long)).

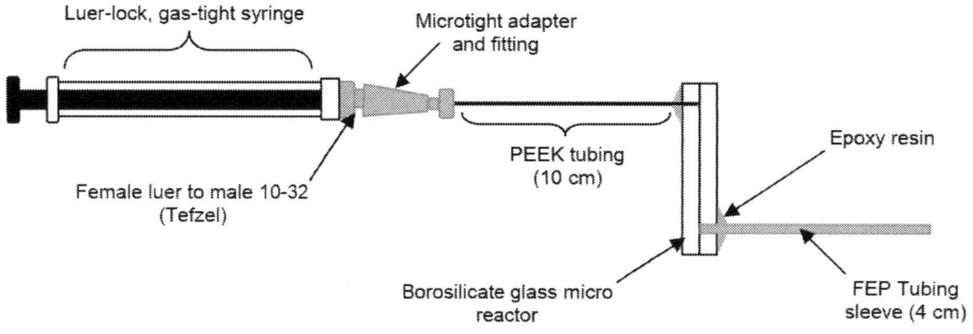

Figure 2. Schematic illustrating the technique employed for the connection of the micro reactor to a syringe driver.

Results and Discussion. To achieve this, solutions of 2-phenylethylamine (0.4 M in MeCN) and 4-bromobenzaldehyde (0.4 M in MeCN) were introduced from separate inlets (A and B) into the micro channel network, under pressure-driven flow (5 to 100 µl min^{-1}), where they were reacted for a specified period of time prior to collection and analysis of the reaction mixture (at outlet C, 0.2 M in MeCN) by GC-MS. As illustrated in Figure 3, employing a flow rate of 100 µl min^{-1} (residence time < 0.2 s) afforded incomplete conversion of 4-bromobenzaldehyde to the respective imine. Further optimization of the reaction conditions however afforded the quantitative synthesis of [1-(4-bromophenyl)meth-(*E*)-ylidene]phenethylamine when operating at a flow rate in the range of 1 to 25 μl min^{-1}, which corresponds to a residence time of between 21.6 and 0.9 sec. Having successfully achieved our first aim, which was to efficiently prepare an imine, the next step of the investigation was to evaluate the synthesis of the respective α-aminonitrile under continuous flow. In order to determine the optimal flow rate, and hence residence time, required to perform the nucleophilic addition step, we initially focused on the use of a pre-formed imine.

Using the reactor illustrated in Figure 1, a solution of [1-(4-bromophenyl)meth-(*E*)-ylidene]phenethylamine (0.2 M in MeCN) was introduced from inlet A and a solution of TMSCN (0.2 M in MeCN) from inlet C, the reactants mixed in the central channel prior to reaction within the catalyst bed, which contained polymer-supported ethylenediaminetetraacetic acid ruthenium (III) chloride (PS-$RuCl_3$) (10 mg, 0.26 mmol g^{-1}). The resulting reaction products were collected at outlet D (0.1 M), concentrated *in vacuo*, prior to dissolution in $CDCl_3$ and analysis by NMR spectroscopy; whereby comparison of the integrals afforded quantification of the percentage conversion. Employing a flow rate of 100 μl min^{-1} afforded 70.0 % conversion to the desired α-aminonitrile; this was subsequently improved upon by reducing the flow rate to 10 μl min^{-1}, and hence increasing the reactant residence time, to afford quantitative conversion of the imine. Having confirmed that both reaction steps could be conducted under continuous flow, the *in-situ* generation of the imine followed by α-aminonitrile synthesis was investigated. As Figure 3 illustrates, quantitative conversion is attainable for both steps and in-line with this observation, the multi-component reaction was performed at a total flow rate of 10 μl min^{-1}.

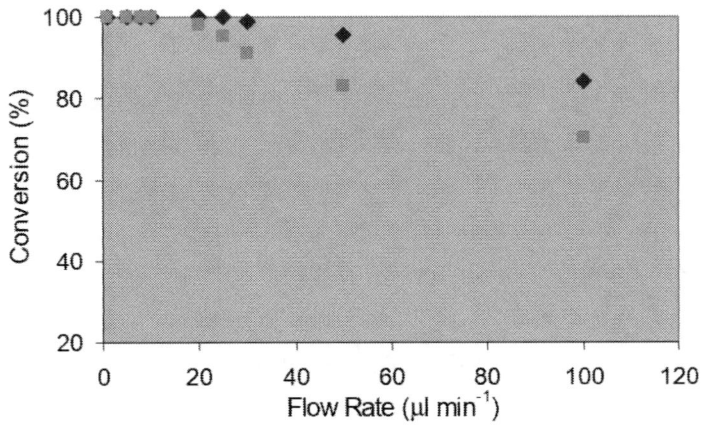

Figure 3. Summary of the reaction conditions evaluated for the synthesis of [1-(4-bromophenyl)meth-(*E*)-ylidene]phenethyl-amine (top data series) and (4-bromophenyl)phenethylaminoacetonitrile (bottom data series).

To conduct a reaction, 4-bromobenzaldehyde (0.4 M) was introduced from inlet A, 2-phenylethylamine (0.4 M) from inlet B and TMSCN (0.2 M) from inlet C (1:1:1). The reaction products were collected at outlet D and concentrated *in vacuo* to afford 2-(4-bromophenyl)-2-(phenethylamino)acetonitrile as a pale yellow solid. Analysis of the material by NMR spectroscopy confirmed quantitative conversion to the desired product and unlike analogous batch reactions, no competing cyanohydrin formation was observed.

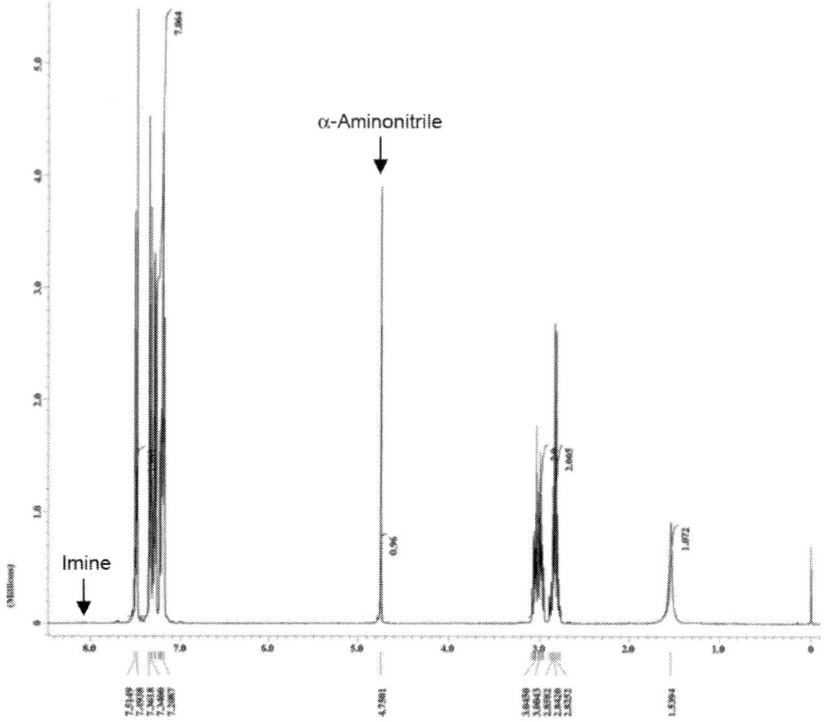

Figure 4. ^{1}H NMR spectra of the α-aminonitrile synthesized under continuous flow.

In order to confirm product purity, the reactor was operated continuously for 2.5 hr to afford sufficient material (0.047 g, 99.6 %) for evaluation by ^{1}H NMR (Figure 4), ^{13}C NMR, IR spectroscopy, MS, ICP-MS and elemental analysis. Importantly, in the absence of the catalyst, no α-aminonitrile formation was observed, simply recovery of the imine.

Having demonstrated the ability to synthesize an α-aminonitrile with excellent selectivity under continuous flow conditions, we subsequently evaluated the preparation and reaction of a series of hydrolytically unstable imines. As Table 1 illustrates, a further 4 amines were investigated ranging from aromatic derivatives to a cyclic aliphatic amine, which afforded an iminium ion as the reactive intermediate. Using this approach it was found that the optimal flow rate for aromatic derivatives (Entries 1-4) was 10 μl min^{-1}, compared to the more reactive pyrrolidine (Entry 5) which afforded quantitative conversion to the α-aminonitrile at a flow rate of 20 μl min^{-1}.

In an attempt to increase the reactor throughput further, a second supported Lewis acid catalyst was investigated, scandium *bis*(trifluoromethanesulfonate) polymer bound (PS-$ScOTf_2$) (10 mg, 0.60 mmol g^{-1}). Using the reaction conditions previously described it was found that comparable conversions were obtained at increased flow rates, affording double the throughput per reactor *i.e.* 0.12 to 0.24 mmol hr^{-1} *cf.* 0.06 to 0.12 mmol hr^{-1} (Table 2); consequently, all further investigations were conducted using PS-$ScOTf_2$ as the catalyst.

entry	product	flow rate, μl min^{-1} [a]	yield, g	throughput, mg hr^{-1}
1	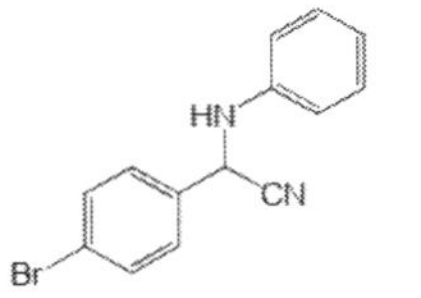	10	0.043 (100) [b]	17.2
2	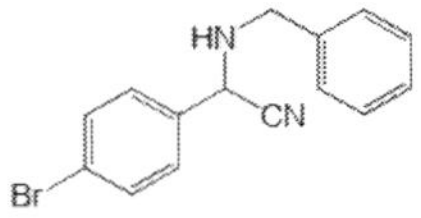	10	0.045 (100)	18.1
3	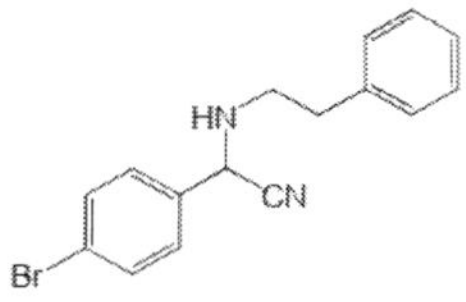	10	0.047 (100)	18.9
4	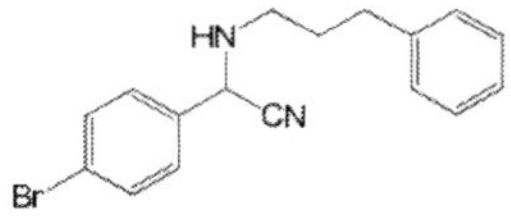	10	0.049 (100)	19.7
5	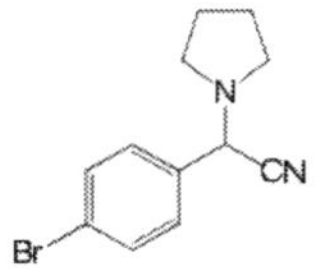	20	0.040 (100)	31.8 [c]

[a] Based on the total flow obtained from 3 fluidic inputs, [b] % conversion determined *via* comparison of the ^{1}H NMR integrals observed for the imine and α-aminonitrile and [c] run time = 1.25 hr.

Table 1. Results obtained for the synthesis of five α-aminonitriles, derived from 4-bromobenzaldehyde (Run time = 2.5 hr), employing PS-$RuCl_3$ as the catalyst.

The complexity of the reaction was subsequently increased through the employment of aliphatic aldehydes as precursors, which unlike the aromatic derivatives illustrated in Tables 1 and 2, have the potential to self condense *via* the aldol reaction to afford a series of reaction products, as depicted in Scheme 2, along with the previously encountered *O*-TMS cyanohydrin/cyanohydrin. Consequently to further demonstrate the synthetic utility of the micro reactor fabricated herein, the reaction of propionaldehyde and 2-phenylethylamine, to afford 2-phenethylamino-butyronitrile was investigated. Employing a total flow rate of 40 μl min^{-1}, the α-aminonitrile was obtained in excellent yield (0.023 g, 99.8 %) unlike the analogous batch reaction where a complex reaction mixture was obtained, requiring additional purification, if the desired product was to be isolated.

entry	product	flow rate, μl min^{-1} [a]	yield, g	throughput, mg hr^{-1}
1	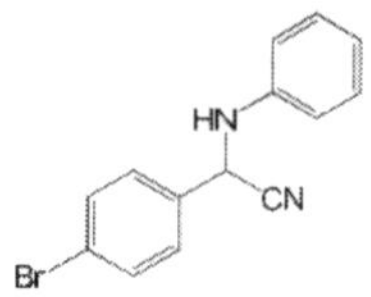	20	0.034 (100) [b]	34.4
2	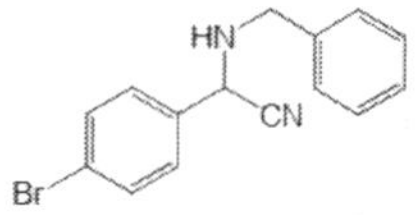	20	0.036 (100)	36.1
3	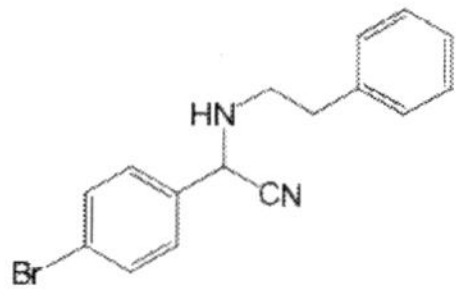	20	0.038 (100)	38.0
4	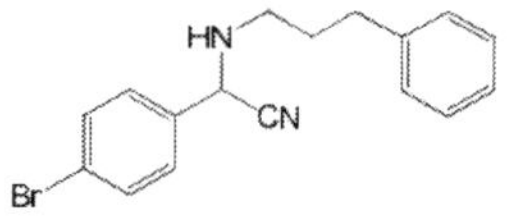	20	0.039 (100)	39.5
5	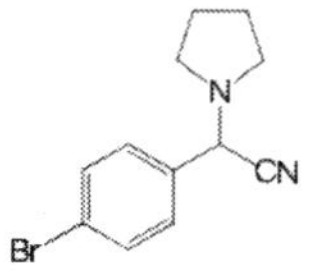	40	0.032 (100)	63.6 [c]

[a] Based on the total flow obtained from 3 fluidic inputs, [b] % conversion determined *via* comparison of the ^{1}H NMR integrals observed for the imine and α-aminonitrile and [c] run time = 0.5 hr.

Table 2. Results obtained for the synthesis of five α-aminonitriles, derived from 4-bromobenzaldehyde (Run time = 1.0 hr), employing PS-ScOTf$_2$ as the catalyst.

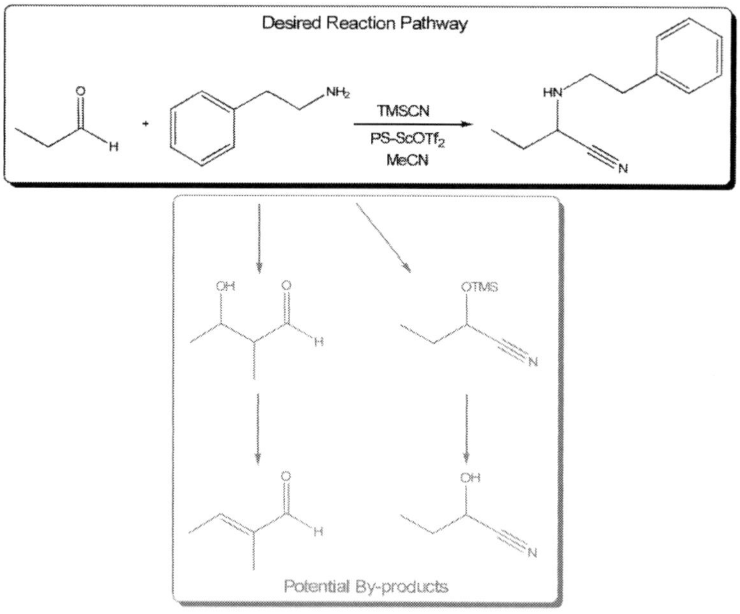

Scheme 2. Schematic illustrating a selection of the reaction products encountered when attempting to synthesize 2-phenethylamino-butyronitrile in a one-pot, batch reaction.

Conclusions. In summary, we report herein a reactor capable of performing multi-component reactions consisting of both solution phase and polymer-assisted steps. This approach, not only affords superior selectivity when compared to analogous batch reactions, due to sequential reactant addition, but also enables the catalyst to be recycled with ease. Furthermore, the technique described affords a simple route to the synthesis of α-aminonitriles without the need for additional purification steps, whilst preventing catalyst leaching (compared to a typical stirred protocol where degradation led to the release of 440 ppm of Ru into the reaction product). With this in mind, further investigations are currently underway within our laboratories in order to increase reactor throughput, product diversity and chemoselectivity of the technique.

Acknowledgements. Financial support provided by the EPSRC (RAIS award) is gratefully acknowledged, as are Dr Steve Clark and Mr Bob Knight for assistance with device fabrication and ICP-MS analysis respectively. The authors would also like to thank Dr Michael Singer (Sigma-Aldrich) for useful discussions with regard to the supported catalyst.

References.

1. Caution! All experiments should be performed within a fume cupboard in order to avoid inhalation of HCN, which may be liberated upon hydrolysis of trimethylsilyl cyanide.

2. A. Strecker, *Liebigs Ann. Chem.* **1850**, *75*, 27.

3. L. M. Weinstock, P. Davis, B. Handelsman, R. Tull, *J. Org. Chem.* **1967**, *32*, 2823.

4. H. Groger, *Chem. Rev.* **2003**, *103*, 2795.

5. a) C. Wiles, P. Watts, *Chem. Commun.* **2007**, 443; b) B. P. Mason, K. E. Price, J. L. Steinbacher, A. R. Bogdan, *Chem. Rev.* **2007**, *107*, 2300; c) K. Geyer, J. D. C. Codee, P. H. Seeberger, *Chem. Eur. J.* **2006**, *12*, 8434.

236a Capillary-Based Microfluidic Reactors for the Production of Size-Controlled Functional Polymer Particles

Christophe Serra[1], Anne Hébraud[1], Michel Bouquey[1], Laurent Prat[2], and Georges Hadziioannou[1]. (1) Ecole de Chimie Polymères et Matériaux - Laboratoire d'Ingénierie des Polymères pour les Hautes Technologies, ULP - CNRS UMR 7165, 25, Rue Becquerel, Strasbourg Cedex-2, F-67087, France, (2) Laboratoire de Génie Chimique - UMR CNRS 5503, Université de Toulouse - INPT, 5 rue Paulin Talabot, Toulouse, 31106, France

We report on the use of microfluidic devices for the synthesis of monodisperse and size-controlled polymer particles (CV<5%), either beads or capsules, in the size range of tens to several hundred of microns.

The emulsification of the monomer phase is achieved in capillary-based microreactors without any surfactant or surface treatment. Polymer beads are obtained after UV irradiation while polymer capsules are directly generated in the microfluidic device through an interfacial polycondensation reaction.

A preliminary study has shown that in the dripping regime the variation of particle diameter with respect to operating parameters (flow rates and viscosities of the dispersed and continuous phases) is fairly well described by an empirical relationship between two dimensionless numbers, the normalized particle diameter and the Capillary numbers ratio [1].

Application of these microsystems concerns O/W or W/O emulsions for the preparation of inorganic-organic hybrid beads as well as hollow and doped capsules. By confining the particles in the microsystem, we have also successfully prepared polymer beads necklaces [2] with different arrangements.

[1] C. Serra, N. Berton, M. Bouquey, L. Prat and G. Hadziioannou, Langmuir, 23 (14) (2007) 7745-7750.

[2] M. Bouquey, C. Serra, N. Berton, L. Prat and G. Hadziioannou, Chem. Eng. J. (In press, doi:10.1016/j.cej.2007.07.043).

103f Application of Microchannel Device for Enrichment of Electrolytes

Ken-Ichiro Sotowa, Keiji Kume, Kaname Takagi, Rie Tomiya, Shigeru Sugiyama, and Keizo Nakagawa. Department of Chemical Science and Technology, The University of Tokushima, 2-1 Minami-Josanjimacho, Tokushima, 770-8506, Japan

For continuous concentration of ions, electrodialysis is one of the commonly used technques in chemical industry. In this technique, an electrolyte solution is introduced to a chamber which is parted with ion exchange membranes. An electric field is applied to the chamber, and then charged ions migrate towards one of the electrodes. The role of the membrane is to define the direction of ion migration, and to prevent treated solutions from mixing. One of the drawback of the electrodialysis is the need to renew the membrane regularly, and thus the operatinal cost can be high.

This study was devoted to explore the possibility of microchannel electrophoresis for continuous ion enrichment. There are already many reports on realizing electrophoresis in microchannels. In these studies, the electrodes are installed at the inlet and outlet of the microchannel, and thus the direction of the electric field was parallel to the fluid flow. Such microdevices can be operated batchwise only, and thus not suited for continuous separation. In our approach, the electrodes are installed on the channel walls, and thus the electric field was normal to the flow direction. The charged ions are attracted to either one of the electrodes, and carried over to the downstream. By separating the fluid flow into two parts in a simple branch, an enriched solution can be easily obtained.

The validity of the proposed approach was examined in an experimental study. As an example, a sodium chloride solution was used as the feed stock, and the degree of ion enrichment was measured for various operating conditions. The results showed that the proposed technique can be used to enrich sodium ions.

Microreactor for Synthesis Via Intermediates with Assembled Units Enabling Rapid Mixing and Heat Transfer and Short Residence Time

*Nobuaki Aoki, Ryouhei Kitajima, Chiharu Itoh, Kazuhiro Mae**

Department of Chemical Engineering, Graduate School of Engineering, Kyoto University
Kyoto-daigaku Katsura, Nishikyo-ku, Kyoto 615-8510, Japan
**kaz@cheme.kyoto-u.ac.jp*

Abstract

This paper describes a reactor for multi-step syntheses via intermediates with improved selectivity of desired products. This reactor is referred as assembled reactor. This reactor consists of mixing, residence time and heat exchange units. Connection of these units forms a module for a single reaction. Repeated joints of modules enable the reactor to be used in multi-step syntheses. Mixing and heat transfer rate of the assembled reactor was evaluated. This reactor enabled rapid mixing under the total flow rates of more than 30 mL/min and heat transfer efficiency of the order of 10 K per 10 ms in the heat exchange of fluids with temperature difference of 20–40 K. This reactor was also applied to the synthesis of bromobenzene via *p*-bromophenyllithium and gave the product yield more than 80%.

Keywords: Synthesis via intermediates, Assembled units, Mixing, Heat transfer, Residence Time

1 Introduction

In organic syntheses, a desired product is often formed via unstable intermediates. The intermediates are changed into by-products in a quite short time. For increasing the selectivity of the desired product, the intermediates should react with reactants immediately after the formation of the intermediates under precisely controlled temperature. To realize such reaction operation, we need a reactor realizing rapid mixing and heat transfer and short residence time. Microreactor has a high mass and heat transfer efficiency due to its large surface area to volume ratio. The miniaturization of a reactor also leads to a reduction of residence time in a reactor. These advantages offer an opportunity to satisfy these requirements. Multistep syntheses including unstable intermediates using microreactors with enhanced desired product yield and selectivity have been carried out. Swern oxidation [1], synthesis of diarylethenes [2], and sequential Br–Li exchange reactions of *p*-, *m*-, and *o*-dibromobenzenes [3,4] are examples of such synthesis. In these syntheses, however, multiple micromixers are connected by tube, and the mixers and tubes are soaked in a cooling bath. Such reactor setup is difficult to achieve short residence time of the order of millisecond and to control temperature independently for each step of multiple reactions. In order for reactions to proceed under optimized conditions giving highest product yield and selectivity, units for mixing, controlling residence time, and heat exchange should be connected without tubes.

In this context, we have developed a reactor for multi-step organic syntheses including the formation of intermediates with improved selectivity of desired products. We call this reactor "assembled reactor." This reactor consists of mixing, residence time (reaction) and heat exchange units. Connection of these three units forms a module for a single reaction. Repeated joints of modules enable the reactor to be used in multi-step syntheses. We can flexibly replace the units or add units according to elementary reactions of a synthesis. This reactor is made of stainless steel (SUS316) and polyimide resin (CEPLA). The polyimide resin acts as heat insulator. The units made of the resin are embedded between the units having inlets of reactants to control feed temperatures of reactant fluid and reaction temperatures independently. The units are connected without tubing. This connection enables a short residence time between consecutive reactions. For using this reactor in industrial production with a high throughput, we designed the channel sizes to be on the order of several hundred micro meters.

We have examined the usefulness of the assembled reactor from the viewpoint of mixing and heat transfer rate and application in organic synthesis including the generation of unstable intermediates. For the study of mixing performance, we used the Villermaux/Dushman reaction. For the study of heat transfer rate, we performed the heat exchange between hot water and cold water with temperature difference from 20–40 K. We also used this reactor in the synthesis of bromobenzene via *p*-bromophenyllithium from *p*-dibromobenzene using the Br–Li exchange reaction.

2 Feature of the Assembled Reactor

We have developed a reactor for the multi-step organic synthesis including the generation of intermediates with improved selectivity of desired products. Fig. 1 shows the overview of the developed reactor and the units that compose the reactor. We call this reactor "assembled reactor." This reactor consists of mixing (units No. 5 and 10 in Fig. 1), residence time (3, 4, 8 and 9), and heat exchange (1, 2, 6 and 7) units. The materials of this reactor are stainless steel (SUS316) and polyimide resin (CEPLA, units No. 3, 6 and 8). The polyimide resin acts as heat insulator. The units made of the resin are embedded between the units having inlets of reactants to control feed temperatures of reactant fluid and reaction temperatures independently. Each unit is connected without tubing. This connection enables a short residence time between each reaction. We can flexibly replace the units or add units depending on reaction system. In the reactor setup shown in Fig. 1, an intermediate is produced through the mixing of the reactant fluids from inlet A (referred as fluid A) and inlet B (fluid B) in the first mixing module (units No. 1–5), and a product is then formed by the reaction of the fluid from the first mixing part and that from inlet C in the second mixing module (6–10). Simultaneously, heat exchange between heat transfer fluid from inlet H1 (referred as fluid H1) and fluid A and that between heat transfer fluid from inlet H2 (fluid H2) and the mixture of fluid A and B (fluid A + B) can be carried out. In other words, two-step introduction of reagents and heat transfer fluid is possible in this reactor setup.

The flow of reactant fluids and heat transfer fluids in the reactor shown in Fig. 1 is illustrated in Fig. 2. Enlarged mixing zone circled in Fig. 2 is depicted in Fig. 3. The mixing principle is the same as the micromixer with needle adjustment for instant mixing and heating under high pressure and temperature [5]. Reactant fluid from inlet A in unit No. 2 or fluid A + B in unit No. 7 flows in the channel around the needle of unit No. 2 or 7. The channel width is 200 μm. At the edge of the needle, fluid A or A + B mixed with fluid B from unit No. 5 or C from unit No. 10, respectively. The channel width for fluid B or C, *X*, is 0.5 or 1.0 mm. This size also relates the thickness of the mixing zone. The diameter of outlet channel of mixing zone is 800 μm. For using this reactor in industrial production with high throughput, we designed the channel sizes to be on the order of several hundred micro meters.

As shown in Fig. 3, heat exchange between fluid A or A + B and H1 or fluid H2 passing inside the needle also proceeds. Through this heat exchange, temperature of fluid A and that of the mixture of fluid A and B can be controlled independently. Fluid H1 passes through the needle of unit No. 2 (outside of the inside channel of unit No. 1) into the inside channel and then goes out from the outlet in unit No. 1. Fluid H2 passes through the needle of unit No. 7 (outside of the inside channel of unit No. 1) into the inside channel and then goes out from the outlet in unit No. 6.

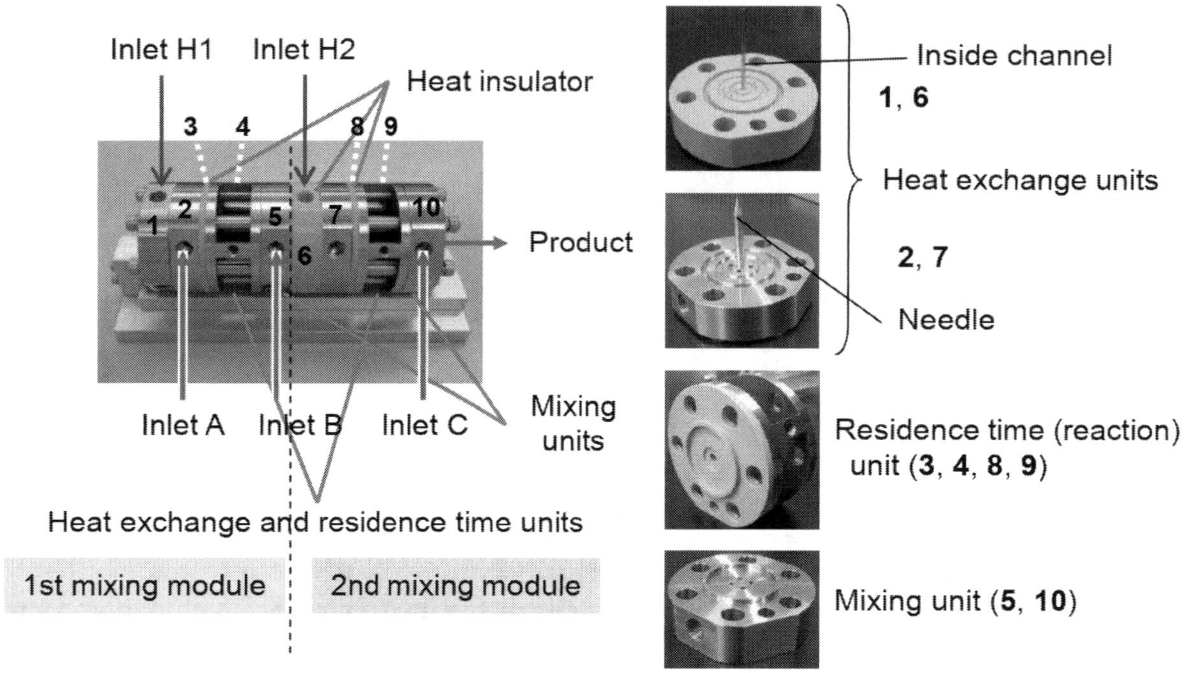

Figure 1. Assembled reactor. The left picture is the overview, and the right pictures are the units composing the reactor.

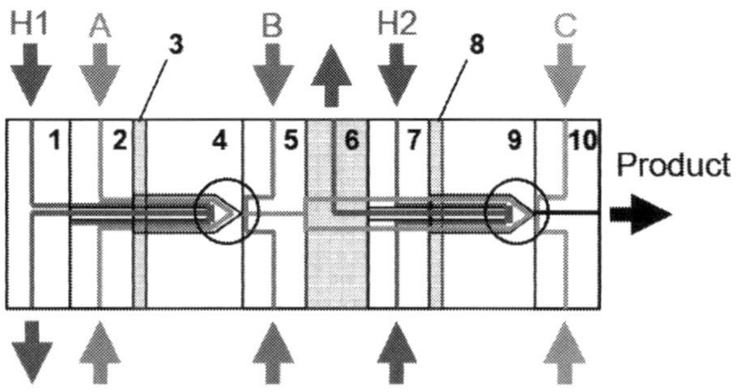

Figure 2. Inside flow of reactant fluids and heat exchange fluids in the assembled reactor.

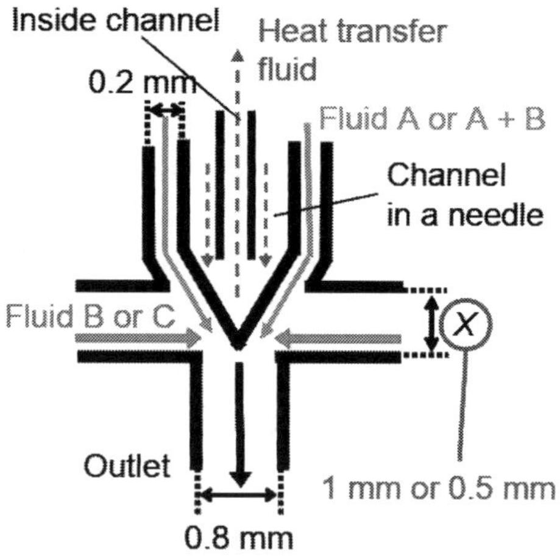

Figure 3. Enlarged drawing of mixing zone.

3 Evaluation Method of Reactor Performance

3.1 Mixing

We evaluated mixing performances of the microreactor described in the previous section using parallel-competitive redox and neutralization reactions called the Villermaux/Dushman Reaction [6–9]. The reactions occur by mixing two reactant fluids of aqueous solutions: One is a solution of diluted strong acid (referred as solution 1); and the other is a buffer solution (mixture of weak acid and strong base) containing KI and KIO_3 (Redox reagents) (solution 2). In our experiments, solution 1 was 0.03 mol/L of HCl, and solution 2 was the mixture of 0.016 mol/L of KI, 0.0032 mol/L of KIO_3, 0.09 mol/L of NaOH, and 0.09 mol/L of H_3BO_3. When boric acid is used as a weak acid for the buffer solution, the reaction formulas are as follows:

$$H_2BO_3^- + H^+ \rightleftarrows H_3BO_3 \qquad (1)$$
$$5I^- + IO^{3-} + 6H^+ \rightleftarrows 3I_2 + 3H_2O \qquad (2)$$
$$I_2 + I^- \rightleftarrows I^{3-} \qquad (3)$$

where reaction (2) is fast, but reaction (1) is instantaneous and is much faster than reaction (2). When mixing is slow, the protons of acid are consumed by reaction (2), and I_2 is produced. The iodine then reacts with iodide ions quickly by reaction (3). The product of reaction (3), that is, triiodide ion has the strong UV light absorbance around 352 nm. Thus, the UV light absorbance at 352 nm depends on mixing performance of the two reactant fluids. Thus, we can use the UV light absorbance at 352 nm as a measure of mixing performance. We measured the absorbance with a UV-Vis spectrometer (Multispec-1500, Shimadzu Co.).

Solution 1 was fed from inlet A and solution was fed from inlet C by using a syringe pump (PHD2000, Harvard Apparatus) at room temperature. Mixing occurred in the second mixing module. The sickness of mixing zone was 0.5 or 1 mm as shown in Fig.3. For comparison, we also evaluated mixing performance of 1/16" union tee with i.d. 0.3 mm. At the outlet of the assembled reactor or the union tee, a PTFE tube with i.d. 0.8 mm was connected. We evaluated mixing performance of the reactors under the total flow rate from 2 to 50 mL/min.

3.2 Heat transfer

For the study of heat transfer rate, we performed the heat exchange between hot distilled water and cold distilled water with temperature difference from 20–40 K. Hot water (50, 60 or 70 °C) was fed from inlet B, and cold water (30 °C) was fed from the inlet of heat transfer fluid 2 (see Fig. 2). Heat exchange occurred in the second mixing module. Flow rate of hot and cold was 50, 75, or 100 mL/min, and the mean corresponding residence time of hot water, τ, was 0.062 s, 0.041 s, 0.031 s, respectively. We used two LC pumps (LC-8A, Shimadzu Co.) for feeding the hot and cold water in the reactor. Temperature of the hot and cold water from the outlets was measured with thermo couples.

3.3 Application to Organic Synthesis

We then applied the assembled reactor to the synthesis of bromobenzene via *p*-bromophenyllithium from *p*-dibromobenzene (substrate) with the Br–Li exchange reaction [3]. Fig. 4 shows the reaction scheme. This reaction is highly exothermic, and increase in temperature reduces the yield and selectivity of bromobenzene (desired product). For this reason, the synthesis is usually carried out at −78 °C in laboratory batch system, although this temperature is inconvenient for industrial production.

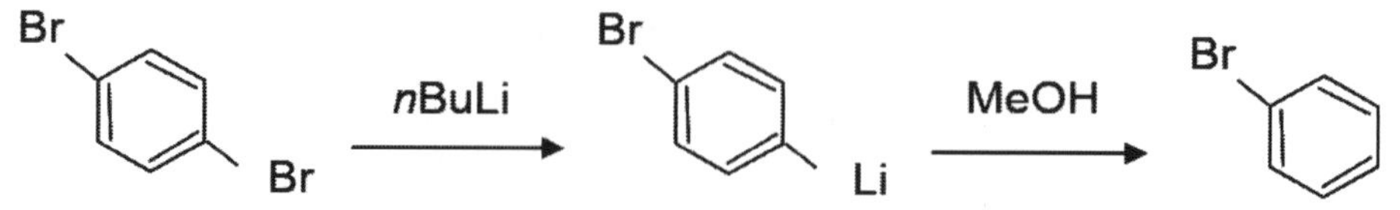

Figure 4. Br–Li exchange reaction of *p*-dibromobenzene using *n*BuLi followed by reaction with MeOH.

In experiments using the assembled reactor, a solution of *p*-dibromobenzene (0.29 M, Aldrich Ltd.) in dehydrated THF and a solution of *n*BuLi (1.61 M) in *n*-hexane were introduced into the reactor from inlets A and B. The stochiometric ratio of *n*BuLi to *p*-dibromobenzene was 1.1. Between the first and second mixing zone, *p*-bromophenyllithium was generated. Dehydrated methanol (neat) was then introduced from inlet C, and the transformation of *p*-bromophenyllithium into bromobenzene proceeded. The reactant fluids were fed in the reactor using three syringe pumps (Pump 11, Harvard Apparatus). Experiments were carried out for several total flow rates summarized in Tab. 1. The flow rate ratio of the three solutions was fixed for all experimental runs. To control the temperature of the system, cold water of 0 °C was introduced at a flow rate of 100 mL/min using two LC pumps (LC-8A, Shimadzu Co.) from inlets H1 and H2 (see Fig. 2). After the system reached steady state, the product solution was taken over 15 s. The amount of bromobenzene was determined using GC (GC-2014, Shimadzu Co.) equipped with a flame ionization detector and a fused silica capillary column (CBP1, 0.25 mm × 25 m).

Table 1. Flow rate for each experimental run in the application of the assembled reactor to an organic synthesis.

Run No.	Flow rate [mL/min]			
	p-dibromobenzene	*n*BuLi	MeOH	Total
1	3.0	0.6	1.5	5.1
2	6.0	1.2	3.0	10.2
3	12.0	2.4	6.0	20.4
4	24.0	4.8	12.0	40.8

4 Results and Discussion

4.1 Mixing Performance

Fig. 5 shows the relation between the absorbance at 352 nm, ABS [-], and total flow rate for the assembled reactor of the thickness of mixing zone, X, = 1 and 0.5 mm and the union tee of i.d. 0.3 mm. The absorbance decreases with improving mixing performance. From the results of X = 1 and 0.5 mm, the thickness of mixing zone has little effect on the mixing performance. The result also indicates that the assembled reactor has lower mixing performance than the union tee at the same flow rate. However, the value of ABS for the assembled reactor is close to 0 in the total flow rate of more than 30 mL/min. Thus, the assembled reactor exhibits high mixing performance in high flow rate range. Therefore, the assembled is useful in industrial production requiring high throughput.

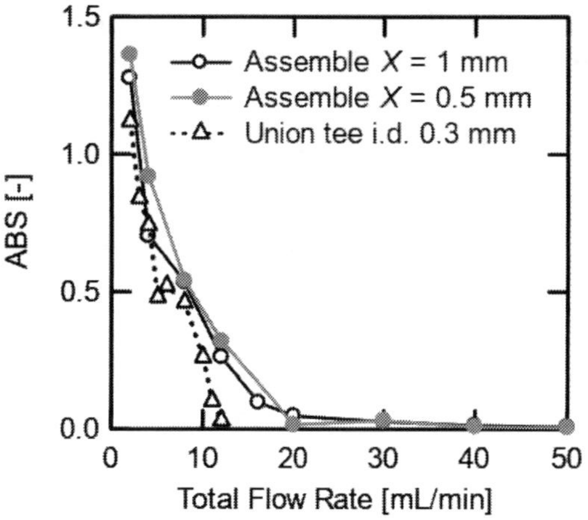

Figure 5. Mixing performance of the assembled reactor.

4.2 Heat Transfer Performance

The results of heat exchange between hot and cold water using the assemble reactor is shown in Tab. 2. The temperature change of hot water in the assembled reactor is denoted by ΔT, and the mean residence time of hot water in the reactor is denoted by τ. The rate of temperature change of hot water, $\Delta T/\tau$, increases with flow rate of hot water. The flow rate of cold water has little effect on $\Delta T/\tau$. The hot water flows in the channel for reactant fluid. This tendency indicates that heat transfer resistance of reactant fluid has larger impact than that of heat transfer fluid in the channel geometry of assemble reactor. Increasing the temperature of hot water, that is, increase in driving force of heat enhances the rate of temperature change of hot water, and the maximum value of $\Delta T/\tau$ was 470 K/s. Thus, the assembled reactor has the heat transfer efficiency of the order of 10 K per 10 ms in the heat exchange of fluids with temperature difference of 20–40 K.

Table 2. Heat transfer performance of the assembled reactor.

hot water (from inlet B)		cold water (from Inlet H2)				
Inlet temp. [°C]	Flow rate [mL/min]	Inlet temp. [°C]	Flow rate [mL/min]	Outlet temp. of hot water [°C]	ΔT (hot water) [K]	$\Delta T/\tau$ [K/s]
50	50	30	50	38.0	12.0	192.3
50	75	30	50	41.5	8.5	204.3
50	100	30	50	43.3	6.7	214.8
60	100	30	50	50.0	10.0	320.5
70	100	30	50	55.9	14.1	451.9
70	100	30	75	55.6	14.4	461.6
70	100	30	100	55.3	14.7	471.2

4.3 Application to Organic Synthesis

Tab. 3 summarizes the result of application of the assembled reactor in the synthesis of bromobenzene from *p*-dibromobenzene using the Br–Li exchange reaction. The yield and selectivity of the desired product bromobenzene tend to increase with total flow rate. The reason for this tendency is that the mixing performance improves and the residence time between the first and second mixing zones decreases with increasing total flow rate. In the synthesis using the assembled reactor, the temperature of cooling water was 0 °C, which is much higher than that in conventional batch synthesis (–78 °C) [3] and preferable in industrial production. The yield of more than 80% indicates that the reaction heat was sufficiently removed to prevent side reactions from proceeding. We can conclude that the assemble reactor enables rapid mixing and heat transfer and short residence time between successive reaction steps and offers a opportunity to be used in multi-step syntheses via unstable intermediates with improved product distribution in industrial scale.

Table 3. Product distribution of the synthesis of bromobenzene via *p*-bromophenyllithium from *p*-dibromobenzene in the assembled reactor.

Run No.	Total flow rate [mL/min]	Conversion of *p*-dibromobenzene [%]	Yield of bromobenzene [%]	Selectivity of bromobenzene [%]
1	5.1	95	74	77
2	10.2	84	63	74
3	20.4	95	82	86
4	40.8	87	78	90

5 Conclusions

We have developed a reactor for multi-step organic syntheses including the formation of intermediates with improved selectivity of desired products, referred as assembled reactor. This reactor consists of mixing, residence time (reaction proceeding) and heat exchange units. Connection of these three units forms a module for a single reaction. Repeated joint of modules enables the reactor to be used in multi-step syntheses. We have examined the usefulness of the assembled reactor from the viewpoint of mixing and heat transfer rate. The results showed that this reactor enables rapid mixing under the total flow rates of more than 30 mL/min and heat transfer efficiency of the order of 10 K per 10 ms in the heat exchange of fluids with temperature difference of 20–40 K. This reactor is also applied to the synthesis of bromobenzene via *p*-bromophenyllithium and gives the product yield more than 80%. The coolant temperature in this synthesis (0°C) is much closer to room temperature than that in conventional batch system (–78°C). Through this study, we have confirmed that the assemble reactor enables rapid mixing and heat transfer and short residence time between successive reaction steps and offers a opportunity to be used in multi-step syntheses via unstable intermediates with improved product distribution in industrial production.

Acknowledgement

This research was partially supported by the New Energy and Industrial Technology Development Organization (NEDO), Project of Development of Microspace and Nanospace Reaction Environment Technology for Functional Materials. We would like to appreciate Dr. Hideharu Nagasawa (FUJIFILM Corporation) for his fruitful discussion on the design of assembled reactor.

References

[1] T. Kawaguchi, H. Miyata, K. Ataka, K. Mae, J.-I. Yoshida "Room-temperature Swern Oxidations by Using a Microscale Flow System" *Angew. Chem. Int. Ed.*, **2005**, *44*, 2413–2416.

[2] Y. Ushiogi, T. Hase, Y. Iinuma, A. Takata, J.-I. Yoshida "Synthesis of Photochromic Diarylethenes using a Microflow System" *Chem. Commun.*, **2007**, 2947–2949.

[3] A. Nagaki, Y. Tomida, H. Usutani, H. Kim, N. Takabayashi, T. Nokami, H. Okamoto, J.-I. Yoshida "Integrated Micro Flow Synthesis Based on Sequential Br-Li Exchange Reactions of *p*-, *m*-, and *o*-Dibromobenzenes" *Chem, Asian J.*, **2007**, *2*, 1513–1523.

[4] H. Usutani, Y. Tomida, A. Nagaki, H. Okamoto, T. Nokami, J.-i. Yoshida "Generation and Reactions of *o*-Bromophenyllithium without Benzyne Formation Using a Microreactor" *J. Am. Chem. Soc.*, **2007**, *129*, 3046–3047.

[5] K. Mae, A. Suzuki, T. Maki, Y. Hakuta, H. Sato. K. Arai "A New Micromixer with Needle Adjustment for Instant Mixing and Heating under High Pressure and High Temperature" *Journal of Chemical Engineering of Japan*, **2007**, *40*, 1101–1107.

[6] P. Guichardon, L. Falk "Characterisation of Micromixing Efficiency by the Iodide–Iodate Reaction System. Part I: Experimental Procedure" *Chem. Eng. Sci.*, **2000**. *55*, 4233–4243.

[7] S. Panić, S. Loebbecke, T. Tuercke, J. Antes, D. Bošković "Experimental Approaches to a Better Understanding of Mixing Performance of Microfluidic Devices" *Chem. Eng. J.*, **2004**, *101*, 409–419.

[8] M. A. Schneider, T. Maeder, P. Ryser, F. Stoessel "A Microreactor-based System for the Study of Fast Exothermic Reactions in Liquid Phase: Characterization of the System" *Chem. Eng. J.*, **2004**, *101*, 241–250.

[9] N. Aoki, K. Mae "Effects of Channel Geometry on Mixing Performance Using Collision of Fluid Segments" *Chem. Eng. J.*, **2006**, *118*, 189–197.

Immobilisation of thermophilic enzymes within miniaturised flow reactors for biotransformations

B. Ngamsom,[a] L. Marle,[a] A.M. Hickey,[b] G.M. Greenway,[a] P. Watts,[a] T. McCreedy,[a] and J.A. Littlechild[b]

[a]Department of Chemistry, Faculty of Science, University of Hull, Cottingham Road, Hull, HU6 7RX, UK
[b]Henry Wellcome Building for Biocatalysis, School of Biosciences, University of Exeter, Stocker Road, Exeter, EX4 4QD, UK

Introduction

The use of enzymes for biotransformation reactions to synthesise a range of chemical and pharmaceutical products has been widely studied. However, their use is often limited to the production of specific products. The limitations of commercially available enzymes with respect to industrial applications can often be attributed to their low stability compared to chemical catalysts, narrow substrate specificity and high cost to produce/ use on a large scale. The use of immobilised enzymes on a small scale with incorporation of screening as an early step in enzyme selection can overcome some of these problems.[1]

After initially being applied to a number of fields, e.g., analysis,[2] catalyst screening[3] and synthetic chemistry,[4-6] microfluidic reactors have now been extended into pharmaceutical and fine chemical studies. The use of enzymes in micro reactors has proved to be successful, for instance in the immobilisation of trypsin in micro reactor channels,[7] alkaline phosphatase and xanthine oxidase on controlled pore glass in a micro reactor[8] and cross-linked enzyme aggregates (CLEAs) of L-aminoacylase.[9]

In this study, immobilisation of an L-aminoacylase from the thermophilic archaeon, *Thermococcus litoralis* onto polymer monoliths within micro reactors was investigated with a view to determining whether the micro reactors held potential for use in industrial biotransformation reactions and screening.

Experimental

The micro reactors used were fabricated in-house using, and consisting of, borosilicate glass containing 4 parallel channels with dimensions 100 μm deep, 300 μm wide and 15 mm long. The channel surfaces were pre-treated with 3-(trimethoxysilyl)propyl methacrylate for covalent attachment of the monolith to the channel walls. Two portions of the monomer solution consisting of GMA and EDMA (3:1; structures of both monomers are shown in Figure 1) were mixed with one portion of the porogenic mixture containing 4:1 cyclohexanol: dodecanol, into which 1 (% w/v) of the initiator DMPAP (with respect to total weight of monomers) was dissolved. Polymerisation was initiated by placing the micro reactors under the UV lamp (365 nm) and irradiating for 30 min. The monoliths were washed with methanol and dried with nitrogen prior to use.

After preparation, the monolithic micro channels were washed with 100 mM Tris-HCl buffer pH 8.0 for 30 min. Recombinant L-Aminoacylase (E.C. 3.5.1.14) from the thermophilic archaeon, *Thermococcus litoralis*, was then immobilised onto the monolithic channels within the micro reactors by passing a solution of L-aminoacylase in Tris-HCl buffer through the channels using a syringe pump at a flow rate of 1 μL min^{-1} for 3 h. Prior to use, the free enzyme was washed off with Tris-HCl, pH 8.0 buffer for 30 min.

The substrates that were to be screened for activity with the L-aminoacylase enzyme were dissolved in 100 mM Tris-HCl buffer pH 8.0. Unless otherwise stated, the preferred substrate was 10 mM *N*-benzoyl-L-phenylalanine. The substrate solution was charged into a 1 mL syringe and supplied to the monolithic micro reactor using a microfluidic pump.

The activity of the immobilised enzyme was investigated at flow rates 1-8 μL min^{-1}, at room temperature and 50°C, using the preferred substrate. The reaction product was analysed by HPLC equipped with a 4.6×150 mm C18 BDS Hypersil column (Thermo Electron Corporation, UK) using 35% acetonitrile-water as the mobile phase at a flow rate of 2 mL min^{-1} at ambient temperature over 10 min.

The enzyme specificity towards other potential substrates was also screened with *N*-acetyl-L-phenylalanine, *N*-chloroacetyl-L-phenylalanine, *N*-acetyl-L-tryptophan, *N*-benzoyl-L-leucine, *N*-acetyl-L-tyrosine and *N*-benzoyl-L-threonine. The reaction was carried out using 10 mM of each substrate and run at room temperature at the flow rate of 1 μL min^{-1}. The reaction products were monitored by HPLC as previously mentioned.

Results and Discussion

Preparation of monolithic micro reactors

Polymer monoliths were formed inside the micro reactor channels (100 μm depth; 300 μm width; 15 mm length) using the polymerisation mixture and photo initiation as described in the experimental section. The SEM images in Figure 1. show the internal morphologies of the monolith which demonstrate that the porous polymers are homogenous across the entire monolith. The complete absence of voids between the channel wall and the monolith was also revealed from the micrograph, demonstrating that the monoliths were formed on the surface of the glass channels. Reactions carried out using the monolithic micro channels at flow rates between 1-8 μL min^{-1} were operational, applying higher flow rates, however, resulted in back pressure and leaking through connectors.

Figure 1. SEM micrographs of the synthesised monoliths.

Enzymatic evaluation within the monolithic micro reaction system

Miniaturised flow reactors containing recombinant *T. litoralis* L-aminoacylase were investigated for their potential utilisation in industrial biotransformation reactions. The enzyme was immobilised on polyglycidylmethacrylate-co-ethylenedimethacrylate monoliths formed within the micro reactors by the reaction between amino groups of the enzyme and epoxy terminal groups on the surface of the monolithic micro channels.

The attachment of the enzyme to the matrix may influence the interaction of the immobilised enzyme with its substrate. The enzyme stereospecificity after being immobilised onto the monolith was examined using 10 mM *N*-benzoyl-D,L-phenylalanine. The maximum substrate conversion was found to be approximately 50%, confirming that the enzyme remained stereospecific after being immobilised onto the monolithic microchannels. The immobilised enzymes were assayed with the preferred substrate, *N*-benzoyl-L-phenylalanine. Using the continuous flow micro reactor system, 100% substrate conversion was obtained at room temperature which is a significantly lower temperature than the optimal condition reported for the conventional batch system of 85°C.[10, 11]

Figure 2 illustrates that the highest enzyme activity at room temperature was achievable when operated at flow rates between 1 to 4 μL min^{-1}, higher flow rates and the operating window could be expanded by performing the reaction at a higher temperature (50°C).

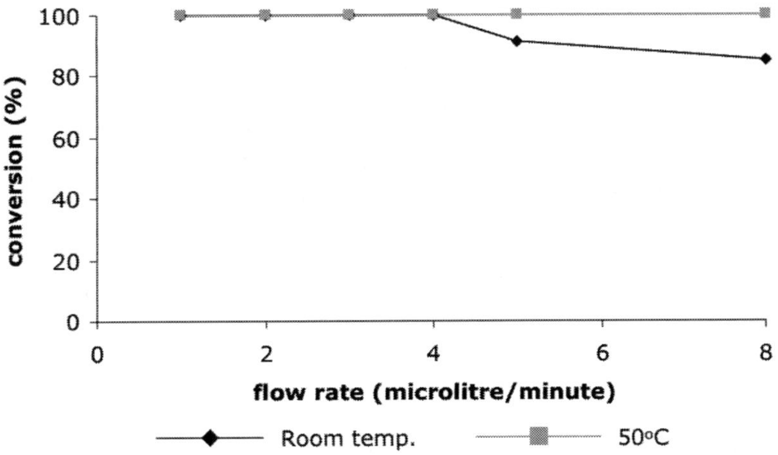

Figure 2. Effect of the flow rate on the activity of the enzyme in converting 10 mM *N*-benzoyl-L-phenylalanine to L-phenylalanine in 100 mM Tris-HCl buffer pH 8.0 at room temperature and 50°C.

The stability of the immobilised enzyme was subsequently tested both during continuous use and when stored in buffer at 4°C. The enzyme remained optimally active even after being continuously used at room temperature for 120 h, Figure 3(a). The thermostability test of the immobilised enzyme at 50°C also exhibited its stability after being exploited for 50 h, Figure 3(b). The storage stability of the immobilised enzyme examined daily during the first two weeks also showed no activity loss (6 hour-continuous run; results not shown here). These results demonstrate that the continuous-flow micro reaction system with this enzyme is a suitable tool for further development.

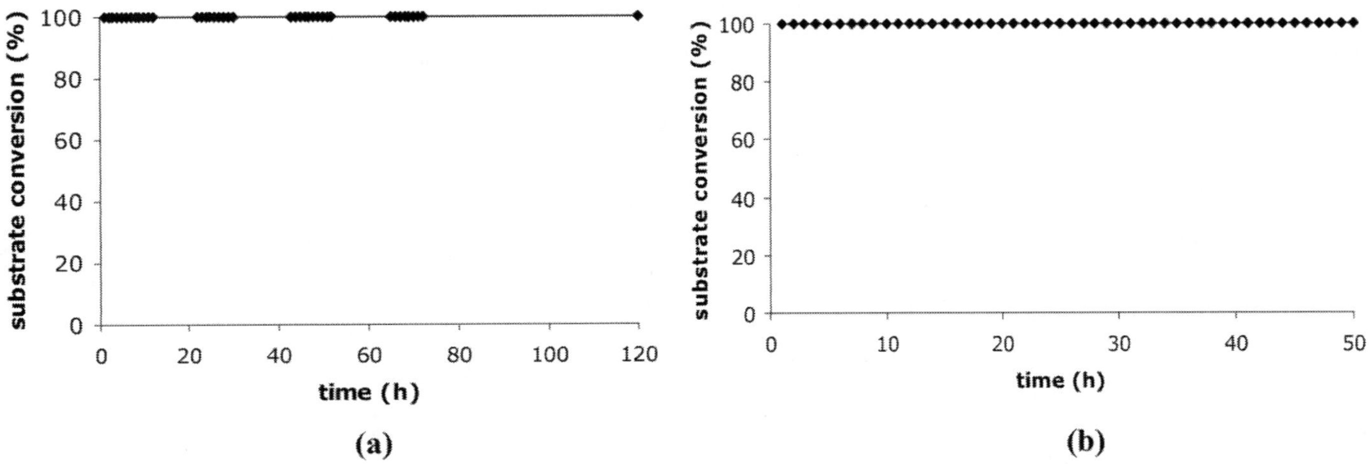

Figure 3. Time courses of the conversion of 10 mM *N*-benzoyl-L-phenylalanine to L-phenylalanine in 100 mM Tris-HCl pH 8.0; reactions performed at a flow rate of 1 μL min^{-1}; (a) at room temperature and (b) at 50°C.

Many biotransformation reactions need to be carried out in organic solvents due to the insolubility of the substrates to be used. Enzyme immobilisation onto solid supports is an approach to stabilise enzymes for use under these conditions. The activity of the immobilised enzyme within the monolithic micro channels in the presence of organic solvents (5-50%) was studied using dimethylformamide, dimethylsulfoxide, ethanol, methanol and acetone, Figure 4. The presence of solvents reduced the activity of the enzyme from 20-80% depending on type and amount of solvent used.

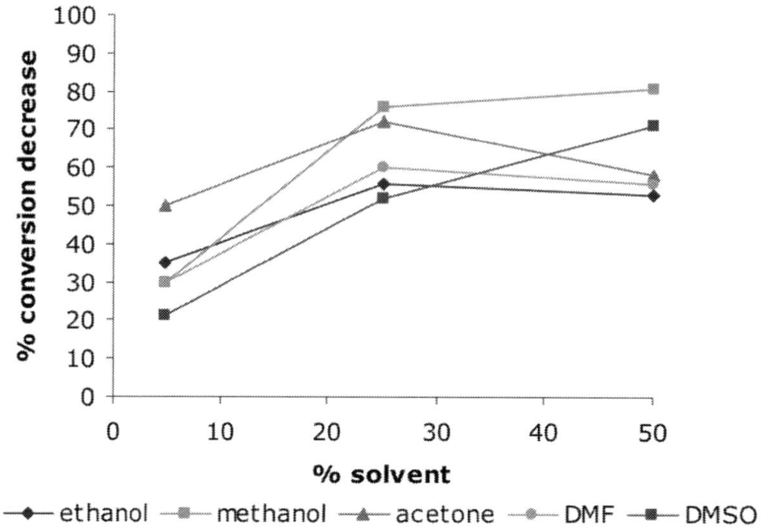

Figure 4. Effect of solvents on substrate conversion; 0% conversion decrease means that the use of solvent in the system does not affect the enzyme's ability to convert 10 mM *N*-benzoyl-L-phenylalanine (preferred substrate) to L-phenylalanine, while 100% conversion decrease indicates no activity towards its preferred substrate in the presence of the studied solvent.

Substrate specificity of the immobilised enzymes was assayed using 10 mM substrates in 100 mM Tris-HCl, pH 8.0 inside the micro reactor using a flow rate of 1 µL min^{-1}. The enzyme appeared to have a preference for *N*-substituted phenylalanine. Maximal substrate conversion was observed when *N*-benzoyl-L-phenylalanine, *N*-acetyl-L-phenylalanine and *N*-chloroacetyl-L-phenylalanine were used as substrates. Among the other substrates examined, the order of preferred activity decreased from *N*-benzoyl-L-threonine (68.3%) > *N*-benzoyl-L-leucine (52.2%) > *N*-acetyl-L-tyrosine (33.3%) > *N*-acetyl-L-tryptophan (7%).

Conclusions

Immobilisation of recombinant L-aminoacylase from the thermophilic archaeon, *T. litoralis* onto polymer monoliths fabricated within micro reactor channels was successfully carried out. The linkages were formed between the amino groups of the enzyme and epoxy terminal on the surface of the monoliths. The immobilised enzyme retained its stereospecific behaviour and remained stable with the optimal activity at both room temperature and 50°C for a substantial period of time. Using the continuous flow micro reactor system, 100% substrate conversion was obtained at room temperature, which is a significantly lower temperature than the optimal condition reported for the conventional batch system. Additionally, the results indicate that substrate screening and higher throughput over a shorter period of time can be achieved using the micro reactors.

Acknowledgements

This work was financially supported by Engineering and Physical Sciences Research Council (EPSRC), UK.

References

1. A.M. Hickey, L. Marle, T. McCreedy, P. Watts, G.M. Greenway and J.A. Littlechild, *Biochem. Soc. Trans.*, **35** (2007) 1621-1623.
2. P.A. Greenwood, C. Merrin, T. McCreedy, G. M. Greenway, *Talanta*, **56** (2002) 539-545.
3. P. Claus, D. Hőnicken and T. Zech, *Catalysis Today*, **67** (2001) 319-339.
4. P. Watts, *QSAR Comb. Sci.*, 24 (2005) 701-711.
5. P. Watts and S. J. Haswell, *Chem. Eng. Technol.* **28** (2005) 290-301.
6. G. N. Doku, W. Verboom, D.N. Reinhoudt and A. van den Berg., *Tetrahedron*, **61** (2005) 2733-2742.
7. D.S. Peterson, R. Rohr, F. Scec and J.M.J. Fréchet, *Anal. Chem.*, **74** (2002) 4081-4088.
8. S. Banu, G.M. Greenway, T. McCreedy and R. Shaddick, *Anal. Chim. Acta.,* **486** (2003) 149-157.
9. T. Honda, M. Miyazaki, Y. Yamaguchi, H. Nakamura and H. Maeda, *Lab. Chip*, **7** (2007) 366-372.
10. H. S. Toogood, I. N. Taylor, R. C., Brown, S. J. C., Taylor, R. McCague and J.A. Littlechild, *Biocatal. Biotransform.*, **20** (2002) 241-249.
11. H. S. Toogood, I. N. Taylor, R.C. Brown, S. J. C., Taylor, R. McCague and J.A. Littlechild, *Extremophiles*, **6** (2002) 111-122.

240e Synthesis of Carbon Nanofibers as Support Layer for Metal Catalysts in a Microreactor for Three-Phase Reactions

Digvijay B. Thakur[1], R. M. Tiggelaar[2], K. Seshan[1], L. Lefferts[1], and J. G. E. Gardeniers[2]. (1) Catalytic Processes and Materials, University of Twente, Meander, Room 360, P. O Box 217, Enschede, Netherlands, (2) Mesoscale Chemical Systems, University of Twente, Meander, Room 146, P. O Box 217, Enschede, Netherlands

The main goal of one of the projects within our microreactor-research-program is to develop a microstructured multiphase microreactor, *viz.* a gas-liquid contactor with specially modified catalytic coatings on its microstructured internals, for studying three-phase (G-L-S) reactions. The microreactor will be fabricated using micromachining (cleanroom) technologies in silicon and vitreous materials (i.e., fused silica and/or Borofloat glass). Figure 1a shows the top view of the reactor channel containing silicon pillars. The schematic cross-sectional view, depicted in figure 1b, explains the working principle of the proposed multiphase microreactor. Gas can be injected into a liquid stream *via* the hollow pillars with porous walls. On the outside of these porous pillars Carbon Nano Fibers (CNFs) will be synthesized that act as a support layer for the catalytic active phase. Figure 2a shows, schematically, an enlarged view of porous pillars with CNFs anchored on the outside. Figure 2b shows an entangled jungle of such a CNF layer which can be further utilized as structural support for metal catalyst [2], as illustrated in figure 2c. This abstract focuses on the synthesis of CNFs on the microreactor interior.

Currently, there is a trend to use structured catalyst supports, i.e., that the catalyst is deposited on a rigid, orderly arranged support. Catalytic support layers based on CNFs offer a novel option to facilitate this task [1, 3]. With their inherent high surface area-to-volume ratio, CNFs provide more catalytic surface area [1, 2], thereby obtaining sufficient activity per unit volume of catalyst, while maintaining a low fluidic resistance (which is important in microfluidic devices). Additionally, diameter and length of the fibers and hence the bulk density of the CNF layer, can be manipulated to tailor porosity and overcome tortuosity problems. This also eliminates the internal diffusion limitations by preventing concentration gradients inside the CNF layer.

A CVD-process is applied to synthesize CNFs using hydrocarbon gas (e.g. ethylene) and a metal catalyst (such as nickel). Preparation of nickel thin-films along with various adhesion layers (Ti, Ta and TiW) is done by evaporation (thickness of films vary from 10 nm to several hundreds of nanometers). Various compositions or layer configurations (dual-layer and sandwich-structured) are currently investigated in order to understand their influence on the morphology and stability of the synthesized CNF layers. The fundamental growth/synthesis of CNFs – structure, morphology, length, diameter etc. – is studied as function of CVD conditions (growth temperature, and time). Different stages in CNF-growth can be discriminated based on the growth time. After 5 min. of reaction, formation of nickel nanoparticles initiated along with very few and thin CNFs as depicted in figure 3a. Whereas, significant growth took place after 15 min. achieving thick layer of CNFs (8 to 15 μm Long) as shown in figure 3b. Longer reaction times (30 min. to 1h.) leads to formation of cauliflower resembling structures along with CNFs as illustrated in figure 3c.

It is expected that this novel approach of constructing microstructured reactors with stable and well-defined carbon nanofibers upon porous silicon pillars as structured catalyst support, results in accomplishing the goal of designing smart and efficient multi-phase microreactors.

References:

[1] M.-J. Ledoux , C. Pham-Huu, Catal. Today 102-103 (2005) 2.

[2] M. L. Toebes, F. F. Prinsloo, J. H. Bitter, A. J. van Dillen, K. P. de Jong, J. Catal. 214 (2003) 78.

[3] J. K. Chinthaginjala, K. Seshan, L. Lefferts, Ind. Eng. Chem. Res. 46 (2007) 3968.

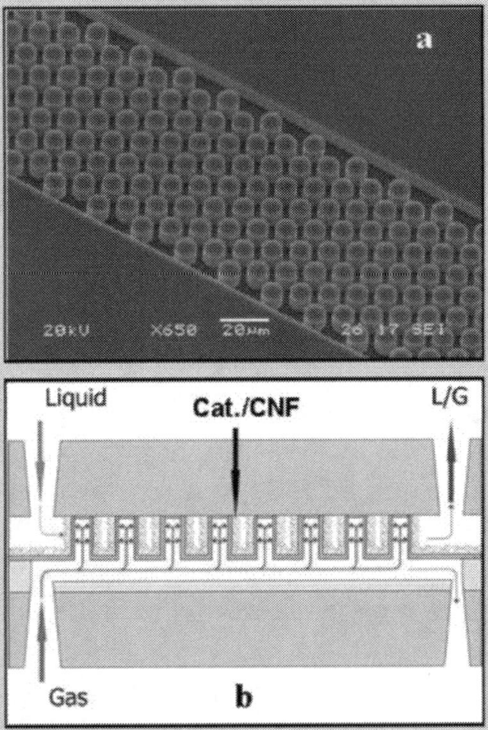

Figure 1: *Multiphase-microreactor with carbon nanofibers*

- **a.** Top view of the fabricated reactor microchannel showing Si-pillars
- **b.** Schematics of three-phase reactor in operation

Figure 2: *Carbon nanofiber as structured support for metal catalyst*

- **a.** Expanded schematics of porous hollow pillars with CNF
- **b.** Layer of CNF grown using C_2H_4 on Ni-film
- **c.** TEM image of 5 wt% CNF supported Ru catalyst [2]

Figure 3: *HR-SEM images of different stages of CNF-synthesis on an evaporated Ni-film (25% C_2H_4 in N_2, 700 °C)*

- **a.** Growth Time: 5 minutes (Formation of Ni particles)
- **b.** Growth Time: 15 minutes (Thick layer of CNFs - 8 to 15 µm long)
- **c.** Growth Time: 30 minutes (Formation of cauliflower resembling structures along with CNFs)

45b Fabrication and Reaction Properties of Mesoporous Silica Thin Films Inside Microreactors

Sho Kataoka, Akira Endo, Atsuhiro Harada, Yuki Inagi, and Takao Ohmori. Energy-Efficient Chemical Systems Group, Research Institute for Innovation in Sustainable Chemistry, National Institute of Advanced Industrial Science and Technology (AIST), Japan, AIST Tsukuba Central 5, 1-1-1, Higashi, Tsukuba-shi, Ibaraki, 305-8565, Japan

Micelle-templated mesoporous silica thin films are formed on the inner walls of microreactors. Microreactors possess a high interfacial area per unit volume and the use of microreactors can be effective for heterogeneous catalytic reactions. However, the methods for immobilizing complex molecular catalysts in a micrometer-scale space are quite limited and homogeneous catalysts are flowed with reactant solutions in many cases. Micelle-templated mesoporous silicas (e.g., MCM-41 and SBA-16) possess highly ordered structures and can serve as a catalyst support inside microreactors. In this study, SBA-16 mesoporous silica films are formed inside microreactors and their structures are fully investigated. Furthermore, Al-doped SBA-16 films are formed as a heterogeneous catalyst inside microreactors since pure SBA-16 has little catalytic activity. By conducting catalytic reactions inside microreactors, the reaction properties of SBA-16 films are discussed.

83c Precipitation of Barium Sulphate in a Hollow Fibers Membrane Device – Numerical Simulation and Experimental Validation

Roland Kieffer, **Catherine Charcosset**, Denis Mangin, and François Puel. LAGEP, CNRS, UCBLyon1, CPE, 43 Bd du 11 Novembre 1918, Villeurbanne, 69 622, France

Precipitation of barium sulphate in a hollow fibers membrane device involves the permeation of KSO4 through the membrane pores to react with BaCl2 flowing in the lumen side. In this paper, the numerical simulation of barium sulphate precipitation in a hollow fibers membrane device is obtained from the Navier-Stokes equations, transport equations for the chemical species and moments of the crystal size distribution (CSD). The numerical simulations allow the determination of the number-averaged particle size, the coefficient of variation of the CSD, the particles flux, as well as the chemical conversion of the product. The effect of the reactants concentration (5-100 mol m-3), the mean velocity at the inlet of the lumen side (0.025-0.2 m s-1), the inner radius of the hollow fiber (10-3 - 10-4 m), and its length (0.1-0.4 m), on barium sulphate precipitation are investigated. The numerical simulations show that narrow and long hollow fibers allow obtaining high particle fluxes and high chemical conversions. Narrow hollow fibers are advised to obtain low coefficients of variation of the CSD, and short hollow fibers to obtain small averaged particles size. For practical applications, a hollow fibers membrane device should be chosen according to these various criteria. Moreover, experimental measurements of barium sulphate precipitation in a hollow fibers membrane device are realized, and scanning electron microscopy is used to determine the crystal size. It is shown that experimental data compared well with model predictions.

Well-Defined Hyperbranched and Linear Polymers by Anionic Polymerization in Continuous Flow

Daniel Wilms, Frederik Wurm, Johannes Klos, Holger Löwe* and Holger Frey*

Institute of Organic and Macromolecular Chemistry, University of Mainz, Duesbergweg 10-14, Mainz, 55099, Germany

Introduction

Continuous flow processes involving micromixers offer intriguing potential with respect to various types of controlled polymerizations since these often proceed strongly exothermic and are diffusion-limited due to their fast kinetics. Nevertheless, only limited efforts have been reported to date towards the actual preparation of polymers in such microdimensional devices.[1-3] Only within recent years, vital progress in the field of polyreactions in microstructured reaction devices has been made by several research groups who carried out free radical,[4] controlled radical,[5] anionic[6, 7] and cationic[8-11] polymerizations using different microreactor designs. The fascinating potential of polymer chemistry in continuous flow has also been pointed out by recent studies focusing on the synthesis of dendrimers[12] and block copolymers[13] in microfluidic devices. Since branched polymers generally possess lower viscosity than their linear counterparts both in solution and in the melt, we considered it a worthwhile challenge to realize a hyperbranching polymerization in a microreactor set-up.[14] On the other hand, ring-opening of highly strained epoxides is a fast and exothermic reaction that qualifies for syntheses in such a device.

Hyperbranched polyglycerols offer several benefits, e.g., a high number of hydroxyl groups as a versatile functionality, excellent biocompatibility[15] and offer the possibility to tailor architecture, functionality and solubility. One of the goals of this work was to develop an alternative approach for the preparation of hyperbranched polyglycerol, avoiding tedious and time-consuming conventional methods.[16, 17]

Our work also involved the extension of the micromixer-based system to living carbanionic polymerizations which have been established as one of the major polymerization techniques since the 1950ies.[18] Only about ten years after the first report on living polymers, pioneering work involving anionic polymerization in continuous flow was accomplished by M. Szwarc et al.[19, 20] and G. V. Schulz et al.[21-23] The respective approach was later extended by A. Müller et al. in elegant work, using a flow-tube reactor for a detailed kinetic investigation of the anionic polymerization of methyl methacrylate and for the preparation of block copolymers.[24-27] Herein we report the development of an approach for living carbanionic polymerization on a synthetic scale, using a microstructured platform under continuous flow conditions.

Experimental

Micromixing device: All polymerizations were carried out in a microstructured reaction set-up provided by the *Institut für Mikrotechnik Mainz (IMM).* It consists of a tempered oil or water bath hosting the mixing devices (stainless steel interdigital SIMM-V2 mixer with an internal volume of 8 µL) that is equipped with two reactant inlets and one outlet. After mixing, the reaction mixture is directly guided into the respective delay tube (60µL, 2 or 6 mL). The rear end of the tubular delay tube leads to the reactor outlet. Flow rates are controlled via HPLC pumps (Knauer WellChrom K-501, inert 10mL pump heads with ceramic inlays).

Synthesis of hyperbranched polyglycerol: Purified glycidol and a solution of the initiator (prepared by partial deprotonation of trimethylol propane (TMP) with potassium methylate) in diglyme were separately pumped into the micromixing device. The oil bath was tempered to the desired temperature (usually 120°C). A delay loop with an inner volume of 6 ml was used. After passing the delay loop, the product was recovered at the device outlet and immediately quenched with methanol. Methanolic solutions of the products were stirred with a cation exchange resin (Lewatit K 1131) at room temperature until neutral. After filtration, the product was redissolved in methanol, precipitated twice into cold acetone and dried in vacuo at 80°C for 12 h.

Synthesis of polystyrene: Before each reaction the solvent was pumped through the microstructured reactor set-up for approximately 15 minutes at 0.5 ml/min via both HPLC pumps, followed by freshly distilled dry and degassed solvent for an additional 10 min at 0.5 ml/min. Purified styrene was kept under argon atmosphere in a flask covered with a septum as a 10-20 weight% solution in dry cyclohexane or THF. In a second flask the desired solvent and the required amount of initiator (*sec*-BuLi) were combined by cryo-transfer of the solvent and injection of the initiator solution via syringe. Both reactant flasks were connected to the HPLC pumps via PTFE tubes transfixed through the septa and the water/oil bath of the reactor plant was tempered to the desired temperature (25°C or 80°C respectively). For all experiments carried out in cyclohexane, a delay tube with an internal volume of 2 ml was used (250 cm length and 1000 µm diameter). For reactions in THF, the polymers were recovered directly behind a short tube (30 cm length and 500 μm diameter, volume approx. 60 μL) connected to the micromixer. Polystyrene samples were subsequently dried in vacuum at 40°C for 12 h.

Block Copolymer Synthesis: THF was used as reaction solvent. A second micromixing device was installed after the living polystyrene anions passed a delay tube. 4-tert-butoxystyrene was pumped into the second micromixer via a third HPLC-pump with different flow rates. By variation of the flow rates of both monomers, different diblock copolymers with varying block lengths were synthesized. Polymers were recovered after precipitation into methanol and drying in vacuo at 40°C for 12 h.

Results and Discussion

Hyperbranched polyglycerols have been prepared in continuous flow up to molecular weights of 1,000 g/mol by reacting a solution of partially deprotonated trimethylol propane (TMP) with glycidol monomer in a reactor equipped with a slit interdigital mixer. Although the technique of slow monomer addition, which is essential for control over the reaction in the conventional approach, is not applied here, the polymers obtained were remarkably narrowly distributed ($M_w/M_n < 1.5$). This result is very promising in terms of process optimization because reaction times can be reduced from several hours to minutes by using the microreactor approach. NMR spectroscopy and MALDI-ToF mass spectrometry (Figure 3) confirmed full incorporation of the initiator core into all species present in the distribution of the hyperbranched structure.

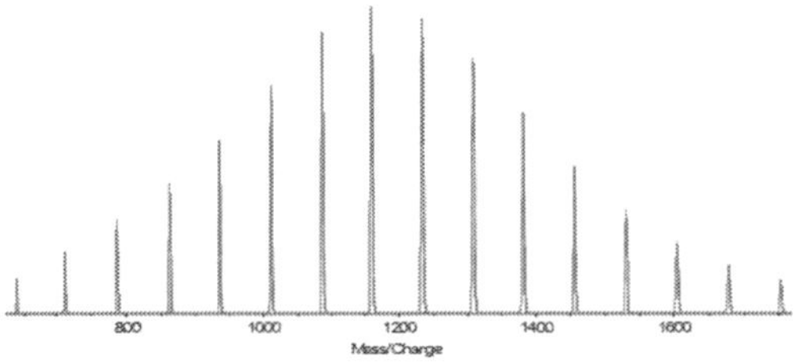

Figure 1. MALDI-ToF spectrum of hyperbranched polyglycerol (M_n = 1,000 g/mol) prepared in the microstructured reaction device. Each peak corresponds to the added masses of the TMP core, the respective number of glycidol units and one lithium ion.

In living anionic polymerizations, fast mixing of initiator and monomer is essential for simultaneous chain growth and hence narrow molecular weight distributions. Microstructured reactors should therefore offer some crucial benefits for the polymerization of vinyl monomers under living conditions as well.

For example, polymerization of styrene by initiation with *sec*-butyllithium in cyclohexane can be carried out at a temperature of 80°C in the microstructured reaction device, permitting a recovery of living anions of the desired chain length after several minutes of reaction time. Comparably mild conditions have to be applied in a conventional reactor for carbanionic polymerizations, leading to prolonged reaction times (usually several hours) and often impeded control over the degree of polymerization and molecular weight distribution. The SEC diagrams obtained from the polystyrene samples prepared in the microstructured reactor under continuous flow conditions are shown in Figure 2, demonstrating monomodal and narrow molecular weight distributions.

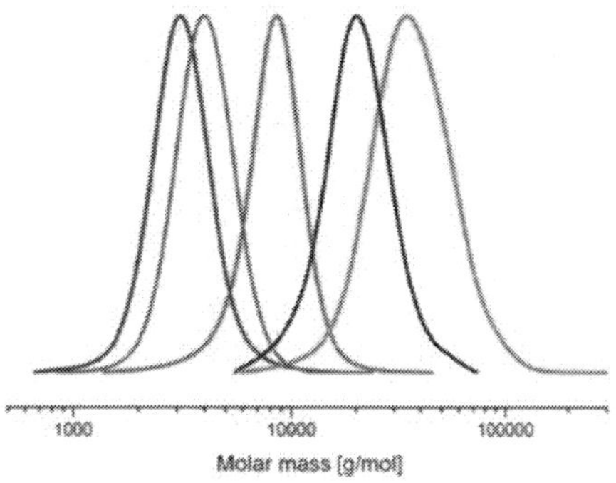

Figure 2. SEC diagrams (measured in chloroform) obtained from polystyrene samples prepared in the microstructured reactor with attached delay tube.

The anionic polymerization of styrene proceeds significantly faster in polar reaction media, e.g., in polar ether solvents like THF. Narrowly dispersed (M_w/M_n = 1.08 – 1.2) polystyrenes were synthesized in THF in a broad range of molecular weights (up to 70,000 g /mol) within seconds using the microstructured reaction device while under these conditions, no control over the polymerization can be achieved in a conventional set-up.

Furthermore, block copolymers consisting of styrene and 4-*tert*-butoxystyrene blocks were successfully synthesized, retaining excellent control over molecular weight and polydispersity by using the set-up illustrated in Figure 3.

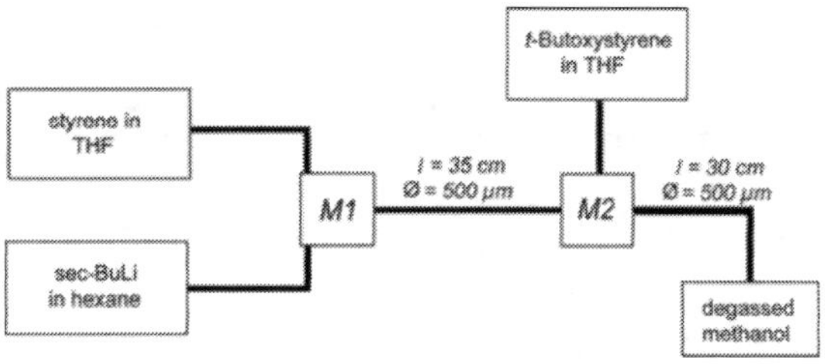

Figure 3. Illustration of the experimental set-up used for the synthesis of poly(styrene-*b*-4-*tert*-butoxystyrene

The obtained block copolymers (Table 1) were characterized in detail by NMR spectroscopy, SEC and MALDI ToF MS.

Table 1. Characterization data of polystyrene-*b*-poly(4-tert-butoxystyrene) samples prepared by living anionic polymerization in a microstructured reaction set-up using THF as solvent.

S:*t*-BuOS[a)]	M_n[b]	M_n[c]	M_n[d]	M_w/M_n[e]
0:20	3,600	3,700	4,200	1.21
5:5	1,400	1,300	1,500	1.18
10:5	1,900	1,900	2,100	1.22
22:12	4,400	4,800	4,700	1.17
80:30	13,600	13,100	13,600	1.15
200:25	25,300	24,700	25,600	1.14

[a]) Block ratio styrene (S) : 4-*t*-butoxystyrene (*t*-BuOS) adjusted by flow rates.

[b]) theoretical value of the number average molecular weight distribution in g/mol.

[c]) number average of the molecular weight distribution in g/mol determined by size exclusion chromatography in tetrahydrofuran.

[d]) number average of the molecular weight distribution in g/mol determined by multi-angle light scattering (MALLS) in tetrahydrofuran.

[e]) polydispersity index determined by MALLS detection in tetrahydrofuran.

Conclusion

We have presented the first report on a continuous process for the preparation of hyperbranched polymers in a microstructured reactor. The highly efficient heat and mass transport in microstructured reaction devices affords a quick pathway to these materials, without having to resort to time-consuming conventional methods. As glycidol, like several other epoxide monomers, is a highly toxic compound, the reaction can be carried out in the safer environment of the contained microreactor device.

In addition, we have successfully developed a versatile process for living carbanionic polymerization with subsequent functional termination and diblock copolymer synthesis using a continuous flow reactor set-up equipped with a micromixing device. This approach provides an elegant pathway to well-defined polymers, avoiding time-consuming and laborious conventional methods that involve individual reactor crafting and prolonged reaction times. We are convinced that these examples of transferring well-established polymer synthesis procedures to a microdimensional device will motivate further studies, including the modification of polymer architectures by further variation of reactor designs.

References

1. Hessel, V.; Serra, C.; Löwe, H.; Hadziioannou, G. *Chem. Ing. Tech.* **2005,** 77, 1693.
2. Steinbacher, J. L.; McQuade, D. T. *J. Polym. Sci. Part A: Polym. Chem.* **2006,** 44, 6505.
3. Wilms, D.; Klos, J.; Frey, H. *Macromol. Chem. Phys.* **2008,** 209.
4. Iwasaki, T.; Yoshida, J. *Macromolecules* **2005,** 38, 1159.
5. Wu, T.; Mei, Y.; Cabral, J. T.; Xu, C.; Beers, K. L. *J. Am. Chem. Soc.* **2004,** 126, 9880.
6. Honda, T.; Miyazaki, M.; Nakamura, H.; Maeda, H. *Lab Chip* **2005,** 5, 812.
7. Miyazaki, M.; Honda, T.; Nakamura, H.; Maeda, H. *Chem. Eng. Technol.* **2007,** 30, 300.
8. Nagaki, A.; Kawamura, K.; Suga, S.; Ando, T.; Sawamoto, M.; Yoshida, J. *J. Am. Chem. Soc.* **2004,** 126, 14702.
9. Iwasaki, T.; Nagaki, A.; Yoshida, J. I. *Chem. Commun.* **2007,** 1263.
10. Iwasaki, T.; Yoshida, J. *Macromol. Rapid Commun.* **2007,** 28, 1219.
11. Ouchi, M.; Inagaki, N.; Ando, T.; Sawamoto, M. *Polym. Prepr. (Am. Chem. Soc., Div. Polym. Chem.)* **2005,** 46, 939.
12. Liu, S. H.; Chang, C. H. *Chem. Eng. Technol.* **2007,** 30, 334.
13. Wu, T.; Mei, Y.; Xu, C.; Byrd, H. C. M.; Beers, K. L. *Macromol. Rapid Commun.* **2005,** 26, 1037.
14. Wilms, D.; Nieberle, J.; Klos, J.; Löwe, H.; Frey, H. *Chem. Eng. Technol.* **2007,** 30, 1519.
15. Kainthan, R. K.; Janzen, J.; Levin, E.; Devine, D. V.; Brooks, D. E. *Biomacromolecules* **2006,** 7, 703.
16. Sunder, A.; Hanselmann, R.; Frey, H.; Mülhaupt, R. *Macromolecules* **1999,** 32, 4240.
17. Sunder, A.; Mülhaupt, R.; Haag, R.; Frey, H. *Adv. Mater.* **2000,** 12, 235.
18. Szwarc, M. *Nature* **1956,** 178, 1168.
19. Geactinov, C.; Smid, J.; Szwarc, M. *J. Am. Chem. Soc* **1962,** 84, 2508.
20. Bhattacharyya, D. N.; Lee, C. L.; Smid, J.; Szwarc, M. *J. Phys. Chem.* **1965,** 69, 612.
21. Hostalka, H.; Figini, R. V.; Schulz, G. V. *Makromol. Chem.* **1964,** 71, 198.
22. Löhr, G.; Schulz, G. V. *Makromol. Chem.* **1964,** 77, 240.
23. Löhr, G.; Schmitt, B. J.; Schulz, G. V. *Z. Phys. Chem.* **1972,** 78, 177.
24. Baskaran, D.; Müller, A. H. E. *Macromolecules* **1997,** 30, 1869.
25. Baskaran, D.; Müller, A. H. E.; Sivaram, S. *Macromolecules* **1999,** 32, 1356.
26. Baskaran, D.; Müller, A. H. E.; Sivaram, S. *Macromol. Chem. Phys.* **2000,** 201, 1901.
27. Königsmann, H.; Jüngling, S.; Müller, A. H. E. *Macromol. Rapid Commun.* **2000,** 21, 758.

Synthetic Study of Biofunctional Molecules by Using Microfluidic System

Koichi Fukase,[a]* Katsunori Tanaka,[a] Shinya Motomatsu,[a,b] Koichi Koyama[b]
[a]*Department of Chemistry, Graduate School of Science, Osaka University, 1-1 Machikaneyama, Toyonaka, Osaka, 560-0043, Japan*
[b]*Kishida Chemical Co., Ltd., Technopark 14-10, Sanda, Hyogo 669-1339, Japan*
koichi@chem.sci.osaka-u.ac.jp

A continuous flow microreactor, which is reported to realize efficient mixing and fast heat transfer is recognized as an innovative technology in recent organic synthesis.[1,2] The flow system also allows the reaction to be quenched immediately after the formation of the unstable product. In this paper, we report successful applications of a continuous microfluidic system to oligosaccharide synthesis and an immunostimulating terpenoid pristane. We also report acetone-based Aldol reaction in aqueous biphasic medium under microfluidic conditions.

In order to investigate the biological function of sialylated oligosaccharides, we have developed an efficient α-sialylation by utilizing the highly reactive sialyl donors having C5-cyclic imides by virtue of the "fixed dipole effect".[3] The scale-up in batch process, however, decreased the yield and selectivity. The decrease in sialylation efficiency might be due to the high reactivity of the donor **1**. For such a case, precise reaction control is very difficult under the conventional batch process conditions, especially when the reaction is scaled-up. Thus, the disorder of the reaction factors in the scaled-up batch reaction, i.e., (i) precise temperature control, (ii) mixing efficiency between acceptor, donor, and Lewis acid, and (iii) reaction time, might lead to the glycal production. In order to circumvent these problems, we used a continuous flow microreactor. An application of the microfluidic system to the glycosylation reaction was first reported by Seeberger on α-mannosylation.[2d] We also have established an efficient microfluidic glycosylation in combination with the affinity separation method.[4]

For the present microfluidic sialylation, a propionitrile solution of sialyl donor **1** and acceptor **2** with various concentrations was mixed with TMSOTf solution in dichloromethane at -78 °C using an IMM micromixer at the flow rate of 1.0 mL/min. After the reaction mixture was allowed to flow at -78 °C for additional 47 seconds through a reactor tube (Φ = 1.0 mm, l = 1.0 m), the mixture was quenched by another flow of excess triethylamine dissolved in dichloromethane by using a T-shaped mixer at -78 °C. When the concentrations of the donor **1**, acceptor **2**, and TMSOTf were adjusted to 0.15 M, 0.1 M, and 0.08 M, respectively, disaccharide **3** was obtained in only 14 % yield and an excess acceptor **2** was recovered (entry 1). However, we were pleased to find that the yield of **3** dramatically increased (88 %) when the concentration of the Lewis acid was increased up to 0.15 M. (entry 2). Finally, the desired α-sialoside **3** was obtained in quantitative yield by increasing the concentration of the donor **1** to 0.2 M (entry 3). The microfluidic reaction successfully controlled the high reactivity of the sialyl donor **1** for α-sialylation, and efficient and large-scale procedure have now been realized.[5]

Optimization of α-(2-6)-Sialylation between Donor **1** and Acceptor **2** Using Microreactor.

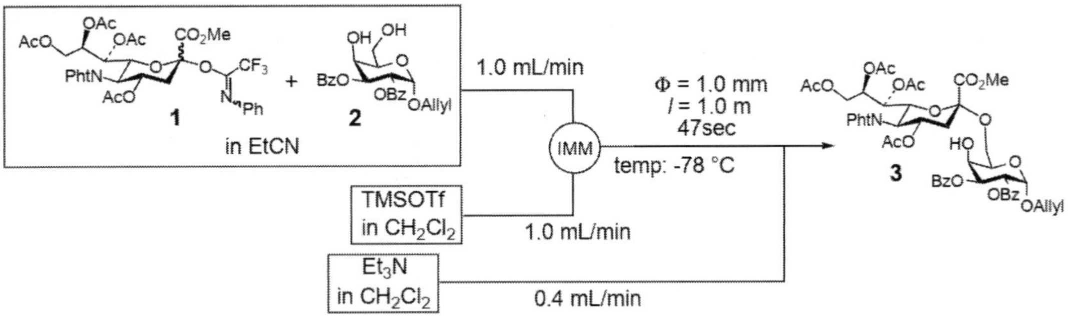

entry	donor **1** (M)	acceptor **2** (M)	TMSOTf (M)	yield of **3**(%)	α : β
1	0.15	0.1	0.08	14	α only
2	0.15	0.1	0.15	88	α only
3	0.2	0.1	0.15	>99	α only

An efficient procedure for the reductive opening of 4,6-*O*-benzylidene acetals was then investigated under microfluidic conditions. Reductive opening of the sugar 4,6-*O*-benzylidene acetals by the combination of acid/hydride reagents is one of the most useful transformations in the field of carbohydrate chemistry, since the benzyl-protected derivatives at either the C4- or C6-hydroxyl can be selectively prepared by the choice of the reagents and/or the solvents systems. For example, the treatment of 4,6-*O*-benzylidene protected glucose derivative **4** with $Et_3SiH/BF_3{\bullet}OEt_2$[6] in dichloromethane provides the 6-*O*-benzyl derivative **7**, while the treatment with $BH_3{\bullet}Me_2NH/BF_3{\bullet}OEt_2$[7] gives the 4-*O*-benzyl derivative **8** in good selectivity.

The typical procedure of the reductive opening of the acetals involves the very slow addition of the acid to a mixture of the substrate and excess hydride reagent at 0 °C, followed by continuous stirring at room temperature for another few hours to ensure the completion of the reaction. Since the reaction is exothermic, it is important to control the addition speed of the acid in order to prevent concomitant acid-catalyzed hydrolysis of the benzylidene groups. However, the yields are not always reproducible, especially when the reaction is performed in a large scale, giving rise to the hydrolyzed by-products, 4,6-*O*-diols.

The reductive opening by a microfluidic system was then examined using the glucose 4,6-*O*-benzylidene acetal **4**. Benzylidene acetal **4** (1.0 M) and Et_3SiH (10 M) dissolved in CH_2Cl_2 were mixed with various concentrations of $BF_3•OEt_2$ in CH_2Cl_2 (0.1 M – 1.0 M) at 0°C by using a micromixer (Techno Applications Co., Ltd., Tokyo, Japan)[8] at the flow rate of 0.5 mL/min. After the reaction mixture was allowed to flow at room temperature for additional 45 seconds through a reactor tube (Φ = 1.0 mm, l = 0.9 m), the mixture was quenched by saturated $NaHCO_3$ solution at 0 °C.[9]

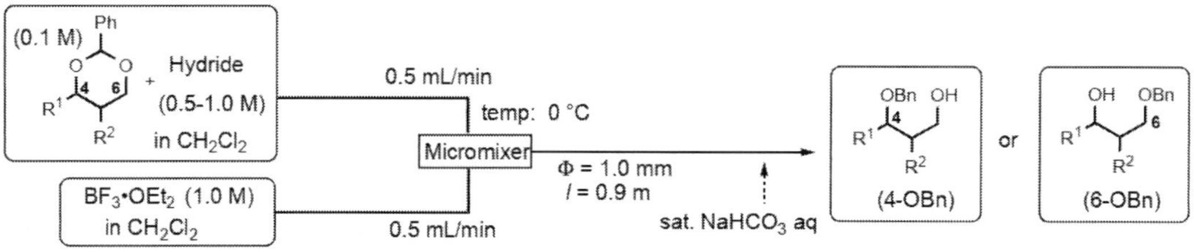

entry	substrate	reducing agent	product	yield (%, microfluidic)[a]	yield (%, batch)[a,b]
1	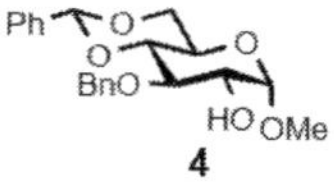 **4**	Et_3SiH (1.0 M)	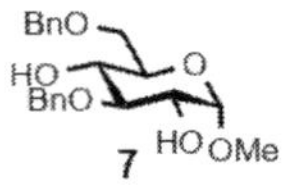 **7**	93[c] (6-OBn)	58
2	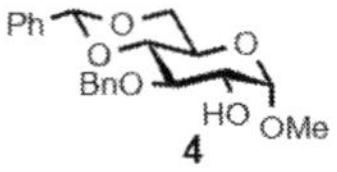 **4**	$BH_3•Et_2NH$ (0.5 M)	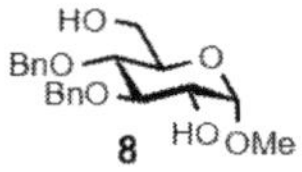 **8**	100 (4-OBn)	90
3	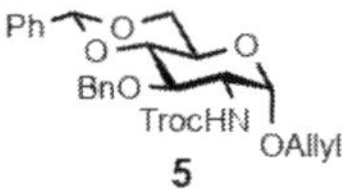 **5**	Et_3SiH (1.0 M)	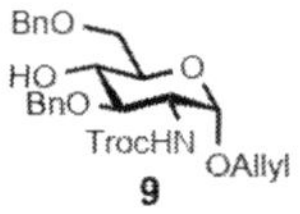 **9**	91 (6-OBn)	83
4	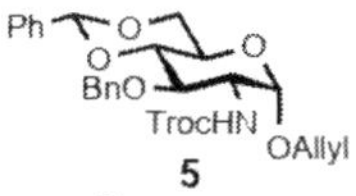 **5**	$BH_3•Et_2NH$ (0.5 M)	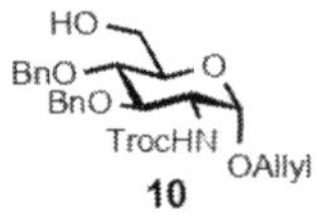 **10**	100 (4-OBn)	86
6	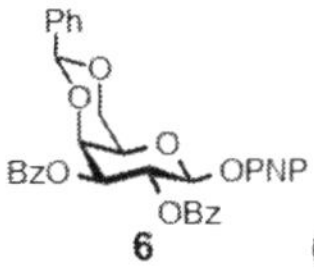 **6**	Et_3SiH (1.0 M) (PNP: *p*-nitrophenyl)	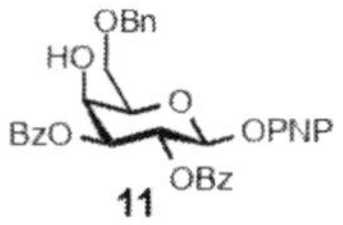 **11**	91 (6-OBn)	62

When 0.1 M solution of BF_3•OEt_2 in CH_2Cl_2 was used, only a hydrolyzed compound was obtained in about 50 % yield accompanied with recovery of the starting material **4**. However, the yield of 6-*O*-benzyl derivative **7** dramatically increased when the concentration of BF_3•OEt_2 was increased up to 1.0 M; the desired **7** was obtained in 93 % yield together with 5% of the 4-*O*-benzyl derivative **8**, and no hydrolyzed by-product was detected (entry 1). Interestingly, the reaction with a more concentrated BF_3•OEt_2 solution (BF_3•OEt_2 : CH_2Cl_2 = 1 : 1) also provided **7** in excellent yield. The reaction was very rapid and **7** was obtained within a minute after mixing with the acid at the micromixer device.

The treatment of **4** with BH_3•Et_2NH selectively provided 4-*O*-benzyl derivative **8** in quantitative yield (entry 2). The benzylidene group of glucosamine derivative **5** was also selectively cleaved under the microfluidic condition by utilizing either Et_3SiH or BH_3•Et_2NH as a hydride source to provide the 6- and 4-*O*-benzyl derivatives **9** and **10** in excellent yields, respectively (entries 3 and 4). The selective reduction of galactose derivative **6** gave the 6-*O*-benzyl derivative **11** in 91 % yield (entry 6). It is worthwhile mentioning that the established microfluidic reaction reproducibly provided the *O*-benzylated compounds **7-11** in greater than 90% yields in all cases.[9]

Pristane is a saturated isoprenoid isolated from the basking shark, *Cetorhinus maximus*. This hydrocarbon oil has been widely used as an adjuvant for monoclonal antibody production in mouse ascites. However, in 2002, the basking shark was listed in Article II of the Washington Convention, and since then, the availability of pristane from natural sources has been very limited. We therefore investigated simple and efficient procedure for the synthesis of pristane, i.e., (1) oxidation of farnesol, (2) alkylation, (3) dehydration, and (4) hydrogenation, as shown in Scheme 1.[10]

Scheme 1

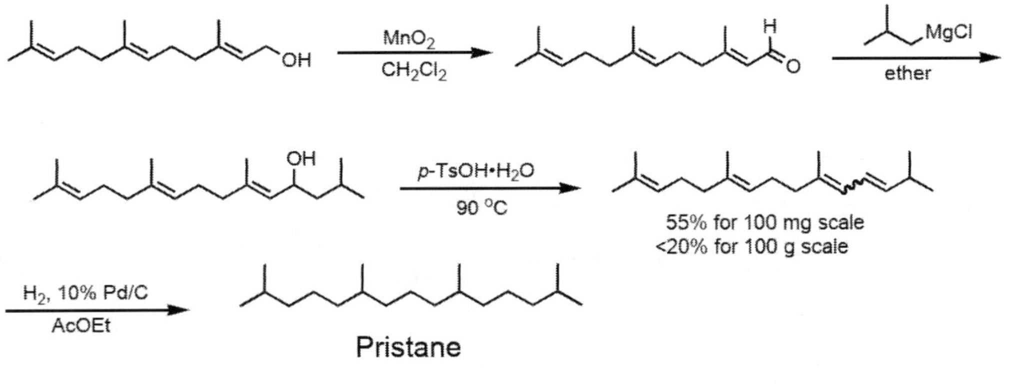

The challenging step in Scheme 1 is the acid-catalyzed dehydration of allylic alcohol. When the reaction was performed in 100 mg scale using a catalytic amount of *p*-TsOH in benzene at 80 °C, the corresponding diene was obtained in 55 % yield. However, when the scale was raised to 100 g, various cation-mediated by-products, such as the cyclized products or the alkyl group-migrated compounds were produced.

We then examined the microfluidic dehydration under the following conditions (Figure 1): Allylic alcohol (1.0 M in THF) was mixed with a solution of *p*-TsOH (various concentrations of 0.1-1.0 M in THF : toluene = 1 : 1) by using an IMM micromixer or Comet X-01 at each flow rate of 0.3 mL/min. After the reaction mixture was allowed to flow for additional 45 seconds through a reactor tube (Φ = 1.0 mm, l = 1.0 m), the mixture was quenched by a saturated $NaHCO_3$ solution at room temperature. Higher concentration and higher reaction temperature gave better result. The desired diene was finally obtained in 80 % yield (total yield from farnesol) at the acid concentration of 1.0 M. It is noted that under the established microfluidic conditions, the formation of other by-products could not be detected by TLC analysis.

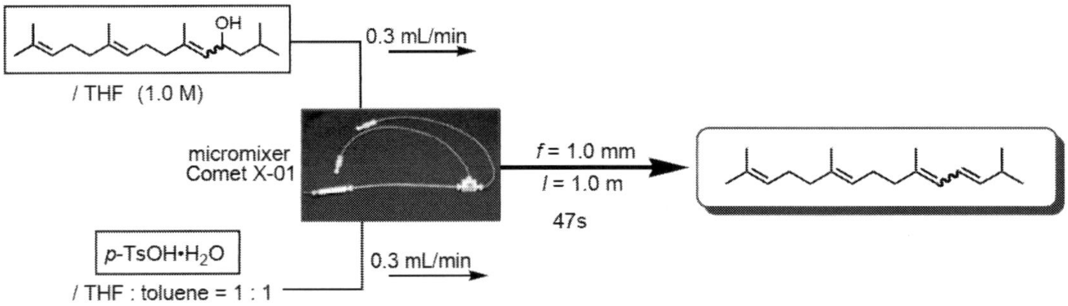

Entry	Conc. of Acid (M)	Temp. (°C)	Yield (%)
1	0.1	rt	< 10
2	0.2	rt	30
3	0.5	rt	55
4	1.0	rt	70
5	1.0	50	85
6	1.0	80	100

Figure 1. Optimization of microfluidic dehydration.

The synthetic pristane by this route is confirmed to induce antibody production as same efficiency as natural pristane in mouse. The present microfluidic process was generally useful for dehydration of β-hydroxyketone, simple alkanol, and allylic alcohol, , giving corresponding olefins quantitatively. The method was successfully applied to a multi-kilogram synthesis of pristane in one batch.

Aldol condensation is one of the most fundamental C-C bond forming reactions. The simplest version of this reaction utilizes the acetone as the precursor of the enolate. This acetone-based Aldol reaction is conveniently performed by slowly adding the aldehydes to biphasic solution of acetone and sodium hydroxide below 10 °C. However, the efficiency of this protocol under the batch conditions significantly decreases when the reaction is performed in a large-scale, leading to the aldehyde- and the acetone-polymerization. The situation becomes miserable particularly when the reaction is applied to the aldehydes containing the α-protons, e.g., the reaction with decyl aldehyde **12** resulted in the production of a large amount of aldehyde-polymer gel, and the desired 4-hydroxy-2-tridecanone **14** is obtained only in less than 20 % yield (Figure 2). The microfluidic conditions led to the high-yielding reaction of acetone enolate with even α-proton-containing aldehydes in biphasic aqueous-acetone system, by minimizing the formation of self-condensation products.[11]

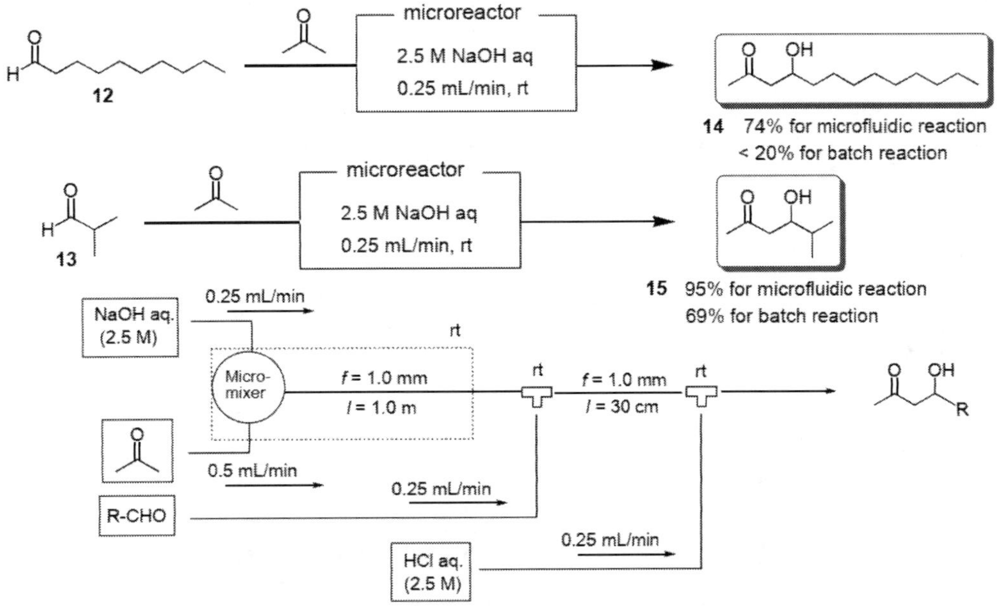

Figure 2. Biphasic Aldol condensation of acetone with aldehydes under the microfluidic conditions.

In summary, we have achieved the efficient glycosylation, dehydration, reductive opening of benzylidene group, and Aldol condensation under the microfluidic conditions. The success of these methods owe to the integration of all favorable aspects of microfluidic reactions, namely, efficient mixing, precise temperature control, and the "easy handling" of the reactive intermediate by controlling the residence time. The efficiency of microfluidic system for both acid-mediated reactions and Aldol condensation in the present study provokes the need to re-investigate the traditional reactions, which have not been throughout utilized for organic synthesis.

1. Representative reviews, see: (a) Ehrfeld, W. ed., *Microreaction Technology*, Springer, Berlin, 1998; (b) Manz, A.; Becker, H. eds., Mycrosystem Technology in Chemistry and Life Sciences, Springer, Berlin, 1998; (c) Ehrfeld, W.; Hessel, V.; Lowe, H. *Microreactors*, Wiley-VCH, Weinheim, 2000; (d) Hessel, V.; Hardt, S.; Lowe, H. *Chemical Micro Process Engineering*, Wiley-VCH, Weinheim, 2004; (e) Yoshida, J.-I.; Suga, S.; Nagaki, A. *J. Synth. Org. Chem. Japan* **2005**, *63*, 511-522.
2. Recent applications, see: (a) Pennemann, H.; Hessel, V.; Loewe, H. *Chem. Engineer. Sci.*, **2004**, *59*, 4789-4794; (b) Jahnisch, K.; Hessel, V.; Loewe, H.; Baerns, M. *Angew. Chem. Int. Ed.* **2004**, *43*, 406-446; (c) Nagaki, A.; Togai, M.; Suga, S,; Aoki, N.; Mae, K.; Yoshida, J.-I. *J. Am. Chem. Soc.* **2005**, *127*, 11666-11675; (d) Ratner, D. M.; Murphy, E. R.: Jhunjhunwala, M.; Snyder, D. A.; Jensen, K. F.; Seeberger, P. H. *Chem. Commun.* **2005**, 578-580.
3. Tanaka, K.; Goi, T.; Fukase, K. *Synlett* **2005**, 2958-2962.
4. Fukase, K.; Takashina, M.; Hori, Y.; Tanaka, D.; Tanaka, K. Kusumoto, S. *Synlett* **2005**, 2342-2346.
5. Tanaka, S; Goi, T.; Tanaka, K.; Fukase, K. *J. Carbohydr. Chem.* **2007**, *26*, 369-394.
6. DeNinno, M. P.; Etienne, J. B.; Duplantier, K. C. *Tetrahedron Lett.* **1995**, *36*, 669-672.
7. Oikawa, M.; Liu, W.-C.; Nakai, Y.; Koshida, S.; Fukase, K.; Kusumoto, S. *Synlett* **1996**, 1179-1180.
8. A new micromixing device, "Comet X-01" (Techno Applications Co., Ltd., 34-16-204, Hon, Denenchofu, Oota, Tokyo, 145-0072, Japan) was used, which exhibited much superior mixing efficiency to the simple T-shape mixer and similar efficiency to the IMM micromixer.

9. Tanaka, K.; Fukase, K. *Synlett* **2007**, 164-166.
10. Tanaka, K.; Motomatsu, S.; Koyama, K.; Tanaka, S; Fukase, K. *Org. Lett.* **2007**, *9*, 299-302
11. Tanaka, K.; Motomatsu, S.; Koyama, K.; Fukase, K. *Tetrahedron Lett.* **2008**, *49*, 2010-2012.

Fouling and Long Term Stability in Micro Heat Exchangers

Juergen J. Brandner[1)], W. Benzinger, U. Schygulla, K. Schubert
Forschungszentrum Karlsruhe GmbH, Institute for Micro Process Engineering IMVT, P.O. Box 3640, DE-76021 Karlsruhe, Germany
[1)] corresponding author, Phone +49 7247 82 3963, Fax +49 7247 82 3186, e-mail: juergen.brandner@imvt.fzk.de

For a couple of years, micro heat exchangers are available not only as tools for laboratory work but also for industrial applications [1]. They are well known for their superior heat transfer properties due to the large surface-to-volume ratio (values up to 30.000 $m^2 \cdot m^{-3}$ have been reported, [1]). Applications in chemical and pharmaceutical industry, automotive industry and many more have been presented [2]. Furthermore, the well-known excellent temperature control of fluids achievable with microstructure heat exchangers made them interesting for applications in bio-technology. But there are still little data available on the long term stability of these devices in conventional chemical processes as well as in bio-technological processing.

In this publication, we will present several examples for micro heat exchangers made of polymer and stainless steel. The polymer devices have been manufactured by micro stereo lithography, a rapid prototyping process. Details are given in [3]. The metal devices consists out of stainless steel foils providing numerous micro channels generated by mechanical micromachining or wet chemical etching. A number of the foils are arranged in a specific way and bonded together, i.e. by diffusion bonding. Details on the manufacturing process can be found in [3].

Single microstructure layer polymer devices have been tested in a special experimental arrangement. They have been manufactured by micro stereo lithography, a rapid prototyping process. Details to this can be found in [3, 4]. The polymer micro channel shell was covered with a non-structured stainless steel foil, sealed by a polymer o-ring. This arrangement was screwed on top of an electrically heated aluminium block. In figure 1 such a test section is shown.

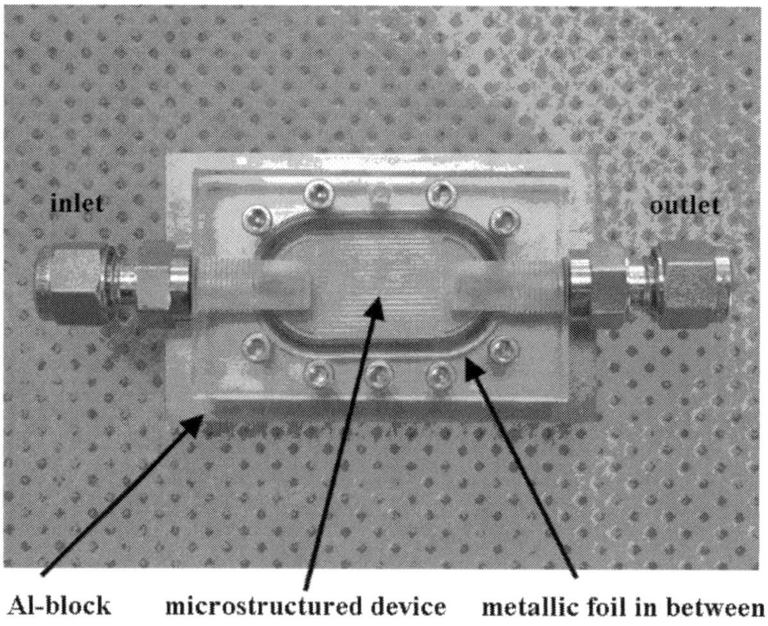

Fig. 1: Test section for fouling tests using calcium nitrate and sodium hydrogen carbonate.

A solution of calcium nitrate and sodium hydrogen carbonate was used as test fluid. The fluid was pumped at a flow rate of 1,5 kg/h through the channels of the microstructures, using hose fittings. The Reynolds number Re was calculated to about $Re \approx 110$, thus the experiments were performed at laminar flow. The solution was heated electrically by the hot stainless steel surface. At the beginning of the experiment the temperature rises from 25°C at the inlet to 86°C at the outlet. The high temperature causes the precipitation of solid calcium carbonate. Because of the fouling layer, the heat transfer capabilities decreases.

It was expected that a calcium carbonate layer would grow more or less homogeneously on the relatively rough stainless steel surface, but not on the smooth polymer micro channel walls. But from the experiments it is obvious that fouling mainly took place at the polymer micro channels and at the outlet manifold of the test configuration. In figure 2, an example for the deposition of the calcium carbonate inside the micro channel arrangement is shown.

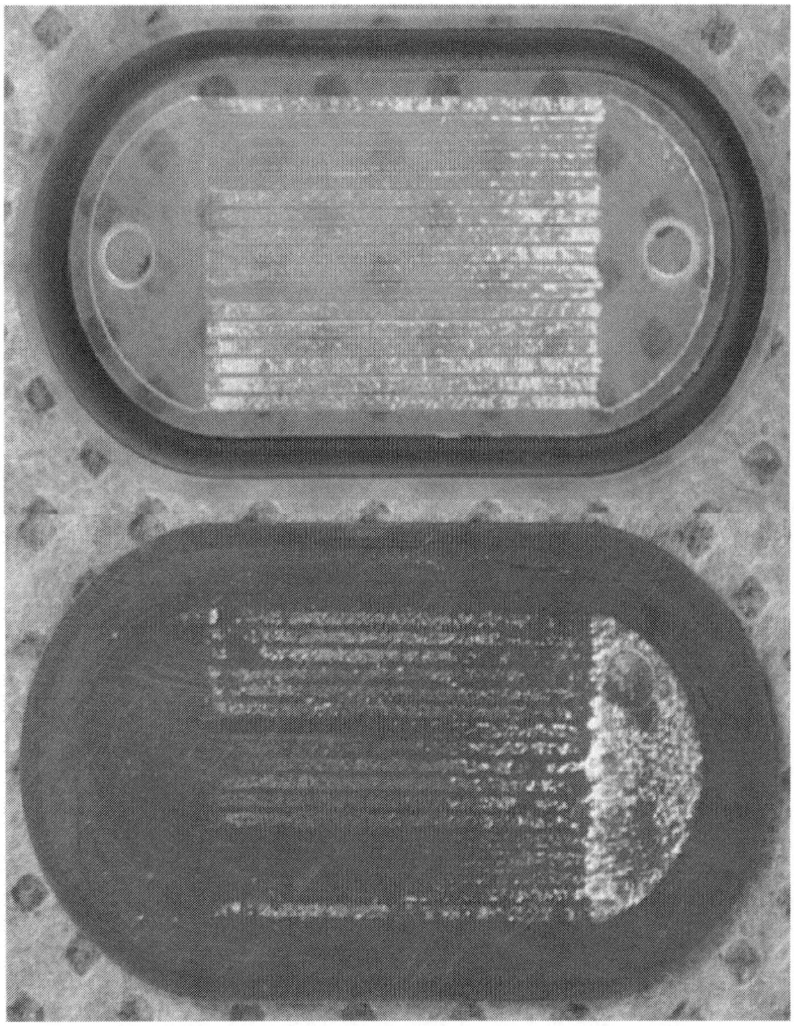

Fig. 2: Polymer micro channel shell and stainless steel heat transfer foil with calcium carbonate deposited by fouling. From this figure it is clearly to see that the main deposition takes place at the end of the micro channels and inside the outlet manifold.

To obtain more data and maybe reduce the fouling on the heat transfer foil side, the heat transfer surface was modified. Within the same test arrangement, foils coated with FEP (fluorinated ethylene propylene) and DLC (diamond like carbon) have been tested. Figure 3 shows the three type of foils used for the experiments.

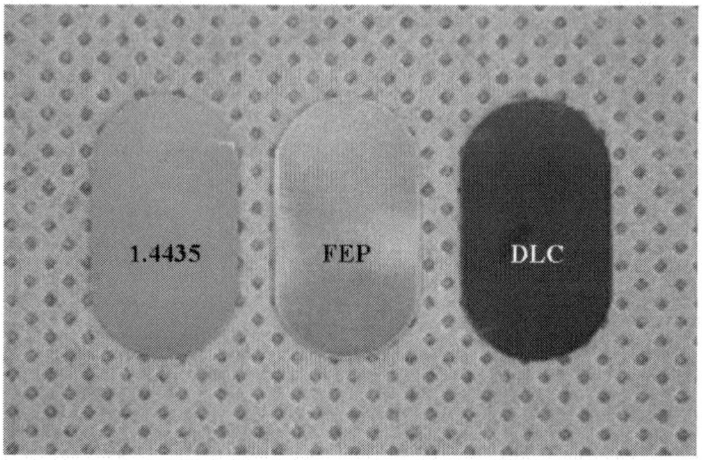

Fig. 3: Three types of foils, used for fouling tests in our arrangement: stainless steel (1.4435, left), stainless steel coated with FEP (fluorinated ethylene propylene, middle), and stainless steel coated with DLC (diamond like carbon, right)

It was expected that the coating of the heat transfer surface might give some better properties of the material against fouling. With simple pressure drop measurements obtained between inlet and outlet of the test section, the increasing pressure drop due to the fouling process was monitored. As shown in figure 4 and combined with the findings from other measurements (heat transfer, mass flow), no really significant difference between the single materials could be observed. From all measurements performed so far, the modification of the surface seems not to have any influence to the fouling properties. More details are given in [5].

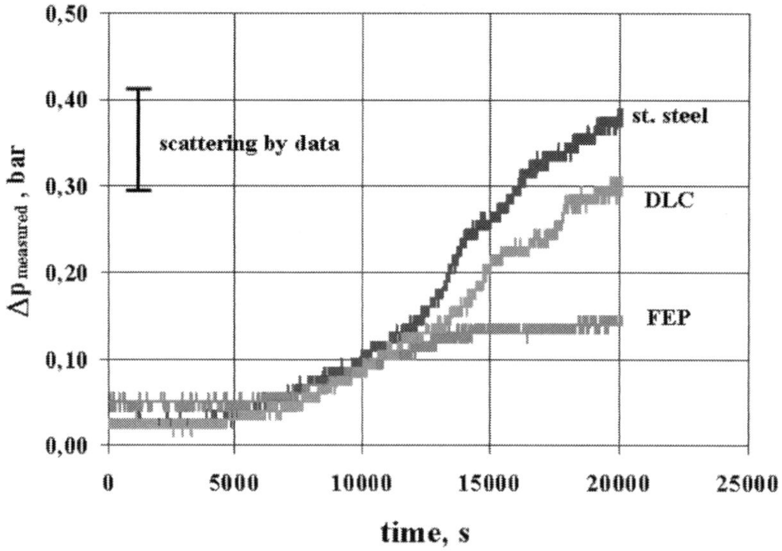

Fig. 4: Pressure drop between inlet and outlet of the test section. As shown in this figure, the difference between the simple stainless steel, the FEP and the DLC coated foil is not in a really significant range but in the range of the data scattering. Thus, no strictly defined or relevant advantage or disadvantage of the coatings could be monitored.

Due to the low differences in measured data (which is in the range of the measurement scattering), no explicit advantage or disadvantage of the coatings could be found yet. Additionally, after disassembling the test section, more or less the same fouling behaviour was visualised inside the micro structure, no matter whether FEP or DLC coating or simple stainless steel was applied for the heat transfer. Therefore, it is assumed that the flow velocity is too low to efficiently draw away calcium carbonate particles and maybe reveal the advantages of the coatings. Some particles are ripped off and re-deposited in the outlet manifold due to a reduction of the flow velocity. Furthermore, the temperature is equalised across to the flow direction due to the low flow velocity values of about 0.3 m/s and the relatively high heat conduction of water. Therefore, precipitation of calcium carbonate crystals occurred in the bulk of the fluid. Experiments at higher Reynolds numbers – especially in the turbulent flow regime – might lead to other results and will be performed in future. More details on this experiments can be found in [5].

To obtain relevant data on multilayer devices two stainless steel crossflow heat exchangers have been investigated for their long term stability. We used an existing test rig for characterisation of microstructure heat exchangers which was described in details before [6, 7]. We used two stainless steel devices in crossflow arrangement, one with rectangular micro channels manufactured by mechanical precision machining, and one with elliptically shaped micro channels manufactured by wet chemical etching. Details on the manufacturing process can be found in [7].

Both devices have been tested for more than 8000 hours each, using deionised water as test fluid. Experimental data on the heat transfer properties and the pressure drop will be given and compared. It was found that the heat transfer capabilities were decreased by almost 50% within the first few hundred hours of testing and then ran into a saturation state, still providing good heat transfer capabilities. Details to this are given in [8]. As an example, the development of the overall heat transfer coefficient k with the time for the two different types of stainless steel crossflow heat exchangers tested are given in figure 5. It is clearly to see that there is an initial decrease of the heat transfer which is then running into saturation. Nevertheless, the heat transfer capabilities are still one order of magnitude above that reached with conventional devices.

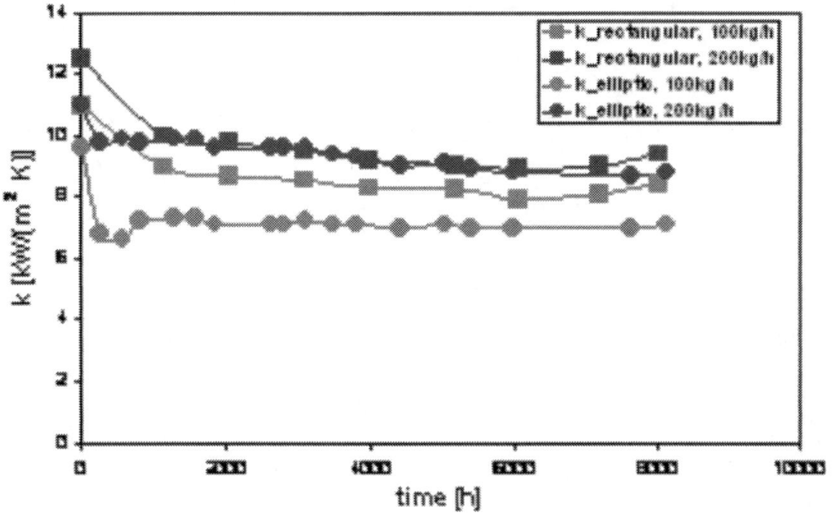

Fig. 5: Overall heat transfer coefficient k for the rectangular and the elliptical channels device, plotted versus the operation time, for two different mass flow values.

As a lab-scale application example, results of tests for the quench of mammalian cell solution will be given. This method was developed to provide a possibility to stop the metabolism of mammalian cells as rapid as possible without damaging the cells, for conventional methods like cooling with liquid nitrogen lead to a large number of disrupted cells and therefore to an increased expenditure for the analysis.

Within a first study, a stainless steel crossflow heat exchanger was tested for its suitability to instantaneously cool down a sample flow of a mammalian cell culture suspension from 37°C to 0°C by the use of a mixture of ethylene glycol and water as cooling fluid. This was necessary to almost stop the cell metabolism and to analyse the metabolites inside the cells. In figure 6, the temperature dependency of molar densities of four metabolites plotted against the temperature are given as examples. Here it is clearly to see that with increasing sample temperature the molar mass of metabolites is decreasing in the most cases. Thus, to observe the intracellular density, a snapshot-like sampling and analysis procedure is necessary. No fouling was observed during the experiments with the cell suspension [7, 9]. It will be tried to engage the method to other bio-tech processes like, e.g., temperature control of fermentation.

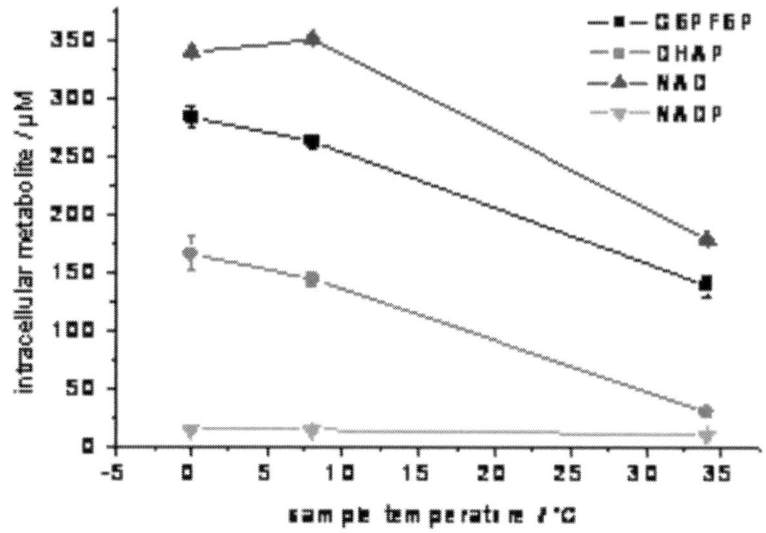

Fig. 6: Dependency of the intracellular density of four cell metabolites against the sampling temperature.

Conclusion

Fouling and blocking in microstructure devices is definitely a point to keep in mind. The microstructures may be in the same size range as blocking particles are. Generation of fouling films inside of microstructure systems may lead not only to less performance with respect to heat transfer but also to complete blocking and, thus, the possibility of a non-working device as can happen in conventionally sized heat exchangers. Unlike those, most microstructure devices can not be disassembled and cleaned as single parts. Thus, on-stream cleaning procedures have to be developed.

There are certain countermeasures for conventional devices dealing with fouling and blocking, such as filters, protective layers, etc. Some of those may also be useful for microstructure devices, some are not applicable. Due to the huge variety of different sizes, types, manufacturing methods and applications of microstructure systems, there are lots of different more or less important points to deal with when talking about long term stability. Like with the corrosion resistance, it has to be taken in account that the restraints and limitations are different from those existing for conventional sized devices, and, thus, new regulatories have to be generated. Moreover, special designs have to be developed and applied to prevent fouling inside microstructure devices.

For fouling and blocking in microstructure devices as well as for their long term stability, new design concepts and manufacturing methods as well as integrated countermeasures could lead to a significant better acknowledgement of this technology in the industry. In many cases, not only a lack of experimental data but also the fear to run into difficulties generated by fouling in microstructures often prevents the use of these high-performance devices. Overcoming these restraints will be one of the most important tasks in the development of microstructure devices for the next few years.

References

[1] Brandner, J.J., Bohn, L., Henning, T., Schygulla, U., Schubert, K., Microstructure heat exchanger applications in laboratory and industry, Heat Transfer Engineering 28 (8-9), 761-771, 2007

[2] Kraut, M., Wenka, A., Bohn, L., Schubert, K., Successful upscale of laboratory micro reactor into industrial scale, Proc. of the 6th ANQUE International Congress of Chemistry, Dec 5-7, 2006, Puerto de la Cruz, Tenerife, Spain, paper T2-L-9

[3] Brandner, J.J., Microfabrication in Metals and Polymers, Advanced Micro- and Nanosystems Vol.5, pp.267-320, Wiley-VCH, Weinheim, Germany, 2006

[4] Brandner, J.J., Hansjosten, E., Anurjew, E., Pfleging, W., Schubert, K., Microstructure devices generation by selective laser melting, Proc. of SPIE Photonics West, Jan 25-27, 2007, San Jose, CA, USA

[5] Benzinger, W. Brandner, J.J., Schygulla, U., Schubert, K., Influence of different surface materials on the fouling process in a microstructured heat exchanger, Proc. ECI Conf. on Heat Exchanger Fouling and Cleaning VII, July 1-7, 2007, Tomar, Portugal

[6] Brandner, J.J., Schubert, K., Fabrication and Testing of Microstructure Heat Exchangers for Thermal Applications, Proc. of ICMM2005, paper 75071, 5th Int. Conf. on Micro and Minichannels, June 13-15, 2005, Toronto, Canada

[7] Brandner, J.J., Bohn, L., Henning, T., Schygulla, U., Schubert, K., Microstructure Heat Exchanger Applications in Laboratory and Industry, Heat Transfer Eng. 28 (8-9), 761-771, 2007

[8] Brandner, J.J., Benzinger, W., Schygulla, U., Schubert, K., Metallic Micro Heat Exchangers: Properties, Applications and long term stability, Proc. ECI Conf. on Heat Exchanger Fouling and Cleaning VII, July 1-7, 2007, Tomar, Portugal

[9] Wiendahl, C., Brandner, J.J., Küppers, C., Luo, B., Schygulla, U., Noll, T., Oldiges, M., A Microstructure Heat Exchanger for Quenching the Metabolism of Mammalian Cells, Chem. Eng. Techn., 2007, 30, No.3, 1-8

COMPACT HEAT EXCHANGE REACTORS

Dr. Mark Wood, Chart Energy & Chemicals, Inc. Wolverhampton, UK
Dr. Zhijun Jia, Chart Energy & Chemicals, Inc. LaCrosse, WI, US
Larry Stryker, Chart Energy & Chemicals, Inc. LaCrosse, WI, US

ABSTRACT

Reactors are at the heart of a chemical process and consequently are a key step to focus upon when considering process intensification (PI). Heat transfer in the reactor can make the difference between a successful process and an out of control process.

Heat removal or addition can be especially difficult for a highly exothermic or endothermic reaction. Coupling the reaction and heat transfer allows temperature profiles to be closely controlled, thereby improving the product quality and yield.

Chart Energy & Chemicals manufactures compact heat exchangers and chemical reactors that are used for PI using either aluminium or stainless steel or special metals

Chart has been manufacturing compact brazed aluminium plate-fin heat exchangers for over 50 years, with these being used as chemical reactors as early as 1979. Over the last 15 years a shim-based, diffusion bonded heat exchanger and reactor, known as ShimTec®, has been developed and widely applied. ShimTec® has options for construction from a wide range of materials depending on temperature, pressure and corrosiveness. Structures are capable of operating under high temperature and pressures.

Manufacturing with other metals, including nickel alloys and high temperature metals, is also done for fuel processing and hydrogen generation fields, which show a lot of activity. In many applications, the processing rate achieved was many times higher than that anticipated.

This paper describes Chart's reactors and includes case studies of their use in the fine chemicals, pharmaceutical and fuel processing fields.

PROCESS INTENSIFICATION

Process Intensification (PI) is a phrase that is gaining wider recognition in the US and worldwide after starting around 30 years ago in the UK[1]. Whilst there is no single definition, proponents agree that the ultimate objective is to reduce the size, and hence cost, of chemical processing plant. A means of achieving PI is to design the process plant to meet the needs of the chemical process, rather than fitting the process into standard equipment.

Chemical reactors are an important unit operation when considering PI as changes made to the reactor can affect the entire plant. Applying PI to reactors has the ultimate objective of allowing the reaction to run stoichiometrically at its inherent kinetic rate, providing the highest possible throughput in a given volume of reactor. Reaction rates can be restricted in existing equipment to achieve safe operation, for example by using lower concentrations, lower temperatures or reduced feed rates to limit the rate of heat release. PI aims to remove these restrictions by matching enhanced heat transfer capability to the heat production or consumption rate. Other aims are to provide effective mixing when it is required and ensuring that the residence time is equal to the actual reaction timescale. These seem obvious targets, but traditional reactors such as stirred vessels and packed columns may not achieve them, particularly for fast or very exothermic reactions.

PI reactors by their very nature tend to be continuous in operation. Whilst this is a normal mode of operation for larger scale processing, it is not normal for fine chemicals and pharmaceuticals production where products are made in batches. Switching from batch to continuous operation is therefore one of the barriers that face the application of PI. When correctly designed, a PI process can achieve any or all of the following benefits:

- Enhanced heat transfer
- Enhanced mixing
- Improved product quality
- Reduced residence time
- Reduced reactant inventory
- Reduced energy consumption
- Inherently safe operation

Not every chemical reaction can have PI applied due to the nature of the kinetics or physical form of the reactants and products. Despite this, a number of improvements could still be made by applying the principles of PI. It has been demonstrated that applying PI can allow reactions to be operated under conditions that simply would not be possible within existing equipment, allowing kinetics to be enhanced and reducing the reaction timescale.

HEAT TRANSFER IN REACTORS

Temperature Effects in Reactors

Temperature can be a critical factor in operating chemical reactions. For exothermic reactions, heat has to be removed quickly as increased temperature can cause undesirable byproduct formation or even runaway reactions. For endothermic reactions, heat has to be provided as and when it is required to enable the reaction to proceed.

Stirred vessels have limited heat transfer capability, which reduces as the vessel is scaled in size due to the volume to surface area ratio decreasing. Additional heat transfer surface can be provided, but even so there is a large difference between operation at different scales, leading to problems when scaling up.

Packed beds have a relatively high heat transfer capability, particularly those in the form of tubular reactors, but the reaction conditions at the centre of the column can be different to the conditions at the wall as the heat has to travel between the point of reaction and the point of heat transfer.

With continuous operation of a fast exothermic reaction, there will be a rapid increase in temperature following mixing of the reactants. This temperature peak can be very undesirable if it causes product degradation or safety issues. Consequently, the heat transfer capability has to be high enough to reduce the temperature peak to a manageable value and/or feed has to be added in discrete quantities along the length of the reactor.

Compact Heat Exchange Reactors

The principle of operating chemical reactions inside heat exchangers is one means of achieving

PI. As heat exchangers are designed to have high heat transfer capability, they are suited to the task of providing and removing heat of reaction as and when required.

Heat exchangers transfer externally generated heat between two or more streams. Heat Exchange Reactors have much the same construction and function, except that the heat is generated or removed internally. Compact heat exchangers have short heat transfer paths and high heat transfer surface areas per unit volume. Plate-fin heat exchangers are one example of compact heat exchangers manufactured by Chart. In these, corrugated fins are used to provide multi-layered units where each layer is only a few mm tall. For plate-fin, UA is about 20 times greater than a shell and tube heat exchanger, allowing close approach temperatures.

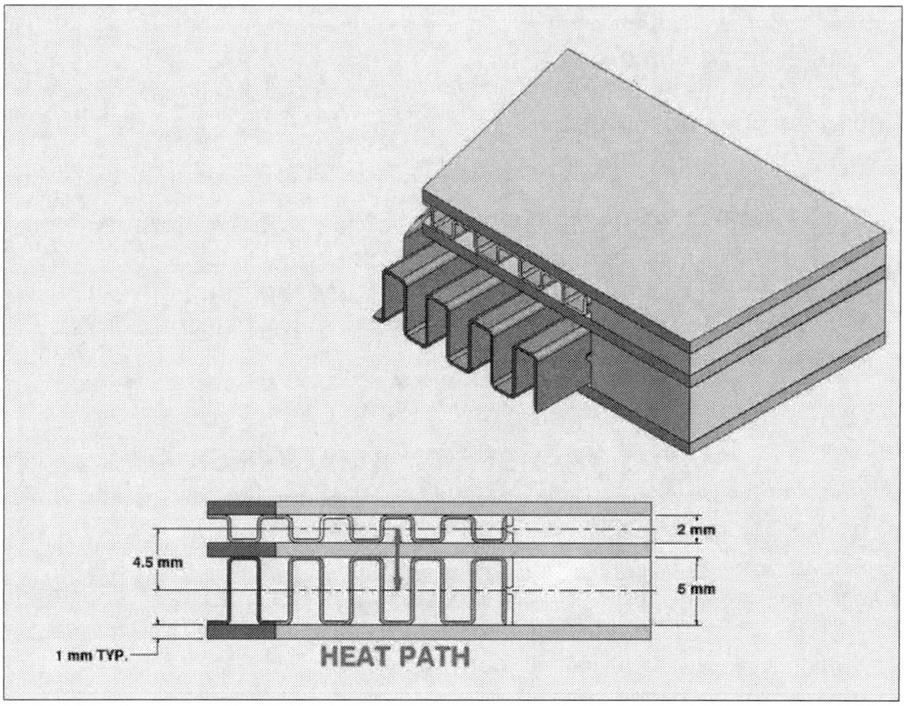

Figure 1. Short heat path in a compact plate-fin heat exchanger

It is important to note that "Compact" refers to the spacing of the channels and distance between heat transfer surfaces, rather than the physical size of the unit, as these exchangers and reactors can be very large.

CHART'S COMPACT HEAT EXCHANGE REACTORS

Chart manufactures two ranges of compact Heat Exchange Reactors, FinTec™ and ShimTec®. FinTec™ is suitable for medium to large scale operation, whilst ShimTec® is suitable for small to medium scale operation. A particular version of ShimTec® has previously been referred to in literature as "Marbond" and "HEX".

Chart has been manufacturing compact brazed aluminum plate-fin heat exchangers for over 50 years, with these being used as chemical reactors as early as 1979. FinTec™ is the name Chart applies to plate-fin reactors. In brazed aluminum, these reactors can have volumes of several cubic metres, making them suitable for bulk chemical processing. Aluminum is an appropriate material of construction for a number of chemical reactions, with operation at temperatures up to 260°C and pressures up to 30barg.

ShimTec reactor

Plate-fin heat exchanger

Figure 2. Different scales of compact heat exchangers manufactured by Chart

ShimTec® is a more recent development, though still dates back more than 15 years. Construction is from thin metal sheets with features cut into them, stacked to provide channels. The robust structure lends itself to being diffusion bonded, resulting in a reactor capable of withstanding extremely high operating temperatures and pressures. Reactors have been operated at well over 800°C and proven to withstand design pressures of 180 bar at lower temperatures. ShimTec® reactors are available in a range of metals including stainless steel and nickel alloys.

Both reactor types are designed around the same principles, with a number of alternating layers so that reaction streams are sandwiched between heat transfer streams. Intermediate layers and channels can be incorporated to allow the controlled addition of additional chemicals into the bulk streams. This addition can be divided up along the length of the reaction path, ensuring that temperature profiles or product compositions are closely controlled.

Both FinTec™ and ShimTec® are readily scaled up as the channel dimensions can be kept constant, regardless of the physical size of the finished unit.

Microchannel reactors (microreactors) and minichannels are widely referred to in literature when considering PI. Like PI itself, there is no specific definition for microchannels, though it is typically seen as referring to channels with a maximum characteristic dimension of around 1mm. One study[2] suggest that microchannels should be defined as those with a hydraulic diameter of < 0.2mm due to the change in flow fields and different manufacturing techniques required for such small channels. Channels from 0.2-3mm are defined as minichannels. Chart has manufactured reactors with microchannels according to this definition, but mostly works in the minichannel range as these are more applicable to chemical production by, for example, reducing the risk of blockage.

CATALYST APPLICATION

Catalytic reactions have been widely operated within Chart's Compact Heat Exchange Reactors. Catalyst has been incorporated within the reactor structures in the following manner:

- Coated on internal surfaces
- Packed
- Pre-applied to a removable insert
- Injected along the reaction path

Open channels with dimensions of several millimetres upwards can be produced to allow packing and insertion of catalyst, which can be removed and replaced when required.

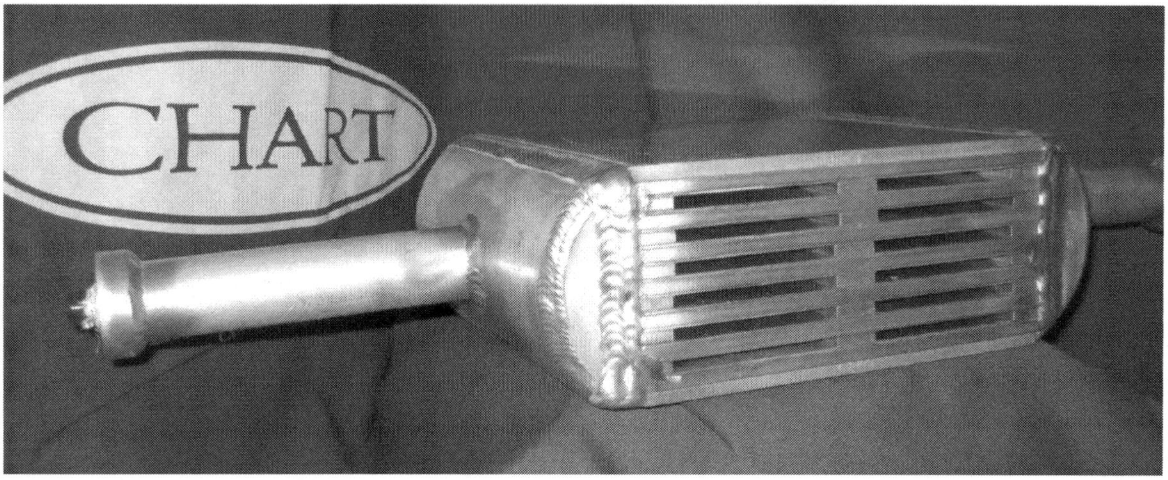

Figure 3. Open channels for packing with catalyst

Catalysts can be coated onto the internal structure when it is known that catalyst replacement will not be required. ShimTec® includes a surface called Pin-Fin which has proven particularly beneficial for coating applications. Pin-fin has a high surface area per unit volume and its mesh-like structure allows process fluids to pass through the catalyst, rather than flowing over the surface as would happen with, for example, coated tubes. Coating catalyst allows the reaction to take place immediately adjacent to the metal structure, providing very short heat paths.

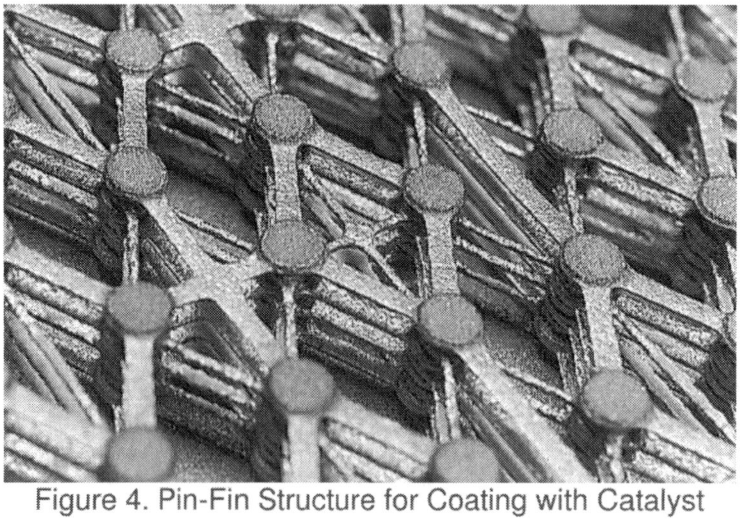

Figure 4. Pin-Fin Structure for Coating with Catalyst

CASE STUDIES

A number of case studies are now presented to give an illustration of the type of reactions carried out in ShimTec® and FinTec™ reactors.

Case A. Exothermic Oxidation with Liquid Phase Catalyst [3]

Current operation is in a semi-batch stirred vessel. It is an exothermic reaction with a liquid-phase catalyst and significant quantities of byproducts are generated. Feed addition rate has to be restricted to control the reaction rate.

A Chart compact heat exchange reactor with high heat transfer capability allowed the operating temperature to be increased to enhance reaction kinetics. Fluids were pre-mixed with a rotor-stator mixer to generate a two-phase dispersion before entering the channels, which helped maintain the dispersion.

Reaction time was reduced from 1 hour to 6 seconds with a 99% reduction in byproduct formation.

Case B. Steam reformer with coated catalyst [4,5]

Two separate projects examining steam reforming. In both, catalyst was coated onto the internal structure. Operation was carried out both within a furnace and with a combustion stream providing heat for the endothermic reforming reaction[5].

Project 1[4] proved the concept of reforming within compact heat exchange reactors. In project 2[5], high conversion and reactor performance allowed the residence time to be reduced 10-fold compared to the anticipated value. A reformer with an active volume of 8 cm^3 produced 170 litres/day of H_2 from autothermal reforming and 150 litres/day of H_2 from steam reforming. The target 5000m^3/day of H_2 could be produced by a reactor with total volume less than 1m^3.

Case C. Liquid phase oxidation with packed catalyst [6]

A three phase catalysed reaction with an adiabatic temperature rise of 180°C. Internal pre-mixing of the oxygen and liquid stream was carried out before contacting the catalyst.

Kinetic control was achieved with removal of mass transfer limitations. Temperature change controlled to ±1°C. 99.5% selectivity and 55% conversion in only 10cm length

Case D. Liquid phase hydroformylation with slurry phase catalyst [7]

Stirred vessel operation is compared with a compact heat exchange reactor. Using the compact heat exchange reactor there was a 10-fold increase in reaction rate, no mass transfer limitation and improved selectivity compared to a stirred batch reactor

Case E . Low concentration hydrogen peroxide production [8]

Hydrogen peroxide is produced in bulk at a high concentration to reduce shipping costs, then diluted on-site for use. It is desirable to produce hydrogen peroxide at low concentrations in situ, but it is not possible to combine hydrogen and oxygen in low concentrations within conventional reactor systems as the mixture becomes flammable and even explosive, with high pressures also required. Operation in a compact heat exchange reactor has allowed this reaction to be operated safely and efficiently at a low pressure. The pilot plant can produce over 2%wt hydrogen peroxide, suitable for many applications and work is ongoing to increase this to 5%wt.

REFERENCES

1. Tsouris and Porcelli. "Process Intensification – Has its time finally come?" CEP October 2003, 50-55.

2. Kandlikar. "Microchannels and Minichannels – History, Terminology, Classification and Current Research Needs". First International Conference on Microchannels and Minichannels, Rochester, New York, 24-25 April 2003.

3. Green, Johnson, Westall, Bunegar and Symonds. "Combined Chemical Reactor/Heat Exchangers: Validation and Application in Industrial Processes." 4th International Conference on Process Intensification for the Chemical Industry, Brugge, 10-12 September 2001.

4. Dicks, Jones, Judd, Pointon, Gough. "Assessment of Advanced Performance and Fabrication Options for a Compact Steam Reformer." DTI Publication URN 01/1163, 2001.

5. "Intensified Technology for Distributed Hydrogen Production" (Hydrofueler) EU Fifth Framework project ref ENK6 CT-2002-00684.

6. Plucinski et al, "Liquid-Phase Oxidation of Organic Feedstock in a Compact Multichannel Reactor" Ind. Eng Chem. Res. 2005, 44, 9683-9690.

7. Enache, Thiam, Dumas, Ellwood, Hutchings, Taylor and Stitt. "Intensification of the solvent-free catalytic hydroformylation of cyclododecatriene: Comparison of a stirred batch reactor and a heat-exchange reactor". Catalysis Today 128, 2007, 18-25.

8. DoE Industrial Technologies Report "Microchannel Reactor System". November 2008 http://www1.eere.energy.gov/industry/chemicals/pdfs/microchannel_reactor.pdf

Improvement of molecular weight distribution in synthesis of liquid oligomers for reactive resins by micro flow-through polymerization

G. Alexander Gross, P. Mike Günther, Hagen Zollfrank, Monika Bauer and J. Michael Köhler

Techn. Univ. Ilmenau, Inst. f. Mikro- and Nanotechnologies, D-98693 Ilmenau, Germany
Fraunhofer-Gesellschaft, Inst. f. Appl. Polymer Techn., Teltow

Abstract

The solution polymerization of a monomer mixture by a radical process was implemented in a micro reaction setup including injectors, a valve and a pressure sensor. The process was realized by application of a segmented flow. The polymerization takes place after activation in a PTFE micro tube. The molecular weight distribution could be deciding improved by enhancing reaction temperature from 78 to 120 °C, enhancing reaction pressure up to six bar und reducing the reaction time from 120 down to 15 minutes.

Keywords

micro fluidics, segmented flow, polymerization, molecular weight distribution, dispersion, continuous process

1. Introduction

Micro reaction technology has been introduced for a lot of different application during the last 15 years. Beside homogenous phase reactions, heterogenous reaction as well as particle synthesis were successfully implemented in micro reactors. The use of micro reactors opened the possibility of conversion of discontinuous batch processes into lab-scale continuous-flow processes [1, 2]. The continuous processes guarantee a high stability of reaction conditions and help to improve the selectivity, the yield and the efficiency of chemical reaction [3]. The precise control of product parameters is one of the most important motivations for the applications of micro reaction technology. This aspect is of particular interest in the synthesis of polymers, but the application of micro reactors for polymerization suffers from the strong increase in molecular weight and the resulting increase of viscosity. This increase in viscosity is a serious problem for micro channels and can be responsible for the arise of very high pressure drops and clogging of the flow-through devices.

Despite this difficulties, there are some special applications of micro reactors for polymerization. The viscosity problem can be avoided if an emulsion polymerization is applied instead of a mass or solution polymerization. So, a continuous-flow polymerization in emulsion could be realized by an micro reactor arrangement consisting on two static micro mixers for generation of emulsion and initiation of the reaction and a following tube coil as residence loop for completion of reaction [4]. Polystyren nano particles with diameters between some dozen and a few hundred nanometers are obtained by this method. The principle can also be used for generation of polymer particles integration smaller metal nano particles [5]. The generation of particle chains [6] and polymer rods by polymerization of micro fluid segments are other examples of polymerization in micro fluidic systems [7]. In this cases the surrounding medium or a carrier liquid support the transport of polymerizing phases inside the fluid channel.

An only moderate increase of viscosity takes place if the polymerization is limited to lower polymerization degrees and carried as solution polymerization. In this case, the polymerization can take place in micro devices without danger of blocking and with moderate flow-induced pressure drops. This gives the possibility for polymerization under unusual conditions and for searching for process parameter sets for improved product quality.

Oligomers of mixed monomer composition are of particular interest for reactive resins. The products should possess molecular weights in the range of about 1.5 kDa and a low dispersity.

Here, the influence of the conversion of the batch polymerization into a micro polymerization process under continuous flow conditions was investigated. The flow-through synthesis of resins from monomer mixtures in ethanolic solution under pressures is reported. The goal of the study was to find a possibility of influencing the distribution of molecular weights by shifting the temperature, pressure, flow rate and residence time in the heating zone.

2. Experimental

The experimental setup was implemented for a segmented-flow process. Two syringe pumps and an injector were coupled fluidically in a modular micro reactor arrangement and connected with a pressure sensor and a valve for pressure control (experimental set-up in fig. 1). The solutions of monomer mixtures and initiator were actuated by PC-controlled fluid actuators.

The acrylate monomer mixture with the initiator in MEK solution were conducted through a PTFE tube (1 mm inner diameter, length up to 40 m) after mixing the in the micro mixing device. The first series of polymerization experiments was started by a thermostating bath (70°C). The flow rate was varied between 2.5 and 10 mL/h. The residence time at elevated temperature was kept at about 2 h.

The second series of experiments was carried out under elevated temperature and pressure, but reduced residence times. A PEEK tube was used for the stabilization of the 0.5-mm-PTFE reaction capillary at the specific reaction conditions. The flow rates were enhanced.

3. Results and discussions

Products with a mean molecular weight Mn of about 2000 Da and a dispersion (Mw/Mn) of about 2.3 are typically obtained in 4-component batch synthesis of mixed oligomers from methylacrylate (57 mol%), ethylhexylacrylate (12 mol%), 2-hydroxyethylmethacrylat (30 mol%) and acrylic acid (1%) in methylethylketone (MEK). This monomer mixture was applied in the described micro reactor reactor arrangement and polymerized in a continuous-flow process.

All product solutions were investigated for molecular weight and molecular weight distribution. The obtained polymers (after removal of residual monomers) show a significant reduction of dispersity (Mw/Mn = 1.83 ..1.88) in comparison with the batch process. In addition, the experiments show a slow tendency of reduction of molecular weight with increasing flow rate (fig. 2).

In order to come to a lower molecular weight without loss of homogeneity, the residence time was shortened down to 15 minutes and the temperature was strongly increased. The changed conditions demand for a pressure-controlled flow-through process, because boiling points of components are exceeded. A 0.5 mm PTFE capillary inside a PEEK tube was applied instead of the larger PTFE tube. The application of a perfluorinated carrier liquid and the realization of a micro segmented flow process guarantee a high mixing quality and a narrow residence time distribution.

The reaction mixture was applied as a multiphase flow (plug-like transport) using perfluorinated hydrocarbons as immiscible carrier liquid. The pressure was enhanced up to 6 bar in order to suppress the boiling of the solvent. So, the solution polymerization in liquid ethanol could be carried out up to temperatures of 120 °C.

In result, a further reduction of dispersity and a strong reduction of molecular weight was achieved (fig. 3) by the changed conditions. The transfer of the (discontinuous) batch process into a continuous micro flow-through process allowed not only a considerable acceleration of the process but lead to a significant change in the product properties.

4. Conclusions

The comparison with a batch process made the advantages of the micro flow through process clear: The dispersity was lowered from 2.28 down to 1.69 and the molecular weight Mw reduced from 4763 down to 1993. The experiments show the high potential of micro reaction technology for the preparation of polymer materials with particular specifications.

Acknowledgement

We like to thank S. Schneider for technical support.

References

[1] W. Ehrfeld, V. Hessel, H. Löwe: Microreactors (Wiley-VCH 2000)

[2] N. Kockmann: Process engineering methods and microsystem technology, in: Micro process engineering (ed. N. Kockmann, Wiley-VCH 2006), 1
cont. Flow in microreactors

[3] V. Hessel, H. Löwe: Chem. Eng. Technol. 28 (2005), 267-284

[4] P.M. Günther, J. Wagner, G.A. Gross, J.M. Köhler: "Generation of dye-doped polymer and composite nanoparticles by use of chip reactors", 9th Internatl. Conf. On Miniaturized Systems μ-TAS (Boston, October 9-13, 2005), 918-920

[5] P.M. Günther, G.A. Gross, J. Wagner, F. Jahn, J.M. Köhler: Introduction of surface-modified Au-nanoparticles into micro flow-through polymerization of styrene, Chem. Eng. J. 135 S (2008), S 126-130

[6] M. Bouquey, C. Serra, N. Berton, L. Prat, G. Hadziioannou: Chem. Eng. J. 15, Suppl1 (2008), S 93-S 98

[7] G.A. Gross, C. Hamann, M. Günther, J.M. Köhler: Formation of polymer and nanoparticles doped polymer minirods by use of microsegmented flow principle, Chem. Eng. Technol. 30, 3 (2007), 341-346

Figures

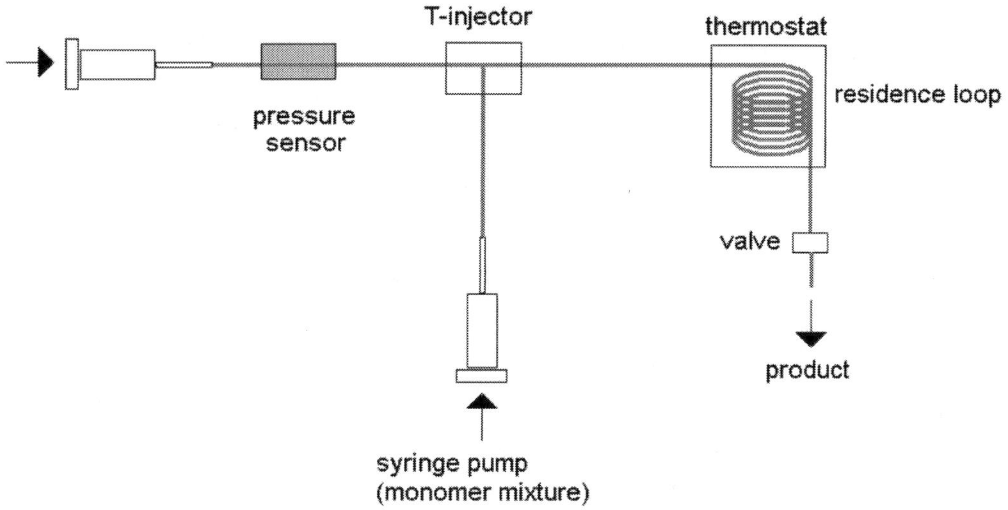

Fig. 1
Experimental set-up for micro flow-through synthesis

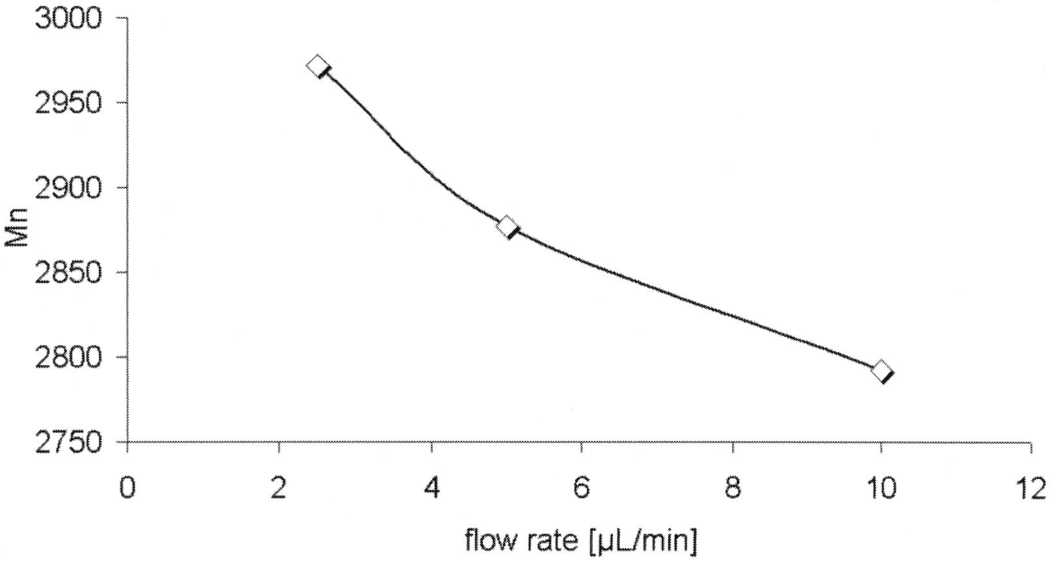

Fig. 2
Dependence of molecular weight of product polymer on flow rate under continuous-flow conditions

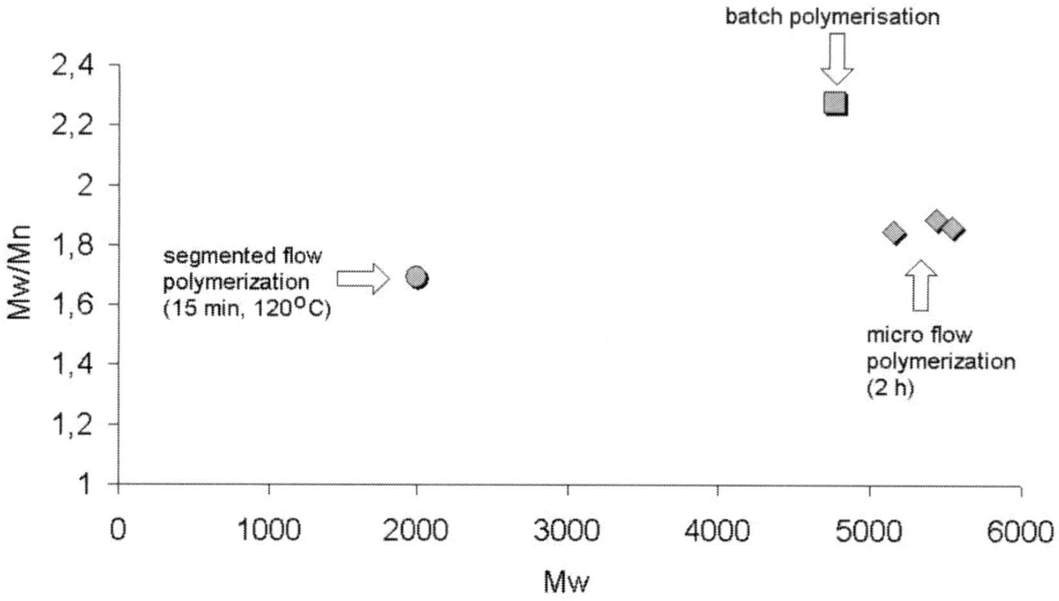

Fig. 3
Improvement of product properties by enhancing process temperature, pressure, lowered reaction time and application of segmented-flow conditions

101b Combining Chemical and Enzymatic Catalysis for the Enantioselective Oxidation of Sulfides

Sanjib Kumar Karmee[1], Christoph Roosen[1], Christina Kohlmann[2], Benjamin Ganchegui[1], Walter Leitner[1], Stephan Lütz[3], and Lasse Greiner[1]. (1) Institut für Technische und Makromolekulare Chemie, RWTH Aachen University,Germany, Worringer Weg 1, Aachen, 52074, Germany, (2) Institute of Biotechnology-2, Forschungszentrum Jülich, Jülich, 52425, Germany, (3) Institute of Biotechnology-2, RWTH Aachen University,Germany, Worringer Weg 1, Aachen, 52074, Germany

Chloroperoxidase (CPO) from Caldariomyces fumago is a versatile and highly selective catalyst for the sulfoxidation of thioethers preferably using hydrogen peroxide as oxidizing agent.[1] However, inactivation of CPO in the presence of high concentration of H_2O_2 is an inherent problem. Various strategies have been advocated viz. alternative oxidants than H_2O_2 [2,3], fed-batch addition[4], electrochemical generation[5], and immobilization of enzyme[6]. We report the synthesis of optically pure sulfoxide combining the direct generation of hydrogen peroxide from hydrogen and oxygen[7] with enzymatic catalysis using CPO (Scheme-1). This reaction is carried out using the biphasic media H_2O/supercritical carbon dioxide. Thus, the otherwise hazardous stoichiometric combination of hydrogen and oxygen for the formation of hydrogen peroxide and the enzymatic conversion can be carried out conveniently in one pot. The results of our investigation towards this cascade reaction will be presented. Scheme-1 1. Van Rantwijk F.; Sheldon R A. Selective oxygen transfer catalysed by heme peroxidases: synthetic and mechanistic aspects. Curr. Opin. Biotechnol. 2000, 11, 554-564. 2. Piero P.; Giacomo C.; Enrico M.; Nicoletta G.; Stefano C. Chloroperoxidase-catalyzed enantioselective oxidation of methyl phenyl sulfide with dihydroxyfumaric acid/oxygen or ascorbic acid/oxygen as oxidants, Biotechnol. Bioeng. 1999, 62 , 489-493, 3. Van de Velde F.; Van Rantwijk F.; Sheldon R. A. Selective oxidations with molecular oxygen, catalyzed by chloroperoxidase in the presence of a reductant. J. Mol. Cat. B: Enzymatic, 1999, 6, 453-461. 4. Van Deurzen M. P. J.; Seelbach K.; Van Rantwijk F.; Kargl, U.; Sheldon R.A. Chloroperoxidase: use of hydrogen peroxide-stat for controlling reactions and improving enzyme performance. Biocatal. Biotransform. 1997. 15, 1-16. 5. Kohlmann C.; Lütz S. Electroenzymatic synthesis of chiral sulfoxides. Eng. Life Sci., 2006, 6, 170–174, 6. Trevisan V.; Signoretto M.; Colonna S.; Pironti V.; Strukul G. Microencapsulated chloroperoxidase as a recyclable catalyst for the enantioselective oxidation of sulfides with hydrogen peroxide, Angew. Chem. 2004, 116, 4189-4191 7. Beckman E. J. Production of H_2O_2 in CO_2 and its use in the direct synthesis of propylene oxide,Green Chem., 2003, 5, 332-336,

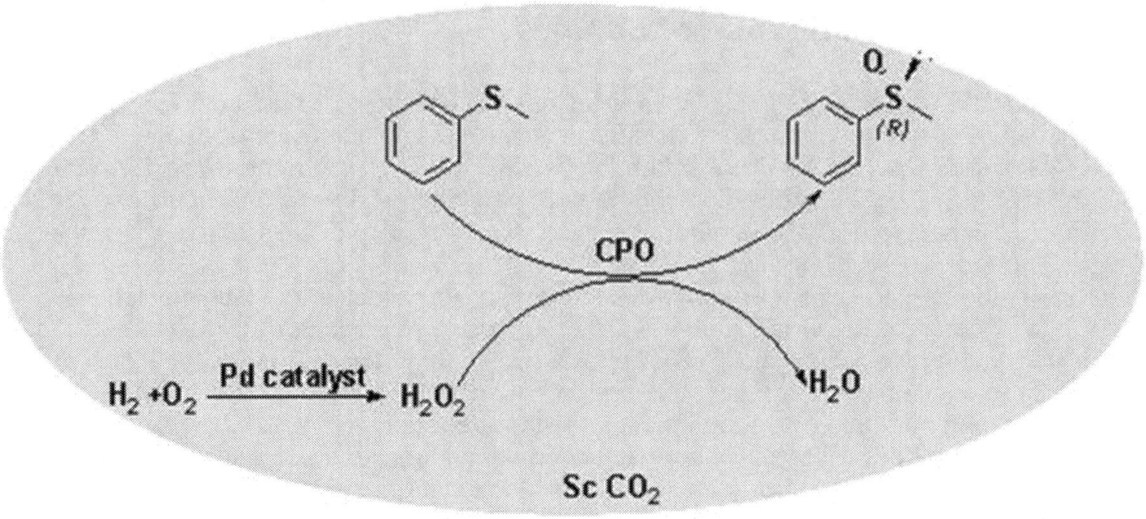
S
O
S
(R)
CPO
H_2 + O_2
Pd catalyst
H_2O_2
H_2O
Sc CO_2

128e Process Simulation of a Pilot Case: Production of Perfume-Containing Microcapsules

Laureano Jiménez Esteller and Rodolfo Vicente Tona. Chemical Engineering Department, University Rovira i Virgili, Av. Països Catalans, 26, Campus Sescelades, Tarragona, 43007, Spain

1. Introduction

On the basis of a radical change in the paradigm microtechnology may help to build the foundations for a new generation of chemical production plants for high added value products. Microtechnology is able to achieve process and products improvements by orders of magnitude, rather than percentage points on key sectors (pharmaceuticals, additives, pigments, coatings, food additives…). The idea behind microdevices is the combination of two concepts:

· *Intensification*: produce more with less energy, less solvent, less inventory and reduced transport, thus reducing the environmental impact.

· *Miniaturization*: supply better products with smaller volumes, higher efficiency and greater precision.

The need for microsystem modeling arises for two reasons: (i) to allow the rapid penetration into the market it is essential that the design, manufacture and test loop is kept to a minimum; (ii) device models can lead to a higher understanding of microdevice operation. To this end, model complexity (many of the new microsystems including several sensor types along with their relevant conditioning and read out electronics) and material properties (surface effects and microscopic thermal properties) are among the most important.

2. Perfume-Containing Microcapsules (PmC)

The industrial case is the synthesis and characterization of perfume-containing microcapsules (PmC). PmC are of great interest to the industrial partner as they offer a mechanism for the efficient deposition of perfumes as well as provide a long-lasting fragrance benefits. Perfume deposition on textiles is an inefficient process (» 90 % of perfume added to a detergent is lost during washing). In addition, perfumes have to be very fabric specific and this limits the possible fragrances. Encapsulation of perfume into a microcapsule helps deposition because, with careful control of the size, PmC become entrapped into the cloth fibers during washing and resist being flushed away.

The production of PmC using microtechnology is at the laboratory stage. The continuous encapsulation technique is *phase inversion precipitation* [1]. The process occurs in a two steps procedure (Figure 1). The polymeric solution and the continuous phase is fed to the first microreactor. The continuous phase is immiscible with the solvent and is not a non-solvent for the polymer, and therefore the product stream is an emulsion containing polymer and perfume microdroplets. In the second microreactor the non-solvent phase is added, and precipitation of the polymer occurs. As a result, microcapsules are produced with the perfume encapsulated. To avoid any operational problem, and according to experimental data, the purity specifications of raw materials is very high (99-99.9 % w/w). In addition, the polymeric phase stream must be free of water [1].

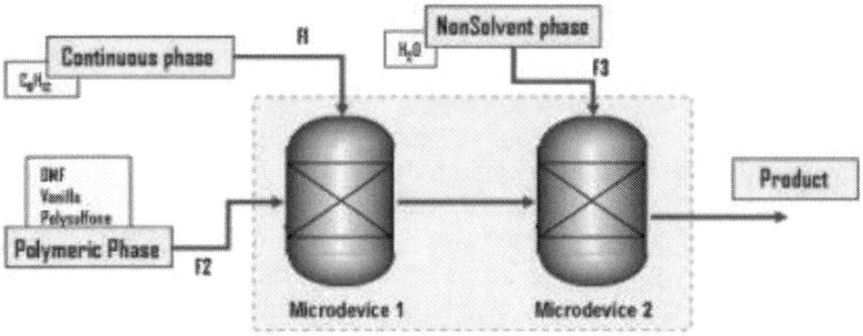

Figure 1. Microcapsule process flowsheet.

3. Integration of Process Modeling Approaches

The MICAP® module is used to perform the simulation of the PmC. The MICAP® tool, developed in Matlab® [2, 3], is used to link the modeling approaches:

1. Molecular simulation to study the micellization process, the precipitation of the polymer, size of porous and the morphology of the microcapsules.

2. Computer fluid dynamics to predict the flow dynamics, the mixing, and mass/heat transfer processes in the microreactor.

3. Process simulation to predict the continuous production of the PmC, including the solvent recovery system.

4. Life cycle assessment to deal with the environmental impact and consider the sustainability.

This work is focused at the process simulation level. Different modeling tools were used, as commercial process simulators are not prepared to efficiently handle microtechnology problems. The tasks are divided in two levels: (a) encapsulation of the MICAP® within Aspen Plus®; (b) PmC separation and recovery of the solvents.

3.1. Encapsulation of MICAP® and Aspen Plus®

Current process modeling and simulating tools offer a broad variety of functionalities to develop reliable models. The process simulator selected is Aspen Plus®. The tool does not offer functionalities for handling microstructured unit operations, and therefore the modeling requires an additional effort to overcame some difficulties [4].

Aspen Plus® and Matlab® were linked using Active X and COM technology [5, 6], thus integrating both tools in the calculation procedure. The problem of this integration is that both calculations tools are involved in several nested loops (Figure 3) giving rise to convergence problems. As Aspen Plus® is a sequential-modular simulator, the modification of the process topology does not affect the overall convergence performance of the system. In this way, the architecture and transfer procedure of the modules is build to promote scalability, re-usability and flexibility of the model

A *Basic Unit Operation* (BUO) was created within Aspen Plus®. The data from MICAP® is send to the BUO and data calculated from Aspen Plus® may be recovered through this BUO. The BUO acts as an encapsulate version of MICAP® within Aspen Plus® and, consequently, as a microdevice unit within the simulator (Figure 2).

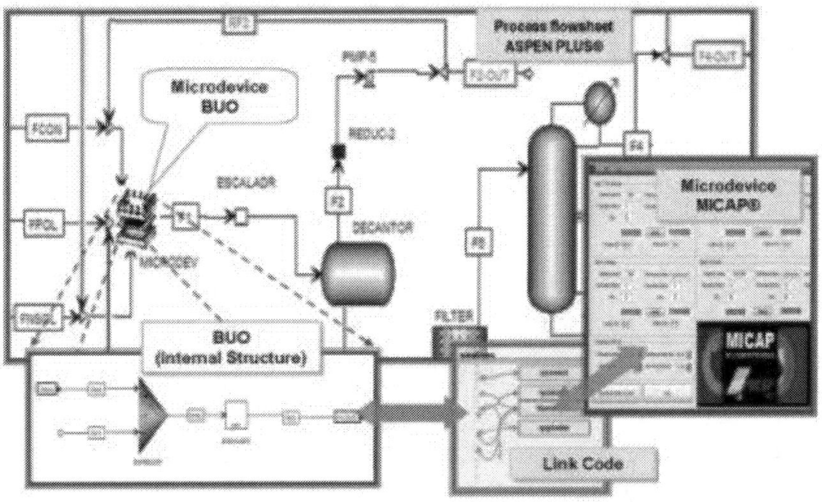

Figure 2. Encapsulation of MICAP® within Aspen Plus®.

4. Perfume-Containing Microcapsules Separation and Solvent Recovery

The microdevice output stream contains the PmC with cyclohexane, DMF, water and perfume (vanilla). Although this separation has not been experimentally tested, it is based in the information available. Figure 3 shows a complete diagram of the process.

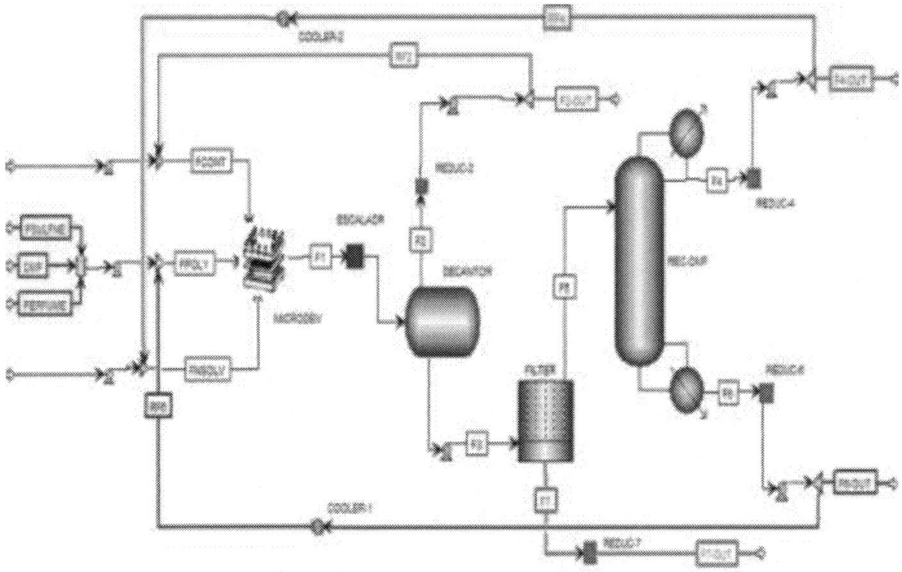

Figure 3. Flowsheet of the process in Aspen Plus®.

4.1. Recovery of cyclohexane

Cyclohexane is highly immiscible with water and DMF [7, 8], and the modeling of the decanter is straightforward. Then, the adoption of microstructured technology does not offer any *a priori* advantage. Therefore, the model proposed a scheme with microdevices working in parallel (N = 238) and the resulting stream is processed in a decanter, where the organic phase (cyclohexane), and the aqueous phase (PmC, DMF, water and perfume) are obtained. The number of microdevies can be modified by the user, and this change of scale allows to combining multiscale process units with microstructured elements.

4.2. Recovery of microcapsules

The average size of microcapsules in the laboratory is over 6 microns with a very small percentage under 4.5 microns (2% in weights). Then, a screen filter to retain particles (efficiency » 95-99 % at 4.5 microns) is enough for PmC recovery. To mimic the production line PmC must be preserved in a gel media (» 20 % liquid). In addition, the porous structure of the microcapsules retain some liquid (» 5 % w/w).

4.3. Recovery of water and DMF

The filtrate liquid contains DMF, water and a small fraction of perfume. The separation of water and DMF mixtures is the testing system for pervaporation units [9, 10]. Due to miscibility, perfume is recovered with the DMF (free-water).

4.4. Minimizing the environmental impacts

A key aspect in the pilot plant tests is the reuse of solvents, leading to a minimization of wastes and raw materials. Nevertheless, a life cycle assessment (LCA) is performed following the *cradle to grave* approach using the Ecoinvent [11] database and SimaPro® [12]. The aggregation of the impact categories into a single value (Eco-indicator 99) improves its value for the decision makers.

Most of the environmental load is assigned to the DMF production, and the residual streams containing DMF. Effort is needed to minimize this stream or select an alternative solvent.

5. Conclusions

It is clear that at present there are many industrial applications of microdevices (biosensors, microprocessors…), the state of the art in chemical engineering is far from the one achieved in other areas. The end goal of the IMPULSE project (“*fab-on-a-chip*”) is far from achieved, as microdevice has a clear advantage in some market niche (*e.g.* microreactors), while separation technology is far from developed.

In addition to model microsystems, experimental data based on real behavior is required to tune the model predictions (drag-out, efficiencies…).

6. Acknowledgements

This work has been funded by the IMPULSE EU project (NMP2-CT-2005-011816).

Literature

[1] Production of capsules using modified lab reactor - Deliverable D3.2f. IMPULSE. ETSEQ, February 2007.

[2] C. Torras, and R. Garcia. The [MICAP] simulator. Registered software T-177/07.

[3] Torras_MICAP_Flyer.pdf and Torras_MICAP_Oral.pdf (available at the IMPULSE website).

[4] Aspen Plus® 12.1. User Guide. Aspen Technology Inc., Cambridge (MA), USA. 2003.

[5] A. Bojarski, L. Jiménez, A. Espuña and L. Puigjaner. European Congress of Chemical Engineering. Copenhagen, 2007.

[6] Matlab® 7.0. External Interfaces. The MathWorks, Inc. Natick (MA), 2007. pp. (8-3) - (8-129). 2007.

[7] M.L. Campbell. "Ciclohexane" in Ullmann's Encylopedia of Industrial Chemistry (6th Ed). John Wiley & Sons, Inc., 2007.

[8] H. Bip and H. Kieczka. "Formamides" in Ullmann's Encylopedia of Industrial Chemistry (6th Ed). John Wiley & Sons, Inc., 2007.

[9] H.K. Lee, J.Y. Kim, Y.D. Kim, J.Y. Shin and S.C. Kim. Polymer, 42 (2001) 3893.

[10] D.Anjali Devi, B. Smitha, S. Sridhar and T.M. Aminabhavi. Separation and Purification Technology, 51 (2006) 104.

[11] Swiss Centre for Life Cycle Inventories, The Ecoinvent database, (2006).

[12] Pré-Product Ecology Consultants, Introduction to LCA with SimaPro® 7, Pré-Product Ecology Consultants, (2006).

Continuous Microreactor Processing, Reactive Ionic Liquids and Microwave Heating – Approaches Towards Intensified Kolbe-Schmitt Synthesis

Ulrich Krtschil[1], Faysal Benaskar[2], Volker Hessel[1,2], Patrick Löb[1]

[1]Institut für Mikrotechnik Mainz GmbH, Mainz, Germany

[2]Eindhoven University of Technology, Eindhoven, The Netherlands

The Kolbe-Schmitt synthesis is a traditionally used carboxylation reaction for phenolic cores at which a carboxylic group is introduced by an electrophilic substitution. High pressure CO_2 or carbonic acid derivates like $KHCO_3$ act as COOH precursor. The Kolbe-Schmitt synthesis is normally carried out batch wise with long reaction times. This, in 1860 by A.W.H. Kolbe developed and 1985 by R. Schmitt improved synthesis route is to date a standard method used for the fabrication of aromatic hydroxy carboxylic acids like acetylsalicylic acid (Aspirin®), p-aminosalicylic acid (PAS) and alkyl p-hydroxybenzoat (paraben). These products are used both as pharmaceuticals (Aspirin®), antimicrobial (paraben) for foods, cosmetics and pharmaceuticals and as a fine chemical (p-hydroxybenzoic acid) for the production of dyes, pesticides and pharmaceuticals and not to forget of liquid crystalline polymers.

This paper will address new approaches to intensify the synthesis using and combining different methods of process intensification: Continuous processing using microstructured devices, microwave heating and the use of reactive ionic liquids.

Due to the commercial importance of the Kolbe-Schmitt synthesis a first explorative work carried out at IMM was engaged in the exemplary implementation of the batch process into a continuous one combined with the application of high temperature and pressure. Here the conventionally heated aqueous Kolbe-Schmitt synthesis was continuously carried out for the synthesis of 2,4-dihydroxybenzoic acid from resorcinol (1). The product is amongst others applied for the fabrication of industrial dyestuffs, chemicals for reprography and pharmaceuticals.

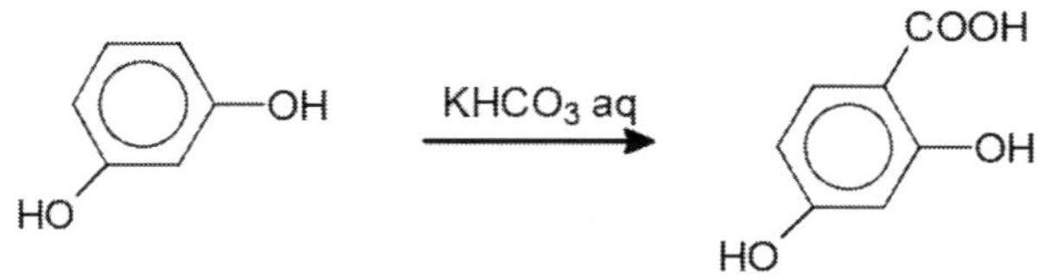

(1)

The experimental setup was composed of a capillary reactor and a cross-flow micro heat exchanger for fast chilling the reaction mixture at the end of the reactor (Figure 1).

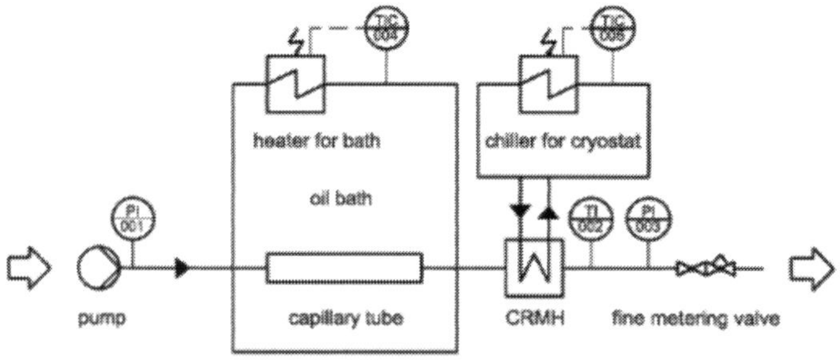

Figure 1. *Flow diagram of the experimental setup for continuously performed Kolbe-Schmitt synthesis using conventional oil bath heating*

A product yield of 43% was obtained at a reaction time of about one minute, by means of oil bath heating at a temperature of 180°C and under pressure of 40 bar. The best yield was 47% (390 s, 140°C). The achieved yields are comparable with experimental results (up to 50 % yield) using a stirred batch; however the latter needs 120 minutes reaction time which demonstrates the process intensification by means of continuous capillary reactor technology [1] (Figure 2). Long residence times, which are practiced in the classic batch synthesis, and higher temperatures promote the decarboxylation reaction. The latter could be suppressed by using the good temperature control in the capillary reactor and the application of high pressure which allows the formidable reduction of the reaction time due to higher reaction temperatures while still staying in the liquid phase.

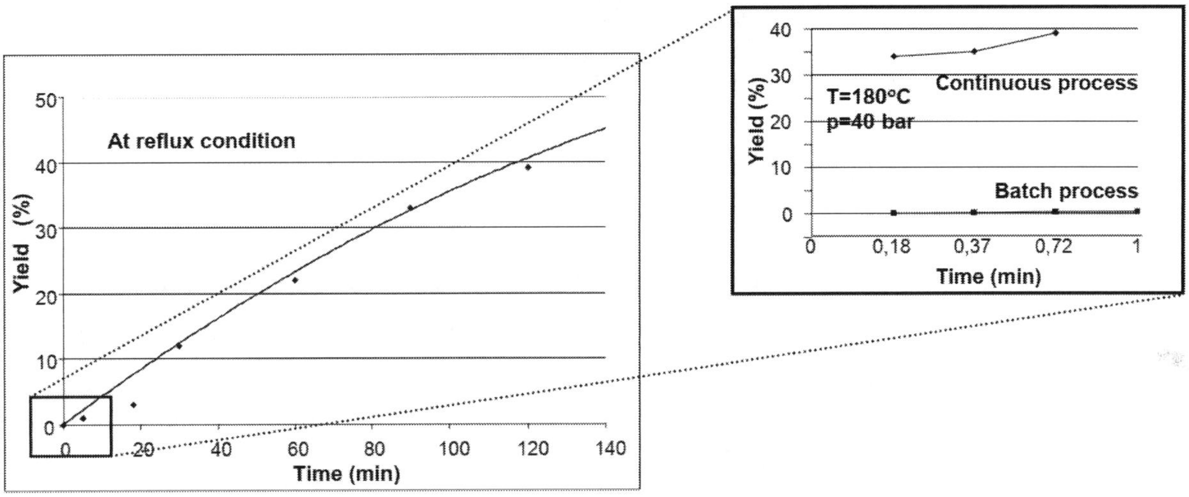

Figure 2. *Comparison of continuous vs. batch process*

A similar indication for the intensification was observed for the synthesis of 2,4,6-trihydroxybenzoic acid from phloroglucinol. But in this case the decomposition reaction became dominant already for lower temperatures over 125 °C, i.e. the effect with regard to the required reaction time was not in the same dimension (Figure 3).

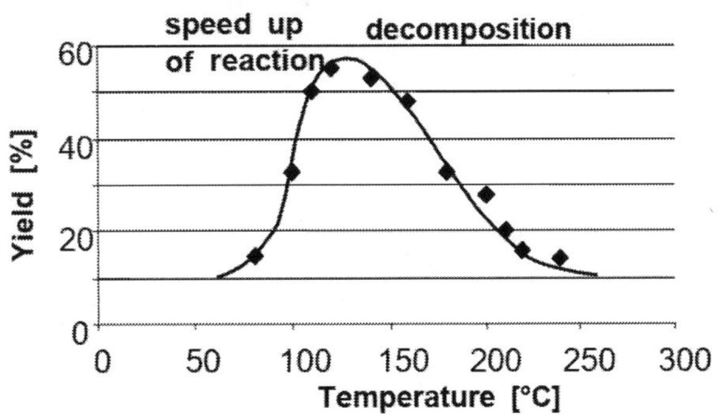

Figure 3. *Speed up and decomposition effects with increasing temperature for the synthesis of 2,4,6 trihydroxybenzoic acid [2]*

The results of the explorative experimental studies show process intensification when applying high temperature and pressure combined with equipment which is able to control the heat of reaction. Another process intensification approach is the application of reactants at higher concentrations or to operate even solvent-free. In the case of the aqueous potassium bicarbonate solution it is not possible to exceed markedly the concentration due to the limits of solubility. In this context, the substitution of the aqueous potassium bicarbonate solution by reactive ionic liquids which offer a carbonate ion may be helpful. In addition, an increase in selectivity can be expected when making use of the fast heating rate of microwaves combined with almost instantly chilling by a microstructured heat exchanger to suppress side reactions and decomposition

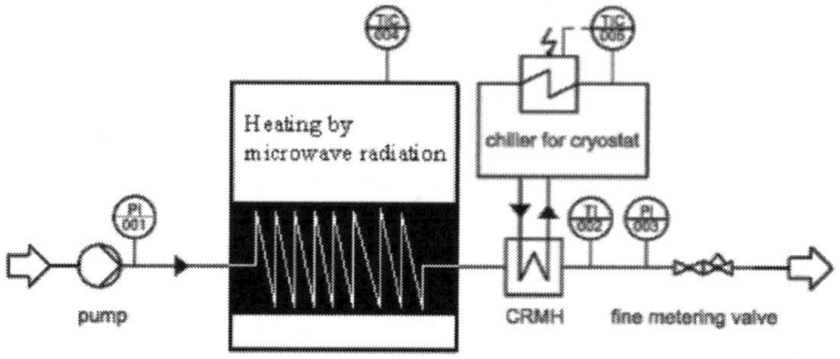

Figure 4. *Flow diagram of the experimental setup for continuously performed Kolbe-Schmitt synthesis using microwave heating [3]*

Pursuant to these ideas appropriate experiments were performed in the context of a student's internship from Eindhoven University of Technology at IMM. A continuous microwave processing (Figure 4) was developed, with a capillary reactor of comparable dimensions to the prior trials described above. In advance it was for appropriate power settings necessary to determine both flow and material dependent heating-up curves (Figure 5) and temperature profiles along the capillary reactor.

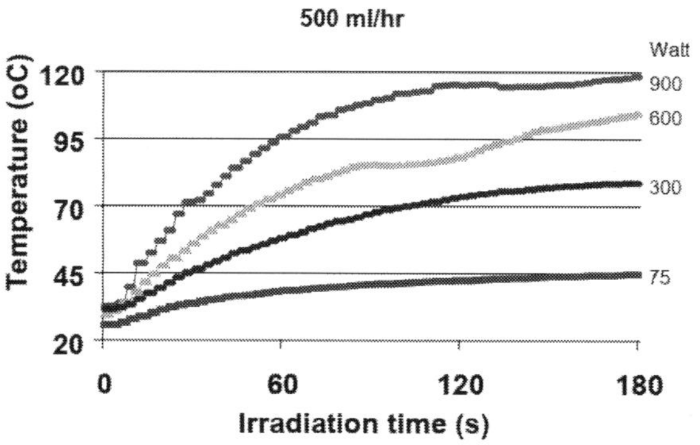

Figure 5. *Outlet temperature as function of irradiation time and power for microwave heating with water [3]*

The microwave heated aqueous Kolbe-Schmitt synthesis was carried out under a pressure of 8 bar at different temperatures. This resulted in considerable higher yields compared to the conventionally heated approach. The best product yield obtained was 52%, which is an improvement of 5% compared with the best value obtained by conventional heating even at decreased reaction time.

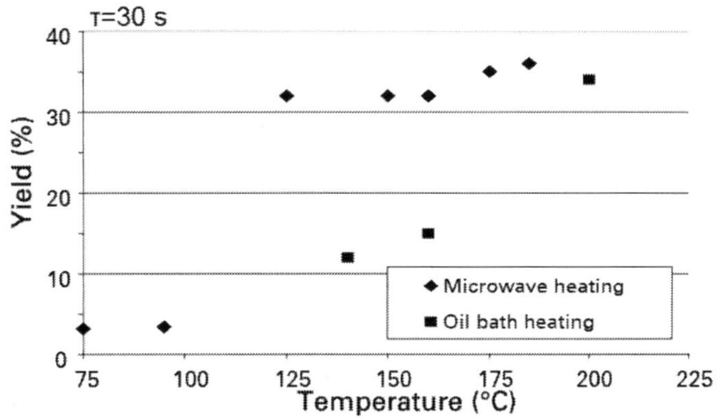

Figure 5. *Comparison of the obtained yields for the aqueous Kolbe-Schmitt synthesis at a residence time of 30 s for oil bath heating and microwave heating, respectively [3]*

Then, ionic liquids were used as CO_2 donating agents, diluted (2.91 mol/L) or concentrated (5.81 mol/L), thus being reactive solvents which allow to perform solvent-less processing. In the diluted case 1-ethyl-3-methylimidazolium hydrogen carbonate was used, while in the concentrated case 1-butyl-3-methylimidazolium hydrogen carbonate was applied. Both ionic liquids were used for the conventionally heated Kolbe-Schmitt synthesis. The yield obtained in the diluted case was 44% at 180°C, 35 bar and at 130 s reaction time. Only a relatively low flow rate could be applied due to the limited available amounts of the ionic liquids. The obtained yield in the concentrated case was 59% at the same conditions, but with a considerably higher space-time-yield of 2800 kg/(h m³) compared to 960 kg/(h m³) in the diluted case (Figure 6).

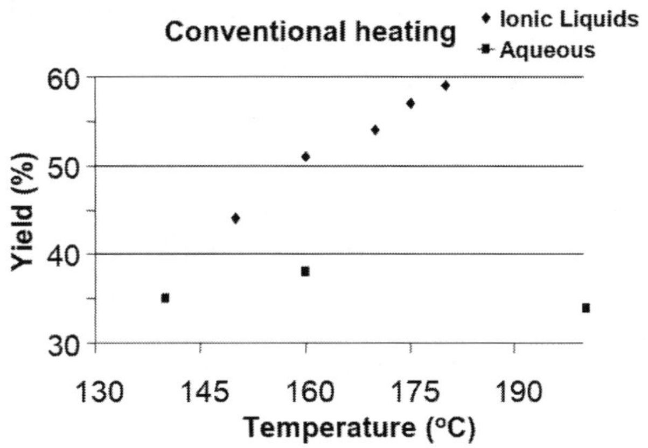

Figure 6. *Yield comparison for conventionally heated Kolbe-Schmitt synthesis using aqueous potassium bicarbonate solution and ionic liquids, respectively [3]*

Finally experiments using a combination of microwave heated Kolbe-Schmitt synthesis and reactive ionic liquid were performed at different temperatures and resulting in a yield of 57 % (i.e. 7 % more than for the standard procedure described in the Organikum [4]) using a reaction time of 130 s, 10 bar and 140°C. These yields are similar to those obtained for the conventionally heated Kolbe-Schmitt synthesis with reactive ionic liquids, but at 40 K to 50 K lower outlet temperatures (Figure 7).

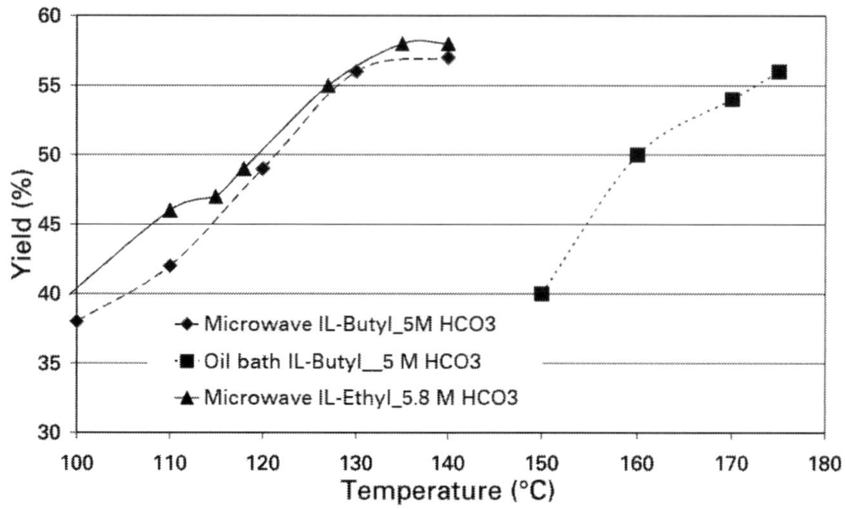

Figure 7. *Comparison of the obtained yields at a residence time of 130 s for oil bath heating and microwave heating, respectively [3]*

Motivated by these optimistic results, more detailed investigations are planned within the recently started project "New process windows for the process intensification of the Kolbe-Schmitt synthesis" which will be performed in the frame of the German Federal Foundation Environment (Deutsche Bundesstiftung Umwelt, DBU) funding cluster "Novel Process Windows". Among others this project aims at a systematic investigation of new synthesized reactive ionic liquids as CO_2 donating agents, the recirculation of the ionic liquid and in-situ remanufacturing of the carbonate ion. In addition, both the synthesis of the ionic liquids and the Kolbe-Schmitt synthesis will be carried out microwave heated. Latest results will be implemented into the presentation.

[1] Hessel, V. et al., Org. Proc. Res & Dev, **2005**, 9, 479-489

[2] Hessel, V.; Hofmann, C., Löb, P., Löwe, H., Parals, M.: Chem. Eng. Technol., **2007**, 30 (3), 355-362

[3] Benaskar, F., Hessel, V., Krtschil, U., Löb, P., Rebrov, E., Wille, G., Presentation at 2nd Symposium on Microwave Accelerated Synthesis, Düsseldorf, Germany, **2007**

[4] Organikum, Wiley-VCH, Weinheim, 22. Edition, **2004**

83f Micro Reactor Application for the Beckmann Rearrangement to Epsilon-Caprolactam

K.T. Zuidhof[1], D.V. Raulji[2], M.H.J.M. De Croon[2], and J.C. Schouten[2]. (1) Department of Chemical Engineering and Chemistry, Eindhoven University of Technology; P.O. Box 513, 5600 MB Eindhoven, the Netherlands, Mr v Busselstraat 64, Mierlo, Netherlands, (2) Chemical Engineering and Chemistry, Eindhoven University of Technology; P.O. Box 513, 5600 MB Eindhoven, the Netherlands, Netherlands

Mass and heat transfer can put significant technological limitations to reactor and process design, which leads to far from ideal solutions for bulk processes. Scaling down to centimeter, millimeter, or even micrometer scale can increase mass and heat transfer in such cases tremendously to allow for a much improved process design with respect to conversion, selectivity, reactor volume, and, in case of strongly poisonous and dangerous chemicals and runaway conditions, safety. When several of these conditions are present, a strong impetus exists to replace the present bulk scale technology by micro reactor technology. The goal of this paper is to report on the design, construction, and testing of a single channel micro reactor integrated with a micro mixer to demonstrate the viability of a miniaturized reactor system for an industrial bulk process, based on micro reaction technological considerations. As model reaction is chosen the Beckmann rearrangement of cyclohexanone oxime to epsilon-caprolactam with oleum as a catalyst and solvent. Epsilon-caprolactam is the main precursor for nylon-6 and is produced with an annual production of 4.4 million tons and may therefore be considered as one of the main bulk chemicals. The admixing of molten cyclohexanone oxime (1 mPas) and the reaction mixture (50-300 mPas), consisting of caprolactam and oleum, is the key step in the industrial process. In current processes, a jet mixer, operated in the turbulent regime ($Re > 5000$), is used with a considerable recycle of the reaction mixture. However, the integration of the reaction and the heat exchange sections by the application of micro technology will certainly decrease residence time and may therefore increase selectivity and safety. Furthermore, the window of operation in the Beckmann reactor is enlarged, which is beneficial for process flexibility. Already in the 1980s BASF described a process for the continuous preparation of caprolactam by Beckmann rearrangement of cyclohexanone oxime, dissolved in a solvent which is inert towards oleum and immiscible with reaction mixture. This is beneficial for the process. Useful solvents are cycloalkanes, which in addition remove water from the cyclohexanone oxime feed. For the reason that two phase flow in micro mixers creates a large interfacial surface area by the formation of small alternating fluid zones, the hydrodynamics and flow patterns of relevant viscous liquids and immiscible cycloalkanes have been investigated by microscopic high speed imaging. A split and recombine mixer shows various flow patterns as a function of different flow rates (Figure 1). At low flow rate, Taylor flow patterns can be distinguished. Higher flow rates result in the formation of a large number of small droplets in liquid slugs in different fashions. Six different patterns can be distinguished. Other mixer designs, including the Y-mixer and the interdigital mixer, give rise to various other flow patterns which will be shown in the paper. The formation of large numbers of small droplets is beneficial for increased mass transfer in micro channels, as the interfacial mass transfer area will increase per volume of reactor. Also fluid viscosity and surface tension effect the observed flow patterns. Image analysis tools have been used to determine droplet size (ca. 30 ìm) distributions, which are critical parameters in mass transport models. At 100% conversion, the Beckmann rearrangement in a stainless steel single channel micro reactor, with a residence time of approximately 4 sec, shows a strong dependency of the selectivity on the mixing intensity and the way of mixing. In Y-mixer reactors (250 ìm channel diameter), used as a benchmark, an increase in reactor length from 20 cm to 50 cm increases the selectivity from 84 to 94%. Increasing the mixing intensity and mass transfer by the application of two phase flow, reaction mixture and inert solvent, cyclooctane, for cyclohexanone oxime, the selectivity ranges from 60 to 99+%. Despite the simple design, a Y-mixer reactor already demonstrates selectivities up to 99.4% to caprolactam, with 100% conversion. However, the selectivity is very susceptible to process conditions, especially temperature. By the application of more sophisticated designs, such as the well known IMM high pressure interdigital mixer, the selectivity is much more stable towards reaction conditions (Figure 2) and, within experimental error, results in 100% selectivity with 100% conversion for high oleum concentrations. To reach high selectivities at low oleum concentration (higher viscosity), further intensified mixing is needed. To this end, the split and recombine mixer has been used. This mixer offers a very elegant method for creation of multi micro zones of oxime and oleum rich phases and shows excellent performance.

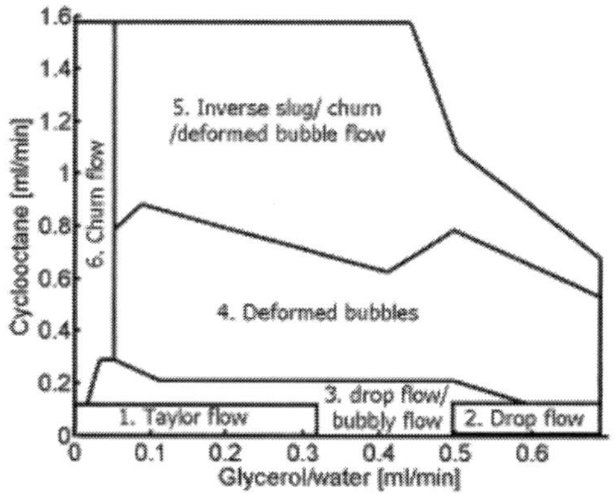

Figure 1: *Flow regime map for glycerol/water and cyclooctane system for a split & recombine mixer.*

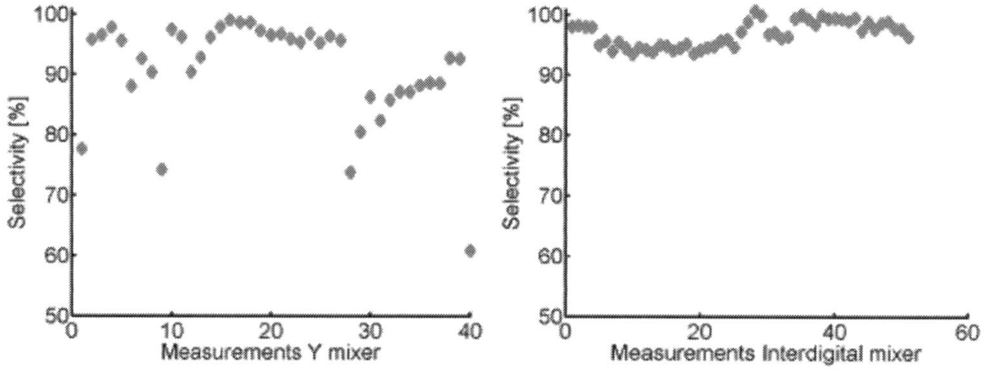

Figure 2: *Selectivities to caprolactam for a number of experiments at comparable operation conditions for a Y-mixer and an interdigital mixer. Notice the more stable operation of the interdigital design.*

83aj Effect of Surface-to-Volume Ratio in Microplasma Reactors

Anil Agiral[1], K. Seshan[2], Leon Lefferts[2], and J.G.E. (Han) Gardeniers[1]. (1) Mesoscale Chemical Systems, MESA+ Institue for Nanotechnology, University of Twente, P.O. Box 217, Enschede, 7500 AE, Netherlands, (2) Catalytic Processes and Materials, MESA+ Institue for Nanotechnology, University of Twente, P.O. Box 217, Enschede, 7500 AE, Netherlands
Non-thermal plasma generated at atmospheric pressure can provide highly energetic electrons which is necessary to decompose molecules. Performing the plasma process in a microreactor can lead to precise control of residence time and extreme quenching conditions, enabling control over the reactants to selectively produce desirable products [1]. Inelastic collisions with processing gas can excite molecules and atoms giving radicals, electronically and vibrationally excited species. Extremely small volume and large surface area of microplasma reactors can be an efficient tool to study plasma-surface interactions. As the surface-to-volume ratio increases, presumably wall collisions will be important and these collisions will affect the rate of decomposition by changing the main pathways by which radicals and activated species are lost in the discharges. High surface-to-volume ratio can be attained in a microplasma reactor by splitting the flow into multiple microchannels so that maintaining the reactor volume the same. Figure 1(a) shows the schematic diagram of the approach. If the same electric field and power are applied to the microreactors, useful information on effect of surface on the decomposition mechanism could be obtained. CO_2 decomposition in plasma reactors have generated considerable interest because of its high energy efficiency [2]. In this study, we discuss the effect of glass surface on the decomposition of CO_2 at atmospheric pressure in glass type microplasma reactor experimentally and by chemical kinetic modeling. We simulate the barrier discharge with CO_2 gas. Modelling employs a well mixed plasma reactor model as available in Chemkin 4.1 [3]. An electron energy balance equates the time rate of change of electron's swarm internal energy to the net flow of electron enthalpy into and out of the reactor, accounting for net chemical production rates, surface losses, collisional losses and power deposition from the externally applied electromagnetic field. Reaction rate coefficients of the electron induced reactions was determined by 0-D solution of the Boltzmann equation using BOLSIG+ [4]. The relationship between reaction rate coefficient and average electron energy was described by fitting the Boltzmann solution results to a three parameters Arrhenius fit. Ionization, vibrational and electronic excitation processes were thought to play an important role in CO_2 plasma. Homogenous and heterogeneous reactions of the species were added to the Chemkin model. In Figure 1(b), calculated energy loss coefficients of important electron impact processes in CO_2 discharge are shown. These results imply that energy spent on dissociative electronic excitations (Electronic1&2) and ionization processes increase with increasing electron energy in the range of 1eV and 8eV. Comparison of the results of one channel microplasma reactor with multiple channels will be discussed to provide information on the surface processes. References [1] T. Nozaki, A. Hattori, K. Okazaki., Catal. Today, 98, (2004), 607. [2] V.D. Rusanov, A.A. Fridman, G.V. Sholin, Sov. Phys. Usp., 24, 447. [3] R. J. Kee, F. M. Rupley, J. A. Miller, M. E. Coltrin, J. F. Grcar, E. Meeks,H. K. Moffat, A. E. Lutz, G. Dixon-Lewis, M. D. Smooke, J. Warnatz, G. H. Evans,R. S. Larson, R. E. Mitchell, L. R. Petzold, W. C. Reynolds, M. Caracotsios, W. E. Stewart, P. Glarborg, C. Wang, C. L. McLellan, O. Adigun, W. G. Houf, C. P. Chou, S. F. Miller, P. Ho, P. Young, D. J. Young, CHEMKIN Release 4.0, Reaction Design, San Diego, CA (2004). [4] G.J.M. Hagelaar, L.C. Pitchford, Plasma Sources Sci. Technol., 14, (2005), 722-733.

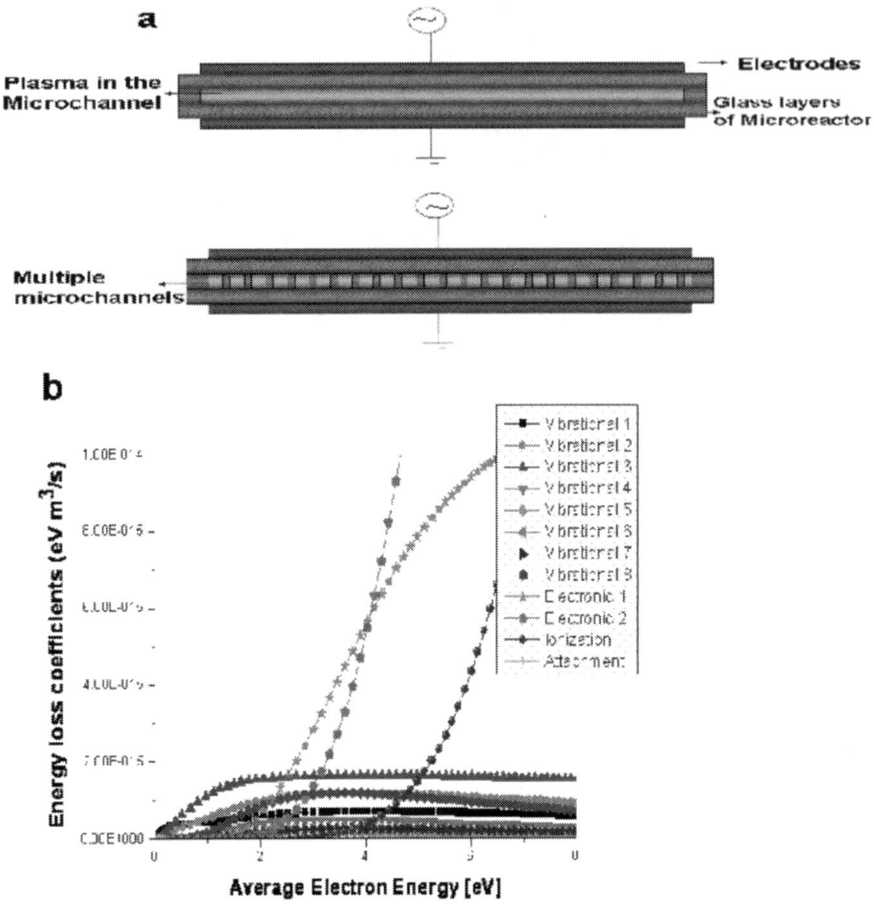

a
Electrodes
Plasma in the Microchannel
Glass layers of Microreactor
Multiple microchannels
b
Energy loss coefficients (eV m³/s)
Average Electron Energy [eV]
Vibrational 1
Vibrational 2
Vibrational 3
Vibrational 4
Vibrational 5
Vibrational 6
Vibrational 7
Vibrational 8
Electronic 1
Electronic 2
Ionization
Attachment

101e Continuous Production of NADPH with Malic Enzyme in Micro Scale

Daniela H. Mueller[1], Pia Mueller[1], Shukrallah Na'Amnieh[2], and Lasse Greiner[1]. (1) Itmc, RWTH Aachen, Worringerweg 1, Aachen, 52074, Germany, (2) X-Zyme, Merowinger Platz 1A, Duesseldorf, 40225, Germany

The reduced cofactor NADPH is an all-purpose redox equivalent for biological systems. NADPH is used in a number of applications as redox equivalent for enzymatic reduction in synthesis as well as enzyme based sensors, and as food additive. Also the production of NADPH from $NADP^+$ leads to an enhancement in value by the factor of 5.

Malic enzyme can be applied successfully for reductive cofactor regeneration.[1,2] For the cell free reductive production we found malic enzyme to be superior to existing systems as formate dehydrogenase and glucose dehydrogenase. A study for production of NADPH was carried out in the micro-enzyme-membrane reactor.[3,4] This reactor was realised with a volume of less than 200μL. That allows evaluation of continuous conditions at early stage process development and also production of the exact amount of cofactor that is needed. We will present results regarding process stability and productivity of this novel enzyme for the continuous on-demand production of NADPH.

[1] Shukrala Naxamnieh, Werner Hummel, and Harald Gröger. Malat-dehydrogenase zur coenzym-regenerierung,(2002). Deutsche Patentanmeldung

[2] Shinichiro Suye and Sadaji Yokoyama; Enzyme Microb. Technol., (1985), 7, 418-424

[3] D. Müller, diploma thesis, FH Aachen (2005)

[4] D. H. Müller, M. A. Liauw and L. Greiner, Chemical Engineering & Technology, (2005), 12, 1569-1571

Process development for fluorination of a steroid with DAST and Deoxofluor

D. Negi, K. Lovis, U. Budde, J. Geisler
Bayer Schering Pharma AG, Müllerstrasse 178, 13353 Berlin

Selective presence of fluorine in some organic molecules is known to alter their physiological properties[1]. In the past few decades there has been increasing interest in the development of pharmaceutical products that contain fluorine in their structure[2]. Consequently, the research on new fluorination processes is also gaining importance in pharmaceutical industry. Fluorination reactions are often very exothermic, and require use of fluorinating agents that need careful handling. Microreactors, in general, can be beneficial in cases where the presence of heat or mass transfer limitations or certain safety aspects pose scale-up problems. For many fluorination reactions, these advantages can be decisive in developing or selecting new processes. Recently microreactors have also attracted interest for fluorination, especially with elemental fluorine as a fluorinating agent[3,4]. Direct fluorination provides an economic method but is also a non-selective process.

The conversion of C-O to C-F bond represents a viable method to produce selectively fluorinated organic compounds. This reaction, also sometimes termed as deoxofluorination, needs fluorinating reagents that are potentially hazardous[5]. Microreaction technology offers a good alternative to benefit this field, mainly by offering improved safety features realized due to smaller reaction volumes and better heat transfer compared to conventional reactors. Furthermore with some fluorinating reagents, scale-up in batch systems is constrained by the recommend limit for the maximum reactor volume. In such cases continuous processing also offers the advantage of higher throughput with a single reactor with much smaller hold-up than the recommended batch reactor size. In this work, we have utilized these advantages in the development of a continuous process for the deoxofluorination of a C=O bond in a steroid (Figure 1).

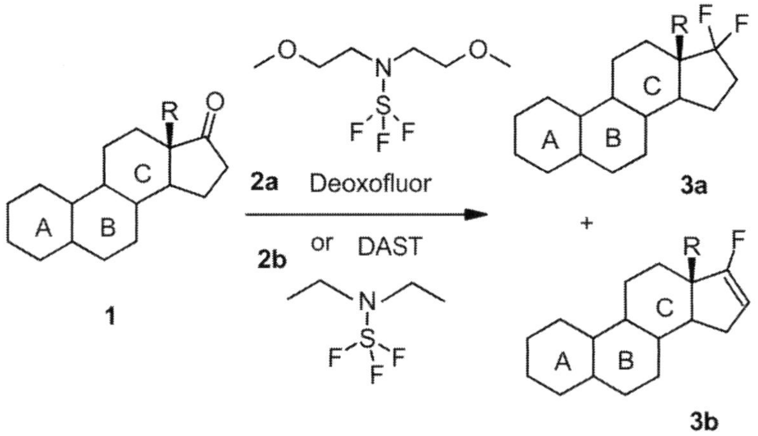

Figure 1. Deoxofluorination of a steroid with DAST or Deoxofluor.

The steroid **1** is fluorinated to get a geminal difluoride **3a**. A portion of the steroid **1** is also converted to yield vinyl fluoride **3b**. After testing some commercially available fluorinating agents, Bis(2-methoxyethyl)aminosulfur-trifluoride (Deoxofluor)[5] **2a** and Diethylaminosulfur-trifluoride (DAST)[6] **2b** were found suitable for this reaction. These fluorinating agents have high thermal potential and are thermally unstable. DAST is known to decompose violently at temperatures above 90°C and it is not recommended for production on an industrial scale[5]. Preliminary reaction studies showed similar results with both fluorinating reagents. Calorimetry was performed with pure reagents and reaction mixtures to identify the boundaries for safe operating conditions. In this work, we implemented a *simulation based experimental design* approach to find the optimum conditions for the continuous process. A kinetic model was derived from batch reactor data, which was then used to simulate the continuous processes. This gives a better understanding of the reaction, and once a suitable model is found, many experimental conditions can be screened reliably based on simulations.

The formal kinetics was first investigated in a 50 mL thermostatted batch reactor with Deoxofluor. The reactant **1** was dissolved in a solvent prior to the reaction. The samples collected from batch reaction were analysed by HPLC to obtain concentration-time profiles of the components **1**, **3a** and **3b**. The effect of temperature and Deoxofluor molar equivalents on the conversion of Steroid **1** and product distribution was investigated. The results of experiments are represented in Figures 2-3. HPLC analysis was also used to calibrate the method used later for online IR monitoring of the continuous process. In batch reactions, it was observed that higher temperature leads not only to faster reaction rate but also to faster decomposition of Deoxofluor as a parallel side reaction. This puts a limit on the maximum attainable conversion of **1** after a reaction time depending on the operating temperature. Since the reaction and decomposition both are exothermic, a precise temperature control is needed to maintain optimum reaction conditions. The product distribution was not found to vary significantly under different reaction conditions.

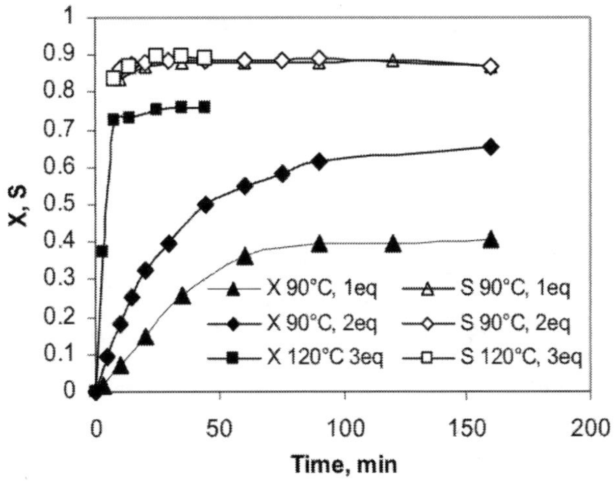

Figure 2: Effect of molar equivalents of Deoxofluor 2a on conversion (X) and selectivity (S). Conversion is defined as moles of steroid 1 reacted per mole of 1 fed. Selectivity is defined as moles of difluoride 3a produced per mole of 1 reacted.

On the basis of the measured concentration-time profile of the steroid components, the reaction was modelled as:

Reaction 1: Steroid **1** + Deoxofluor **2a** $\xrightarrow{k_1}$ geminal difluoride **3a** + byproducts

Reaction 2: Steroid **1** + Deoxofluor **2a** $\xrightarrow{k_2}$ vinyl fluoride **3b** + byproducts

Reaction 3: Deoxofluor **2a** $\xrightarrow{k_3}$ byproducts

The concentration-time profile of active Deoxofluor in the reaction mixture could not be directly measured. An estimation was made for the decomposition of Deoxofluor by heating it at a temperature and then reacting it with a measured amount of steroid **1**. The conversion of steroid gave an approximate idea of the degree of decomposed Deoxofluor as a basis of Reaction 3. The model parameters of reactions 1-3 were estimated by fitting the model to the batch data using the program Berkeley Madonna (Version 8.0.1). Figure 3 shows a comparison of experimentally measured batch reaction data with simulated data using the above kinetic model. Further simulations were carried out using the model to find the optimum conditions of temperature, reagent amount and residence time for the flow reaction, assuming ideal plug flow. Selected reaction conditions were verified by experiments in flow reactor.

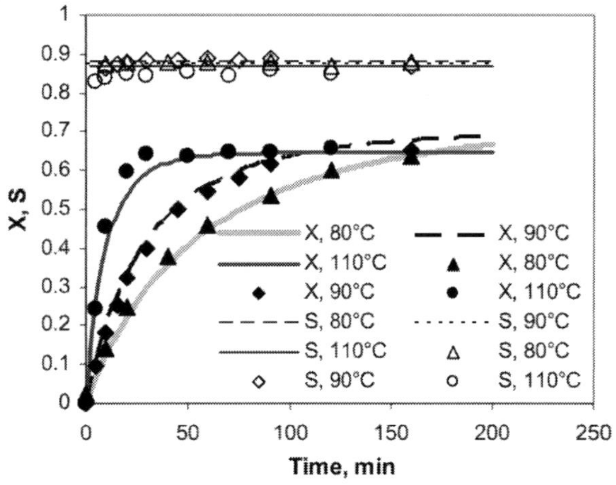

Figure 3: Comparison of simulations and experimental data. Symbols represent experimental data, lines represent simulated data.

Figure 4 shows the process flowsheet for this reaction. The reactants were pumped by rotary piston pumps through a micromixer into a residence time module. The feed vessel and lines carrying the steroid **1** were preheated to avoid the crystallization of the steroid. The residence time module was immersed in a thermostatted oil bath. Depending on the operating temperature and other specific requirements (visual flow observation, ease of cleaning etc.), a metal or fluoropolymer tube with submillimeter to a few millimetres diameter and a specific length was used as residence time module for the reaction. Inline monitoring of the product stream was performed by FT-IR spectroscopy using a specially fabricated miniaturized ATR probe immersed in the product stream (Fig. 4b). Figure 5 shows the typical IR bands of the Steroid **1** that were selected for monitoring the product stream.

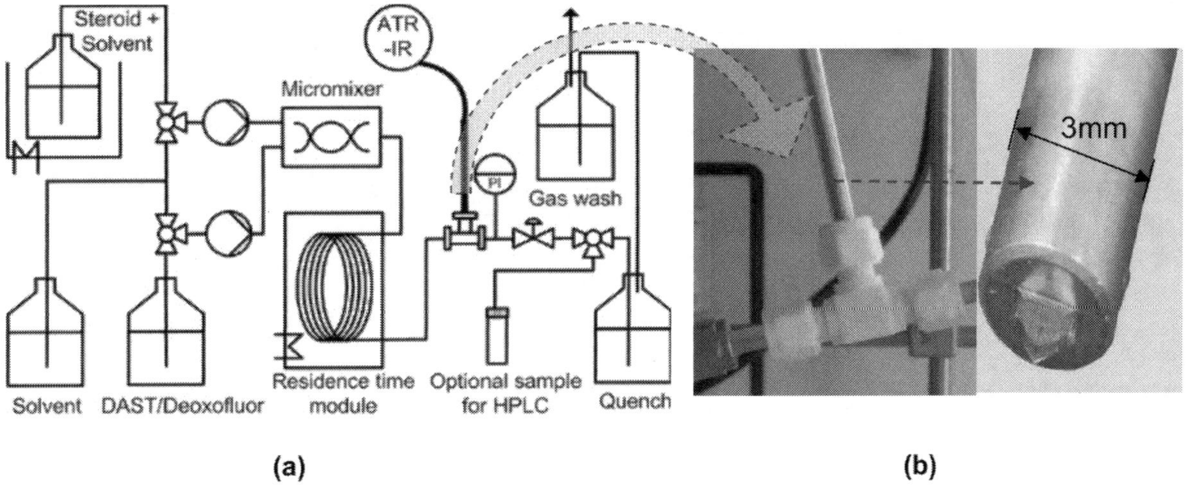

Figure 4. Schematic diagram of the continuous process for deoxofluorination of a steroid (a) and ATR probe with measurement cell (b).

Figure 6 shows a comparison of experimental values obtained in flow reaction and simulation profiles for the conversion of Steroid **1**. The throughput of the laboratory process with DAST was increased fifteen times by applying a combined scale-up and numbering-up concept. Kilograms of the fluorinated steroid intermediate were produced by this process for further processing for clinical tests.

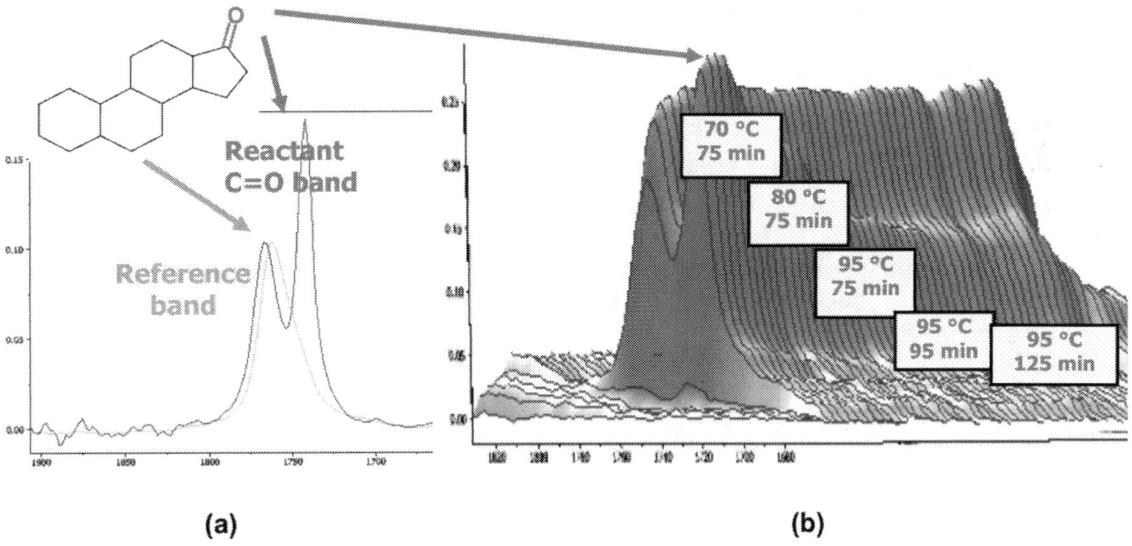

Figure 5. Online monitoring of Steroid 1 conversion by FT-IR spectroscopy in a flow reactor with DAST as Fluorinating reagent. (a) shows the IR bands that were considered for monitoring. The green (smaller) band corresponds to a reference group in the steroid molecule that remains unchanged during the fluorination. The IR band marked with red represents the carbonyl band that is fluorinated and its intensity reduces as the reaction conversion increases. (b) displays some examples of these bands collected from different experimental conditions.

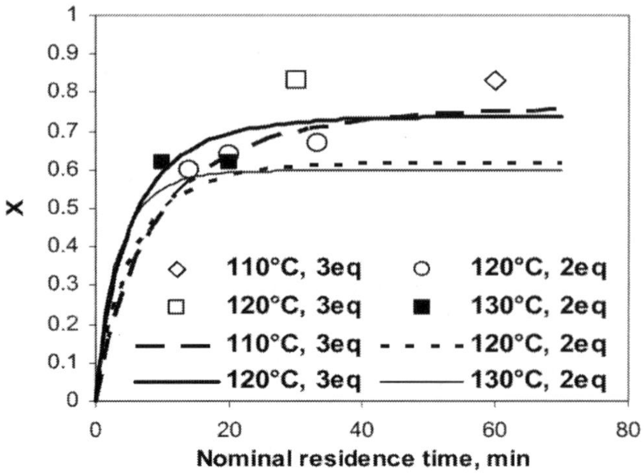

Figure 6. Conversion in continuous process with different parameters. Symbols represent experimental data, lines represent simulated data.

References

1. Strunecka, A., Patocka, J., Connet, P. Fluorine in Medicine. *J. App. Biomed,* **2004**, 2, 141-150.

2. Böhm, H.J., Banner, D., Bendels, S., Kansy, M., Kuhn, B., Müller, K. Obst-Sander, U. Stahl, M. Fluorine in Medicinal Chemistry. *ChemBioChem,* **2004**, 5, 637-643.

3. Jähnisch, K., Baerns, M., Hessel, V., Ehrfeld, W., Haverkamp, V., Löwe, H., Wille, Ch. and Guber, A. Direct fluorination of toluene using elemental fluorine in gas/liquid microreactors. *J. Fluor. Chem.,* **2000,** 105, 117–128.

4. Chambers, R.D., Spink, R.C.H. Microreactors for Elemental Fluorine. *Chem. Comm.* **1999**, 10, 883-884.

5. U.S. Patent 6,242,645. **2001**

6. U.S. Patent 3,976,691. **1976**

SIPROCESS - A Micro Process System for Process Development and Production

Thomas Bayer, Daniel Kadzimirsz, Markus Kinzl, Oliver Lade
Siemens AG, Industriepark Hoechst, Geb. G811, 65926 Frankfurt

Abstract

The batch-to-continuous-approach and its subset micro process technology are technologies with great efficiency potential for the syntheses of fine chemicals and active pharmaceutical ingredients. Micro process technology provides the chemist new possibilities in mixing and heat-transfer, which in lab-scale for many examples have already brought to use by now.

Still, process development for continuous processes - with or without micro structured components - has been considered to be complex in set-up and use. Therefore, the absolute number of continuous processes in fine chemical and pharmaceutical industries is limited and only few production plants use micro process technology today.

The micro process system SIPROCESS successfully meets the challenge. Siemens designed SIPROCESS for chemical syntheses in process development laboratories and to produce first quantities of a definite substance.

Siprocess – What is it about?

Siprocess is a modular, open, and automated micro process system for the synthesis of chemical substances. Siprocess consists of process modules in which process functions (e.g. metering, chemical reactions, mixing) can be carried out. The system can be configured individually using the following six different module types:

- Flow Distribution Module (FD) for feeding of alternatively raw materials, solvents or cleaning agents, respectively
- Pump Module (PU) for exact metering of liquids with negligible pulsation
- Reaction Module (RE) for mixing and subsequent chemical reaction in a temperature controlled micro structured reactor
- Pressure Control Module (PC) for the control of the process-related system pressure
- Sample Module (SA) for sampling and quenching

All modules are specified for the same flow and pressure range. Fluid-contacting parts are made of materials with a high chemical resistance (cf. Table 1).

Table 1: Technical features of Siprocess

Reaction temperature	-30° to 150°C
Process-related pressure	Up to 15 bar
Flow rate per pump module	0.1 to 50 ml/min at η = 1mPas
Material in fluid carrying pathways	Alloy C-4 (2.4610), ceramics, perfluorinated polymers

The modules are attached to racks in which they can be positioned relative to each other and connected to the power supply, the communication bus, the purge air as well as to the temperature control liquids (cf. Figure 1).

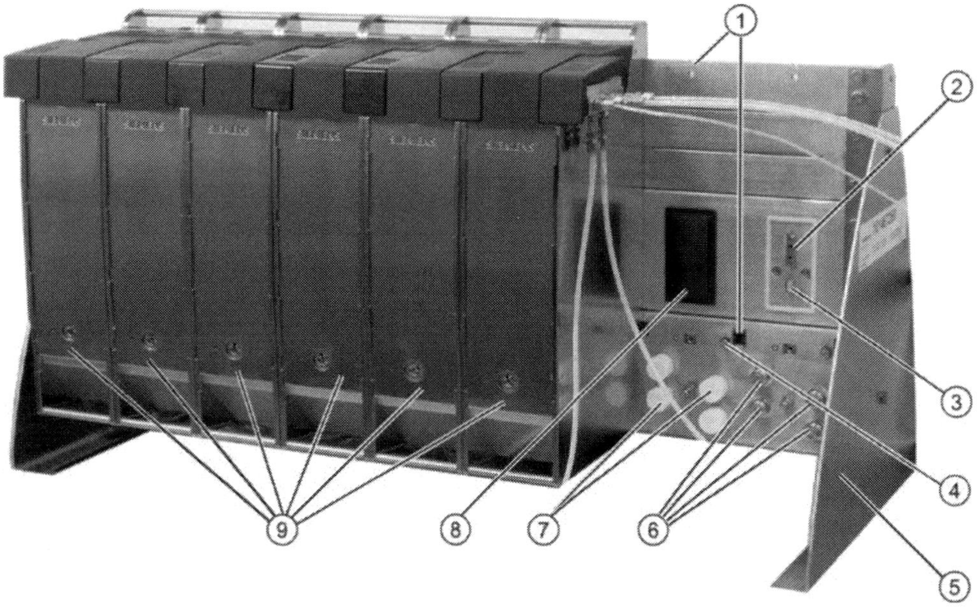

Figure 1: Siprocess rack equipped with 6 modules. (1) Positioning aids for modules, (2) Voltage supply 24 V DC/42 V DC, (3) Connection for communication bus, (4) Compressed air outlet for modules, (5) Stand (can be screwed on), (6) Connections for temperature control liquids, (7) Insulating caps of the connections for temperature control liquids, (8) Covering cap for voltage supply and communication bus connection, (9) Process modules

For applications in the lab, Siprocess is favourably operated with one or two racks, each of which can accommodate up to 8 process modules. Connectors are used for joining the liquid flows between the modules. The system can be adapted to specific tasks thanks to an easy rearrangement of the modules on the rack. Liquids can be routed out of and into the lab environment by means of connection adapters. This makes Siprocess open and flexible for individual configurations.

Siprocess is operated utilizing the industrial approved distributed control system (DCS) Simatic PCS7 on a PC basis. The automatic recording of measured values and of process alarms, operator acknowledgments and interventions is achieved. Data exporting is possible, too. The central automation controls all function in which more than one module is involved.

Nevertheless, closed-loop and open-loop control of the module functions are carried out in a distributed configuration in the relevant modules in accordance with the set points specified by the automation software of the DCS. The modules are equipped with sensors and actuators for this purpose. They communicate with the automation software by means of a communication bus. The automation hardware and software is located in an electric control cabinet along with the power supply. Software for module specific functions, e.g. the closed loop pressure control in the PC-module, runs on the distributed controller of the module.

Characterization of pressure losses

The pressure loss of the system is dominated by the micro structured mixer in the reaction module (figure 2). The remaining pressure losses of the other modules are negligible.

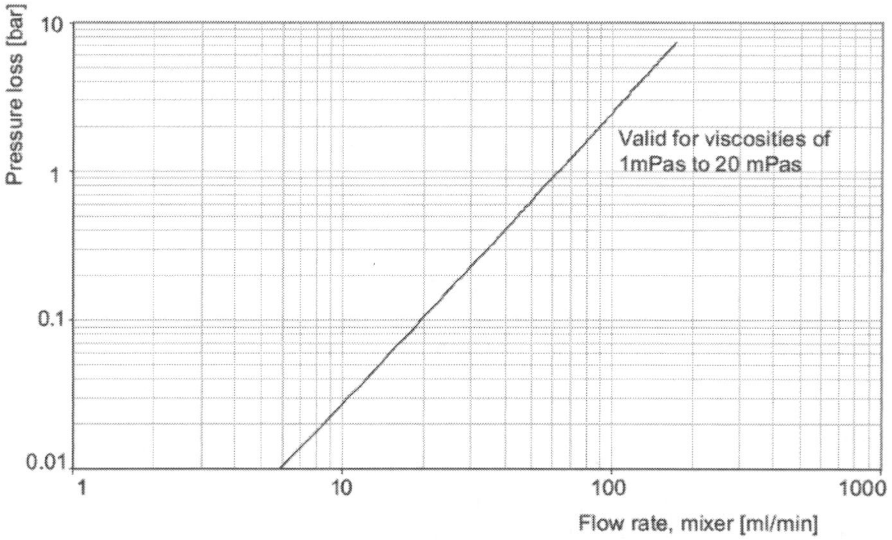

Figure 2: Pressure loss of the micro mixer in the RE module

Customer benefits

The benefit in using SIPROCESS is a remarkable reduction of process development time. This results from continuous processing in an automated system and a very fast set-up of the modular system within less than a day.

Example 1: Determination of kinetic parameters

To determine kinetic parameters a typical experimental setup could have three sample positions along the flow path. At each position up to 14 analytical samples can be taken. This allows 42 samples obtained at different reaction conditions within only 3.5 h (figure 3).

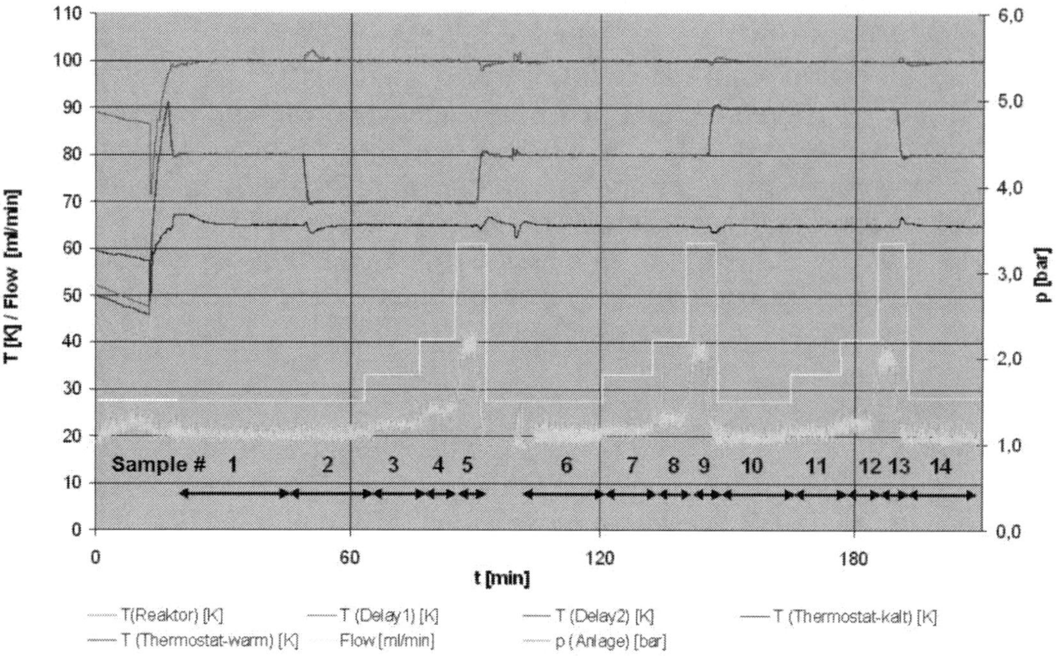

Fig. 3: Determination of kinetic parameters in SIPROCESS

Example 2: Production of first quantities

In the last years several industrial relevant chemical reactions of different classes' have been successfully performed in SIPROCESS, e.g. metal organic chemistry (avoidance of deep temperature), nitration (reduced hold-up). In general higher yields and/or space-time-yields were observed. In some cases product properties resulted never achievable in stirred vessels in batch mode. SIPROCESS is used not only for process development, but also for the production of first quantities. With a throughput of 6 kg/hr several kilogram of material can be synthesized. For a lot of customers this is sufficient to perform further experiments.

Running the system over a year (8,000 hours) with a throughput of 9 kg/hr more than 70 tons of could be processed. In the case of an active ingredient (5-10% in solution) several tons could be produced. Even in shorter campaigns lots of some hundred kilograms per week can be prepared. Therefore SIPROCESS bridges the gap between lab and pilot plant without scale-up risks.

45a Model-Based Design of Hierarchically Structured Porous Catalysts

Gang Wang, Department of Chemical and Biological Engineering, Rensselaer Polytechnic Institute, 110, 8th Street, Troy, NY 12180 and Marc-Olivier Coppens, Howard P. Isermann Department of Chemical Engineering, Rensselaer Polytechnic Institute, 110 8th Street, Troy, NY 12180.

Nanoporous catalysts like zeolites have an extremely large internal surface area, which contributes to the high intrinsic catalytic activity. However, their small pore size leads to slow molecular transport and pore blocking, limiting the efficient use of the catalytic materials. This indicates that, apart from the nanopores (i.e., micropores and narrow mesopores) where reactions actually occur, a "distribution" network of large mesopores and macropores is needed for molecules to quickly move in and out of the catalyst. The question is how this network should be designed to achieve optimal catalyst performance. Such a design is of considerably practical relevance, since it has become technically possible to control the hierarchical structure of porous catalysts, containing both nanopores and a distribution network, thanks to tremendous progress in (nano)materials science.

Inspired by the fractal design of leaves and lungs, which differs so strongly from that of typical catalysts, we studied whether broad, perhaps scaling hierarchical pore networks would perform better than those where the large pore channels are all of the same size. This optimization problem differs from those in which the nanopore size is varied, and is highly practical in view of the fact that the features at the nanoscale typically determine the (given) intrinsic kinetics, which we would like to preserve. Therefore, the texture at all length scales larger than the narrowest pores should be optimized, with regard to efficient molecular transport through the distribution network.

Remarkably, the optimal catalyst performance is found to be essentially the same regardless of whether a broad distribution of large pore channel diameters is allowed (bimodal) or not (bidisperse). This is because, in the optimal hierarchical structures, irrespective of the distribution of large diameters around the optimal average channel diameter, the minimized diffusion limitations exist to the same extent in a principle direction (e.g., a radial direction for a spherical catalyst particle), while they essentially disappear in a direction perpendicular to this principle direction. It is also found that one Thiele modulus governs the optimal catalyst performance when this Thiele modulus is defined in a way analogous to the generalized Thiele modulus (Froment and Bischoff, 1990), but using the molecular diffusivity in the large pore channels, rather than the effective diffusivity in the nanopores. This study calls for the synthesis of hierarchically structured porous catalysts with an optimized, approximately bidisperse pore size distribution, thanks to its easier synthesis and an identical performance as compared to those with a bimodal pore size distribution, especially given the allowed variance around the average.

Synthesis of Porous Silica with Unique Morphology in Narrow Passages Using the Ice Templating Method

S. R. Mukai[1], S. Murata[1], K. Onodera[1], A. Eguchi[1], H. Wada[2] and K. Kotera[2]
[1]Chemical Process Engineering, Graduate School of Engineering, Hokkaido University, N13W8, Kita-ku, Sapporo, Japan
[2]Shinwa Chemical Industries, LTD., 50-2 Kagekatsu-cho, Fushimi-ku, Kyoto, Japan
E-mail: smukai@eng.hokudai.ac.jp

Introduction

Due to the rapid progress in microchemistry, nano-structured porous materials are now frequently used in microchannels. In order to reap the benefits of microsystems in which porous materials are involved, generally the diffusion length within the porous material must be kept as short as possible. In porous materials having the form of particles, this can be easily achieved by reducing their size. However, such fine particles are likely to cause an extremely high resistance to flows when they are packed in microchannels.

Such a situation can be avoided by changing the morphology of the porous material. For example, fibrous porous materials with moderate resiliency will show high performance without causing a severe resistance to flows if they are packed in microchannels properly. Monolithic honeycombs with straight microchannels which are formed by thin porous walls may even do a better job. However, porous materials with such morphologies are difficult to synthesize and/or difficult to pack in microchannels.

Recently, we introduced the "Ice Templating" method, a method which allows the direct synthesis of nano-structured porous materials with unique morphology, such as microfibers and monolithic microhoneycombs, in channels.[1)-3)] In this method, materials are synthesized by freezing their parent hydrosol or hydrogel unidirectionally. Ice crystals which grow within the precursor during freezing act as the template, therefore unlike other methods, which use sacrificial templates to mold materials to have a desired morphology, the template is extremely inexpensive. Moreover, the template can be easily removed just by simply thawing and drying the material. However, synthesis of porous materials in narrow channels using this method has not been attempted yet.

In this work, we attempted to synthesize porous silica with unique morphologies in narrow passages using the ice templating method. The characteristics of the resulting material were also analyzed. Based on the knowledge that we obtained through such studies, we also attempted to synthesize a typical microdevice which requires a porous material in its microchannels. We selected a device for solid phase dynamic extraction (SPDE) which is practically a syringe needle including an adsorbent in its microchannel.

Experimental

Commercial sodium silicate solutions (Wako Pure Chemical Industries) were diluted with ion-exchanged and distilled water, and their SiO_2 concentrations were adjusted to desired values. Then the pH of the solutions was adjusted using an ion-exchange resin to control their gelation time. The solutions were introduced into tubes made of teflon (*o. d.* 1000 μm, *i. d.* 500 μm) and stainless steel (*o. d.* 700 μm, *i. d.* 480 μm) and were aged for different periods. Samples were also prepared using polypropylene tubes (*o. d.* 16 mm, *i. d.* 14 mm) in order to investigate the influence of the size of the channel the silica samples were synthesized in. All of the tubes were dipped at a constant rate of 6 cm/h into a cold bath maintained at 77 K. After the tubes were completely frozen, they were taken out from the bath and were thawed at 323 K. Then the water in the tubes was exchanged with *t*-butanol, and the samples were freeze-dried.

The morphology of the silica samples was directly observed using a scanning electron microscope (SEM). The porous properties of the samples were evaluated through nitrogen adsorption-desorption experiments. First, nitrogen adsorption-desorption isotherms of the samples were measured at 77 K. Next, BET surface areas of the samples were calculated from the obtained isotherms using the BET method.

Next, porous silica was synthesized in the channel of a syringe needle (22 gauge) following the same procedure as above, and a microdevice for SPDE was obtained. The performance of the obtained microdevice was evaluated using nitrogen containing 20 μg/ml each of ethyl acetate, isobutanol, toluene and methanol. 100 ml of the sample gas was passed through the device using a handy type gas sampler. Next the device was set to a syringe containing 1 ml of nitrogen (desorption gas), and the gas was injected into a gas chromatograph via the device. From the results of gas chromatography, the recovery percentage of each component was calculated.

Results and Discussion

In the ice templating method, the morphology of the resulting sample can be controlled by adjusting the firmness of the precursor prior to freezing. When a fairly soft hydrogel is frozen, a monolithic microhoneycomb is obtained, and when a fairly firm hydrogel is frozen, the gel is molded into the form of fibers. The firmness of the precursor hydrogel can be controlled, for example, by adjusting the aging time prior to freezing.

Figure 1 compares a microhoneycomb synthesized in a polypropylene tube (*o. d.* 16 mm, *i. d.* 14 mm) with that synthesized in a teflon tube (*o. d.* 1000 μm, *i. d.* 500 μm). The

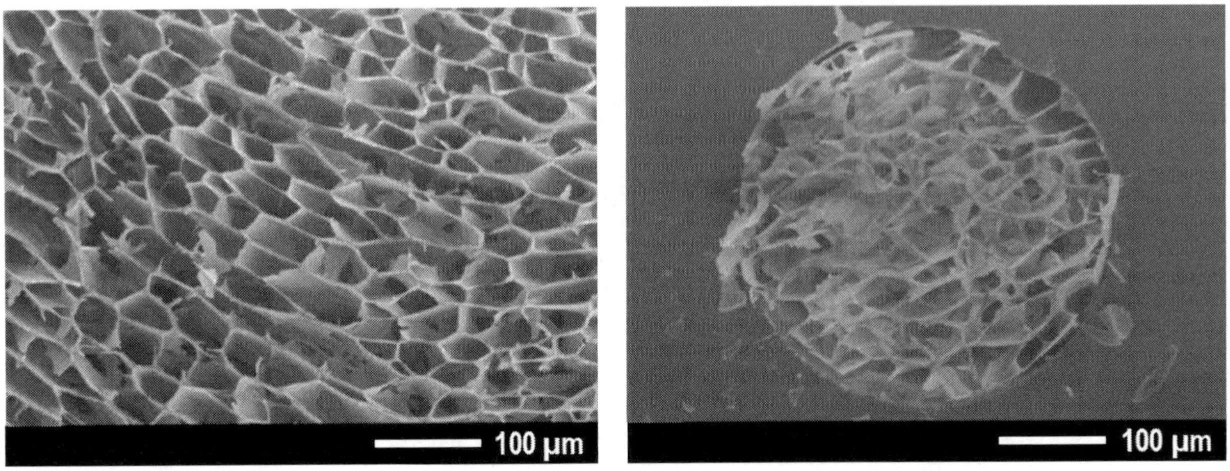

Fig. 1 Typical Monolithic Microhoneycombs of Silica Synthesized in (Left) a P. P. tube (*o. d.* 16 mm, *i. d.* 14 mm) and (Right) a Teflon Tube (*o. d.* 1000 μm, *i. d.* 500 μm)

Fig. 2 Typical Fibers of Silica Synthesized in (Left) a P. P. tube (*o. d.* 16 mm, *i. d.* 14 mm) and (Right) a Stainless Steel Tube (*o. d.* 700 μm, *i. d.* 480 μm)

morphology of microhoneycombs synthesized in stainless steel tubes with narrow passages could not be observed, as it was difficult to directly observe the microhoneycomb without taking it out of the tube, and it was also difficult, if not impossible, to take the microhoneycomb out of the tube without damaging its structure. Both microhoneycombs basically have the same structure, which indicates that microhoneycombs can also be synthesized in narrow passages.

Figure 2 compares fibers synthesized in a polypropylene tube (*o. d.* 16 mm, *i. d.* 14 mm) with those synthesized in a stainless steel tube (*o. d.* 700 μm, *i. d.* 480 μm). Fibers were successfully obtained in both types of tubes, but the shapes of the cross sections of the fibers

Fig. 3 Typical Morphology of Silica Synthesized in a Stainless Steel Tube (*o.d.* 700 µm, *i. d.* 480 µm)

were quite different. Moreover, only fibers were obtained in the large polypropylene tube, but in the stainless steel tube with a small diameter, a large part of the silica had a different morphology. A typical micrograph of such silica is shown in Fig. 3. The resistance to flows of porous materials with this unique morphology may not be as low as materials molded into the form of microfibers or microhoneycombs, but are thought to be lower than materials in the form of particles.

In the ice templating method, the morphology of the resulting sample depends on how the template ice crystals grow within the precursor. When a large tube is used for synthesis, an array of fairly uniform needle-shaped ice crystals grow within the silica precursor, and mold the silica into fibers or a microhoneycomb. However, when a tube with a small inner diameter is used, and especially when the wall of the tube is made of a material which quickly conducts heat, the way the ice crystals grow within the precursor is thought to be quite different. This is probably due to the fact that in a tube, the ratio between the surface area of the channel and the volume of the channel increases with the decrease in diameter, and therefore heat removal from the channel wall becomes more dominant during the unidirectional freezing of smaller tubes.

Figure 4 compares the nitrogen adsorption-desorption isotherms of silica synthesized in a polypropylene tube (*o. d.* 16 mm, *i. d.* 14 mm) with that of silica synthesized in a stainless steel tube (*o. d.* 700 µm, *i. d.* 480 µm). One feature of silica obtained through the ice templating method is that it has developed micropores. Silica with BET surface areas over 1000 m^2g^{-1} can be easily synthesized. Although the adsorption capacities of silica synthesized in stainless steel tubes with small diameters tended be slightly lower than those synthesized in

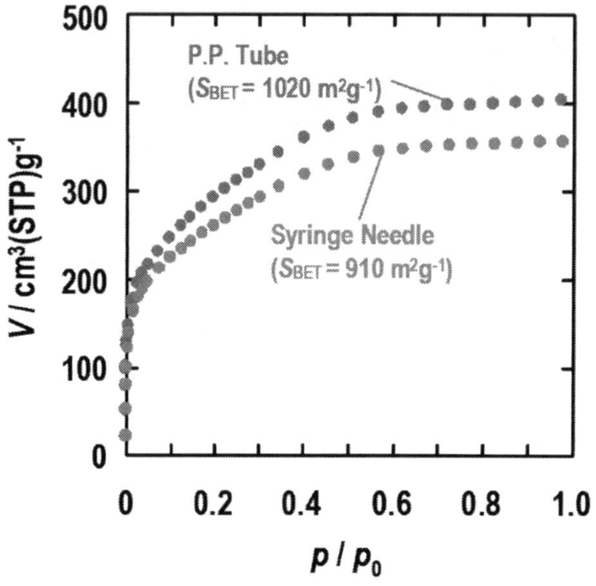

Fig. 4 Nitrogen Adsorption Isotherms (77 K) of Silica Synthesized in Passages of Different Sizes Using the Ice Templating Method

larger tubes, they also were found to have developed micropores within their structure.

Next, we attempted to synthesize a microdevice for SPDE by synthesizing microporous silica directly in the microchannel of a syringe needle using the ice templating method. In a microdevice for SPDE, the adsorbent must have a high adsorption capacity, must show high adsorption-desorption rates and also must not cause significant resistance towards flows. Therefore, silica synthesized using the ice templating method is thought to be suitable for this microdevice.

A microdevice for SPDE is typically used according to the following procedure. First a certain amount of gas which contains a small amount of target substances is passed through the microdevice using a syringe or a handy type gas sampler. During this process the substances will be concentrated in the adsorbent. Next the microdevice is set to a syringe filled with a small amount of desorption gas, and then the gas in the syringe is injected into a gas chromatograph via the microdevice as in a typical procedure of gas chromatography analysis. The target substances desorb from the adsorbent by the heat from the injection port and enter the column along with the desorption gas and therefore can be quickly analyzed. Therefore, detection and accurate quantification of trace level amounts of substances can be conducted using this microdevice.

It was found that a flow rate over 40 ml/min could be achieved at a back pressure of 100 kPa (gauge) when using the obtained microdevice for sampling. This flow rate indicates

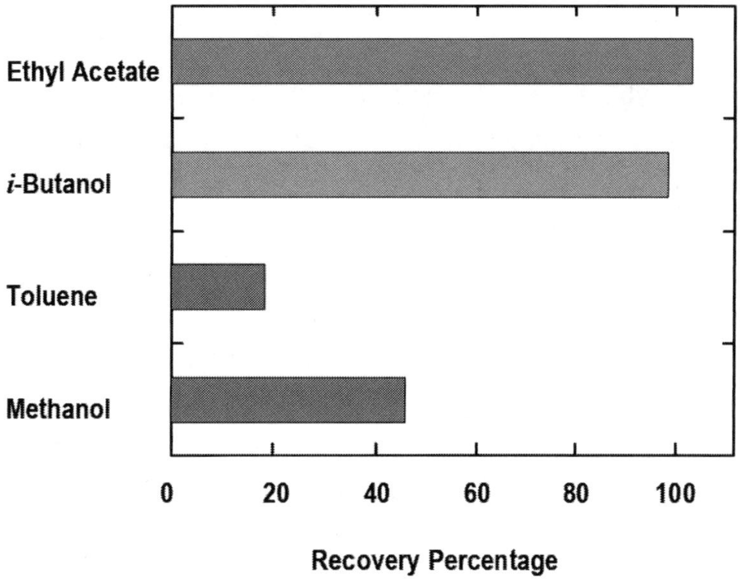

Fig. 5 Performance of the Microdevice for SPDE Synthesized Using the Ice Templating Method

that 100 ml (a typical amount of gas sampling) of the target gas can be sampled within a few minutes. Such fast sampling is thought to be enabled by the unique morphology of the adsorbent included in the microdevice.

Finally, the performance of the synthesized microdevice was evaluated. Figure 5 summarizes the results of performance evaluation tests, where the recovery percentages of target substances are listed. Here, a recovery percentage close to 100 % means that the target substance was efficiently concentrated in the microdevice, and was promptly released during injection. The concentration of such substances can be accurately analyzed. On the other hand, a low recovery percentage indicates that the adsorbent in the microdevice either doesn't have the sufficient adsorption capacity for the target substance, or the adsorbed substance cannot easily desorb from the adsorbent. As expected, recovery percentages of hydrophobic substances, such as toluene, and substances with small molecular weights, such as methanol, were low. The recovery percentage of such substances didn't increase even when the desorption-injection process was repeated, which indicates that the silica in the microdevice didn't have the sufficient adsorption capacity for such substances. On the other hand, it was found that substances which have a hydrophilic part within its molecule, such as ethyl acetate and isobutanol, can be accurately analyzed. Therefore, it can be concluded that the ice templating method is a powerful method to synthesize porous materials for a microdevice for SPDE which is designed to concentrate such substances.

Conclusions

In this work, the synthesis of porous silica with unique morphology in narrow passages was attempted using the ice templating method. It was found that microfibers and monolithic microhoneycombs similar to those obtained in larger passages can be synthesized. However, it was found that in certain cases, the morphology of the resulting material was significantly different from those obtained in larger passages. The high surface area to volume ratio of narrow passages is thought to affect the manner of heat removal during freezing. This difference is thought to be significant when the walls of the passages easily conduct heat.

Porous silica synthesized in passages of various sizes was found to have short diffusion paths and morphologies which do not cause severe resistance to flows. They were found to be suitable for microdevices for SPDE which require adsorbents having the following features: high adsorption capacities, high attainable adsorption-desorption rates, and low resistance to flows.

References

1. Mukai, S. R., Nishihara, H. and Tamon, H. (2003), " Porous Properties of Silica Gels with Controlled Morphology Synthesized by Unidirectional Freeze-Gelation," *Microporous and Mesoporous Materials, 63*, pp. 43-51
2. Mukai, S. R., Nishihara, H. and Tamon, H. (2004), "Formation of Monolithic Silica Gel Microhoneycombs (SMHs) Using Pseudosteady State Growth of Microstructural Ice Crystals," *Chemical Communications*, pp. 874-875
3. Nishihara, H., Mukai, S. R., Yamashita, D. and Tamon, H. (2005), " Ordered Macroporous Silica by Ice-Templating," *Chemistry of Materials*, 17, pp. 683-689

83I Visualisation of Flow in Microchannel Array Water Evaporators by High-Speed Videography

Stefan Maikowske[1], Juergen J. Brandner[1], Ruediger Lange[2], and Klaus Schubert[1]. (1) Institute for Micro Process Engineering (IMVT), Forschungszentrum Karlsruhe, Hermann-von-Helmholtz-Platz 1, D-76344 Eggenstein-Leopoldshafen, Germany, (2) Department of Chemical Engineering, Dresden Universtiy of Technology, D-01062 Dresden, Germany

In thermal and chemical process engineering evaporation is an important and well-known unit operation. It is done by conventional heat exchangers in most instances. Transition from water to vapour with these conventional devices is well known and understood, as well as flow patterns like bubbly flow, liquid ring flow, slug flow or annual flow in conventional channels. Thermal and chemical process engineering in an industrial environment requires heat exchangers or evaporators with exactly controlled phase transitions which generate strictly defined vapour quality. In many cases this will be obtained with conventional devices. In some applications the use of conventional devices is dangerous or may create some major process difficulties like runaway of the process or decomposition of liquid mixtures by exceed of a certain temperature limit. Since a couple of years microstructure heat exchangers became more and more important in thermal and chemical process engineering. The channels integrated into these devices are in the size range of a few hundred micrometers with related hydraulic diameters. The main advantage of microstructure devices compared to conventional heat exchangers is the very high surface-to-volume-ratio of up to several 10,000 $m^2 \cdot m^{-3}$, which is about one or two orders of magnitude higher than that of conventional devices [1, 2]. Therefore the heat transfer capability of microstructure heat exchangers is respectively much higher than the heat transfer capability of conventional heat exchangers. Due to the small hold-up and the overpressure resistance, which depends on material and manufacturing technique, microstructure heat exchangers may be well suited for high-pressure applications or dangerous fluids [3]. The evaporation process of liquids, e.g. water, in microstructure heat exchangers is not well-investigated yet. The transition from water to vapour in micro channels is accompanied by several phenomena like pulsation at sub audio frequencies, vapour clogging or the occurance of vapour slugs in micro channels [4]. These phenomena are widely observed but not investigated in detail. A technique to explore phase transition in micro channels (especially micro channel evaporators) is the investigation by high-speed videography [4, 5]. This method is used to study phase transition at the Institute for Micro Process Engineering (IMVT) of the Forschungszentrum Karlsruhe. A special test device which allows to look inside the micro channels was developed for this purpose. The device is heated electrically by conventional resistor cartridges, and the temperature inside the device is monitored by Type K thermocouples. A detailed description can be found in [6]. A photo of the device is shown in Fig. 1. In combination with a microscope, this test device allows a detailed view inside the micro channels. Therefore, it is possible to observe the flow pattern and the phase transition directly inside the micro channels applying visual methods. Optical analysis of flow pattern and phase transition inside micro channels is supported by digital high-speed videography. This method enables to slow down the recorded phase transition process for a staggering observation of phase transition in microstructure channels. It is also used to visualize the non-uniformity of flow in parallel micro channels. Information about the void fraction is approachable by extraction of pixel intensity from single pictures or picture strings by digital image processing. Exploration of vapour quality is assisted by a special laser photometer which was assembled at the IMVT. The functional principle of this laser photometer is based on measuring of total reflection by the outlet vapour steam. The work presented here focuses on exploration of phase transition in micro channels. Fundamental issues and dependency of phase transition to, e.g., geometric parameters in micro channels obtained by investigation with high-speed videography will be given. This method of investigation is also suited to provide information about formation and constituents of multiphase flow in micro channels, which will be shown in examples. This information is required to describe several processes in micro heat exchangers, e.g. phenomena during evaporation. Theory of known fluid dynamics will be used and, if necessary, modified for description of flow arrangement and regime. Another point of interest is exploration of vapour quality depending on microstructures, their dimensions and arrangement. It is required to investigate the composition of vapour at the outlet of the microstructure devices. Especially the content of liquid droplets and their size define the character of the outlet steam, e.g. saturated steam or superheated steam. A method of measuring the steam quality will be described, and first results will be given. These fundamental issues will lead to the development of special design rules for micro heat exchangers to generate well-defined vapour quality with stable and controllable phase transition. The design rules will also allow to control or to minimize phenomena like pulsation at sub audio frequencies, vapour clogging and the occurance of vapour slugs in microchannels.

References [1] Brandner, J. J., Anurjew, E., Bohn, L., Hansjosten, E., Henning, T., Schygulla, U., Wenka, A., Schubert, K., Concepts and realization of microstructure heat exchangers for enhanced heat transfer. Experimental Thermal and Fluid Science 30, pp. 801–809, 2006. [2] Wörz, O., Jäckel, K. P., Richter, Th., and Wolf, A., Microreactors, A New Efficient Tool for Optimum Reactor Design, Proc. 2nd Int. Conf. on Microreaction Technology, pp. 183-185, 1998. [3] Schubert, K., Brandner, J. J., Fichtner, M., Lindner, G.,Schygulla, U., and Wenka, A., Microstructure Devices for Applications in Thermal and Chemical Process Engineering, Microscale Thermophysical Engineering, Vol. 5, No. 1, pp. 17-39, 2001. [4] Henning, T., Brandner, J. J., and Schubert, K., Comparison of microchannel array evaporators by high-speed videography. S. G. Kandlikar, ed., Proc. 3rd Int. Conf. on Microchannels and Minichannels (ICMM2005), Toronto/ON, Canada, June 13-15, pp. 128-136, 2005. [5] Henning, T., Brandner, J. J., and Schubert, K., 2004. High-speed imaging of flow in microchannel array water evaporators. S. G. Kandlikar, ed., Proc. 2nd Int. Conf. on Microchannel and Minichannels (ICMM2004), Rochester/NY, June 17-19, pp. 483-490, 2004. [6] Henning, T., Brandner, J. J., and Schubert, K., 2004. Characterisation of electrically powered micro heat exchangers. Chem. Eng. J., 101, pp. 339-345, 2004.

Fig.1: Photo of a test device to observe flow pattern, phase transition and appearing phenomena during phase transition in micro channels. The wires in front of the device are thermocouples to measure temperature inside the microstructure device.

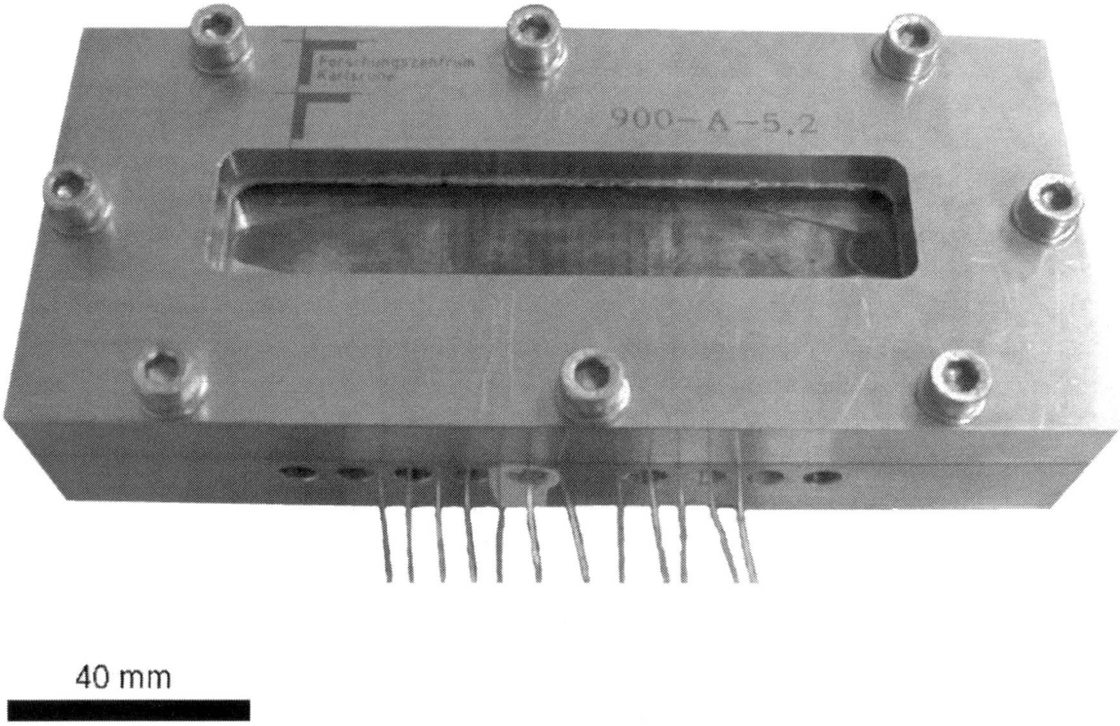

65g Experimental Investigation of the Laminar/turbulent Transition of Liquid Flows in Rectangular Micro Channels Using µpiv

Wolf Wibel, Institute for Micro Process Engineering, Research Centre Karlsruhe, P.O. Box 3640, Karlsruhe, DE-76021, Germany and Peter Ehrhard, Biochemical and Chemical Engineering, University of Dortmund, Emil-Figge-Str. 68, Dortmund, D-44221, Germany.

The advances of micro-fabrication techniques allow the manufacturing of micro heat exchangers or micro reactors. These micro devices are characterized by a large surface-to-volume ratio and, hence, provide the transfer of large heat fluxes or offer large catalytic surfaces for reactions. The design and optimization of such micro devices heavily relies on correlations for pressure drop and heat transfer, as well as on the laminar/turbulent transition. This presentation concentrates onto rectangular stainless steel micro channels with a hydraulic diameter of dh = 133 µm. Three aspect ratios are engaged, namely 1:1, 1:2 and 1:5, whereas the hydraulic diameter is kept constant. The roughness depth of the channel walls is k ≈ 1 - 2 µm in general, specific channels show a roughness depth of k ≈ 20 µm. Filtered and degassed de-ionized water is driven at pressure differences up to 20 bar through the channels to achieve turbulent micro flows at Reynolds numbers up to Re = 4000. The measuring techniques allow a highly-accurate determination of several parameters like mass flow as well as of the time-resolved velocity field (µPIV). For all measured quantities - consistently - the laminar/turbulent transition for smooth channels is found in the range Re ≈ 1900 - 2200, in agreement with findings for macroscopic channels. The influences of channel walls with increased roughness result in a transition to turbulent flow at a lower Reynolds number (Re ≈ 900). Integral measurements, as e.g. the total pressure-difference between inlet and outlet plenum, suffer from a superposition of developing, laminar and turbulent flow regions along the micro channel. Local pressure measurements within small micro channels appear likewise difficult without major disturbances. Even though the pressure-difference measurements between inlet and outlet plenum can be performed at identical micro channels of different length to isolate the pressure drop in fully-developed regions, the measurement of local quantities remains superior to determine the precise site and parameter range of the laminar/turbulent transition. For this reason this presentation concentrates on the results of the µPIV measurements within micro channels. As an example, figure 1 presents the measured local turbulence intensities in the centre of an 1:1 aspect ratio micro channel, close to the channel outlet. The turbulent fluctuations u' and v' are in the axial and wall-normal direction. Starting at Re ≈ 1500 a characteristic increase of the axial fluctuations to a maximum of about 12 % is visible. With further increasing Reynolds numbers the axial fluctuations drop to a constant value of about 4 % beyond Re ≈ 2200. This maximum of the axial fluctuations is characteristic for the laminar/turbulent transition range, where an intermittency of laminar and turbulent states occurs. In contrast, the wall-normal fluctuations in the channel centre do not exhibit such large amplitudes, but a smooth transition (without maximum) is found for increasing Re. The third-direction (likewise wall-normal) fluctuations cannot be measured by the µPIV technique, though due to symmetry they should behave like the given wall-normal fluctuations. Further investigation results given in the presentation include the evaluation of the intermittency factor and the probability density function (PDF) of the velocities at specific local positions. Figure 1 (attachement): Turbulence intensities in the centre of the 1:1 aspect ratio micro channel as function of the Reynolds number.

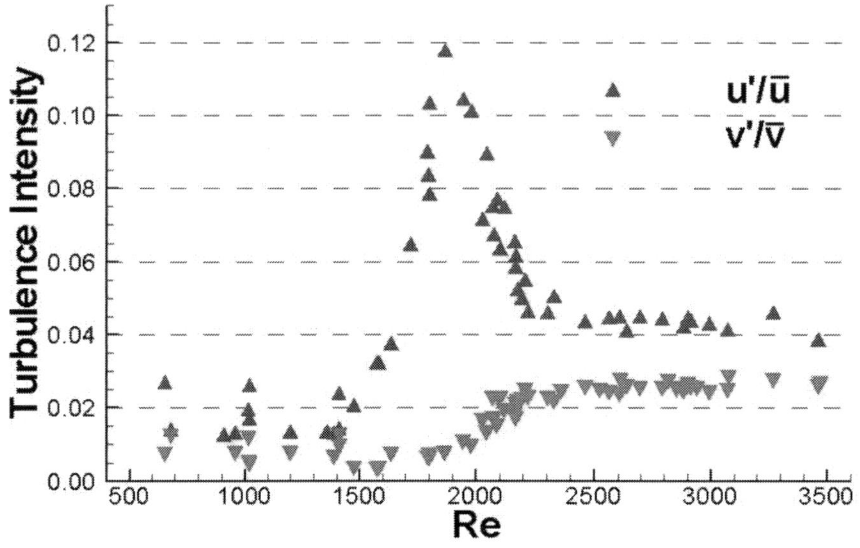

83p Visualisation of Packed Bed Behaviour Using 3D Confocal Microscopy and Microfluidics

Michael S. Shapiro[1], Steve J. Haswell[2], Gary J. Lye[3], and Daniel G. Bracewell[3]. (1) Department of Biochemical Engineering, University College London, Torrington Place, London, WC1E 7JE, United Kingdom, (2) Chemistry, University of Hull, Hull, HU6 7RX, United Kingdom, (3) Biochemical Engineering, University College London, Torrington Place, London, WC1E 7JE, United Kingdom

A microfluidic ion exchange chromatography column was constructed using a conventional adsorbent from the biopharmaceutical sector within a glass chip. The design allows direct visualisation of events in the bed using 3D confocal microscopy and fluorescently labelled proteins. The channels within a glass chip were fabricated using photolithography and isotropic etching. The design includes a 1 cm long microfluidic column where compressible, porous agarose beads keystoned and packed. This type of microfluidic system allowed for the visualisation and study of; bead arrangement (packing), fluid flow, and adsorption within the packed bed.

flnitial studies with the work the 3 dimensional images produced of the packed bed were used to analyse the degree of packing. These pictures showed that the height (150 μm) of the column was such that 2 bead layers were formed, as the depth of the column was double the diameter of an average bead. The voidage of the system, was calculated from the confocal microscopy pictures, to be 0.55, demonstrating that there are more gaps present in the system than in a standard column. This value is greater than is seen in a standard column (approximately 0.4). which has a voidage of approximately 0.4.

Further studies used the system to visualise flow and adsorption in the bed. This was done by generating breakthrough curves. Fluorescently labelled lysozyme was pumped into the channel and the breakthrough was recorded using a fluorescent imaging system. The system allowed the visualisation of flow through the chromatography bed, displaying unconventional flow profiles. A high degree of channelling occurred through the packed system implied that the profile deviated from plug flow conditions. Images of the breakthrough front exiting the packed bed will be shown. Useful breakthrough curves were achieved using 4 linear velocities from 60 – 270 cm h^{-1}. Capacities were calculated from the breakthrough curves and compared to previously published data as well as experimental data performed on a 30 ml column at laboratory scale. Maximum capacities in microfluidic system were recorded in the range of 100 – 140 mg ml^{-1} matrix which was found to be close to values published previously in the literature for laboratory scale columns. This represented a reduction in scale from 500 – 20000 fold.

The microfluidic ion exchange column was also used to separate 3 fluorescently labelled proteins (based on the separation of egg white) using the same fluorescent imaging system as before. At the laboratory scale, lysozyme passes straight through the column, while ovalbumin and conalbumin bind to the chromatography matrix and are eluted using a salt gradient. The same linear velocities as before were used. Offline MALDI Mass Spectrometry analysis demonstrated that the separation occurs at the microscale in the same manner.

Further work will explore the use of these microfluidic columns to predict the performance of larger scale columns to help facilitate more rapid bioprocess development.

Synthesis of Photochromic Diarylethenes Using a Microflow System

Yousuke Ushiogi,[†] *Tomoyuki Hase,*[§] *Yoshiharu Iinuma,*[§] *Atsushi Takata,*[§] *and Jun-ichi Yoshida*[‡]
[†]*The Research Association of Micro Chemical Process Technology (MCPT), Kyoto, Japan*
[§]*Yamada Chemical Co., Ltd. Kyoto, Japan*
[‡]*Department of Synthetic Chemistry and Biological Chemistry, Graduate School of Engineering, Kyoto University, Kyoto, Japan*

1. Introduction

Diarylethenes have received significant research interest from both academia and industry, because some of them exhibit remarkable changing of color by reversible switching of two distinct isomers (photochromism), which is accomplished by absorption of different colors of light as exemplified by Scheme 1.[1]

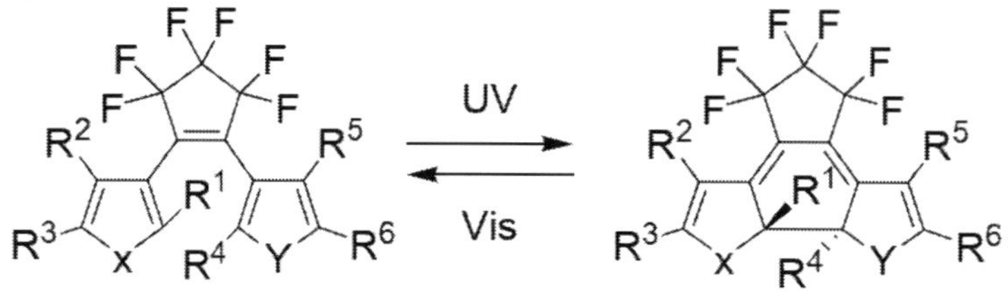

Scheme 1. Photochromism of Diarylethenes.

The most promising photochromic diarylethene candidates include 1,2-diarylhexafluorocyclopentenes, which have been synthesized by the reaction of aryllithium compounds with octafluorocyclopentene in a macrobatch reactor at low temperatures such as –78 °C in order to avoid undesirable side reactions (Scheme 2).[2] The requirement of such low temperatures, however, has been an obstacle to industrial-scale applications. Herein we report that synthesis of diarylethenes can be accomplished at or above 0 °C using a microflow system[3, 4] by virtue of effective temperature control and residence time control.[5]

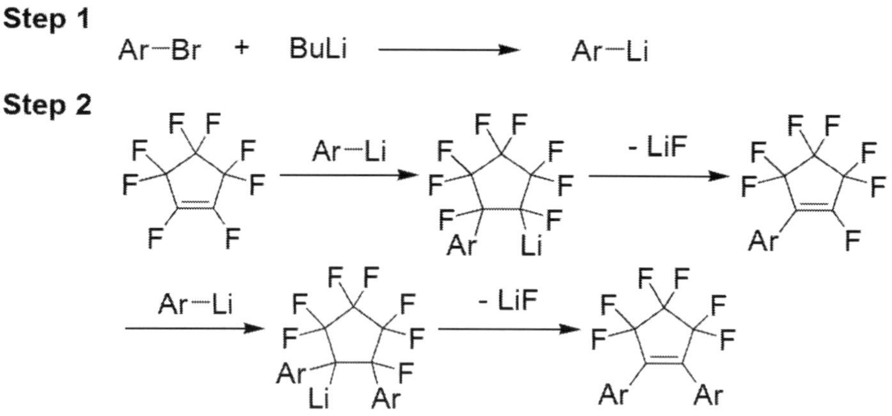

Scheme 2. Synthesis of Diarylethenes.

The first step of the synthesis of diarylethenes is the halogen-lithium exchange reaction of an aryl bromide with butyllithium. The reaction should be carried out at low temperatures such as –78 °C because aryllithium compounds are often unstable and decompose at higher temperatures. In the next step, the aryllithium compounds are allowed to react with octafluorocyclopentene. Usually, two moles of the aryllihitum compound react with one mole of octaflurocyclopentene presumably via the two-fold addition-elimination sequence to give the desired diarylethenes (Scheme 2). Therefore, the method is suitable for the synthesis of symmetrical diarylethenes.

2. Synthesis of Diarylethenes Using a Microflow System

We constructed a microflow system consisting of two T-shaped micromixers and two microtube reactors shown in Figure 1.[6] We chose to use stainless steel as a material for the mixers and reactors because of high heat-exchange ability and mechanical strength from a view point of industrial applications.[6c] In the first mixer **M1**, a solution of an aryl bromide and a solution of butyllithium were mixed, and the mixture was allowed to react in the first microtube reactor **R1**. The resulting aryllithium was mixed with octafluorocyclopentene in the second micromixer **M2** and the mixture was allowed to react in the second microtube reactor **R2**. The temperature of the whole system was controlled using a cooling bath.

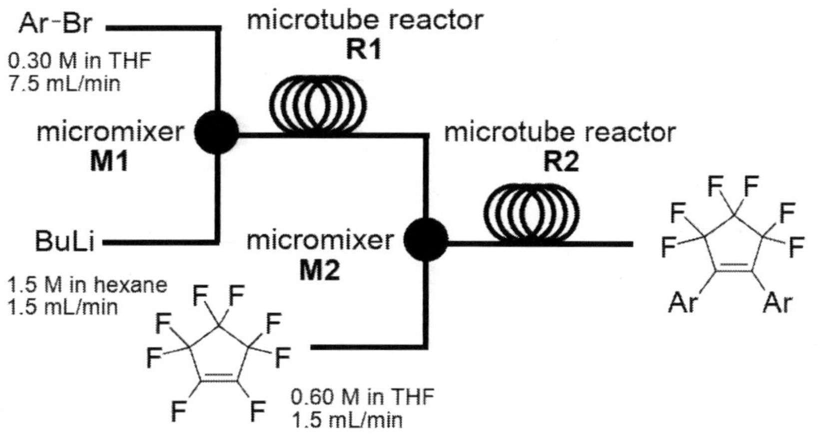

Figure 1. Synthesis of symmetrical diarylethenes using the microflow system.

The reaction temperature was optimized using 3-bromo-2-methyl-5-phenylthiophene as an aryl bromide. As shown in Figure 2, the yield of the desired diarylated product **1** increased with the temperature presumably because the reaction rate increased with the temperature. The highest yield was obtained at 0 °C. Above 0 °C, the yield decreased significantly probably because of the decomposition of the aryllithium intermediate. It is noteworthy that the reaction in a macrobatch system at 0 °C gave a complex mixture. Low temperatures such as –78 °C are essential for the success of the reaction in a macrobatch system.

In order to test the applicability of the present method, various diarylethenes were synthesized using the corresponding aryl bromides and octafluorocyclopentene at 0 °C in the microflow system and the results are summarized in Figure 3. Although the yield depends on the nature of the aryl group, the results indicate that the use of microflow systems serves as a powerful method for synthesis of diarylethenes of various structures (**1** — **8**).

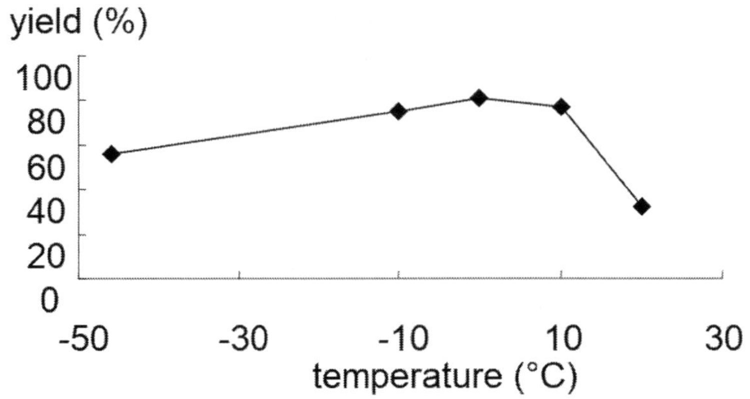

Figure 2. Temperature effect for the synthesis of diarylethene **1** in the microflow system (residence time, **R1**: 3.4 sec, **R2**: 2.9 sec).

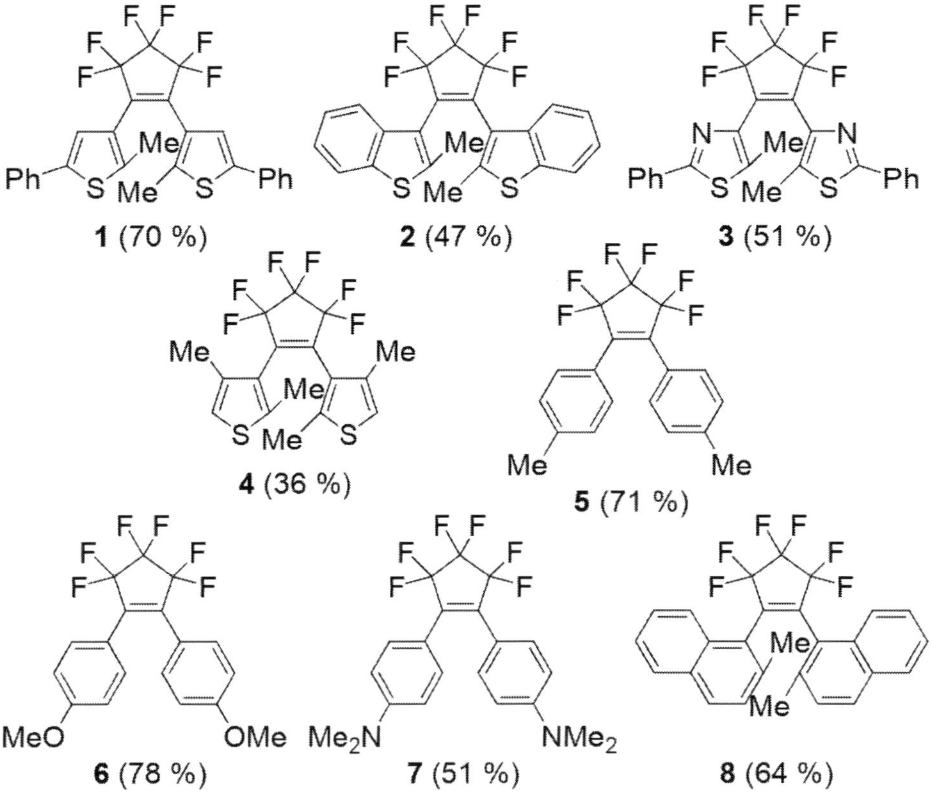

Figure 3. Synthesis of diarylethenes from aryl bromides and octafluorocyclopentene (isolated yields) using the microflow system at 0 °C.

3. Dithiazolylethene and Unsymmetrical Diarylethene

The following observation for 4-bromo-5-methyl-2-phenylthiazole demonstrates a striking advantage of the microflow system (Scheme 3). The macrobatch reaction at –78 °C gave monoarylated compound **9** as a major product even with the use of two equivalents of the aryllithium reagent. The elimination of LiF from the first adduct (Scheme 1) seems to be slow at –78 °C presumably because of nitrogen coordination to Li. LiF elimination seemed to take place during the work-up. In order to synthesize diarylated compound **3**, monoarylated compound **9** was isolated and then was subjected to the second reaction with the aryllithium at –78 °C. In the case of the microflow system, however, the reaction gave diarylated compound **3** as the major product in one-step. The result can be attributed to a higher temperature (0 °C) for the microflow system, where the aryllithium intermediate survived within a short residence time and the LiF elimination took place spontaneously. The present observations suggest a new potentiality of microflow systems; *sequential transformations, which require quite different temperatures in the case of macrobatch systems, can be achieved at the same temperature by virtue of short residence time inherent in microflow systems.*

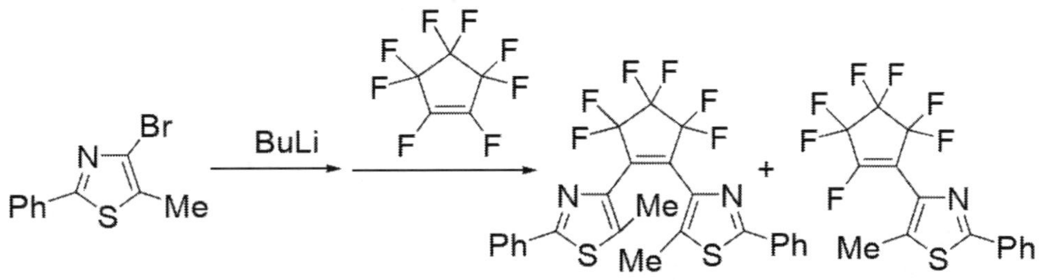

		3	**9**
Batch System	-78 °C, 1 h + 2h	10 %	52 %
Microflow System	0 °C, 3.4 s + 5.7 s	51 %	11 %

Scheme 3. Dithiazolylethene.

It is also noteworthy that the monoarylated product can be obtained selectively in the microflow system by employing one equivalent of the aryllithium. Thus, selective synthesis of unsymmetrical diarylethenes can be easily accomplished using two different aryl bromides sequentially. Figure 4 illustrates the synthesis of unsymmetrical diarylethene **10** (53% yield). The ring-closed isomer of **10** obtained by photo-irradiation exhibited a different color from those of the corresponding symmetrical compounds **1**[7] and **3**[8] (Figure 5). The result indicates a possibility of fine-tuning of physical properties by introducing different heteroaromatic substituents on the alkene framework by taking advantage of the microsystem.

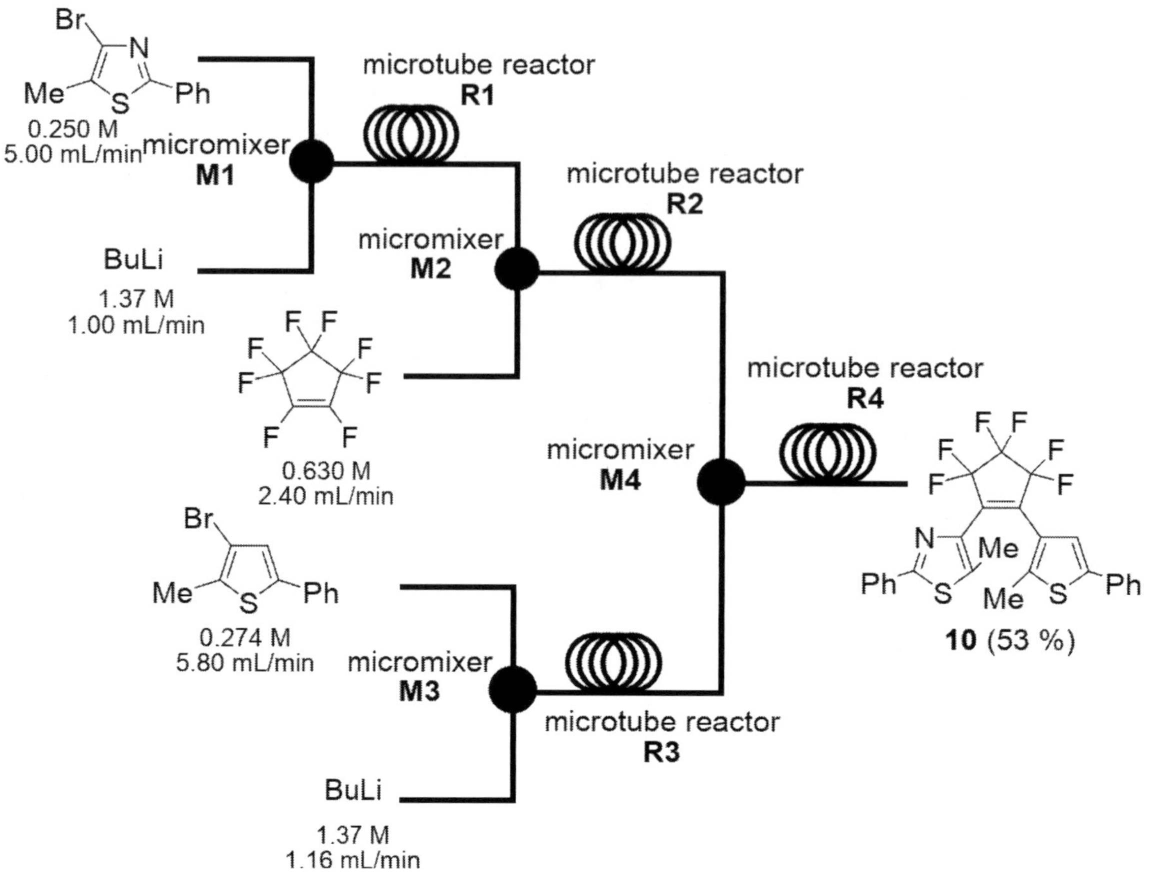

Figure 4. Synthesis of unsymmetrical diarylethene 9 using microflow system.

1 **10** **3**

Figure 5. Colors of the ring-closed isomers of diarylethenes.

Recently, we constructed a pilot plant system to optimize the reaction conditions in continuous microflow synthesis (Figure 6). The pilot plant has four reaction units, each of which consists of a micromixer and a microtube reactor. The inner temperature of the units is controlled by circulating a coolant in the units. Further work using this pilot plant is currently in progress.

Figure 6. A pilot plant.

4. Conclusion

In conclusion, we have developed an effective method for the synthesis of photochromic diarylethenes using microflow systems, which enables reactions at much higher temperatures. Successful synthesis of unsymmetrical diarylethenes[9] using two different aryl bromides in one flow speaks well for the potential of microflow systems in making new functional materials, which are otherwise difficult to synthesize in a conventional way. Because several micro chemical plants in pilot scale have already built,[10] this process seems to be easily transformed to the system in an industrial scale by numbering-up and continuous operation.

5. Reference

(1) Reviews on organic photochromism: for example: (a) Irie, M. *Chem. Rev.* **2000**, *100*, 1685. (b) Berkovic, G.; Krongauz, V.; Weiss, V. *Chem. Rev.* **2000**, *100*, 1741. (c) Kawata, S.; Kawata, Y. *Chem. Rev.* **2000**, *100*, 1777. (d) Feringa, B. L.; van Delden, R. A.; Koumura, N.; Geerts, E. M. *Chem. Rev.* **2000**, *100*, 1789. (e) Bouas-Laurent, H.; Dürr, H. *Pure Appl. Chem.* **2001**, *73*, 639. (f) Raymo, F. M.; Tamosulo, M. *Chem. Soc. Rev.* **2005**, *34*, 327. Some recent papers on diarylethenes: (g) Lucas, L. N.; de Jong, J. J. D.; van Esch, J. H.; Kellogg, R. M.; Feringa, B. L. *Eur. J. Org. Chem.* **2003**, 155. (h) Morimoto M.; Kobatake, S.; Irie, M. *J. Am. Chem. Soc.* **2003**, *125*, 11080. (i) de Jong, J. J. D.; Lucas, L. N.; Hania, R.; Pugzlys, A.; Kellogg, R. M.; Feringa, B. L.; Duppen, K.; van Esch, J. H. *Eur. J. Org. Chem.* **2003**, 1887. (j) de Jong, J. J. D.; Lucas, L. N.; Kellogg, R. M.; van Esch, J. H.; Feringa, B. L. *Science* **2004**, *304*, 278. (k) Choi, H.; Lee, H.; Kang, Y.; Kim, E.; Kang, S. O.; Ko, J. *J. Org. Chem.* **2005**, *70*, 8291. (l) Higashiguchi, K.; Matsuda, K.; Tanifuji, N.; Irie, M. *J. Am. Chem. Soc.* **2005**, *127*, 8922.

(2) Reaction of fluoroolefins with organolithium compounds: Dixon, S. J. *J. Org. Chem.* **1956**, *21*, 400. Reaction of octafluorocyclopentene: Hanazawa, M.; Sumiya, R.; Horikawa, Y.; Irie, M. *J. Chem. Soc., Chem. Commun.* **1992**, 206.
(3) Reviews: For example, (a) Haswell, S. J.; Middleton, R. J.; O'Sullivan, B.; Skelton, V.; Watts, P.; Styring, P. *Chem. Commun.* **2001**, 391. (b) Jensen, K. F. *Chem. Eng. Sci.* **2001**, *56*, 293. (c) Fletcher, P. D. I.; Haswell, S. J.; Pombo-Villar, E.; Warrington, B. H.; Watts, P.; Wong, S. Y. F.; Zhang, X. *Tetrahedron* **2002**, *58*, 4735. (d) Jähnisch, K.; Hessel, V.; Löwe, H.; Baerns, M.; *Angew. Chem. Int. Ed.* **2004**, *43*, 406. (e) Yoshida, J. *Chem. Commun.* **2005**, 4509. (f) Doku, G. N.; Verboom, W.; Reinhoudt, D. N.; van den Berg, A. *Tetrahedron* **2005**, *61*, 2733.
(4) Some recent examples: (a) Salimi-Moosavi, H.; Tang, T.; Harrison, D. J. *J. Am. Chem. Soc.* **1997**, *119*, 8716. (b) Chambers, R. D.; Spink, R. C. H. *Chem. Commun.* **1999**, 883. (c) de Bellefon, C. ; Tanchoux, N. ; Caravieilhes, S. ; Grenouillet, P. ; Hessel, V. *Angew. Chem. Int. Ed.* **2000**, *39*, 3442. (d) Watts, P.; Wiles, C.; Haswell, S. J.; Pombo-Villar, E.; Styring, P. *Chem. Commun.* **2001**, 990. (e) Hisamoto, H.; Saito, T.; Tokeshi, M.; Hibara, A.; Kitamori, T. *Chem. Commun.* **2001**, 2662. (f) Sands, M.; Haswell, S. J.; Kelly, S. M.; Skelton, V.; Morgan, D. O.; Styring, P.; Warrington, B. *Lab. Chip* **2001**, *1*, 64. (g) Wiles, C.; Watts, P.; Haswell, S. J.; Pombo-Villar, E. *Lab. Chip* **2001**, *1*, 100. (h) Suga, S.; Okajima, M.; Fujiwara, K.; Yoshida, J. *J. Am. Chem. Soc.* **2001**, *123*, 7941. (i) Wiles, C.; Watts, P.; Haswell, S. J.; Pombo-Villar, E.; *Chem. Commun.* **2002**, 1034. (j) Fukuyama, T.; Shinmen, M.; Nishitani, S.; Sato, M.; Ryu, I. *Org. Lett.* **2002**, *4*, 1691. (k) Ueno, M.; Hisamoto, H.; Kitamori T.; Kobayashi, S. *Chem. Commun.* **2003**, 936. (l) Garcia-Egido, E.; Spikmans, V.; Wong, S. Y. F.; Warrington, B. H. *Lab. Chip* **2003**, *3*, 73. (m) Lai, S. M.; Martin-Aranda, R.; Yeung, K. L. *Chem. Commun.* **2003**, 218. (n) Mikami, K.; Yamanaka, M.; Islam, M. N.; Kudo, K.; Seino, N.; Shinoda, M. *Tetrahedron Lett.* **2003**, *44*, 7545. (o) Kobayashi, J.; Mori, Y.; Okamoto, K.; Akiyama, R.; Ueno, M.; Kitamori, T.; Kobayashi, S. *Science* **2004**, *304*, 1305. (p) Horcajada, R.; Okajima, M.; Suga, S.; Yoshida, J. *Chem. Commun.* **2005**, 1303. (q) Ducry, L.; Roberge, D. M. *Angew. Chem. Int. Ed.* **2005**, *44*, 7972. (r) Nagaki, A.; Togai, M.; Suga, S.; Aoki, N.; Mae, K.; Yoshida, J. *J. Am. Chem. Soc.* **2005**, *127*, 11666. (s) He, P.; Watts, P.; Marken, F.; Haswell, S. J. *Angew. Chem. Int. Ed.* **2006**, *45*, 4146. (t) Midorikawa, K.; Suga, S.; Yoshida, J. *Chem. Commun.* **2006**, 3794.
(5) (a) Kawaguchi, T.; Miyata, H.; Ataka, K.; Mae, K.; Yoshida, J. *Angew. Chem. Int. Ed.* **2005**, *44*, 2413. (b) Nagaki, A.; Kawamura, K.; Suga, S.; Ando, T.; Sawamoto, M.; Yoshida, J. *J. Am. Chem. Soc.* **2004**, *126*, 14702.
(6) Reactions of organolithium compounds in microsystems: (a) Schwalbe, T.; Autze, V.; Hohmann, M.; Stirner, W. *Org. Process Res. Dev.* **2004**, *8*, 440. (b) Zhang, X.; Stefanick, S.; Villani, F. J. *Org. Process. Res. Dev.* **2004**, *8*, 455. (c) Yoshida, J.; Okamoto, H. *Advanced Micro & Nanosystems Volume 5, Micro Process Engineering, N. Kochmann Ed.* Wiley, **2006**, 439. (d) Usutani, H.; Tomida, Y.; Nagaki, A.; Okamoto, H.; Nokami, T.; Yoshida, J. *J. Am. Chem. Soc.* **2007**, *129*, 3046. (e) Nagaki, A.; Tomida, Y.; Usutani, H.; Kim, H.; Takabayashi, N.; Nokami, T.; Okamoto, H.; Yoshida, *J. Chem. Asian J.* **2007**, *2*, 1513.
(7) Irie, M.; Lifka, T.; Kobatake, S.; Kato, N. *J. Am. Chem. Soc.* **2000**, *122*, 4871.
(8) Uchida, K.; Ishikawa, T.; Takeshita, M.; Irie, M. *Tetrahedron* **1998**, *54*, 6627.
(9) For example: Uchida, K.; Irie, M. *Chem. Lett.* **1995**, 969.
(10) (a) Iwasaki, T.; Kawano, N.; Yoshida, J. *Org. Process Res. Dev.* **2006**, *10*, 1126. (b) H. Wakami and J. Yoshida, *Org. Process Res. Dev.* **2005**, *9*, 787. See also ref 6c.

46d A Multistep Process for the Synthesis of Pharmaceutical Intermediates by An Ozonolysis–Reduction–Sequence Using Microchemical Engineering with An Integrated Miniaturized Fiber–Optical Diamond ATR Sensor in A Multipurpose Plant

S. Hübner[1], K. Jähnisch[1], U. Bentrup[1], N. Steinfeldt[1], L. Küpper[2], U. Budde[3], K. Lovis[3], T. Dietrich[4], and A. Freitag[4]. (1) Leibniz–Institut für Katalyse e.V. an der Universität Rostock, Richard–Willstätter–Str. 12, Berlin, D-12489, Germany, (2) infrared fiber sensors, Im Gillesbachtal 33, Aachen, D-52066, Germany, (3) Global chemical development, Bayer Schering Pharma AG, Müllerstrasse 178, Berlin, Germany, (4) mikroglas chemtech GmbH, Galileo-Galilei-Str. 28, Mainz, D-55129, Germany

Microstructured reactors have recently gained increasing importance as useful devices for chemical reactions and processes. Improved temperature management, shorter diffusion paths and higher specific interfacial areas increase mass and heat transfer rates. Furthermore, due to the low hold–up in microdevices the hazard of highly exothermic and explosive reactions is considerably reduced. Within a research project called ZOHIR these advantages were utilized in a multistep process using microchemical engineering for the preparation of intermediates of vitamin D analogs. Scheme 1 shows the reaction sequence, which was successfully performed in a microplant. A general flow chart for this two step process is depicted in Figure 1.

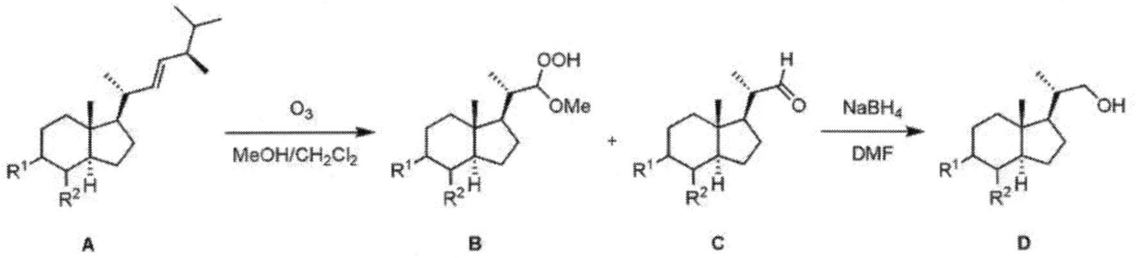

Scheme 1. Reaction scheme for the preparation of intermediates of vitamin D analogs in a two–step process using microstructured reactors.

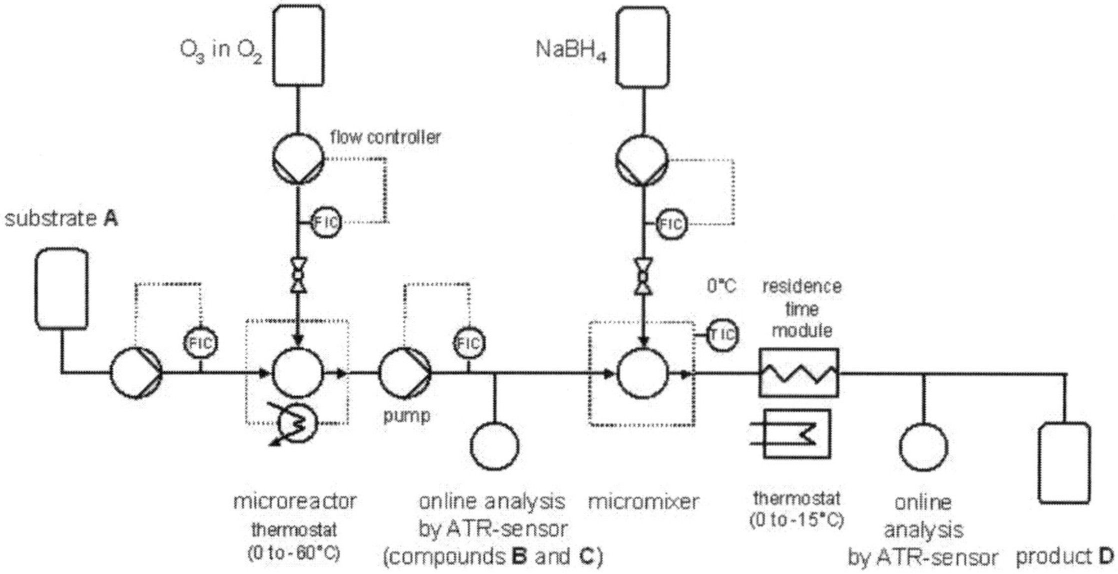

Figure 1. Flow chart for the ozonolysis–reduction–process in microstructured devices.

By a pump educt **A** is fed to the microreactor, where it reacts with ozone in a gas/liquid–reaction. The ozonolysis of the double bond in **A** yields hydroperoxide **B** and aldehyde **C**. The reaction mixture is then pumped out of the reactor and analyzed online by a miniaturized Fibre–optical Diamond ATR sensor, which was developed in the course of this project. The subsequent reduction to alcohol **D** by sodium borhydride is performed in a mixer/residence time module combination without isolation.

Concerning the feasibility and optimization the two reaction steps in microreactors, ozonolysis and reduction have initially been investigated separately. For the ozonolysis, the miniaturized ATR sensor was used for the online–monitoring of the formation of hydroperoxide **B** and aldehyde **C**. It showed that already about 30% of aldehyde **C** is formed along with hydroperoxide species **B** during this step. These results were confirmed by quantitative offline HPLC–analysis of aldehyde **C**.

Additionally, for the ozonolysis step different microstructured reactors have been tested (Figure 2). Surprisingly, best results were obtained with a 5–channel–microreactor (Figure 3) giving quantitative conversion up to a flow rate of 0.26 mmol/min. In comparison for the same flow rate, lower conversions were detected for the 1–channel–micromixer and microreactors which are especially designed for liquid/gas–reactions, such as a falling film microreactor or a cyclone mixer.

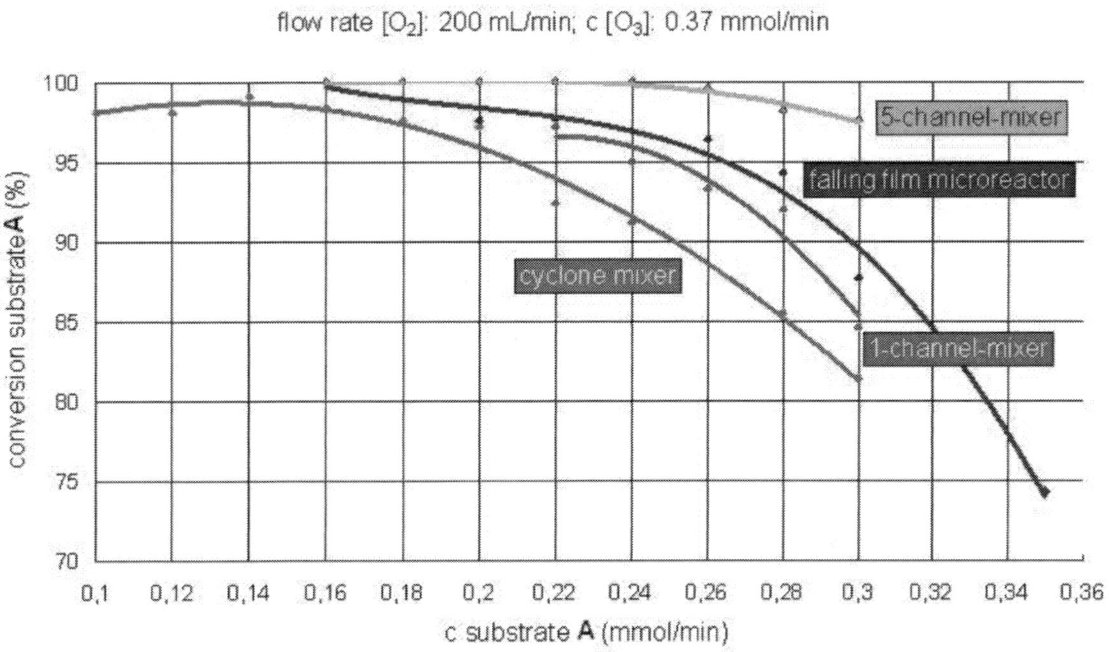

Figure 2. Conversion of compound A in the ozonolysis–step of the multistep process for different flow rates and microreactors.

The reduction to alcohol **D** was also successfully performed in a 5–channel–microreactor (Figure 3). In combination with a residence time module, quantitative conversion was reached up to flow rates of 0.6 mmol/min with a threefold excess of $NaBH_4$.

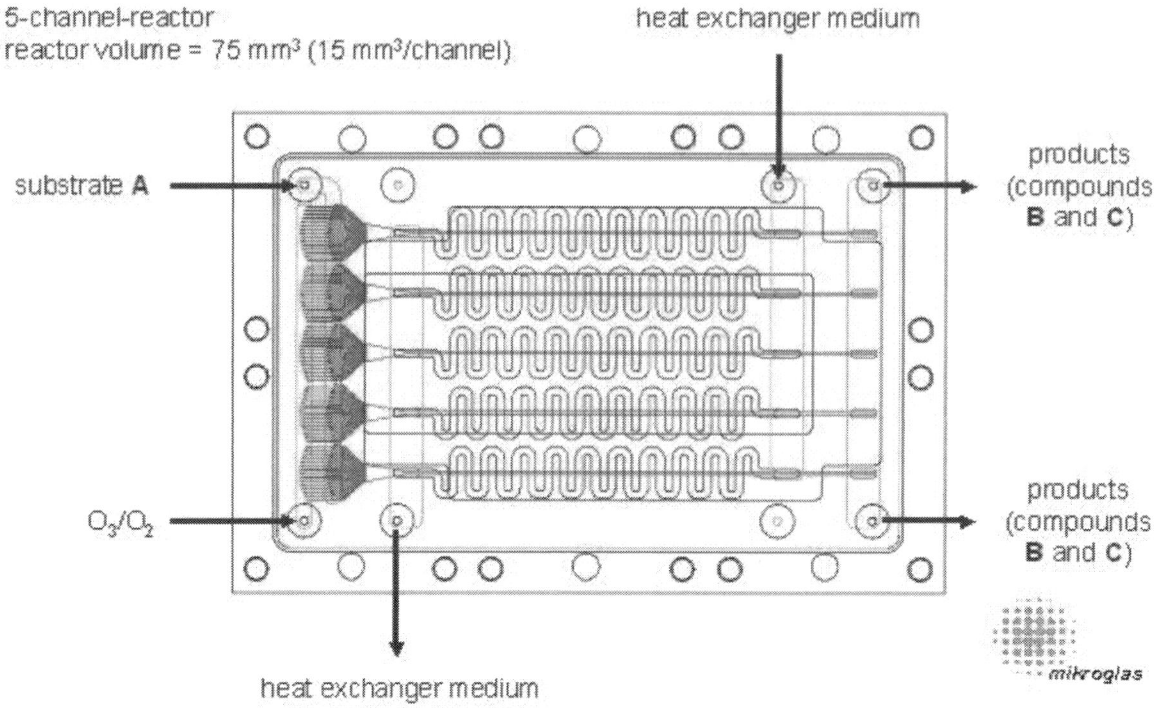

Figure 3. 5–channel–microreactor utilized in the ozonolysis–step of the multistep process (source: mikroglas chemtech GmbH).

The overall process was then performed with two 5–channel–micromixers according to the flow chart given in Figure 1 and the desired product alcohol **D** was isolated in 72% yield.

In conclusion, we were able to run this two step reaction sequence in a continuous manner. The utilization of microstructured devices allows for the realization of a safe and highly controlled multistep process with online–analysis via FTIR. A 5–channel–microreactor turned out to be most suitable for the gas/liquid–reaction (ozonolysis). Additionally, the *in situ* reduction to alcohol **D** leads to a higher efficiency of the process.

THERMAL MASS FLOW SENSORS FOR MONITORING ESTERIFICATION REACTIONS IN RESIDENCE TIME MICRO REACTORS

T. Jacobs[1], M. Kaspereit[3], K.-P. Zeyer[3], A. Kienle[2 3], P. Hauptmann[1]

[1]Institute of Micro and Sensor Systems, University of Magdeburg
[2]Chair of Automation / Modeling, University of Magdeburg
[3]Max Planck Institute for Dynamics of Complex Technical Systems, Magdeburg, Germany

Abstract

In this contribution we present a novel residence time micro reactor with integrated thermal mass flow sensor network for the inline analysis of homogeneously catalyzed esterification reactions on micro scale. The micro reactor design was optimized for low throughput micro plants used in chemical micro process engineering. The sensor network consists of six thermal mass flow sensors based on the thermo-transfer principle (calorimeter) for monitoring relative changes in the isobaric heat capacity during chemical conversion. Multiple in series connected sensors enable the separation of flow pulsation effects even in the two phase regime. Differences in sensor transduction behavior were determined using non reactive liquids with well known and thermal properties. As chemical model system for the characterization of the thermal sensor network we chose the homogeneously catalyzed synthesis of butyl acetate. The esterification was processed in a micro plant with varying reactor capillary length in between each pair of sensors. Focus was on the change in isobaric heat capacity during chemical conversion as a function of residence time and phase transition effects. Furthermore, the achievable measurement sensitivity and the minimization of flow pulsation effects with differential sensor signal analysis was investigated.

Introduction

Recently, in chemical micro process engineering [1,2] the performance of homogeneously as well as heterogeneously catalyzed esterification reactions in micro plants became an interesting field [3,4]. Esters are used as high quality organic solvents with significant industrial relevance, e.g. used for lacquer production. Strong mineral or organic acids are used as liquid catalyst and acidic ion exchange resins immobilized in the micro channel as solid catalyst. Micro residence time reactors or micro fixed bed reactors are state of the art for processing this class of reactive systems in micro plants. Compared to classical batch processing, a micro plant enables fast mixing, precise temperature control, a well defined residence time and interfacing area to the solid or liquid catalyst. The micro reactor can be operated in steady state mode with continuous educt flow or non-steady state mode, where one educt is fed in pulsed mode. When educts and products have different adsorptivities to the solid catalyst, the device behaves like a chromatographic reactor. Inline analysis of the static and dynamic reactor behavior is fundamental for the development, optimization and control of novel integrated chemical processes.

Actually in micro residence time reactors, there is a lack in sensitive, robust, scalable and low priced inline analytics for monitoring chemical reactions. General process parameters of interest are conversion rate, reaction kinetics, phase changes, residence time, dispersion effects and separation efficiency. On micro scale quite expensive flow through chambers with fiber optics attached to spectrometers (UV-VIS-NIR) and light microscopes with attached digital camera are applied [5,6]. Scaling up for multiple micro reactors or channels in parallel is a problem.

In this contribution we present a novel residence time micro reactor with integrated thermal mass flow sensor network based on the thermo-transfer (calorimeter) principle for low mass flow rates [7,8,9]. This class of micro sensors is well established in measuring mass flow rates in micro fluidic systems. It is known, that the measurement signal is also sensitive to changes in isobaric heat capacity. Furthermore, the reactive system is only in contact with a monolithic capillary, which enables high chemical and pressure resistance. Insofar, it emerges as an attractive sensor principle for inline monitoring of chemical reactions in micro reactors by means of relative changes in isobaric heat capacity. For the discrimination between sensor signal changes due to chemical conversion and flow pulsation in the two phase regime more than two sensors are to be connected in series. Differences in physical sensor characteristics were characterized using non-reactive liquids with known thermal properties. Focus was on the propagation of flow pulsation along the micro channel and the capability of pulsation suppression when using multiple sensors in series. As a reactive model system we analyzed the homogeneously catalyzed synthesis of butyl acetate with sulfuric acid as catalyst, (1).

$$\text{1-butanol} + \text{acetic actid} \overset{H_2SO_4}{\rightleftharpoons} \text{butyl acetate} + \text{water} \quad (1)$$

Thermal mass flow sensor network

Thermal mass flow sensors in the presented micro reactor consist of two ohmic resistor coils around the outer surface of a cylindrical capillary made of stainless steel. Sensors are optimized for low mass flow rates, thus using constant power mode (CPM) of operation. Both coils are connected in series to a constant current source for heat generation and act simultaneously as a heater and temperature sensor. Heat is conductively and convectively transferred to the moving fluid and housing material. The fluid is heated up with an almost linear increase in temperature. Local temperature maxima exist at the wall nearby the end of each heating coil. Due to heat conduction and comparably low flow rates temperature gradients decrease rapidly in flow direction, when the fluid passes the heater. Changes in mass flow rate and thermal properties of the liquid, predominantly isobaric heat capacity, lead to a characteristic deformation of the temperature field along the sensor capillary. The deformation nearby the capillary wall can be measured by means of a change in ohmic resistance R(T) of both heater coils. R(T) can be sufficiently approximated with (2), while α_{th} is the linear temperature coefficient of the resistor material and R_{ref} is the resistance at reference temperature (T_{ref}).

$$R(T) = R_{ref}(1 + \alpha_{th}(T - T_{ref})) \quad (2)$$

A simple equation for the relationship between sensor signal U, mass flow rate m and the isobaric heat capacity c_p is given in (3). The conversion factor K_s is dependent on the physical properties of the sensor, e.g. heating power and geometry. In general, the relationship (3) is non-linear and determined during calibration procedure with reference liquids in the operation range of the sensor.

$$U = K_s \cdot c_p \cdot m' \quad (3)$$

The resulting sensor signal U is proportional to mass flow rate and also to changes in isobaric heat capacity of the fluid. Therefore, in a process application at least two sensors in series connected are necessary to distinguish in between changes in isobaric heat capacity and mass flow rate. Measurement sensitivity of the sensor is mainly limited by parasitic heat conduction along the capillary wall, housing material and wires for interconnection. Sensitivity increases almost linear with heating power, which involves a higher over temperature of the liquid or enables the operation at higher flow rates. A compromise between the required measurement sensitivity and the tolerable heating power has to be found.

The sensor network, Fig. 1, consists of six in series connected sensor elements (μ-Flow L01 series, Bronkhorst, Netherlands, [9]) with an inner diameter of 1 mm. Heater/ resistor coils of all sensors are connected to a constant current source. The dissipated heat for each element ranges typically from 100 to 200 mW. Fig. 2 shows the novel micro reactor made of stainless steel with integrated sensor network.

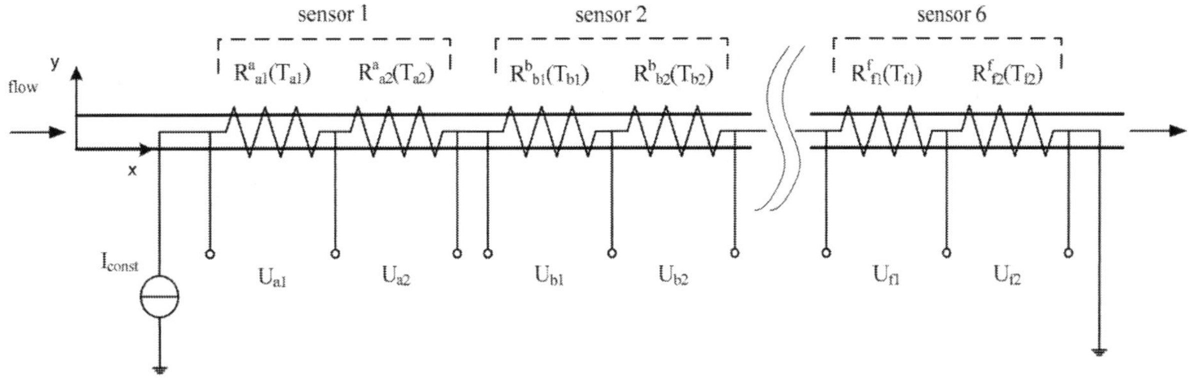

Fig. 1. Sensor network based on thermal mass flow sensors using the thermo-transfer principle

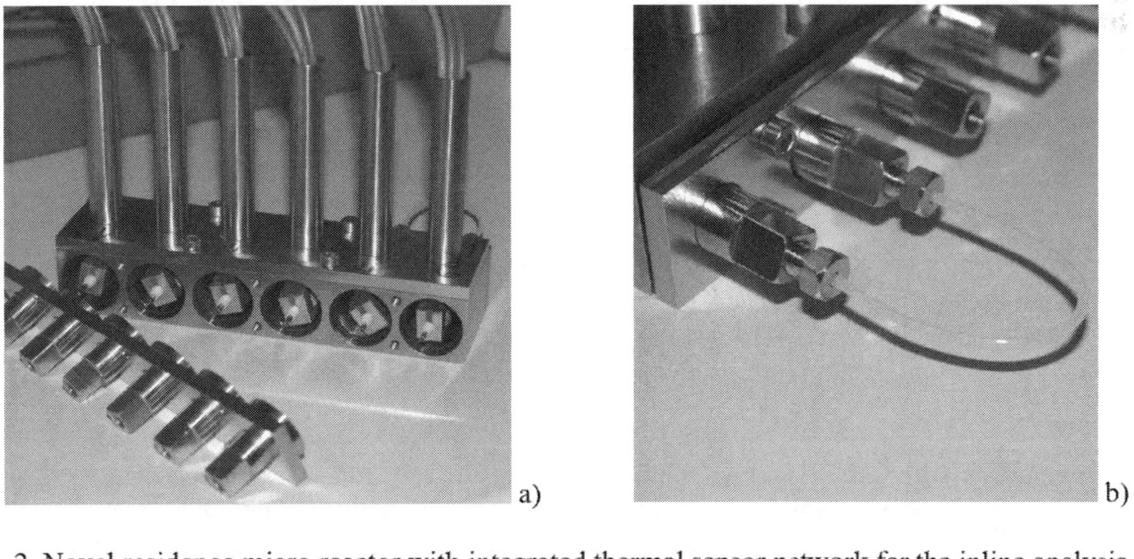

Fig. 2. Novel residence micro reactor with integrated thermal sensor network for the inline analysis of homogeneously catalyzed esterification reactions at low mass flow rates (scale: g/h)

Experimental

The micro reactor was integrated in a micro plant, Fig. 3, that consists of two mass flow meter (mini CORI-FLOW from Bronkhorst, Netherlands) combined with two micro annular gear pumps (mzr-4605 from HNP, Germany) for educt feed, alternatively a syringe pump system (neMESIS from Cetoni, Germany) and a T-type mixer in combination with PTFE-capillaries (inner diameter 0.75 mm) for interconnection and as heat exchanger. Residence time was adjusted using different reactor capillary lengths in the range of 10 cm up to 320 cm, where chemical equilibrium could be premised [3]. When processing the esterification reaction a long PTFE-capillary (inner diameter 0.17 mm), alternatively a pressure regulator (EL-PRESS from Bronkhorst, Netherlands), was used as flow restrictor at the outlet. The complete reactor with directly attached mixer was placed in a thermostat (RE307 from Lauda, Germany) for precise temperature control.

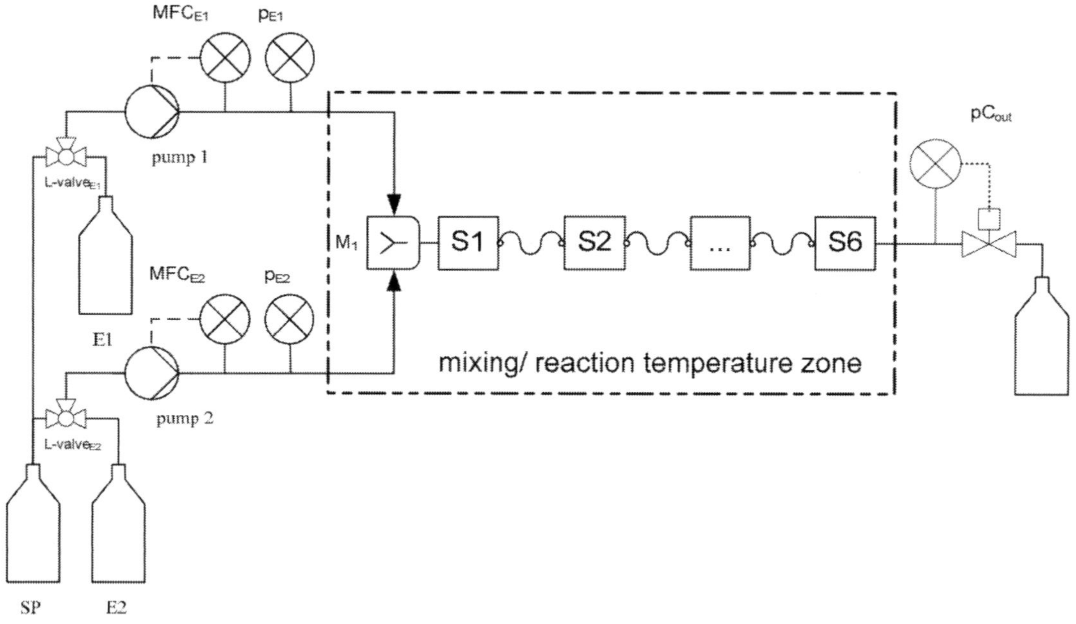

Fig. 3. Micro plant with integrated micro residence time reactor/ thermal sensor network for processing and analysis of esterification reactions at low mass flow rates

Sensor signals, respectively resistor voltage drop, were measured using 4-wire connection technology with a commercially available multimeter combined with a multiplexer (3706 series from Keithley, Germany). A high precision current source (HM7044 from Hameg, Germany) was used for heating power control. Data acquisition was done using a custom made LabView-Programm running on a standard PC. All chemicals used within the experiments were of analysis grade.

For the analysis of the propagation of flow pulsation along and the compensation of different sensor characteristics, sensors signals were evaluated, when flushing the micro reactor with pure 1-butanol at room temperature and ambient pressure. With regard to the amount of dissipated heat sensor signals were investigated in a reasonable mass flow range between 2 and 32 g/h. For each flow rate sensor signals were acquired for at least 10 min with a sample rate of 1 Hz.

The homogeneously catalyzed esterification reaction was processed at 80 °C with equimolar educt feed and 1 % sulfuric acid premixed with acetic acid [3]. An over pressure of approximately 10 bar was applied relative to ambient conditions. Before starting a synthesis the setup was flushed several times with 1-butanol to ensure well defined start conditions. With regard to the total internal volume of the micro reactor the overall flow rate was adapted, that sensor signals are at the upper end of the linear operation range (~3 g/h), when treating it as a simple mass flow sensor. Samples were collected at the end of the capillary to visually verify, whether one phase or two phase regime is present. Sampling rate was fixed to 5 Hz. Experiments were aborted when no significant sensor signal changes could be observed.

Results

When comparing measurement signals of the thermal sensor network for 1-Butanol flowing through the micro reactor, differences in sensor signal can be treated as a constant offset voltage that can be easily compensated. Within the measurement sensitivity no flow rate dependence could be observed. This offset can be mainly attributed to fabrication tolerance, especially the overall resistance of each coil and thermal losses due to the housing material. Fig. 4 shows voltage drops for each resistor coil and the differential voltage drop as a function of mass flow rate. Every measurement data point represents an average over a time period of 5 min. Sensor S1 is not evaluated due to an abnormal sensor signal behavior, that could be related to a mechanical problem with the fixture.

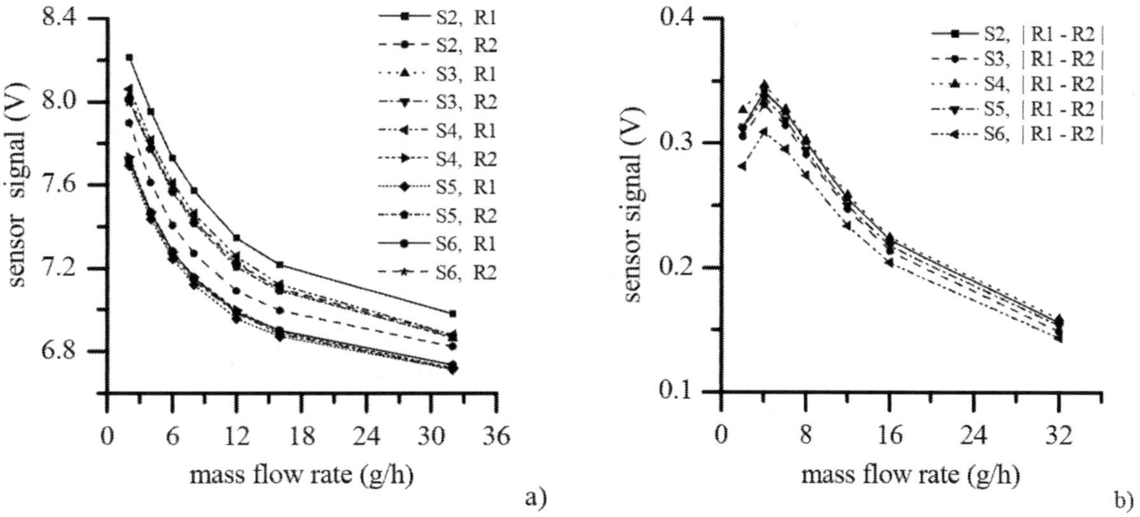

Fig. 4. Differences in sensor transduction behavior as a function of mass flow rate for 1-Butanol, 25 °C, a) single resistor voltages, b) differential resistor voltage for each sensor element

Measurement sensitivity is much higher, when evaluating voltage drops of single resistors, Fig. 4 a), compared to voltage differences of both resistor coils for each sensor element, Fig. 4 b). As expected, sensitivity decreases rapidly with increasing mass flow rate and goes to a steady state value, when the temperature field inside the sensor becomes almost constant. Fig. 5 shows the thermal sensor network response due to flow pulsation that is induced by the rotation of both annular gear pumps.

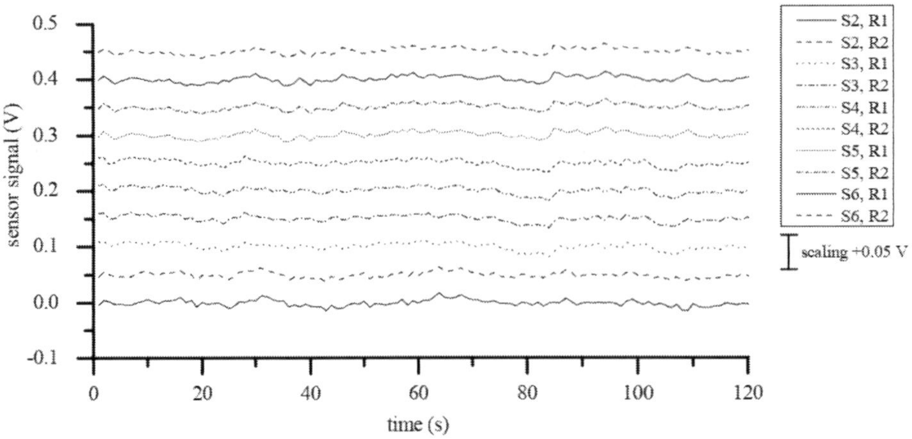

Fig. 5. Thermal sensor network response due to propagation of flow pulsation along the reactor capillary for 1-Butanol, 25 °C, overall mass flow rate is 6 g/h

When using pumps with low residual pulsation all sensors are affected simultaneously by flow alternations. No phase shift or significant damping of the pulsation amplitude could be observed. Remaining fluctuations in the differential sensor signal can be mainly attributed to electrical noise of sensor electronics. Insofar, differential sensor signal evaluation enables an inherent discrimination between flow and heat capacity induced changes. When the reactive system undergoes a phase transition at least two sensors at the beginning of the capillary, where one phase regime is present, are mandatory for the compensation of flow pulsation. Fig. 6 shows a comparison of single sensor signals during esterification reaction for two different capillary lengths, respectively two stages of conversion.

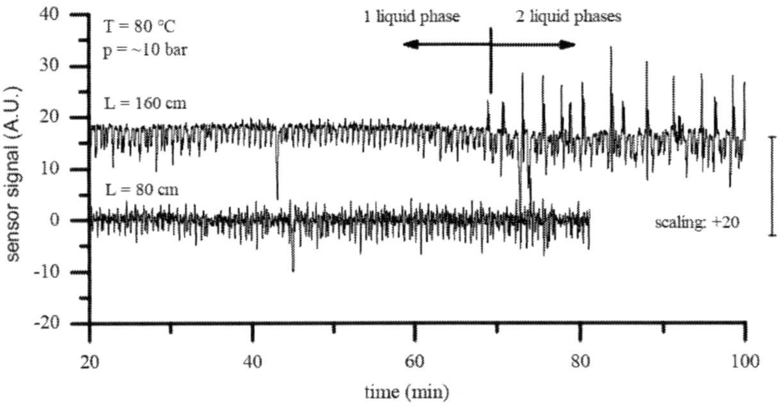

Fig. 6. Single sensor signal during esterification reaction, oscillations in sensor signal due to a phase transition at a certain residence time, respectively capillary length

It is known that during synthesis of butyl acetate at 80 °C a phase transition occurs at a certain mole fraction. This effects leads to strong oscillations in sensor signal that occur at a certain residence time in the micro reactor. Strong positive signal shifts referred to the base line result from the water-rich phase passing the resistor coils. Compared to butyl acetate, acetic acid and 1-butanol, the isobaric heat capacity of water is approx. two times higher. Flow pulsation still can be monitored in parallel using sensors that are at the beginning of the reactor capillary, i.e. evaluating sensor signals, where one phase regime is still present.

Summary

Based on our first results, the sensor network with thermal mass flow sensors using the thermo-transfer principle can be applied for inline analysis of reaction kinetics and phase transition of esterification reactions in residence time micro reactors. Maximum measurement sensitivity can be achieved, when analyzing single resistor voltage signals. Differences in sensor characteristics result in an offset voltage, that must be compensated before starting the chemical reaction using a non reactive liquid, e.g. one of the educts. When the reactive system becomes 2-phase, the influence of flow pulsation effects on the sensor network response can be minimized based on sensor signal evaluation of sensors that are in contact with single phase. The allowed maximum temperature or amount of dissipated heat, e.g. maximum achievable measurement accuracy, that is tolerable for a reactive system must be specified for each process.

References

[1] H. Pennemann, V. Hessel and H. Löwe, "Chemical microprocess technology from laboratory-scale to production", Chemical Engineering Science, vol. 59, 2004, pp. 4789-4794.

[2] K.-F. Jensen, "Microreaction engineering - is small better?", Chemical Engineering Science, vol. 56, 2001, pp. 293-303.

[3] A. A. Kulkarni, K.-P. Zeyer, T. Jacobs, A. Kienle, "Miniaturized Systems for Homogeneously and Heterogeneously Catalyzed Liquid-Phase Esterification Reaction", Industrial & Engineering Chemistry Research, vol. 46, nr. 16, pp. 5271-5277, 2007.

[4] A.A. Kulkarni, K.-P. Zeyer, T. Jacobs, M. Kaspereit and A. Kienle, "Feasibility studies and the dynamics of catalytic liquid phase esterification reactions in a micro plant", Chemical Engineering Journal, vol. 135, supplement 1, 2008, pp. S270-S275.

[5] W. Ferstl, S. Loebbecke, J. Antesa, H. Krausea, M. Haeberl, D. Schmalz, H. Muntermann, M. Grund, A. Steckenborn, A. Lohf, J. Hassel, T. Bayer, M. Kinzl and I. Leipprand, "Development of an automated microreaction system with integrated sensorics for process screening and production", Chemical Engineering Journal, vol. 101, 2004, pp. 431-438.

[6] M. Günther, S. Schneider, J. Wagner, R. Gorges, Th. Henkel, M. Kielpinski, J. Albert, R. Bierbaum and J.M. Köhler, "Characterisation of residence time and residence time distribution in chip reactors with modular arrangements by integrated optical detection", Chemical Engineering Journal, vol. 101, 2004, pp. 373–378.

[7] F.P. Incropera and D.P. DeWitt, "Fundamentals of heat and mass transfer", Wiley, 1996.

[8] M. Ashauer, H. Glosch, F. Hedrich, N. Hey, H. Sandmaier and W. Lang, "Thermal flow sensor for liquids and gases based on combinations of two principles", Sensors and Actuators A, vol. 73, 1999, pp. 7-13.

[9] J. Lötters, "Liquid flow sensor for nano- and micro-flow ranges", Sensors Review, vol. 25, 2005, pp. 20-23.

236f Microreactor Synthesis of Unimodal Molecularly Imprinted Polymer (MIP) Beads Used for Explosives Detection

Dirk Roeseling, Tobias Türcke, Stefan Loebbecke, and Horst Krause. Energetic materials, Fraunhofer Institute for Chemical Technology (ICT), Joseph-von-Fraunhoferstr. 7, Pfinztal, 76327, Germany

Microreactor synthesis of unimodal molecularly imprinted polymer (MIP) beads used for explosives detection

D. Roeseling, T. Tuercke, S. Loebbecke, H. Krause

Fraunhofer Institute for Chemical Technology, 76327 Pfinztal, Germany

Introduction

Molecularly imprinted polymers (MIP) have received an increasing interest in the field of extraction, enantiomer separation, product isolation or enrichment of species from diluted mixtures [1-2]. A quite new field is the use of MIP's as sensing materials for the detection of organic compounds and especially of explosives like trinitrotoluene (TNT), dinitrotoluene (DNT) and dinitrodimethylbutane (DMNB) [3-4]. Sensor networks based on MIP technology are currently developed as a new kind of monitoring devices in complex areas surveillance.

The synthesis of molecular imprinted polymers is usually carried out by suspension polymerization in batch vessels in presence of a template [4]. Subsequent removal of the template reveals binding sites that are specific towards the analyte via either covalent or non-covalent interactions. For the use as a chromatographic material MIP beads must have a narrow size distribution to avoid diffusion problems. Especially inefficient mixing during suspension polymerization often leads to a broad particle size distribution of the polymer beads. Some methods for preparing uniform spherical MIP's, using either expensive solvents, monomers or complex shape polymerizations, are described in literature [5-6].

In recent years microfluidic structures have received an increasing interest as reactors for the synthesis of monodisperse polymer particles [6-7].

This paper reports on the use of microreaction technology for the synthesis of MIP particles with narrow size distribution to be utilized as sensors for the gas phase detection of TNT.

Experimental

The syntheses of particulate MIP particles have been performed in a T-junction microreactor made of glass (micronit microfluidics, The Netherlands) by generating a segmented flow of droplets in a polyvinylalcohol/water mixture used as mobile phase. The droplets are fed with all relevant reagents for the polymerization reaction via a microchannel of 50 µm x 50 µm size (internal diameter) and teared off by the mobile phase in a 200 µm x 200 µm microchannel (Fig. 1). Both fluids (droplet phase and mobile phase) are introduced by syringe pumps (Sykam S1610).

Once the droplets are formed, the segmented flow is released into a wider channel of

500 µm x 500 µm diameter in which either thermally induced polymerization via heating or photochemically induced polymerization is conducted, in the latter case by employing an UV lamp (DYMAX 50 Blue).

The droplet phase consists of a combination of the following functional monomers: acrylamide, methacrylamide, methacrylic acid, 2-hydroxypropylmethacrylate or buntanediol-monoacrylate (all purchased from Aldrich), ethylene glycol dimethacrylate (Fluka) as cross linker, chloroform or toluene (Roth) as porogene and solvent and TNT (Dynamite Nobel) as template molecule. In case of photochemical initiation Irgacure 819 (Ciba Speciality Chemicals) is used as initiator whereas V65 (Wako Chemicals) is used as thermal initiator.

The formed MIP beads were filtrated, dried and the particle size distribution was measured by laser scanning spectrometry (Mastersize S, Malvern Instruments).

The TNT or DNT content of the MIP particles was analyzed by GC (HP 5890) before and after treatment with explosives test vapours, containing typically 500 ppm TNT.

Results and Discussion

We were able to generate MIP particles with a variable size between 3 µm and 150 µm. Control on particle size was achieved by deliberately varying flow rates of the droplet and mobile phase within a range from 4 µL/min up to 5000 µL/min. In comparison to previous suspension polymerizations performed in batch vessels (Fig. 2) we were able to synthesize MIP particles of uniform shape and narrow size distribution (Fig. 3). For the use as an explosives sensor two different particle sizes are of great importance. For example, particles even or smaller than 3 µm can be incorporated onto the surface of a mass sensitive quartz crystal microbalance (QCM) to detect explosives vapour directly from gas phase. Larger MIP particles with a mean size of 20 µm can be used as a stationary phase in columns either to extract explosives out of liquid solutions or to enrich traces of explosives from larger gas volumes.

Conclusion

Unimodal molecularly imprinted polymer beads were successfully produced in a continuously operated microreactor. By varying the flow conditions different particle sizes were obtained. The application of the generated MIP particles for the detection of explosives was shown by GC analysis. Both the synthesis of MIP particles suitable for the detection of other explosives and the generation of smaller MIP particles are currently under further investigation.

References

[1] G. Wulff, A. Sarhan, Angew. Chem. Int. Ed. 1972, 11(2), 341-346

[2] B. Sellergren, Molecularly Imprinted Polymers – Man-made Mimics of Antibodies and their Applications in Analytical Chemistry, Elsevier, Amsterdam, 2001

[3] F. L. Dickert, O. Hayden, Trends in Anal. Chem. 1999, 18 (3), 192-199

[4] G. Bunte, J. Hürttlen, H. Pontius, K. Hartlieb, H. Krause, Anal. Chim. Acta. 2007, 591, 49-56

[5] Z. Chen, R. Zhao, D. Shangguan, G. Liu, Biomed. Chromatogr. 2005, 19, 533-538

[6] M. Zourob, S. Mohr, A.G. Mayes, A. Macaskill, N. Pérez-Moral, P.R. Fielden, N.J. Goddard, Lab Chip 2006, 6, 296-301

[7] T. Takahashi, Y. Takizawa, T. Nisisako, T. Torii, T. Higuchi, Proceed. 8th Int. Conf. on Miniaturized Systems in Chemistry and Life Sciences µTAS 2004, Malmö, Sweden, pp 85-86

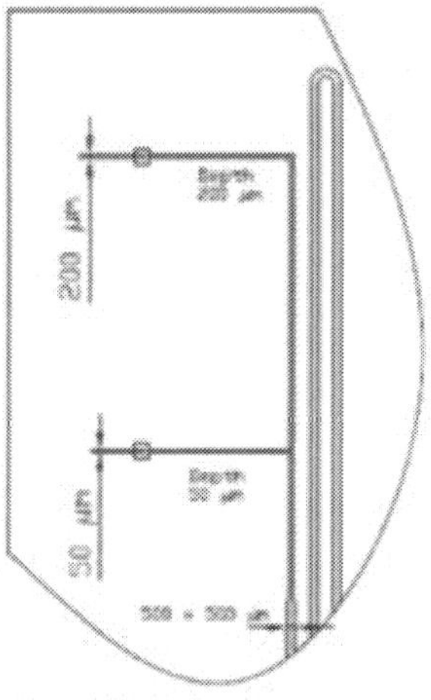

Fig. 1: Microreactor design

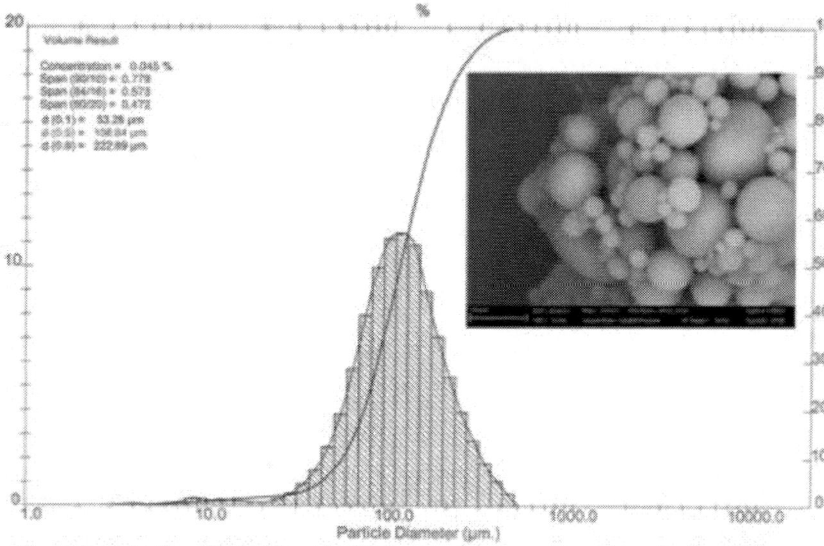

Fig.2: Particle size distribution of a batch vessel suspension polymerisation MIP.

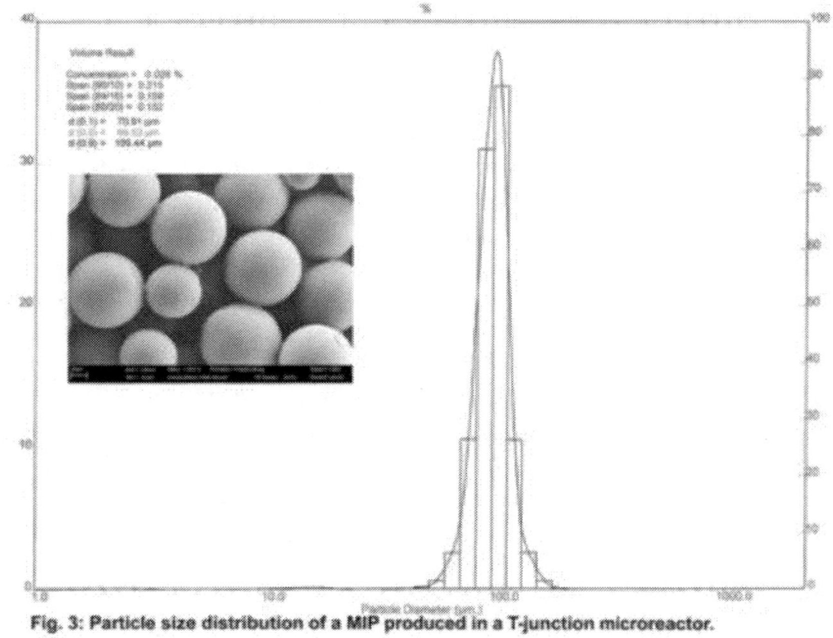

Fig. 3: Particle size distribution of a MIP produced in a T-junction microreactor.

235c Multiphase Packed-Bed Microreactors – Tracer Dispersion and Hydrodynamic Characterization

Nathalie Marquez, Michiel T. Kreutzer, Michiel Makkee, and Jacob A. Moulijn. Delft University of Technology, Julianalaan 136, Delft, 2628 BL, Netherlands

In this contribution we characterize multiphase flow in micro channels, packed with very fine catalyst particles for the application of catalyst performance testing. We analyze the hydrodynamics of the catalyst bed by using accurate measurements of residence times and tracer-pulse broadening.

Motivation:

Miniaturization of catalyst performance testing is primarily motivated by the elimination of temperature and concentration gradients in the catalyst bed. Micro reactors represent a cheaper, faster, and safer alternative to get kinetic data. In the open literature, no dispersion or RTD data are available for the multiphase flows in micro packed-beds.

The objective of this work is to describe the interplay between diffusion, gas-liquid interaction, and convection for a surface tension-dominated laminar flow in these small systems. From a design perspective it is vital to determine the extent of the reactant dispersion, especially at high conversion and in short reactors. Moreover, we need to know the difference in residence time of the gas and liquid: stripping of reactants and products can significantly impact conversion levels and mask the proper interpretation of kinetic experiments.

Van Herk et al. reported initial RTD measurements that indicated some of the major challenges in scaling down trickle-bed reactors. The flow pattern and behaviour in miniaturized beds are different from those observed in commercial or large-scale laboratory trickle beds. At these small scales gravity ceases to be significant and identical results can be obtained for upward- and downward-flow. Indeed, no trickling is observed and all liquid hold-up is static by the classic definition: no dripping is observed upon switching off the liquid and gas feed. The behaviour of these micro packed-beds can not be predicted using correlations from the trickle bed literature.

Experimental Approach:

We report the detailed dispersion measurements using non-adsorbing particles of 50 and 100 μm in a bed with an internal diameter of 2 mm. We avoid significant wall effects using particles that are at least an order of magnitude smaller than the bed diameter.

Gas and liquid superficial velocities are varied from 0.1 mm/s to 10 mm/s. Previous work in optically accessible micro-channels has focussed on using fluorescent dyes under a microscope (Trachsel et al.). We have developed an experimental set-up based on refractive index that allows us to use a wide range of fluids and tracers. We use a miniature gas-liquid separator that works on the capillary forces of a filter with 10 μm pores (Figure 2). By applying a mild under-pressure downstream of the filter, it siphons off all liquid for a wide range of gas and liquid throughputs. The effluent of the filter is immediately connected to the analysis apparatus.

We used different tracer molecules (cumene, methyl stearate, n-hexane, and methanol) in various solvents (n-tetradecane, 1-nonanol, ethanol, and water) and nitrogen as the gas phase. This approach allows us to independently vary the tracer diffusion coefficient, the solvent viscosity, and the gas-liquid interfacial tension.

In the current micro packed-bed with Reynolds numbers smaller than 1 and for volumetric gas/liquid feed ratios between 0.25 and 50, the velocity of the gas is different than the velocity of the liquid, indicating that one phase can significantly by-pass the other depending on the conditions.

The tracer curves are interpreted using the Piston-Dispersion-Exchange model that identifies stagnant zones and the rate of mass exchange between the stagnant liquid and the moving liquid. The data are interpreted in terms of molecular diffusion and hydrodynamic dispersion by considering the relative importance of interfacial stresses and viscous stresses.

Conclusions and Significance:

By varying the volumetric gas/liquid feed ratios, we are able to identify a window of operation where broadening of the RTD data due to evaporation to a fast flowing gas stream and subsequent condensation is minimal. The manner of introduction of the multiphase flow into the bed has an enormous impact on the bed behaviour. The tracer curves and post-processing, thereof, taught us how to avoid stagnant liquid zones and associated back mixing.

The results obtained in this work provide us a valuable insight into the performance of miniaturized multiphase packed-bed reactors.

References:

- Van Herk, D.; Kreutzer, M.T.; Makkee, M. and Moulijn, J.A. (2005) "Scaling down trickle bed reactors", Catalysis Today 106, 227-232

- Trachsel, F.; Günther, A.; Khan, S. and Jensen, K.F. (2005) "Measurement of residence time distribution in microfluidic systems", Chemical Engineering Science 60, 5729-5737

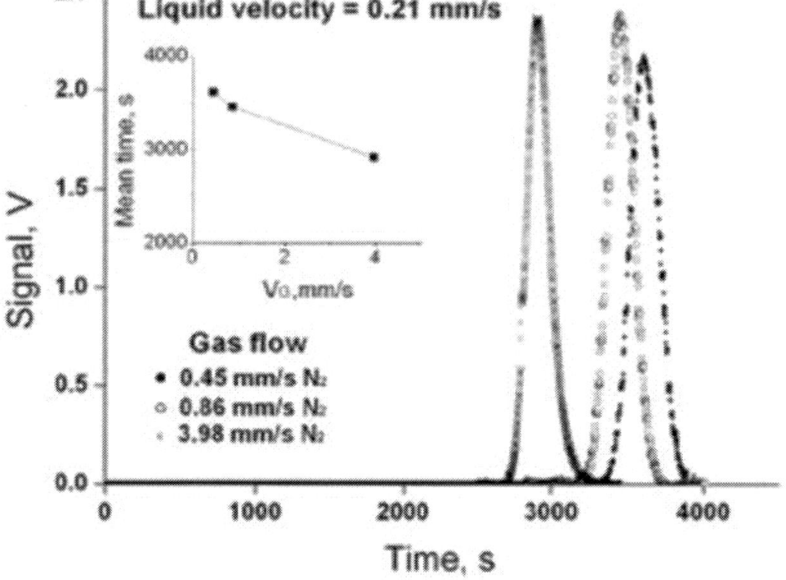

Figure 1. RTD curves showing that the gas velocity does not have a significant effect on tracer dispersion and mean residence time.

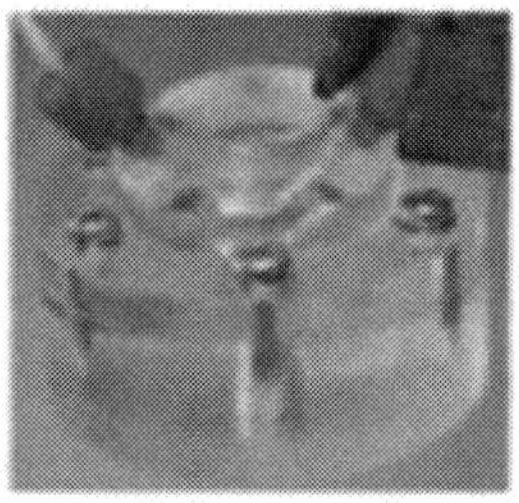

Figure 2. Gas-liquid separator.

83r Synthesis of Carbon Nanofibers on Silicon Substrates Using Sol-Gel Prepared Porous Alumina Layer

Digvijay B. Thakur[1], K. Seshan[1], R. M. Tiggelaar[2], J. G. E. Gardeniers[2], and L. Lefferts[1]. (1) Catalytic Processes and Materials, University of Twente, Meander, Room 360, P. O Box 217, Enschede, Netherlands, (2) Mesoscale Chemical Systems, University of Twente, Meander, Room 146, P. O Box 217, Enschede, Netherlands

The main goal of one of the projects within our microreactor-research-program is to develop a microstructured multiphase microreactor, *viz.* a gas-liquid contactor with specially modified catalytic coatings on its microstructured internals, for studying three-phase (G-L-S) reactions. The microreactor will be fabricated using micromachining (cleanroom) technologies in silicon and vitreous materials (i.e., fused silica and/or Borofloat glass).

Currently, there is a trend to use structured catalyst supports, i.e., that the catalyst is deposited on a rigid, orderly arranged support. Catalytic support layers based on Carbon Nano Fibers (CNFs) offer a novel option to facilitate this task [1, 3]. With their inherent high surface area-to-volume ratio, CNF's provide more catalytic surface area [1, 2], thereby obtaining sufficient activity per unit volume of catalyst, while maintaining a low fluidic resistance (which is important in microfluidic devices). Additionally, diameter and length of the fibers and hence the bulk density of the CNF layer, can be manipulated to tailor porosity and overcome tortuosity problems. This also eliminates the internal diffusion limitations by preventing concentration gradients inside the CNF layer.

The attachment of CNFs to the flat surface (silicon or glass substrates) upon which they are grown is extremely important. Enhanced mechanical stability of CNFs inside microchannels in actual fluid flow conditions can directly address the critical issues like channel blocking due to loose CNFs as well as loss of catalytic activity over the period of operation.

Various approaches can be utilized to achieve improved mechanical stability of CNFs upon flat surfaces like silicon substrates. One of them is to utilize a porous ceramic layer coating to deposit the growth catalyst (e.g. nickel) necessary for growing CNFs. Figure 1 schematically illustrates the cross sectional view of microchannel filled with CNFs anchored to the channel walls using porous alumina layer.

A sol-gel dip-coating method [4, 5] was utilized to prepare alumina layer on silicon substrates using a boehmite sol. During dip-coating procedure various parameters *viz.* sol viscosity, dwell-time and withdrawal-speed as well as calcination temprature were monitored to obtain stable and uniform mesoporous alumina layer on silicon substrates. Prior to dip coating the silicon substrates were pretreated thermally (at 1100 °C for 15 h.) in stagnant air to vary the roughness via producing oxide layer (of about 600 nm) and investigate its effect on the mechanical stability of the deposited alumina layer compared to untreated smooth surface of silicon substrate. Further nickel, as a CNF growth catalyst, was deposited by impregnation on the stable alumina layer using nickel nitrate solution. Following calcination and reduction (with H_2) pretreatment, these alumina coated and nickel impregnated silicon substrates were subjected to synthesize CNF via catalytic vapor deposition (CVD) process using hydrocarbon gas (e.g. ethylene). The CNFs were grown with diameters in the range of few nanometers to 50 nm maintaining the underlying alumina layer intact. HR-SEM images shown in figure 2 illustrate different views involving porous alumina layer and grown CNFs.

It is expected that the mesoporous alumina layer deposited on silicon substrates can enhance the attachment and thus the mechanical stability of the CNFs to the walls of microchannels. This will further facilitate to realize the ultimate goal of designing smart and efficient multi-phase microreactors using CNFs as structured catalyst support.

References:

[1] M.-J. Ledoux , C. Pham-Huu, Catal. Today 102-103 (2005) 2.

[2] M. L. Toebes, F. F. Prinsloo, J. H. Bitter, A. J. van Dillen, K. P. de Jong, J. Catal. 214 (2003) 78.

[3] J. K. Chinthaginjala, K. Seshan, L. Lefferts, Ind. Eng. Chem. Res. 46 (2007) 3968.

[4] C. J. Brinker, A. J. Hurd, J. Phys. III France 4 (1994) 1231.

[5] J. –W. Lee, C. -W. Won, B. –S. Chun, H. Y. Sohn, J. Mater. Res. 8-12 (1993) 3151.

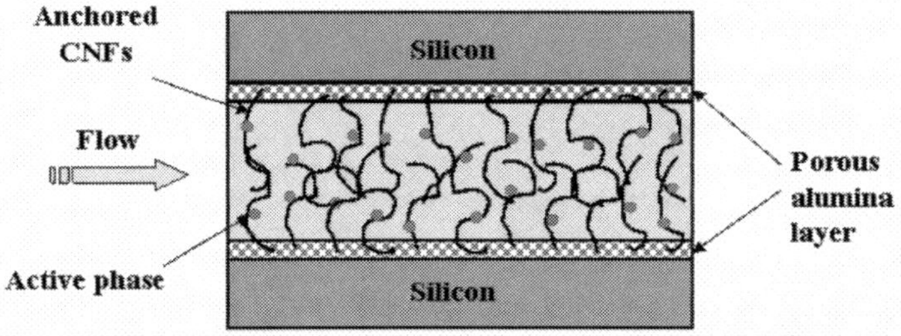

Figure 1: *Schematic cross-section of microreactor filled with CNFs anchored to channel walls using porous alumina layer*

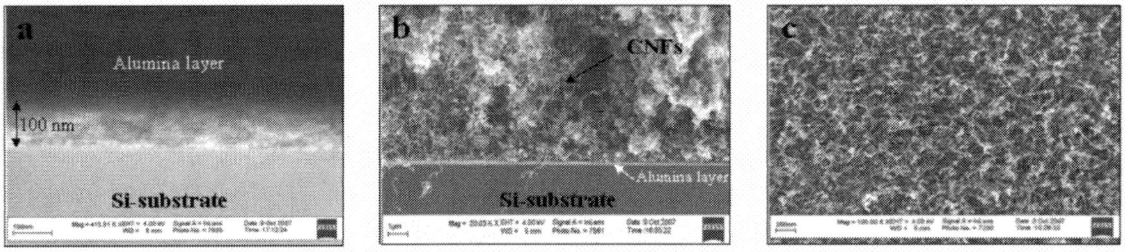

Figure 2: *HR-SEM images* **a)** *Cross sectional view of sol-gel dip coated mesoporous alumina layer on Si-substrate*
b) *Cross sectional view of CNFs grown on Si-substrate*
c) *Top view of CNFs covering Si-substrate*

Intermediate Reynolds Number Reactive Boundary Layers in Catalytic Micro-Combustion Flows

Suzanne A. Smyth and Dimitrios C. Kyritsis

University of Illinois at Urbana-Champaign, Urbana, Illinois 61801-2906, USA

Abstract

Boundary layer flows over catalytically combusting flat plates were investigated for the range of Reynolds numbers that pertain to micro-combustion applications with the use of Gas Chromatography/ Mass Spectroscopy gas analysis and infrared thermography.
Three separate phases of combustion were observed and described in terms of surface temperature and species distributions. High temperature zones, and correspondingly fast reactant conversion, are found in the vicinity of the leading edge, where the streamwise Reynolds number is smaller than an order of 10. For this range of Reynolds numbers, the classical boundary layer assumptions are invalid as both streamwise and transverse gradients are important. Consequently, boundary layer interruption is presented as an approach to increase reactor miniaturization.

1. Introduction

Further miniaturization of portable devices is significantly limited by the size and weight of current power generation methods, namely modern chemical batteries. A potential solution to this problem lies in the high-energy densities of liquid hydrocarbon fuels, which provide energy density an order of magnitude higher than the one of state-of-the art chemical batteries even if a modest efficiency of conversion to electricity is assumed (on the order of 10%) [1]. The process of extracting power from these fuels with large-scale thermodynamic cycles has been well researched. Unfortunately, as these systems are miniaturized, sustaining small-scale combustion becomes a considerable challenge [2-5]. Much of current research has been focused on the feasibility of small scale devices, while the related fluid mechanics has escaped considerable attention. At the small scales desired, reasonably defined Reynolds (Re) and Peclet (Pe) numbers are found within 10 and 100 [1-7]. This is an intermediate range where common assumptions, such as those governing boundary layer formation, may be invalid. These Reynolds numbers fall well within the laminar regime for which the heat/mass transfer coefficients are expected to be relatively low. Catalytic combustion is a frequent mode of combustion in "micro"-burners [1,3,4,6-11] due to the short residence times required. Since in low Pe flows catalytic reactions may not be kinetically limited, the study of gaseous phase phenomena is especially important. Furthermore, it is likely that the flow-catalytic surface interaction will differ significantly from the results of previous studies that were either performed at significantly higher Re [12,13], or targeted the structure of boundary layers over catalytic surfaces and were therefore performed in vacuo [14].

Boundary layer flow over a flat catalytic plate was studied for the set of conditions that are appropriate for micro"-combustion applications. Methane and air flows over thin Pt flat plates were experimentally studied for equivalence ratios and Reynolds numbers relevant to small-scale devices. The surface temperature distributions were measured using infrared thermography and the species concentration profiles of major reactant and products were measured with gas chromatography/mass spectroscopy.

2. Apparatus and Techniques

A schematic of the experimental apparatus is shown in Fig. 1. A radiative heater from ZIRCAR Ceramics, Inc. was used to preheat the incoming methane air mixture to 500 °C. A ceramic shell contained a heating coil with a diameter of 8 cm surrounding a 31 mm OD-19 mm ID steel tube. The reactant gases passed through the tube and flowed over both sides of a flat 25 mm square coupon of Pt foil. The 25.4 μm thick foil was positioned 1 mm above the exit of the heater.

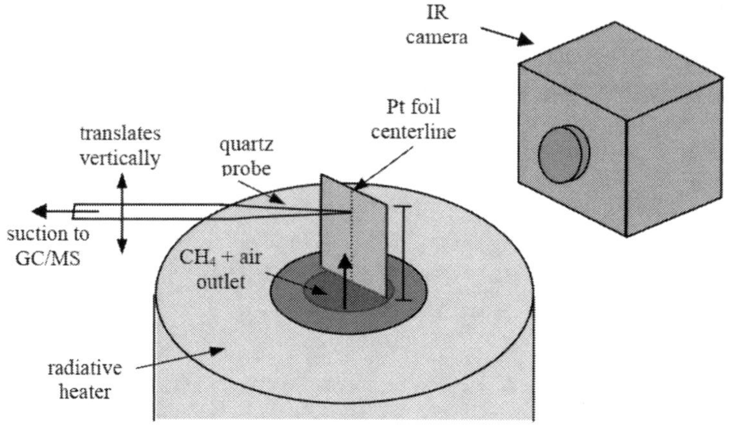

Figure 1. Experimental apparatus

Catalytic combustion was established for five values of the equivalence ratio (φ) ranging from 0.90 to 1.43. Two different Re are used in the analysis. First, the boundary layer flow is described as a function of Re_x, i.e. the value of Re based on the distance x from the leading edge of the flat plate. However, a value of Re based on a length scale for the global size of the device is also necessary. The tube exit diameter (d) was chosen for this reason and results are also reported in terms of the "jet Reynolds number," Re_d. This d was conveniently on the same order as the total length of the catalytic foil (19 and 25 mm respectively). Three intermediate values of Re_d between 76 and 101 were investigated. It is also noted that the length l of the potential core of the jet emerging from the heater approximately scales as $l/d \sim 0.04{*}Re_d$ [15]. Therefore, the value of Re_d was a compromise between the need to stay in the range of $Re_d \sim$ 1-100 that is relevant for micro-combustion studies and the need to have a sizeable portion of the test section in the potential core of the jet, to ensure constant free-stream composition.

Measurements of species concentration were performed above the catalytically reacting flat plate with an Agilent 6890N Gas Chromatograph (GC) and a 5973N Mass Spectometer (MS). The quartz probe, used for GC sampling, had a 200 μm diameter orifice and was positioned 130 μm from the foil surface. The probe was vertically translated along the centerline of the Pt coupon in order to provide concentration profiles. The GC/MS apparatus was equipped with an Agilent HP-PLOT Molesieve column in order to separate oxygen, nitrogen, methane, and carbon monoxide. In order to measure carbon dioxide, an Agilent HP-PLOT Q column was used. The calibration of the GC/MS involved relating the areas of species peaks with the known concentrations of a controlled sample gas.

Surface temperature measurements were made with a PV320L2 infrared imager of Electrophysics with a 50 mm germanium lens that was transparent in the wavelength range of 2-14 μm. Calibration of the infrared camera was performed in order to connect grayscale readings of the infrared camera to quantitative temperature measurements. K-type thermocouples were spot-welded to the catalytic foil surface for this purpose. The flat plate was then imaged and the resultant grayscale intensity at each thermocouple location was correlated to the thermocouple temperature measurement with a fourth order polynomial.

3. Results and Discussion

3a. Temperature measurements

The centerline surface temperature profiles are shown in Figure 2 based on streamwise distance from the leading edge of the catalytic plate. Temperature profiles are shown for $0.9<\varphi<1.43$ and incoming flow jet Reynolds numbers $76< Re_d < 101$. It is observed that the catalytic reaction occurs in three distinct phases. For values of the streamwise Reynolds number smaller than 15-20, a sharp increase in temperature is seen leading to a maximum at approximately Rex ~ 15, labeled as "phase I" in Fig.2. As the free-stream equivalence ratio increases from 0.9 to 1.43, the peak surface temperature increases by more than 350°C and notably peaks at significantly rich equivalence ratios. This is clearly not the case for larger scale reactors [16-18]. Also, these measured temperature profiles differ significantly with the ones of [11] that were acquired for significantly larger Reynolds numbers (in terms of both flow velocities and length scales of the involved hardware). In this the initial region where the temperature maximums are reached, both transverse gradients are on the same order as streamwise ones due to strong diffusional transport [19]. The maximum temperature is consequently determined by a combination of thermo-chemistry and the complicated local transport field. These effects are not seen with larger scale flows over catalytically combusting hardware [11, 16-17] where the effects of the leading edge are practically negligible compared to the overall size of the reactor.

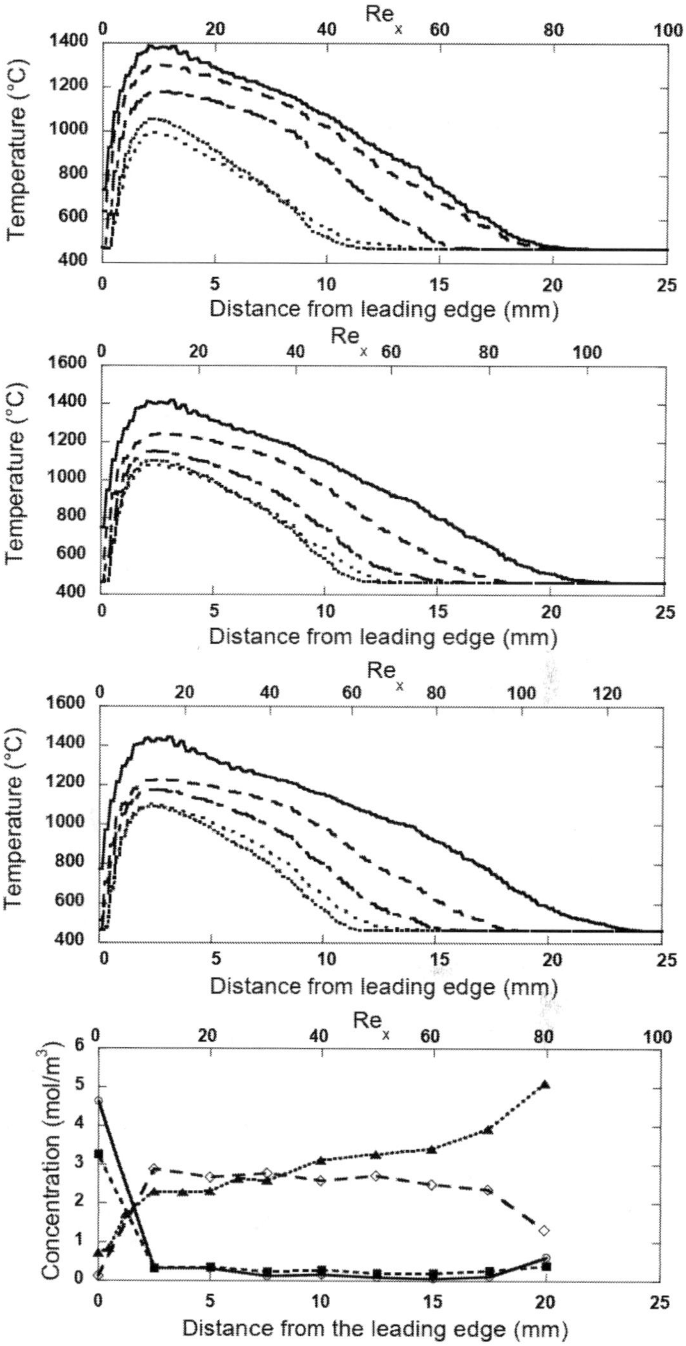

Figure 2. Temperature centerline profiles for equivalence ratios 1.43 (—), 1.25 (---), 1.11 (- —), 1.0 (- - -), and 0.9 (··) and constant jet Reynolds numbers a) 76, b) 88, and c) 101 and the concentration profiles for an equivalence ratio of 1.43 and a Reynolds number of 76 for CO (--◊--), CO_2 (·▲·), CH_4 (--■--), and O_2 (-○-)

Immediately following the temperature peak, a "phase II" consisting of a plateau-like region of nearly constant or slowly decreasing temperature can be observed. The temperatures experienced in this zone are comparable to the peak temperature and extend up to Re_x ~ 40-50. The decrease in temperature can be attributed to boundary layer development that reduces mass transfer to the surface and thus intensity of the catalytic reaction and consequently, surface temperature. The extent of "phase II" is strongly affected by equivalence ratio with longer plateaus associated with richer mixtures. In fact, for leaner mixtures and higher Re_d, the border between "phase II" and "phase III," where the reaction is gradually being completed, is not clear. In the third and final zone, labeled "phase III," temperature decreases more rapidly and ends with surface temperatures comparable to the free stream gas temperature (500 °C) signifying that no further exothermic reactions occur.

The importance of "phase I" in reactant conversion is further supported by species measurements, as shown in Fig. 2d, which reports the species profiles of methane, oxygen, carbon dioxide and monoxide for Re_d = 76 and φ = 1.43. Species measurements are discussed in detail in section 3, while Fig. 2d is provided in order to show the correspondence with the temperature distribution. A sharp decrease in the reactant concentrations and subsequent formation of products immediately following the leading edge indicate a strong reaction in this vicinity. Next, the concentrations of carbon monoxide and carbon dioxide flatten out, which corresponds to the temperature plateu of "phase II" described above. Finally the reaction is completed with the continued conversion to CO_2. Again contrary to what is reported for conditions that do not pertain to micro-combustion (larger Re_d, leaner mixtures – [11, 19]), reactant conversion is not gradual in the streamwise direction, but presents a sharp maximum at the vicinity of the leading edge.

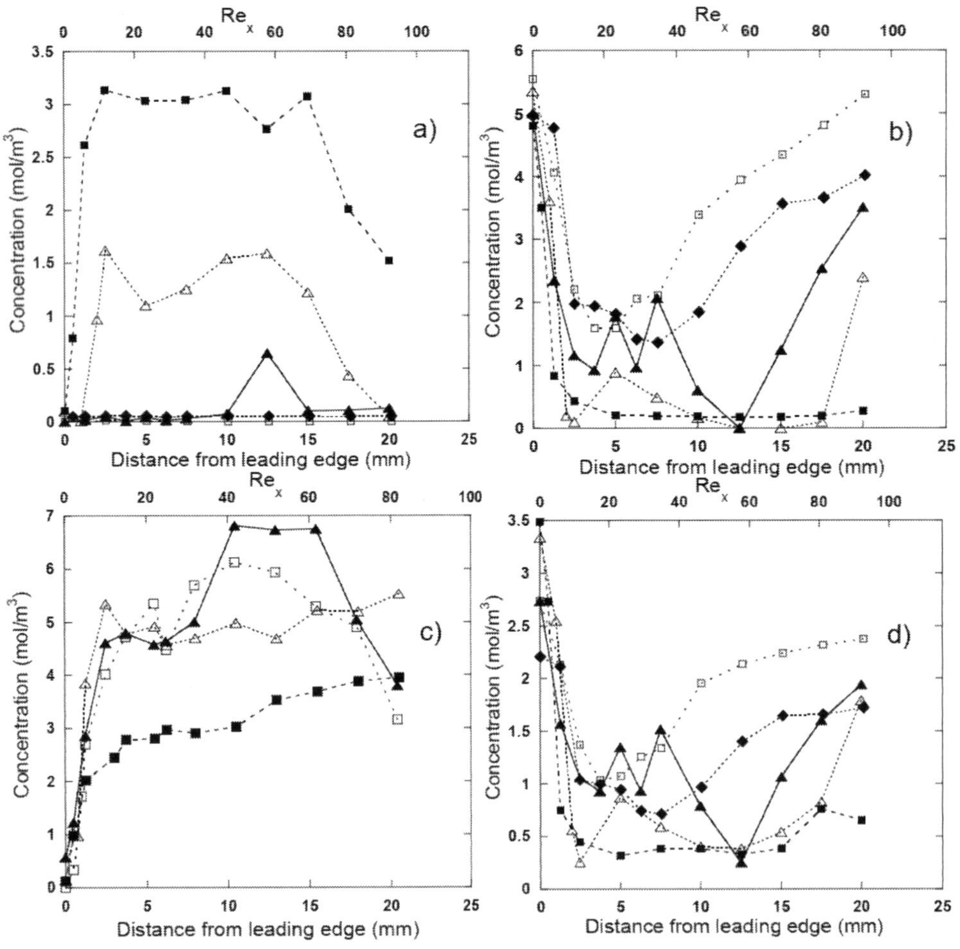

Figure 3. Species profiles for a constant jet Reynolds number of 88 for a) CO, b) O_2, c) CO_2 d) CH_4 for equivalence ratios of 1.43 (--■--), 1.25 (¨Δ¨), 1,11 (-▲-)1.0 (- -□- -), and 0.9 (¨♦¨)

3b. Species measurements – Equivalence ratio effects

Detailed results of species measurements are presented in Fig. 3, which presents centerline species concentration distributions of CH_4, O_2, CO_2 and CO as a function of φ for Re_d = 88. The structure of the profiles is in agreement with the temperature measurement findings that suggest a very intense reaction in the vicinity of the leading edge, and indicate a very fast depletion of the reactants and a corresponding fast generation of products for Re_x < 20.

In order to examine the distributions of the several species individually, one can start from the CO profile in Fig. 3a. Expectedly, it is shown that CO production is favored for richer mixtures, whereas CO is not detectable for lower values of φ. In all cases, there is a region of sharp increase in CO concentration in the vicinity of the leading edge. As the distance from the leading edge increases, the concentration of CO is diminishing as conversion to CO2 and completion of the reaction are approaching. It is also interesting to notice that the maximum of CO formation is observed closer to the leading edge as φ increases.

The effect of increasing equivalence ratio on the conversion of reactants is also confirmed by the results shown for CH_4 and O_2 in Figs. 3b and d. A drastic depletion of these reactants is shown for $Re_x < 20$. This is again in contrast to the gradual conversions shown in experimental and computational studies of large-scale burners [11,19]. Here, chemical reaction occurs mainly in the vicinity of the leading edge where the assumptions of the boundary layer theory ($Re_x >> 1$, streamwise gradients negligible compared to traverse ones) are not valid and the transport of reactants to the surface is intense. Figure 4 provides the location of 90% reactant conversion and 90% of maximum CO generation as a function of the equivalence ratio. The results indicate that these locations move closer to the leading edge for richer mixtures.

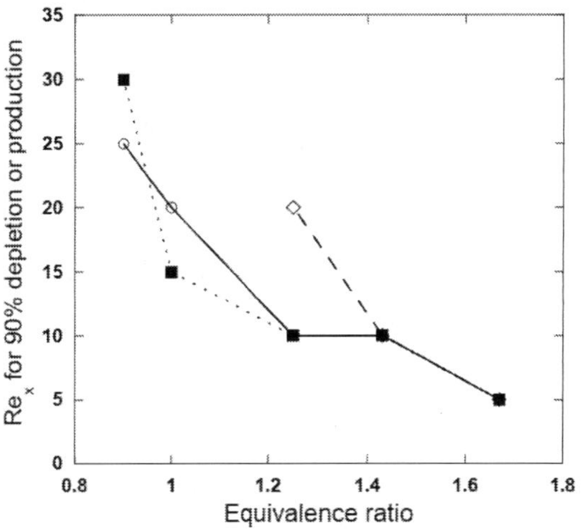

Figure 4. Distance from the leading edge at which the depletion of CH4 (--■--) and O2 (_○_) reaches 10% of their initial maxima and production of CO (_ _◊_ _) reaches 90% of its maximum value for a Re_d of 76 and different equivalence ratios.

A "replenishment" of reactants in the "phase II" region is also seen in the results of Fig. 3b and d. For mixtures close to stoichiometric, both reactants (CH_4 and O_2) reach a minimum, after which their concentrations increase again, recovering values close to those of the free-stream. It is also worth noting the significant qualitative similarity between CH_4 and O_2 distributions in this context. As distance from the leading edge increases, the reactants are depleted more slowly and as the reaction enters "phase III" where it gradually approaches extinction, diffusion of fresh reactants from the free stream leads to an increase in their concentration. This is an important finding in terms of micro-reactor design in that it shows that after a fast "depletion" in the vicinity of the leading edge, the catalytically coated plate does not equally effectively convert the reactants for distances greater than $Re_x \sim 20$. An ideal mode of operation in this context would be continuous boundary layer interruption which would cause very efficient completion of the reactants. This finding is supported by the results of [7] where a practical catalytic micro-combustor of impressively small length was demonstrated that operated through continuous boundary layer interruption with catalytically coated wires.

A phenomenon similar to reactant "replenishment" is shown in the CO_2 distribution in Fig. 3c. Here, two qualitatively different behaviors can be observed. For rich mixtures, a continuous increase in CO_2 formation is observed as the distance from the leading edge increases, whereas for mixtures close to stoichiometric, the CO_2 concentration profile exhibits a maximum. Also as described above, intense CO_2 production is observed in the vicinity of the leading edge. The CO_2 results have to be interpreted on the basis of the CO results of Fig. 3a, keeping in mind that the CO-to-CO_2 conversion is the last step of the hydrocarbon oxidation. Rich mixtures produce significant amounts of CO early in the process (Fig. 3a) that are consequently gradually converted to CO_2, accounting for the continued rise in CO_2 concentration. For leaner mixtures, CO production in the initial steps is less, and as a result,

CO_2 production follows the decrease in the surface temperature and the subsequent reactant "replenishment" indicated by Figs. 2, 3b and d.

4. Implications for micro-reactor design

The results of Figs. 3 and 4 indicate a strong depletion of reactants and formation of products in "phase I," i.e. near the vicinity of the leading edge. The reactants are consumed to a significant extent in streamwise distances that correspond to Re_x on the order of 10 and smaller. This is a region where the assumptions of boundary layer theory (as outlined e.g. in [20]) are not valid. In flows that develop over large length scales (and therefore large Re_x), transport in the vicinity of the leading edge can be neglected compared to the overall transport over a substantial streamwise length. The particular characteristic of micro-combustion flows is that most of the catalytic reaction that converts the reactants occurs in this region in this initial "phase I" region where classical boundary layer assumptions do not hold and both streamwise and traverse gradients are important.

If the flow is left to evolve and reach values of Re_x that are much larger than 10, a classical boundary layer is established as evidenced by the slowly varying temperature and species profiles in "phase II". However, form the point of view of micro-reactor design there is no need for this slowed reaction to occur. In fact, "micro"-combustor design should proceed with geometries that include continuous boundary layer interruption so that the reactants are converted very efficiently in "phase I" zones before "phase II" develops as is shown in Fig.5. This is the reason that the devices proposed by [11, 21] achieve full reactant conversion in lengths that are on the order of 1000 shorter than typical catalytic monoliths [16-18]. In such configurations, boundary layer interruptions were not designed but occurred by staggering catalytically coated wires in a disordered manner. Designed boundary layer interruption based on the results reported here can lead to reactor miniaturization without the pressure loss introduced by staggered wires that block the flow.

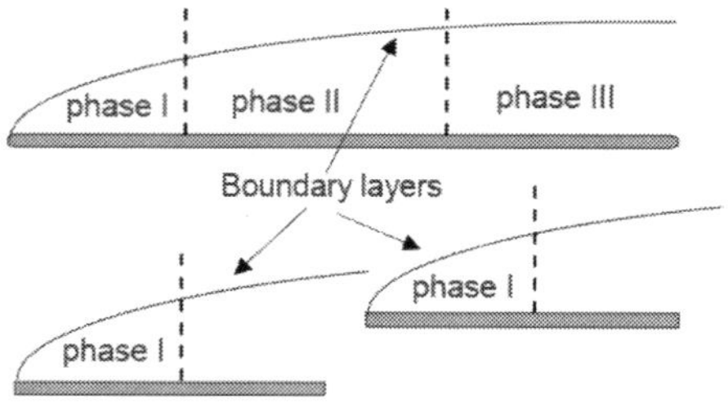

Figure 5. Schematic presenting the concept of boundary layer interruption and establishment of "phase I" zones.

5. Conclusions

For the intermediate Reynolds numbers that pertain to micro-combustion devices, flat plate boundary layer flows that are catalytically reacting develop in a manner significantly different from the boundary layer flows in large-scale catalytic reactors. The main characteristic of these flows is that intermediate Reynolds numbers cause the catalytic reactions to occur in the vicinity of the leading edge, in a region of the flow where the classical boundary layer assumptions do not hold. In particular, in the vicinity of the leading edge, i.e. for Re_x smaller than an order of 10, fast depletion of the reactants and fast formation of carbon oxides was observed, with simultaneous development of high surface temperatures. The equivalence ratio has a strong effect on both the maximum temperature reached and the intensity of the depletion of reactants. The location of 90% reactant conversion shifts closer to the leading edge with increasing equivalence ratio. As the Re_x increases, a second phase takes place where classical boundary layers develop and species and temperature profiles vary less drastically. For equivalence ratios in the vicinity of stoichiometric,"reactant replenishment" was observed, i.e. a gradual increase of the reactant concentration near the catalytic plate because of diffusion from the free-stream. The catalytic combustion is completed in a third phase where the surface temperature is decreased and the combustion gradually extinguishes. The intense surface reaction in the vicinity of the leading edge can be exploited in order to miniaturize micro-burners through carefully tailored boundary-layer interruption arrangements.

6. Acknowledgements

The authors would like to gratefully acknowledge the support of NSF under grant CTS 04-48968CAR (Dr. Phillip Westmoreland, contract monitor).

7. Nomenclature

d: tube exit diameter
l: length of potential core
Re_d: Reynolds number based on tube exit diameter (jet Reynolds number)
Re_x: Reynolds number based on the distance from the leading edge of the foil
x: streamwise direction
φ: equivalence ratio

8. References

1. D.C. Kyritsis, S. Roychoudhury, C. Mc Enally, L.D. Pfefferle and A. Gomez, Exp. Fl. & Therm. Sci., 28 (2004), 97-104.
2. A.C. Fernandez-Pello , Proc. Combust. Inst., 29 (2002) 883-889.
3. D.C. Kyritsis , B. Coroton, F. Faure, S. Roychoudhury and A. Gomez, Comb. & Flame, 139 (2004) 77-89.
4. J. Vican , B.F. Gajdeczko, F.L. Dryer, D.L. Milius , I.A. Aksay , and R.A. Yetter , Proc. Comb. Inst., 29 (2002) 909-916.
5. W.A. Sirignano, T.K. Pham, and D. Dunn-Rankin, Proc. Comb. Inst., 29 (2002), 925-931.
6. J.M. Ahn , C. Eastwood, L. Sitzki, P.D. Ronney, Proc. Comb. Inst., 30 (2005), 2463-2472.
7. A. Gomez, J. Berry, S. Roychoudhury , J. Huth, "Palm power: Using combustion at small scales and a free piston Stirling engine to replace batteries", presented at the 8th International Conference on Energy for a Clean Environment, Lisbon, Portugal, June 27-30, 2005
8. D.G. Norton and D.G. Vlachos, Chem. Eng. Sci. 58 (2003), 4871-4882.
9. C. Miesse , R.I. Masel , M. Short, M.A. Shannon, Comb. Theory & Modeling, 9 (2005), 77-92.
10. K. Maruta, T. Kataoka, N. Kim, S. Minaev, R. Fursenko, Proc. Comb. Inst. 30 (2005), 2429-2436.
11. C. Appel, J. Mantzaras, R. Schaeren, R. Bombach, A. Inauen, B. Kaeppel, B. Hemmerling, and A. Stampanoni, Comb. & Flame, 128 (2002), 340-368.
12. L.D. Pfefferle, T.A. Griffin, M. Winter, D.R. Crosley, and M.J. Dyer, Comb. & Flame 76 (1989), 325-338.
13. R.J. Cattolica and R.W. Schefer, Comb. Sci. & Tech. 30 (1983), 205-212.
14. N. Khadiya and N.G. Glumac, Comb. Sci. & Tech., 159 (2000), 147-167.
15. S. Akaike, and M. Nemoto, *ASME J. Fl. Eng.,* 110 (1988), 392-398.

16. R.A. Dalla Beta, J.C. Schlatter, M. Chow, D.K. Yee, and T. Shoji, Proc. Int. Workshop on Cat. Comb., (1994), 154-157.
17. S.T. Kalaczkowski, Trans. Inst. Chem. Eng., 73 (1995), 168-190.
18. D.B. Fant, G.S. Jackson, H. Karim, D.M. Newburry, P. Dutta, K.O. Smith, R.W. Dibble, J. Eng. Gas Turb. & Power, 122 (2000), 293-300.
19. L.L. Raja, R.J. Kee, O. Deutschmann, J. Warnatz, L.D. Schmidt, Catal. Today, 59 (2000), 47-60.
20. D.E. Rosner (2000) Transport Processes in Chemically Reacting Flows, Dover Publications Inc., Mineola NY.
21. D.C. Kyritsis, I. Guerrero-Arias, S. Roychoudhury, and A. Gomez, Proc. Combust. Instit., 29 (2003) 965-972.

Generation and Reactions of ortho-Bromophenyllithium without Benzyne Formation Using Microflow System.

Aiichiro Nagaki, Hirotsugu Usutani, Yutaka Tomida, Heejin Kim, Naofumi Takabayashi, Toshiki Nokami, Hideho Okamoto, and Jun-ichi Yoshida

Department of Synthetic Chemistry and Biological Chemistry, Graduate School of Engineering, Kyoto University, Kyoto 615-8510, Japan

1. Introduction

Chemical synthesis in microreactors has received significant research interests both from academia and industry.[1,2] Among characteristic features of microreactors, residence time control in the microchannel is especially important.[3] The length of time that the solution remains inside the reactor, i.e. residence time can be greatly reduced by reducing the microchannel length and increasing the flow speed. Based on residence time control, a highly unstable intermediate can be transferred to another location and is used in the follow-up reaction. Fast heat transfer, which minimizes local deviation of temperature, is another important feature of micro flow systems. We report here that these features are extremely useful in analyzing and controlling reactions involving ortho-bromophenyllitium as highly unstable intermediates.[4]

2. Trapping reaction with methanol of *o*-bromophenyllithium using microflow system

Lithiation of 1,2-dibromobenzene to generate ortho-bromophenyllithium is usually carried out at low temperatures, such as -100 °C in order to avoid benzyne formation (Scheme 1).[5,6]

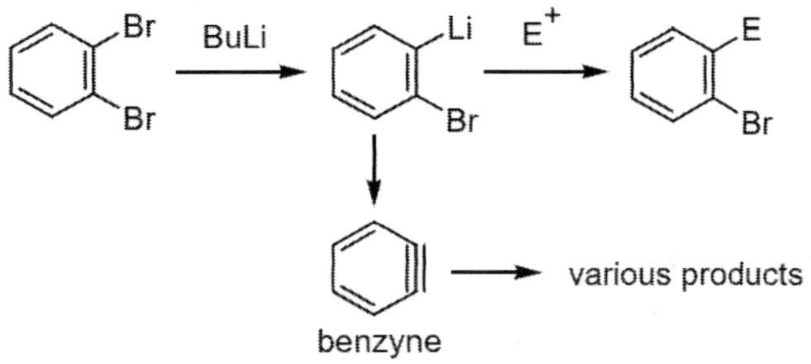

Scheme 1. Generation and reactions of *o*-bromophenyllithium

We began with optimizing the conditions of the Br-Li exchange reaction[7] using a microflow system involving two micromixers (**M1** and **M2**) and two microtube reactors (**R1** and **R2**) shown in Figure 1.[8] *o*-Bromophenyllithium was trapped by the reaction with methanol and the yield of bromobenzene was determined.

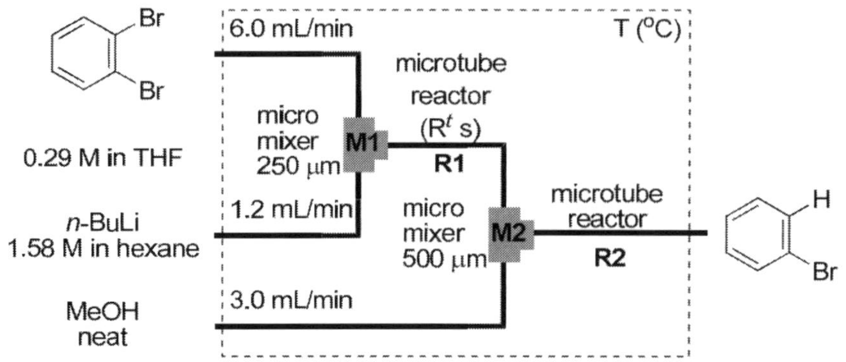

Figure 1. Microflow system for Br-Li exchange reaction.

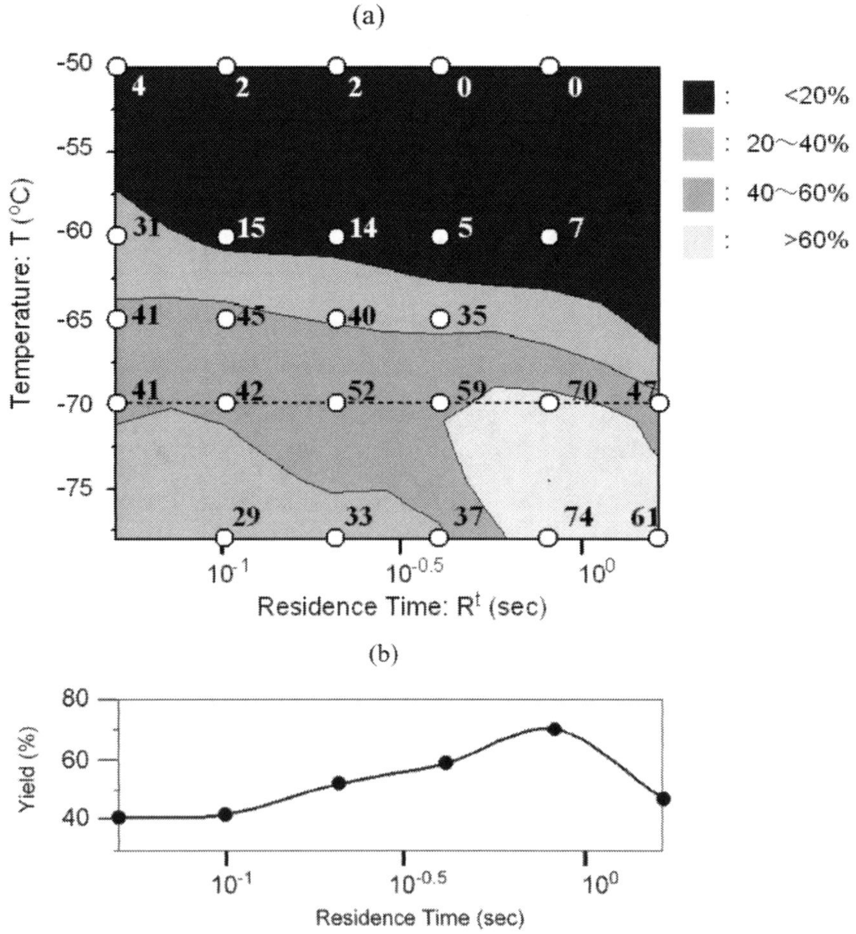

Figure 2. Effects of temperature and residence time on the yield of bromobenzene. (a) Counter plot with scatter overlay of the yields (%). (b) Crosssection at -70 °C.

The reactions were carried with varying the residence time and temperature in the microtube reactor **R1**. As profiled in Figure 2(a), the yield was very low at temperatures higher than -60 °C, presumably because of the decomposition of *o*-bromophenyllithium intermediate to benzyne. At lower temperatures, the yield increased with the decrease of the temperature because of slower benzyne formation. Further decrease of temperature, however, resulted in the decrease of the yield, because Br-Li exchange reaction did not complete at lower temperatures. The effect of the residence time is interesting as shown in Figure 2(b). At -70 °C, the yield increased with the residence time and became the maximum at residence time of 0.8 s. This phenomenon can be explained in terms of the progress of Br-Li exchange with the residence time. Further increase of the residence time caused decrease of the yield, presumably because of the benzyne formation. It is important to note that the present temperature-residence time profile is quite effective to unveil the features of Br-Li exchange reaction together with the stability of *o*-bromophenyllithium, although the efficiency of the mixing may also affect the residence time. Therefore, microflow systems serve as a powerful tool for mechanistic studies on reactions involving highly unstable intermediates.

3. Reaction with *o*-bromophenyllithium of electrophiles using the microflow system

Using the optimized conditions (-78 °C, 0.8 s), the reactions with various electrophiles were examined.

Table 1. Reactions of ortho-bromophenyllithium with electrophiles

electrophile	product	% yield	electrophile	product	% yield
MeOTf	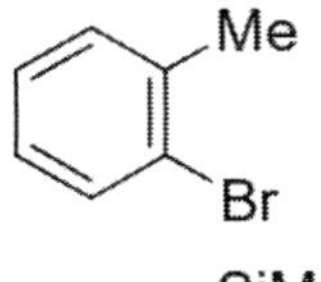	79[a]	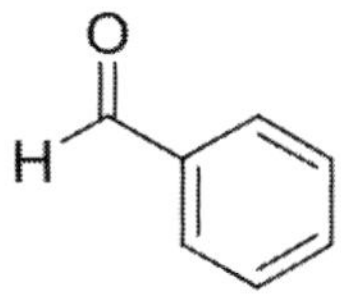	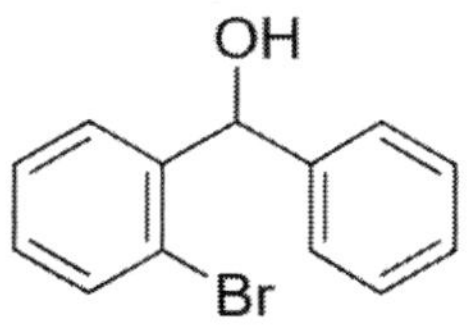	75[a]
HMe_2SiCl	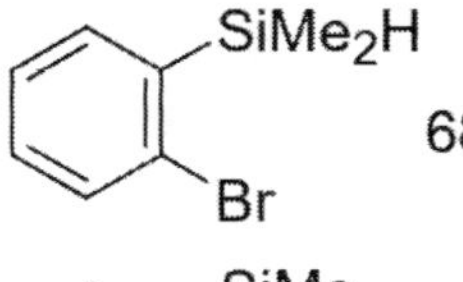	68[a]	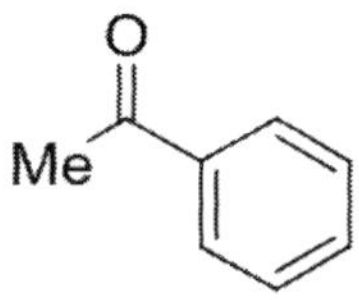	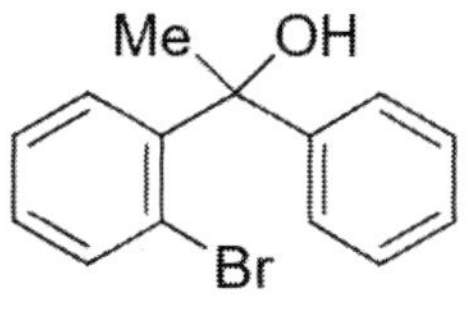	81[a]
Me_3SiOTf	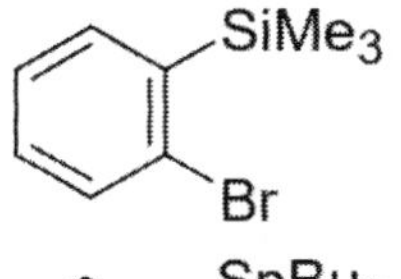	69[a]			
Bu_3SnCl	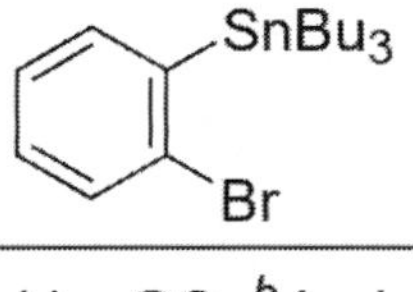	81[b]			

[a] Determined by GC. [b] Isolated yield

Methyl triflate, chlorodimethylsilane, trimethylsilyl triflate, chlorotributylstannane, aldehydes, and ketones were effective to give the corresponding substituted bromobenzene derivatives (Table 1).

4. Sequential introduction of two electrophiles using the microflow system

With successful Br-Li exchange reaction of *o*-dibromobenzene followed by the reaction with an electrophile in the microflow system in hand, we next examined the sequential introduction of two electrophiles into dibromobenzene using a micro flow system consisting of four T-shaped micromixers (**M1**, **M2**, **M3** and **M4**) and four microtube reactors (**R1**, **R2**, **R3**, and **R4**) as shown in Figure 3.

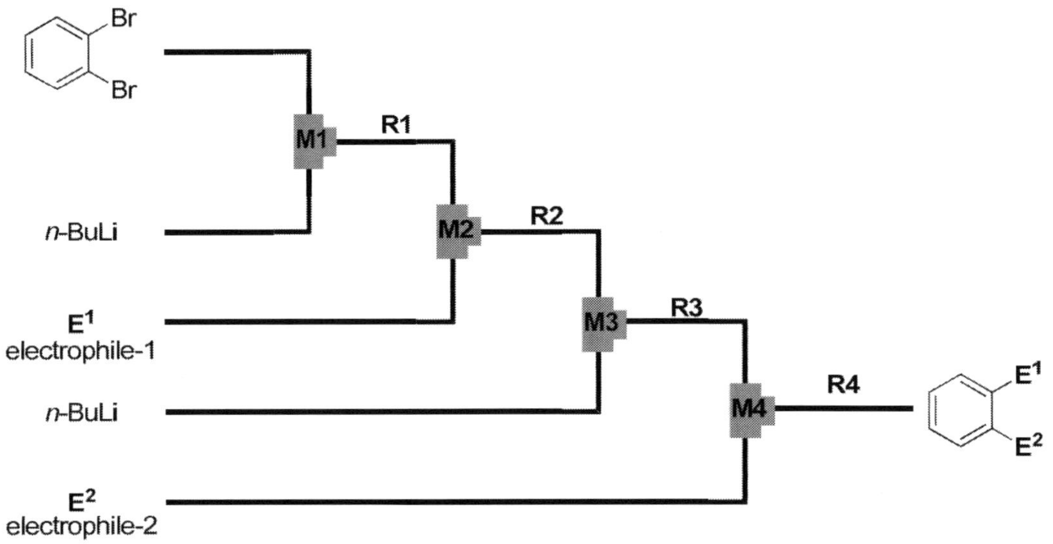

Figure 3. Sequential reactions of *o*-dibromobenzene

Therefore, sequential use of two bromine atoms in *o*-bromobenzene with Br-Li exchange reactions followed by electrophiles using the microflow system consisting of four micromixers and four microtube reactors was accomplished (Table 2).

Table 2. Sequential reactions of *o*-dibromobenzene with two electrophiles using the microflow system.

electrophile-1	electrophile-2	product	yield (%)[a]
MeOTf	Me_3SiCl	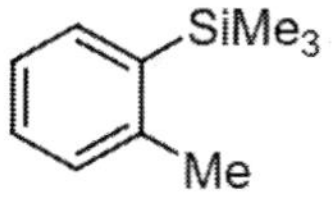	67
MeOTf	Bu_3SnCl	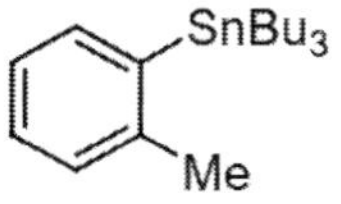	62
MeOTf	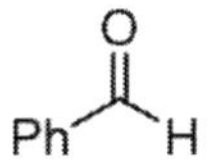	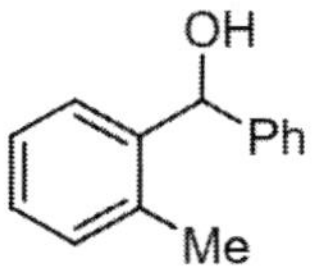	61
MeOTf	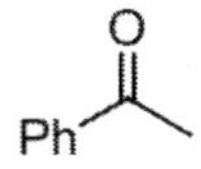	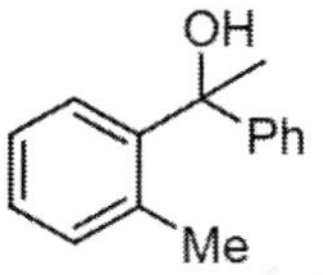	53
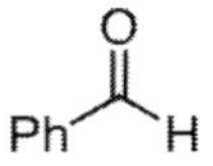	Me_3SiCl	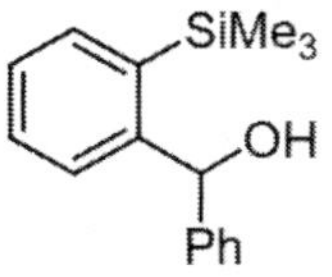	74
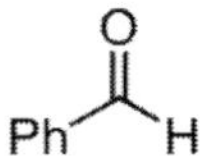	Bu_3SnCl	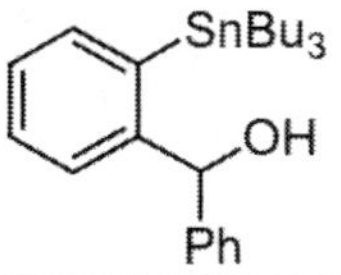	58

[a] Determined by GC

4. Conclusion

We have revealed that microflow systems serve as a powerful tool for mechanistic studies on reactions involving highly unstable intermediates such as *o*-bromophenyllithium, which is known to decompose to benzyne very quickly even at -78 °C. It is also noteworthy that an efficient way of controlling the generation and reactions of highly reactive intermediates can be accomplished by virtue of precise residence time control and temperature control inherent in microflow systems. The method adds a new dimension in organolithium chemistry.

5. Reference

(1) Reviews: For example (a) Fletcher, P. D. I.; Haswell, S. J.; Pombo-Villar, E.; Warrington, B. H.; Watts, P.; Wong, S. Y. F.; Zhang, X. *Tetrahedron* **2002**, *58*, 4735. (b) Jähnisch, K.; Hessel, V.; Löwe, H.; Baerns, M. *Angew. Chem. Int. Ed.* **2004**, *43*, 406. (c) Doku, G. N.; Verboom, W.; Reinhoudt, D. N.; van den Berg, A. *Tetrahedron* **2005**, *61*, 2733. (d) Yoshida, J. *Chem. Commun.* **2005**, 4509.

(2) Recent our examples: (a) Nagaki, A.; Togai, M.; Suga, S.; Aoki, N.; Mae, K.; Yoshida, J. *J. Am. Chem. Soc.* **2005**, *127*, 11666. (b) Suga, S.; Nagaki, A.; Yoshida, J. *Chem. Commun.* **2003**, 354. (c) Yoshida, J.; Nagaki, A.; Iwasaki, T.; Suga, S. *Chem. Eng. Technol.* **2005**, *28*, 259.
(3) (a) Kawaguchi, T.; Miyata, H.; Ataka, K.; Mae, K.; Yoshida, J. *Angew. Chem. Int. Ed.* **2005**, *44*, 2413. (b) Nagaki, A., Kawamura, K., Suga, S., Ando T., Sawamoto, M., Yoshida, J. *J. Am. Chem. Soc.* **2004**, *126*, 14702.
(4) (a) Nagaki, A.; Tomida, Y.; Usutani, H.; Kim, H.; Takabayashi, N.; Nokami , T.; Okamoto, H.; Yoshida, J. *Chem. Asian J.* **2007**, *2,* 1513. (b) Usutani, H.; Tomida, Y.; Nagaki, A.; Okamoto, H.; Nokami, T.; Yoshida, J. *J. Am. Chem. Soc.* **2007** *129*, 3046.
(5) (a) Chen, L. S.; Chen, G. J.; Tamborski, C. *J. Organometal. Chem.* **1980**, *193*, 283. (b) Prabhu, U. D. G.; Eapen, K. C.; Tamborski, C. *J. Org. Chem.* **1984**, *49*, 2792.
(6) *o*-Bromophenylmagnesium reagents, which are more stable than the corresponding lithium reagent, have been reported in the literature: (a) T. Ghosh, Hart, H.; *J. Org. Chem.* **1988**, *53*, 3555. (b) Boymond, L.; Rottländer, M.; Chiez, G.; Knochel, P. *Angew. Chem. Int. Ed.* **1998**, *37*, 1701. (c) Inoue, A.; Kitagawa, K.; Shinokubo, H.; Oshima, K. *J. Org. Chem.* **2001**, *66*, 4333. (d) Krasovskiy, A.; Knochel, P. *Angew. Chem. Int. Ed.* **2004**, *43*, 3333.
(7) Halogen-lithium exchange reactions of simple aromatic halides in micro systems: (a) Schwalbe, T.; Autze, V.; Hohmann, M.; Stirner, W. *Org. Process Res. Dev.* **2004**, *8*, 440. (b) Zhang, X.; Stefanick, S.; Villani, F. J. *Org. Process Res. Dev.* **2004**, *8*, 455.
(8) Although it is difficult to determine the mixing time for the T-shape mixers, it seems to be much less than 0.8 sec at the present flow rate, because most of the starting material (*o*-dibromobenzene) was consumed within this residence time.

A Continuous Flow Microwave-Assisted Microreactor System

Mitsuhiro Matsuzawa, Shigenori Togashi, and Shinji Hasebe
Hitachi, Ltd. Mechanical Engineering Research Laboratory,
832-2, Horiguchi, Hitachinaka, Ibaraki 312-0034, Japan

1. Introduction

Recently, chemical reaction processes with microwaves have attracted attention, and a lot of microwave effects in organic synthesis have been reported[1], e.g., reducing chemical reaction times from hours to minutes, reducing side reactions, and increasing yields. However, most chemical reactors using microwaves that have already been developed are batch-types, and there are few flow-types. Accordingly, the objective of this study is to develop a flow-type microwave-assisted chemical reactor, and moreover, we try to apply the reactor to microreaction technology. We designed the reactor using electromagnetic simulation and examined its energy absorption efficiency by conducting a water heating test. We also examined the yields and the reaction times of the Suzuki-Miyaura coupling reaction[2] under microwave heating and conventional (oil-bath) heating.

2. A continuous flow microwave-assisted microreactor system

Figure 1 shows the developed microwave-assisted microreactor system. It consists of five parts: a microwave generator, waveguide (which transfers microwaves), pump unit, control unit, and monitor. A quartz helical tube, in which a reagent solution flows and is heated by microwaves, vertically penetrates the waveguide. A stab tuner projects down from the upper surface of the waveguide, and it is possible to adjust the impedance in the waveguide by changing the tuner's length and position to optimize the microwave energy absorption efficiency of the reagents.

The waveguide was designed on the basis of analyses of electric field intensity and energy absorption efficiency by electromagnetic simulation. Figure 2 shows the electric field intensity distribution in the waveguide. When water was run through the quartz helical tube, the energy absorption efficiency of the water was 95% when the tuner's length and position were adjusted, whereas when there was no tuner, it was 63%.

3. Water heating test

Figure 3 shows the results of water heating test. The rising temperature (the difference between the temperatures of the inlet and the outlet of the reactor) was measured by changing the water flow rate through the quartz helical tube and the power of the microwave irradiation. The solid line shows the results obtained from the electromagnetic simulation (energy absorption by water is 95%), and the marks show the experimental measurements. The simulation corresponds with the measurements within ±10% accuracy.

4. Evaluation experiment

We examined Suzuki-Miyaura coupling reactions under microwave irradiation using the developed microwave-assisted microreactor. For comparison, we also examined the reactions under conventional (oil-bath) heating. The outlet temperature was kept at 95℃ in both methods. The reaction time under microwave heating reduced to about an eighth of that under oil-bath heating (Fig. 4).

5. Conclusion

We developed a prototype of a continuous flow microwave-assisted microreactor system. The waveguide was designed on the basis of electromagnetic simulations analyzing electric field intensity and energy absorption efficiency. Evaluation experiments showed that the simulation corresponded with the measurements within ±10% accuracy. We found that in the developed reactor, the reagent efficiently absorbed microwave energy and the energy absorption efficiency of the water was 95%. Moreover, by examining Suzuki-Miyaura coupling reactions, we found that the reaction time reduced to about an eighth of that when using conventional heating.

This research was financially supported by the "Development of Microspace and Nanospace Reaction Environment Technology for Functional Materials" project run by the New Energy and Industrial Technology Development Organization (NEDO), Japan.

References

1. C. O. Kappe (2004), "Controlled Microwave Heating in Modern Organic Synthesis," *Angew. Chem. Int. Ed*, 43, pp. 6250-6284.
2. N. Miyaura, A. Suzuki (1995), "Palladium-Catalyzed Cross-Coupling Reactions of Organoboron Compounds," *Chem. Rev.*, 95, pp. 2457-2483.

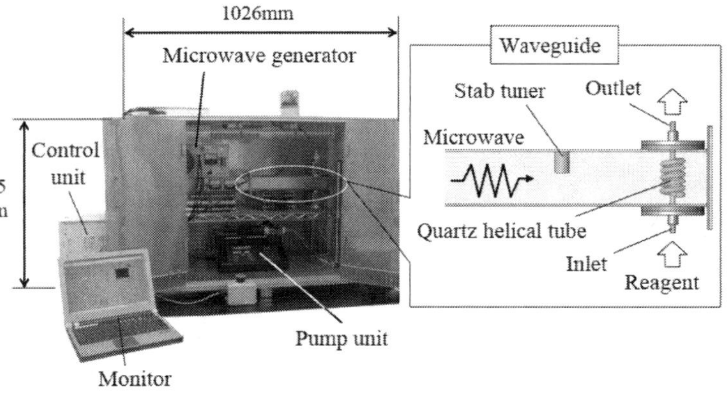

Fig.1 A continuous flow microwave-assisted microreactor system

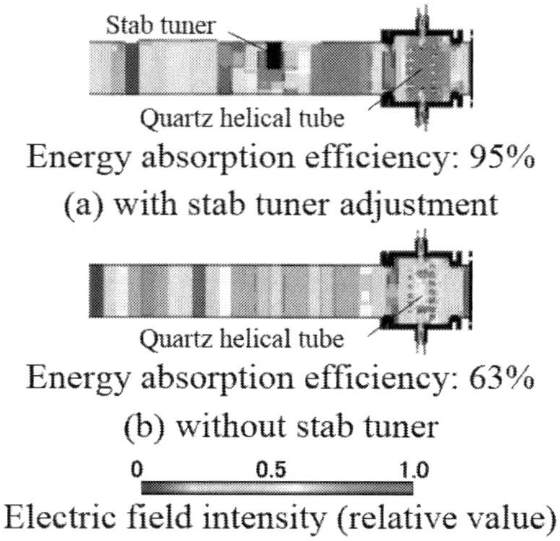

Fig.2 Electric field intensity distribution and energy absorption efficiency

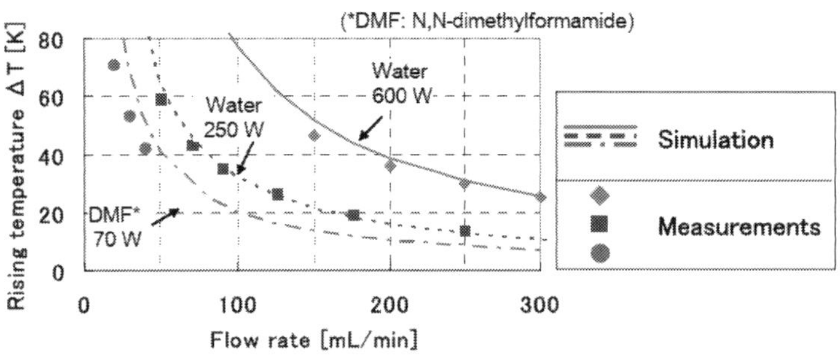

Fig. 3 Heating test results

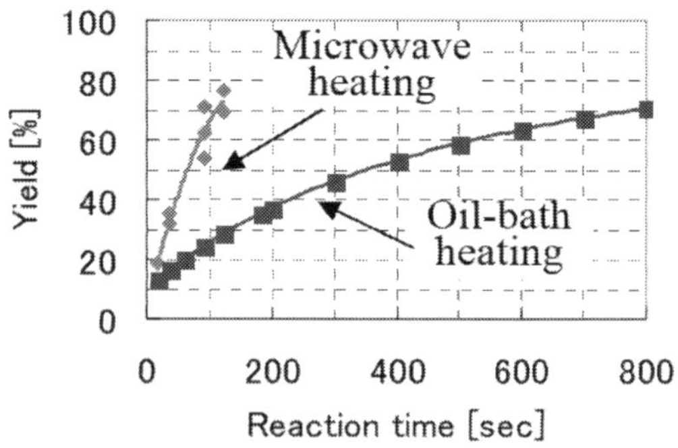

Fig. 4 Evaluation of yields and reaction time

Sequential Integration of Br-Li Exchange Reactions Using Microflow Systems

Yutaka Tomida, Aiichiro Nagaki, Hirotsugu Usutani, Heejin Kim, Naofumi Takabayashi, Toshiki Nokami, Hideho Okamoto, and Jun-ichi Yoshida

Department of Synthetic Chemistry and Biological Chemistry, Graduate School of Engineering, Kyoto University, Kyoto 615-8510, Japan

1. Introduction

Chemical synthesis in microreactors has received significant research interests both from academia and industry.[1,2] Among characteristic features of microreactors, residence time control in the microchannel is especially important.[3] The length of time that the solution remains inside the reactor, i.e. residence time can be greatly reduced by reducing the microchannel length and increasing the flow speed. Based on residence time control, a highly unstable intermediate can be transferred to another location and is used in the follow-up reaction. Fast heat transfer, which minimizes local deviation of temperature, is another important feature of microsystems. We here report our study on sequential introduction of two electrophiles into *p*-, *m*-, and *o*-dibromobenzenes based on Br-Li exchange reactions using micro flow systems (Scheme 1).[4]

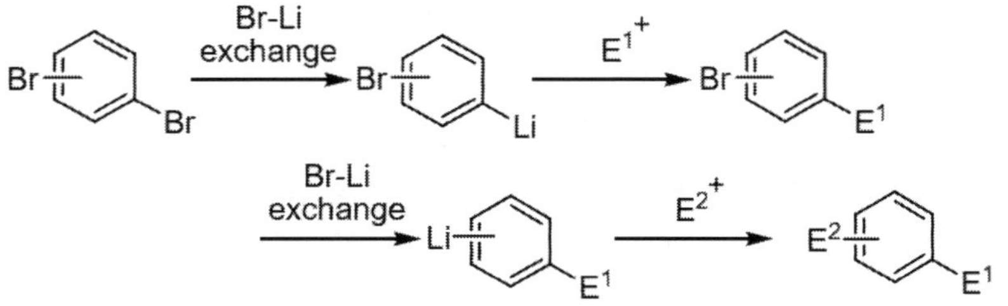

Scheme 1. Sequential introduction of two electrophiles into dibromobenzenes using Br-Li exchange reaction.

2. Br-Li exchange reaction of *p*-dibromobenzene followed by the reaction with electrophiles

First we examined on Br-Li exchange reaction[5] of *p*-dibromobenzene to generate *p*-bromophenyllithium, which is the first and key step of the desired transformation. It is well known that such reactions are usually carried out at -78 °C in laboratory synthesis, although this temperature is not convenient for industrial synthesis.

We employed a microflow system consisting of two T-shaped micromixers (**M1** and **M2**) and two microtube reactors (**R1** and **R2**) shown in Figure 1. A solution of *p*-dibromobenzene (0.29 M) in THF (flow rate: 6 mL min^{-1}, 1.74 mmol min^{-1}) and a hexane solution of *n*-BuLi (1.58 M) (flow rate: 1.2 mL min^{-1}, 1.9 mmol min^{-1}) were introduced to **M1** (ϕ = 250 μm) by using syringe pumping technique. The mixture was passed through **R1** (ϕ = 500 or 1000 μm, length = L cm, residence time = R^t sec) and was introduced to **M2** (ϕ = 500 μm), where methanol (neat, flow rate: 6 mL min^{-1}) was introduced. The resulting mixture was passed through **R2** (ϕ = 1000 μm, length = 50 cm, residence time = 1.79 sec). The temperature of the system was controlled by adjusting the bath temperature. The Br-Li exchange reaction and the quenching with MeOH were conducted at the same temperature. The residence time was adjusted by changing the length of the microtube reactor **R1** with a fixed flow rate.

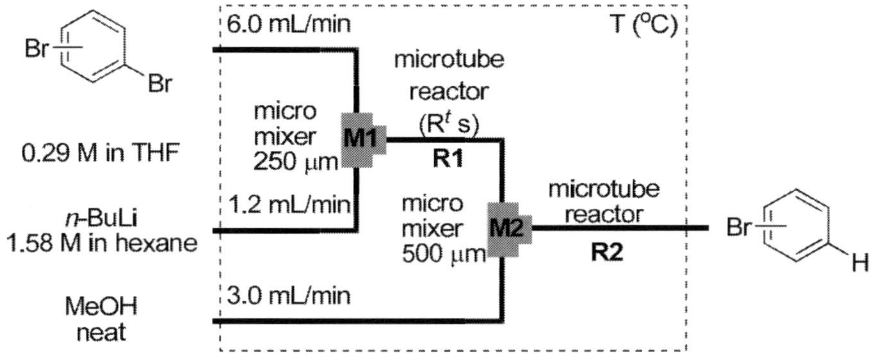

Figure 1. A microflow system for Br-Li exchange reaction of dibromobenzenes followed by the reaction with MeOH.

The results obtained with varying temperature (T) and residence time (R^t) in **R1** are summarized in Figure 2. It is interesting to note that bromobenzene was obtained in good yields above -48 °C. The reaction can be conducted even at 20 °C, demonstrating a significant advantage of microflow systems. Extremely fast heat transfer of the microflow system seems to be responsible for preventing the side reaction. In other words, Br-Li exchange reaction can be inherently conducted at 20 °C, although the reaction in macrobatch systems suffers from insufficient heat removal and over-cooling is necessary.

At -40 °C ~ -78 °C, the yield increased with the temperature. Presumably the Br-Li exchange reaction is slow in this range and temperature increase caused the increase of the reaction rate. In this range, the yield also increased with the residence time. Br-Li exchange reaction progressed with the increase of the residence time.

Under the optimized conditions, the reactions of *p*-bromophenyllthium with various electrophiles were examined using the microflow system at 20 °C. Iodomethane, chlorotrimethylsilane, chlorotributylstannane, benzaldehyde, and acetophenone were effective. The reactions gave the corresponding *p*-substituted bromobenzene derivatives at 20 °C in good yields as shown in Table 1.

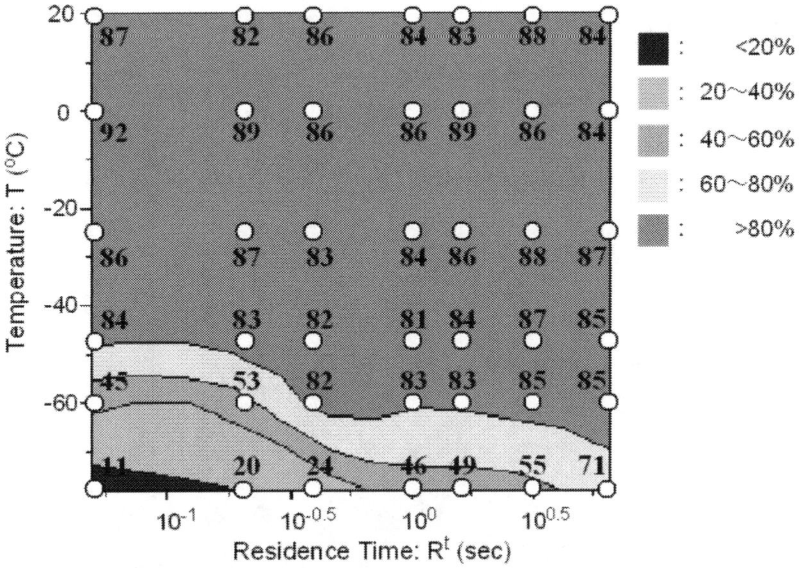

Figure 2. Effects of temperature and residence time on the yield of bromobenzene in Br-Li exchange reaction of *p*-dibromobenzene with *n*-BuLi using the microflow system.

Table 1. Br-Li exchange reaction of *p*-dibromobenzene with *n*-BuLi followed by the reaction with electrophiles using the microflow system.

electrophile	product	yield (%)[a]	electrophile	product	yield (%)[a]
MeI	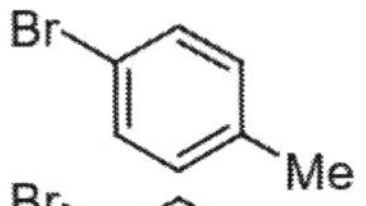	87	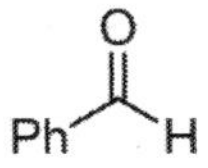	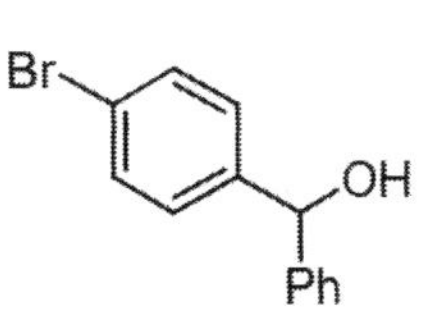	73
Me_3SiCl	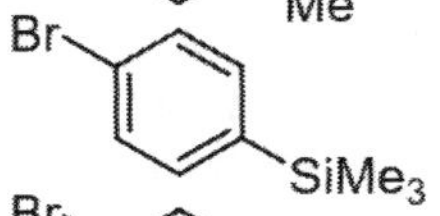	85			
Bu_3SnCl	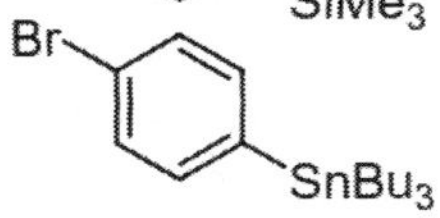	86	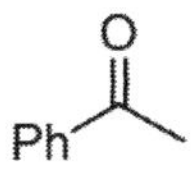	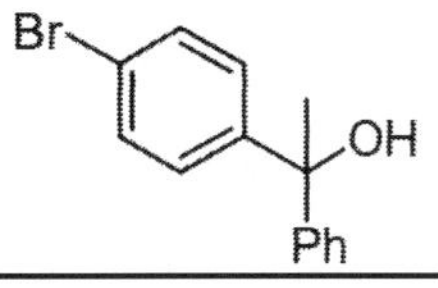	87

[a] Determined by GC

3. Br-Li exchange reaction of *m*-dibromobenzene followed by the reaction with electrophiles

Next, we examined on Br-Li exchange reaction of *m*-dibromobenzene to generate *m*-bromophenyllithium. The microflow reactions were conducted with a system shown in Figure 1. The results obtained with varying temperature (T) and residence time (R^t) in **R1** are summarized in Figure 3. It is interesting to note that bromobenzene was obtained in good yields above -48 °C. The reaction can be conducted at 20 °C. Extremely fast heat transfer of the microflow system seems to be responsible. At -40 °C ~ -78 °C, the yield increased with the temperature. Presumably the Br-Li exchange reaction is slow in this range and temperature increase caused the increase of the reaction rate. In this range, the yield also increased with the residence time. Br-Li exchange reaction progressed with the increase of the residence time.

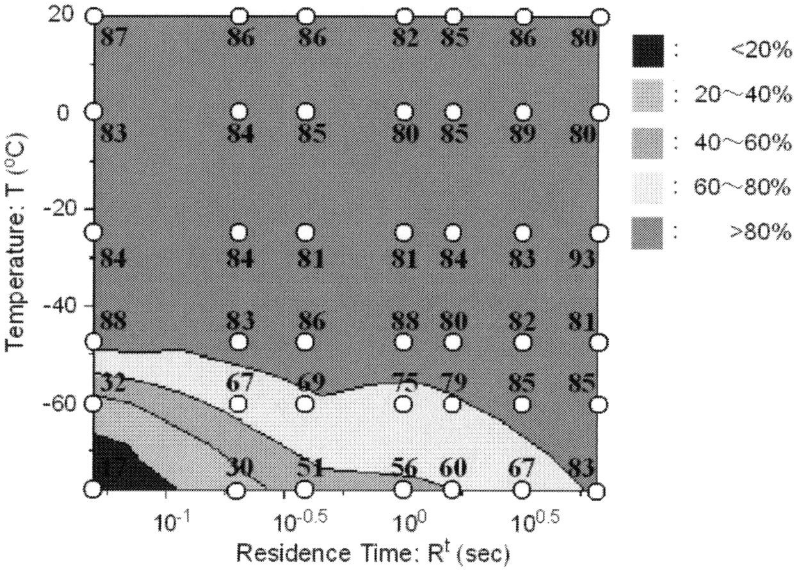

Figure 3. Effects of temperature and residence time on the yield of bromobenzene in Br-Li exchange reaction of *m*-dibromobenzene with *n*-BuLi using the microflow system.

These tendencies are quite similar to those for *p*-bromobenzene case. Therefore, it is reasonable to consider that *m*-bromophenyllithium and *p*-bromophenyllithium have similar thermal stability.

Table 2. Br-Li exchange reaction of *m*-dibromobenzene with *n*-BuLi followed by the reaction with electrophiles using the microflow system.

electrophile	product	yield (%)[a]	electrophile	product	yield (%)[a]
MeI	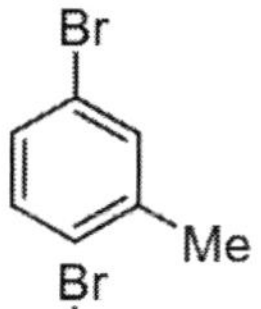	81	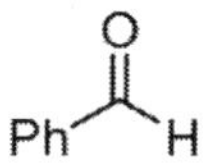	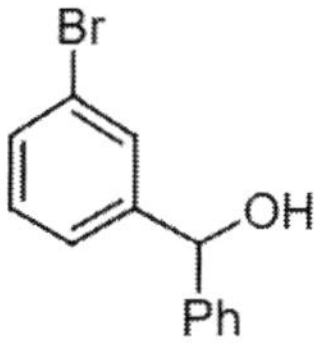	86
Me_3SiCl	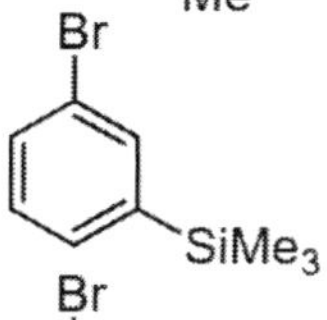	93			
Bu_3SnCl	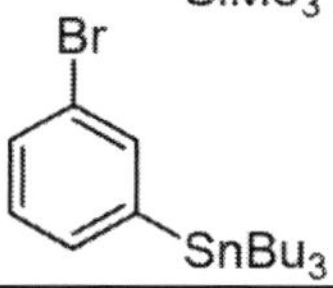	94	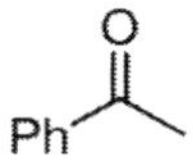	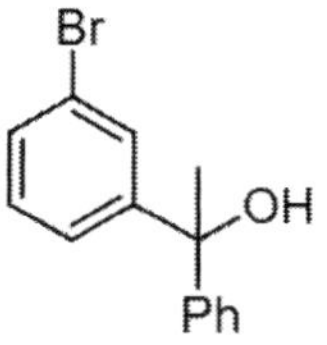	90

[a] Determined by GC

Under the optimized conditions, the reactions of *m*-bromophenyllithium with various electrophiles were examined in the microflow system. Iodomethane, chlorotrimethylsilane, chlorotributylstannane, benzaldehyde and acetophenone as well as methanol were effective and the reactions gave the corresponding *m*-substituted bromobenzene derivatives in good yields at 20 °C as shown in Table 2.

4. Br-Li exchange reaction of *o*-dibromobenzene followed by the reaction with electrophiles

The Br-Li exchange reaction of *o*-dibromobenzene[6,7] was successfully conducted in a microflow system at -78 °C without elimination of LiBr to form benzyne. The details will be described in a separate presentation. Using the optimized conditions, the reactions with various electrophiles gave the corresponding substituted bromobenzene derivatives in good yields.

5. Sequential introduction of two electrophiles into dibromobenzenes using the microflow system

With successful Br-Li exchange reaction of dibromobenzenes followed by the reaction with an electrophile in the microflow system in hand, we next examined the sequential introduction of two electrophiles into dibromobenzene using a microflow system consisting of four T-shaped micromixers (**M1**, **M2**, **M3** and **M4**) and four microtube reactors (**R1**, **R2**, **R3**, and **R4**) as shown in Figure 4.

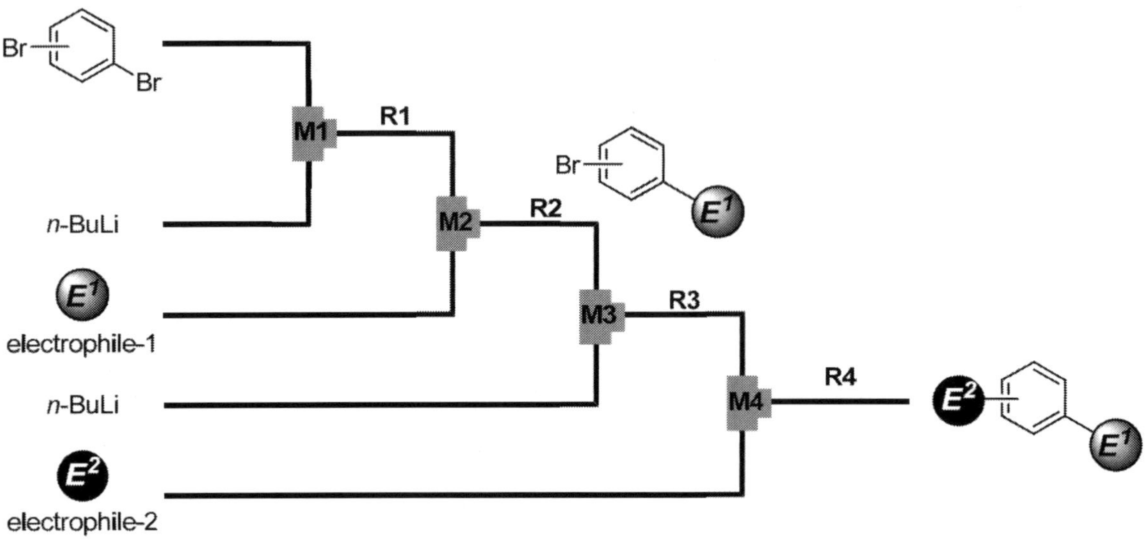

Figure 4. A microflow system for sequential introduction of two electrophiles into dibromobenzenes.

Therefore, sequential use of two bromine atoms in bromobenzenes with Br-Li exchange reactions followed by electrophiles using the microflow system consisting of four micromixers and four microtube reactors was accomplished (Figure 5).

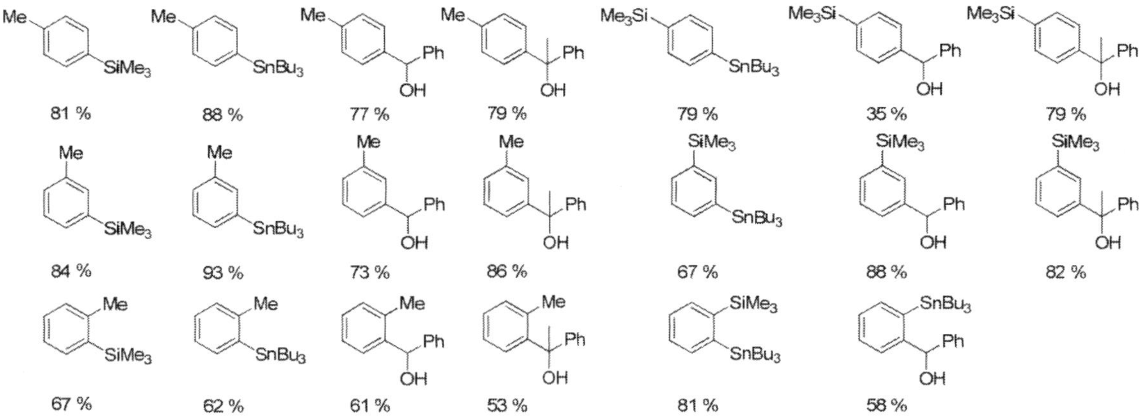

Figure 5. Sequential introduction of two electrophiles into dibromobenzenes using Br-Li exchange reaction.

6. Conclusion

In conclusion, sequential introduction of two electrophiles to *p*-, *m*-, and *o*-dibromobenzenes based on Br-Li exchange reactions has been accomplished using a microflow system consisting of four micromixers and four microtube reactors at much higher temperatures than those for conventional macro batch systems by virtue of residence time control and temperature control. The method serves as a straightforward and powerful access to disubstituted benzene derivatives. The results obtained in this study speak well for potential of microflow systems for the integration of chemical reactions involving highly reactive intermediates. It is hoped that numbering-up of the microflow system could be accomplished to increase the productivity and that the present method will be utilized in the industrial production of fine chemicals.

7. Reference

(1) Reviews: For example (a) Fletcher, P. D. I.; Haswell, S. J.; Pombo-Villar, E.; Warrington, B. H.; Watts, P.; Wong, S. Y. F.; Zhang, X. *Tetrahedron* **2002**, *58*, 4735. (b) Jähnisch, K.; Hessel, V.; Löwe, H.; Baerns, M. *Angew. Chem. Int. Ed.* **2004**, *43*, 406. (c) Doku, G. N.; Verboom, W.; Reinhoudt, D. N.; van den Berg, A. *Tetrahedron* **2005**, *61*, 2733. (d) Yoshida, J. *Chem. Commun.* **2005**, 4509.

(2) Recent our examples: (a) Nagaki, A.; Togai, M.; Suga, S.; Aoki, N.; Mae, K.; Yoshida, J. *J. Am. Chem. Soc.* **2005**, *127*, 11666. (b) Suga, S.; Nagaki, A.; Yoshida, J. *Chem. Commun.* **2003**, 354. (c) Yoshida, J.; Nagaki, A.; Iwasaki, T.; Suga, S. *Chem. Eng. Technol.* **2005**, *28*, 259.

(3) (a) Kawaguchi, T.; Miyata, H.; Ataka, K.; Mae, K.; Yoshida, J. *Angew. Chem. Int. Ed.* **2005**, *44*, 2413. (b) Nagaki, A., Kawamura, K., Suga, S., Ando T., Sawamoto, M., Yoshida, J. *J. Am. Chem. Soc.* **2004**, *126*, 14702.

(4) (a) Nagaki, A.; Tomida, Y.; Usutani, H.; Kim, H.; Takabayashi, N.; Nokami , T.; Okamoto, H.; Yoshida, J. *Chem. Asian J.* **2007**, *2,* 1513. (b) Usutani, H.; Tomida, Y.; Nagaki, A.; Okamoto, H.; Nokami, T.; Yoshida, J. *J. Am. Chem. Soc.* **2007** *129*, 3046.

(5) Halogen-lithium exchange reactions of simple aromatic halides in micro systems: (a) Schwalbe, T.; Autze, V.; Hohmann, M.; Stirner, W. *Org. Process Res. Dev.* **2004**, *8,* 440. (b) Zhang, X.; Stefanick, S.; Villani, F. J. *Org. Process Res. Dev.* **2004**, *8*, 455.

(6) (a) Chen, L. S.; Chen, G. J.; Tamborski, C. *J. Organometal. Chem.* **1980**, *193*, 283. (b) Prabhu, U. D. G.; Eapen, K. C.; Tamborski, C. *J. Org. Chem.* **1984**, *49*, 2792.

(7) *o*-Bromophenylmagnesium reagents, which are more stable than the corresponding lithium reagent, have been reported in the literature: (a) T. Ghosh, Hart, H.; *J. Org. Chem.* **1988**, *53*, 3555. (b) Boymond, L.; Rottländer, M.; Chiez, G.; Knochel, P. *Angew. Chem. Int. Ed.* **1998**, *37*, 1701. (c) Inoue, A.; Kitagawa, K.; Shinokubo, H.; Oshima, K. *J. Org. Chem.* **2001**, *66*, 4333. (d) Krasovskiy, A.; Knochel, P. *Angew. Chem. Int. Ed.* **2004**, *43*, 3333.

Optical characterization of viscosity-dependent residence time distribution in the static micromixer Statmix6

J. Michael Köhler [1], Beate Schleiff [1], Steffen Schneider [1], Dusan Boskovic [2], G. Alexander Gross [1]

1) Techn. University Imenau, Inst. f. Micro and Nanotechnologies, D-98693 Ilmenau, Germany
2) Fraunhofer Institute for Chemical Technologies, Pfinzthal, Germany

Abstract

The residence time behaviour of the static chip micro mixer Statmix6 (IPHT Jena) was characterized by a pulse trace experiment applying a micro photometric measurement with two flow-through detectors. The residence time distribution (RTD) shows a characteristic shift with viscosity and flow rate. The change in RTD function can not be explained by reconsidering the Reynolds number exclusively. At lower flow rates, the shift of RTD tailing with flow rate is nearly independent on viscosity. In a transition region, the RTD tailing is reduced strongly, but the characteristic Re number of transition depends on viscosity. The RTD behaviour of the micro mixer was compared with the behaviour of a PTFE tube (1 mm inner diameter) under same experimental conditions and with similar mean residence times (about 0.5 to 40 s). In the tube, a reduction of tailing with increasing flow rate was observed too. But, the change of RTD function with flow rate and their viscosity-dependence differs considerably of the functions found in the static micro mixer. The experimental results were interpreted by changes in the flow regime under different local convection mechanism inside the micro device and in the tube. Such changes in mechanisms seem to exist in case of the complex geometry of micro mixer as well as in case of the simple tube. But the characteristic transition regions in the shape of RTD-function are strongly influenced by the structure of device and by the viscosity of the liquid.

Keywords

micro reactor, micro mixer, residence time distribution, Reynolds number, viscosity flow rate

1. Introduction

The characterization of residence time distribution is one of the most important aspects for the optimization of single micro reactors and whole micro reaction arrangements. In contrast to conventional reactors neither the model of a continuous-flow stirred tank reactor nor the model of a plug flow reactor is applicable. Homogenous fluids show in micro tubes with constant cross section a nearly laminar flow regime. This results in wide residence time distributions and is connected with the typical large fluidic dispersion observed normally in micro fluidic systems [1].

The residence time behaviour becomes still complex due to complex channel geometries. The superposition of convection in the micro channels on the one hand with different conditions of diffusion due to different thicknesses of laminar streaming structures on the other hand leads to complex conditions for the trajectories of molecules through a micro reactor and can cause very different residence time distributions.

There has been several approaches for description of residence time behaviour of micro reactors. Although the theoretical basics for the description of fluid transport and diffusion are given by the Navier Stokes equations and Ficks laws. An analytical description is impossible due to the geometrical problems. Therefore, FEM simulations or semi empirical approximations must be applied for the description of residence time distributions under micro fluidic conditions [2, 3].

Branched micro channels represent a particular complex case of micro reactor geometry. The residence time behaviour of static micro mixers using a multi lamination strategy – inter digital structures or split-and -recombine-structures can not be approximated by the behaviour of a common micro channel with the same internal volume easily.

Here, the flow rate-dependent RTD of the static micro mixer Statmix 6 (IPHT Jena) is investigated and compared with the RTD of a static micro mixer of Microglas (Mainz) and with the RTD of a simple PTFE tube. The Statmix 6 device consists on a series of split-and-recombine-units and possesses an internal volume of about eight microliters [4]. The construction follows the principle of Si/glass/Si mixers developed by N. Schwesinger and coworkers [5], but the device is realized by a glass/Si/glass sandwich construction. The junctions for splitting and recombining the fluid streams are oriented in lateral and perpendicular direction alternatively in order to ensure optimal conditions for multi lamination. The residence time behaviour of such type of micro mixers is of particular interest in case of the application of two- and three-step static micro mixers working with the same principle. In these devices, liquids from three or four fluid inlets are mixed together in a “2+1", in a “2+1+1" or in a “2+2" order. Each mixing step is implemented by a series of 4 to 6 split-and-recombine units and the liquid is guided immediately from the last junction of the first mixing unit to the first junction of the following one. Short mean residence times take place due to very low internal volumes of the mixing units and due to avoiding parasitic channel volumes between the mixing units. Very short time intervals between the mixing steps can be realized if higher flow rates are applied. This is of importance for the product formation in fast two or three step chemical reactions, for example in the formation of colloidal solutions of binary or surface-functionalized metal nano particles [6, 7].

It was expected, that the distribution of residence times of the micro mixers is not only dependent on the flow rate but additionally influenced by the viscosity of the streaming liquid. Therefore, the flow rate effects on RTD were investigated with a series of liquids with stepwise varied viscosities.

2. Experimental

2.1. Microreactors and experimental arrangement

The both investigated micro mixers are distinguished by construction and internal volumes. The internal volume of the microglas mixers amounts 15 :L (including fluid ports) and is about two times larger than the internal volume of the Statmix 6 device. The microglas mixer uses the multi lamination principle by subdividing the inlet channels in to a series of 15 small channels [8]. Small lamellas of streaming liquid were achieved in the Statmix 6 device by a 8 step lamination structure. Details about the construction of the devices were described earlier [4, 9].

The measurement arrangement consists of a syringe pump for the rate-controlled fluid actuation, an injector for dye tracer solution using a capillary loop with precisely defined injection volume of 1 :L, two micro flow-through photometers and the microreactor (Fig. 1). The both photometers are used for the determination of tracer profile before and after the micro reactor.

The arrangement allowed a flow rate variation in a large range. Flow rates between 500 and 3000 :L/min were applied in the experiments reported here. RTDs for all flow rates and viscosities were calculated using two different models. The mean hydrodynamic residence time was in the range of 0.75 to 50 s. An aqueous solution of Cochenille red A (E 124, absorbance 2.5 at an optical path length of 1 mm) was used for tracing.

The parasitic volumes of tubings between the first measurement position and the fluid port of the mixing devices and between the fluid outlet and the second measurement position were about 6 :L. So, this volumes can not be neglected. But, it can be expected that the observed residence time behaviour is dominated by the reactors itselfs.

2.2. Microphotometric flow-through measurements and signal processing

The both micro flow-through photometers are identically in construction. They consist on a LED, an optical aperture, a detection unit and electrical interconnectors. The tube is fixed by mechanical meanings in the centre of the optical channels. The light path between the exit of LED and the entrance into the optical detector amounts about 3 mm only. This short distance ensures that the light is efficiently conducted through the liquid channel and the uptake of light from the environment is minimized. The size of the aperture is 50 :m. The mechanical construction of the device is adapted to tubes with an inner diameter of 1 mm and an outer diameter of 1/16 inch.

The whole device is mounted on the tube and protected against environment light by housing. The device can be operated with a minimal time constant of 0.2 ms. So, absorbance data can be taken with frequencies up to 10 kHz.

3. Results and discussions

The investigations address the influence of viscosity and flow rate on the residence time distribution of a split and recombine mixer. Therefore, the flow of water and water/glycerol mixtures (up to 50 weight percent) were studied in the flow rate range between 50 and 3000 :L/min. The residence time behaviour of the micro devices was characterized by the residence time distribution in dependence on normalized time. The normalized time is given by the quotient of measurement time and the hydrodynamic residence time. So, experiment over a larger spectrum of flow rates can easily be compared.

The residence time distribution function (RTD function) was determined by analyzing the change of the tracer profile between the first and the second photometer. The RTD function of the fluid system between the first and the second measurement position (photometers) can be derived from the both tracer profiles by use of a dispersion model [3]. The obtained RTD describe the fluidic behaviour between the both photometers despite a certain fluidic dispersion between the injection position of the tracer and the position of the first photometer.

The RTD function of the static micro mixer STATMIX 6 shows a clear dependence on flow rate. At lower flow rates (50- 400 :L/min) the RTD of a water/glycerol mixture containing 10% glycerol is asymmetric, resulting in an early maximum and a significant RTD tailing (Fig. 2a). At higher flow rates (800 :L/min and above), the RTD more symmetrically and the tailing is considerably reduced. The differences between the RTD at higher and lower flow rates were reduced if the viscosity was enhanced (Fig. 2b).

The transition between both types of RTD is well reflected by the dependence of the RTD tailing on flow rate. We use a tailing factor describing the normalized time upto 99% of the integral of the RTD. The tailing is slightly reduced with increasing flow rates in the lower flow rate range, and followed by a steep decrease at enhanced flow rates (Fig. 3a). This behaviour was qualitatively independent on the viscosity, but the steep decrease shifts to higher flow rates if the viscosity was enhanced. This behaviour reflects a dependence of transport mechanism on Reynolds number. This conclusion is correct for lower and mediate viscosities (0 - 30 % glycerol in water). At higher viscosities (40 and 50 % glycerol) the transition between the both types of RTD functions was observed at lower Reynolds numbers as expected (Fig. 3b).

The static mixer of microglas showed a more complex residence time behaviour: On the one hand, there have been found two types of RTD function analogously to the IPHT mixer. On the other hand, there is a significant non-monotonous dependence of RTD function on flow rate (fig. 4). At lower viscosities (0 - 20% glycerol) , this non-monotonous behaviour is reflected by the appearance of a maximum in the RTD tailing factor in front of the steep decrease with increasing flow rate (fig. 5a). The transition between the more asymmetric and the more symmetric RTD (higher and lower tailing) is nearly independent on the viscosity in the investigated concentration range. The dominating role of the flow rate and the nearly neglectable influence of the viscosity for this transition is clearly reflected by the comparison between the flow-rate dependency and the Reynolds number dependency of the tailing factor (Fig. 5b). A non-monotonous characteristics of the RTD function was explained by N. Kockmann by flow-rate-dependent changes in the structures of the streaming fields inside micro fluidic channels [10].

This changes and the complex residence time distribution are not originally caused by specific complex structures of mixing devices. They were qualitatively observed in different micro reactors [11]. They were also observed in more simple fluidic systems like a PTFE tube with an ideal circle-shaped cross section (Fig. 6). This fact confirms the assumption that not the complex fluidic structure of the micro mixers is responsible for qualitative changes in RTD functions and non-monotonous dependencies of tailing on flow rate but the spontaneous change of transport mechanisms inside a single micro channel.

4. Summarizing conclusions

The RTD behaviour of the static micro mixer STATMIX6 (IPHT) and of the microglas mixer can be characterized in detail by a micro fluidic arrangement with a tracer injection unit and two fast measuring micro photometers in dependence on flow rate and viscosity. Both devices show a change in the RTD function in dependence on flow rate. Higher symmetry in RTD functions were found at higher flow rates, larger RTD tailing at lower flow rates. The influence of flow rate and viscosity can be described by the Reynolds number in case of the IPHT static micro mixer, whereas a different behaviour at equal Reynolds numbers was found in case of the microglas mixer and in case of a simple PTFE tube. The investigations show that viscosities as well as flow rates must be reconsidered if the residence time behaviour of a micro flow channels or micro mixing devices should be described adequately.

Acknowledgement

We like to thank J. Albert, G. Mayer, M. Sossna and Th. Henkel (IPHT Jena) for preparation of micro mixers and for discussions. Financial support of the federal ministry for education and research (project Serizell) is gratefully acknowledged.

References

[1] N. Kockmann. in: Micro process engineering (Wiley-VCH, Weinheim 2006, ed. by N. Kockmann), 71
[2] G. Goranovic, H. Bruus: Simulations in Microfluidics, in: Microsystem engineering of Lab-on-a-chip devices (Wiley-VCH, Weinheim 2004, ed. by O. Geschke, H. Klank, P.

Telleman), 79
[3] D. Boskovic, S. Loebbecke: Chem. Eng. J. 135, Suppl. 1 (2008), S 138- 146
[4] G.A. Gross, T. Thelemann, S. Schneider, D. Boskovic, J.M. Köhler: Chemical Engineering Science, submitted
[5] T. Kirner, J. Albert, M. Günther, G. Mayer, K. Reinhäckel, J.M. Köhler: "Static micro mixers for modular chip reactor arrangements in two step reactions and photochemical activated processes", Chem. Engn. J. 101 (2004), 65
[6] N. Schwesinger, T. Frank, H. Wurmus:J. Micromech. Microengin. 6,1, (1996), 99-102
[7] J.M. Köhler, M. Held, U. Hübner, J. Wagner: Formation of Au/Ag nanoparticles in a two step micro flow-through process, Chem. Eng. Technol. 30, 3 (2007), 347-354
[8] J.M. Köhler, H. Romanus, U. Hübner, J. Wagner: Formation of star-like and core-shell nanoparticles during two- and three-step preparation in batch and in microfluidic systems, J. of Nanomaterials (2007), Article ID 98134
[9] www.microglas.de
[10] N. Kockmann: Transport Phenomena in Micro Process Engineering (Springer, Berlin-Heidelberg 2008)
[11] G.A. Gross, S. Schneider, B. Schleiff, J.M. Köhler: Internatl. Conf. on Micro and Nanochannels ICNMM (Darmstadt 2008)

Figures

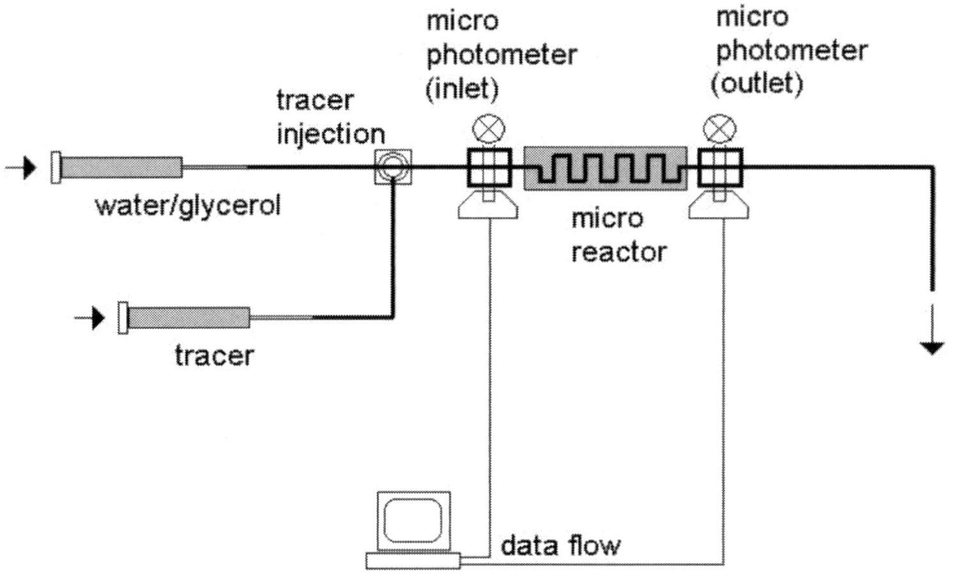

Fig. 1
Experimental set-up for residence time characterization

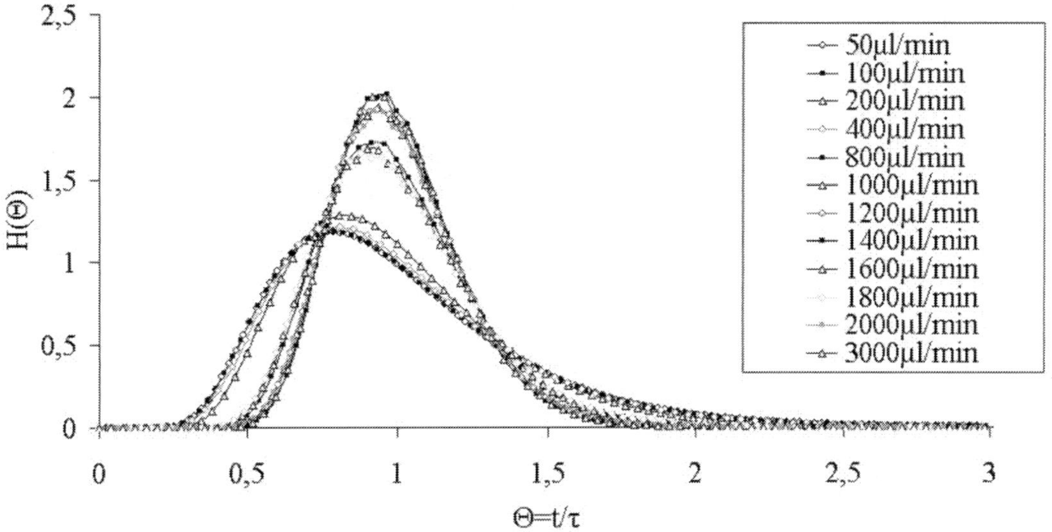

a

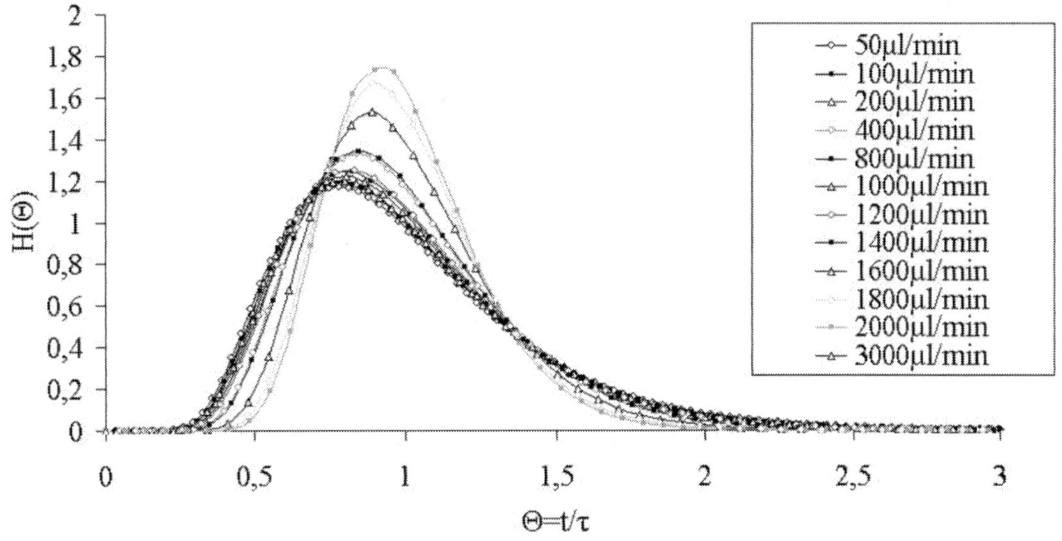

b

Fig. 2
Normalized RTD functions of the static micro mixer STATMIX6 (IPHT Jena) for different flow rates a) in case of 10% aqueous mixtures of glycerol, b) in case of 50% aqueous mixtures of glycerol

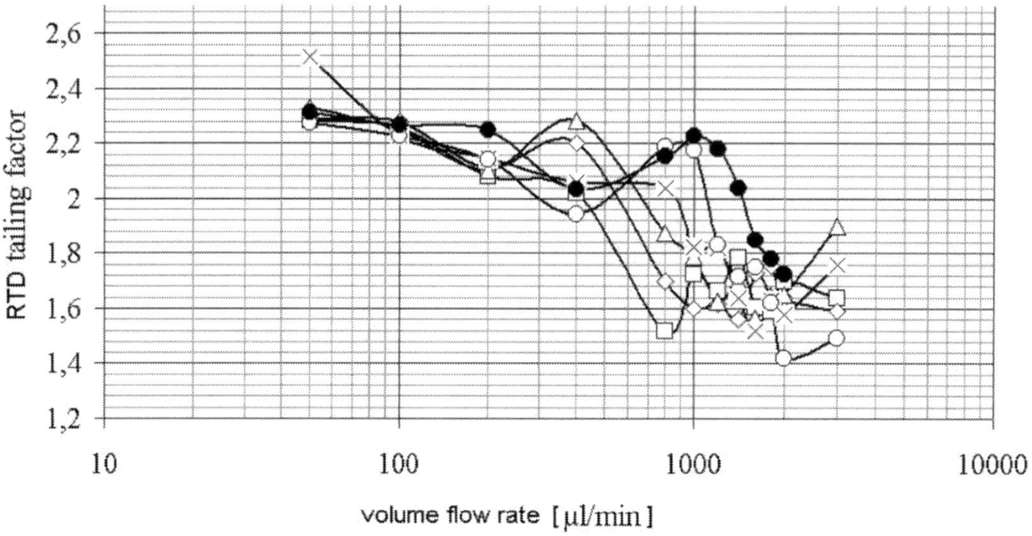

a

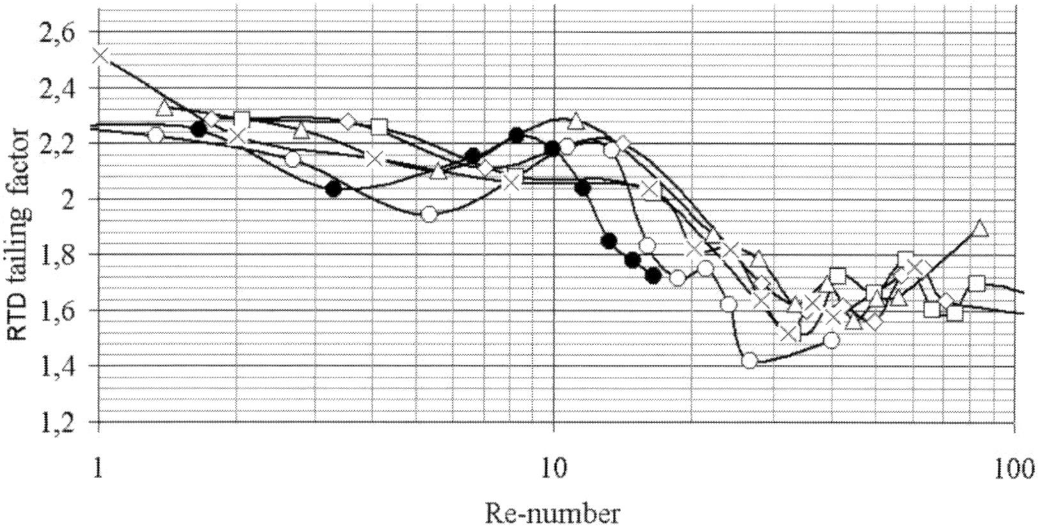

b

Fig. 3
RTD Tailing effect of different water/glycerol mixtures in the static micro mixer STATMIX6 (IPHT Jena): a) dependence on volume flow rate, b) dependence on Reynolds number

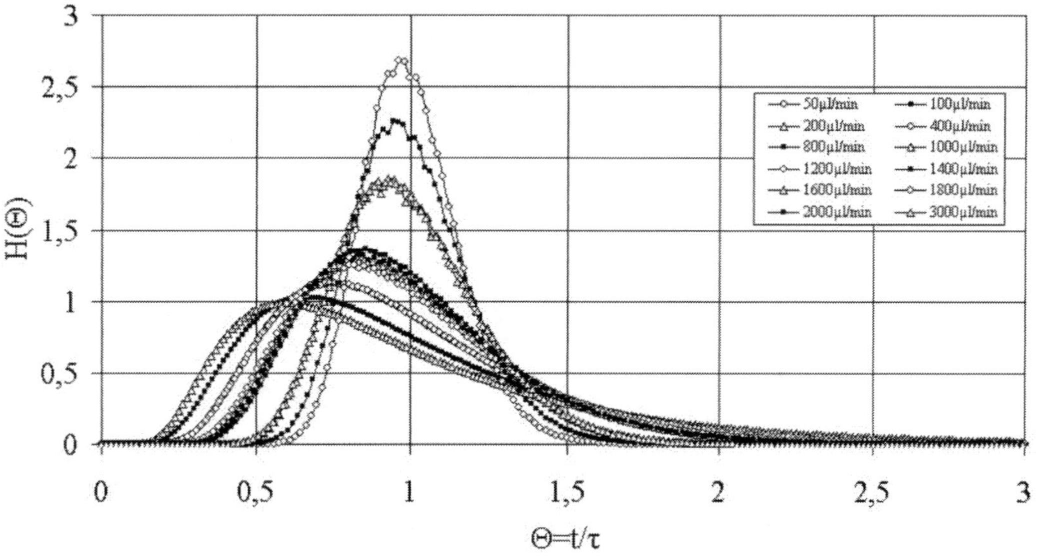

Fig. 4
Normalized RTD functions of a static micro mixer (microglas) for different flow rates (10% glycerol in water)

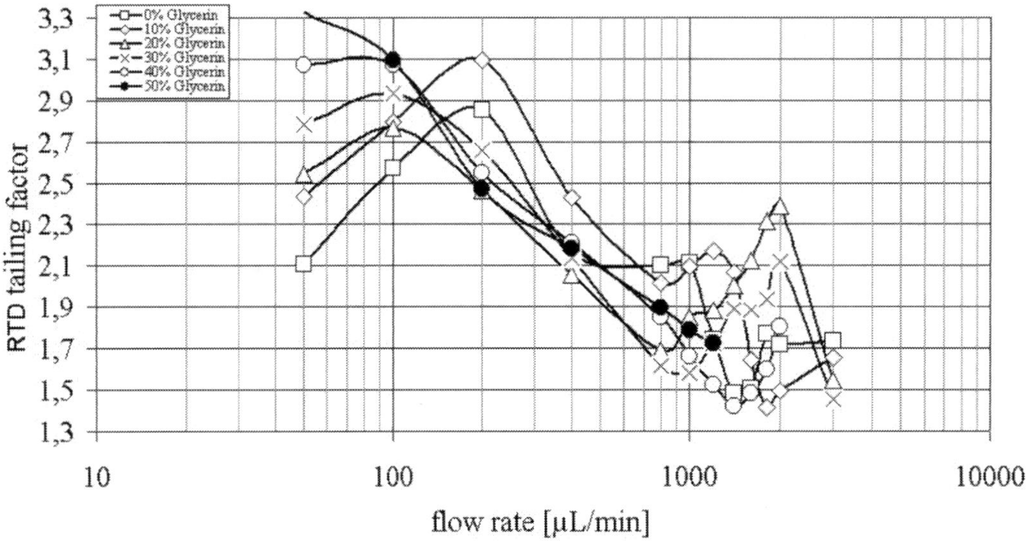

a

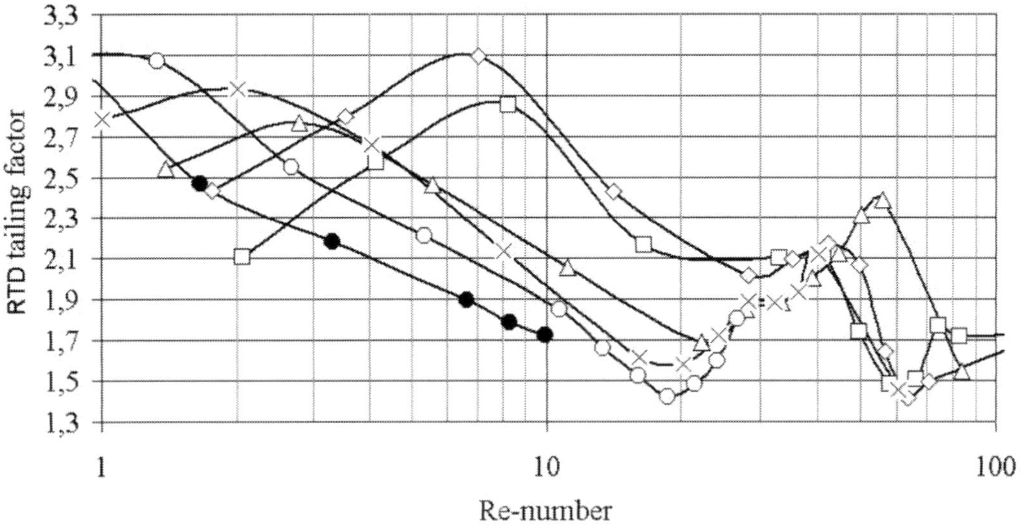

b

Fig. 5
RTD Tailing effect of different water/glycerol mixtures in the static micro mixer (micro glass: a) dependence on volume flow rate, b) dependence on Reynolds number

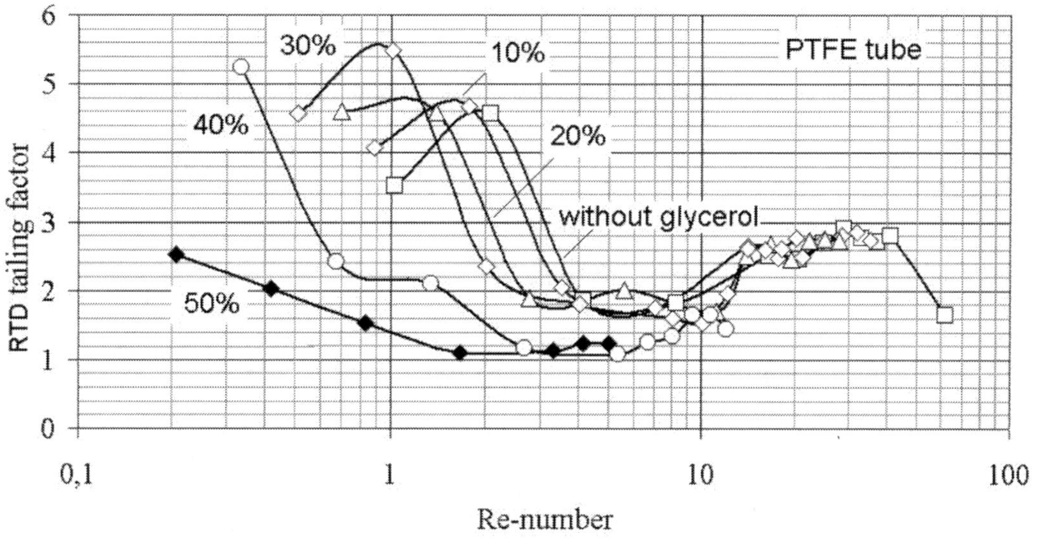

Fig. 6
RTD Tailing effect of different water/glycerol mixtures in a PTFE tube of 100 mm length (0.5 mm internal diameter) as reference object in the experimental set-up for RTD characterization in dependence on Reynolds number

Effect of surfactants on the formation of Au/Ag core/shell nanoparticles in micro segmented flow synthesis

A. Knauer, H. Romanus and J.M. Köhler
Techn. Univ. Ilmenau, D-98693 Ilmenau, Germany

Abstract

The influence of surfactants on the synthesis of core/shell nanoparticles in micro flow-through synthesis using the segmented flow principle was investigated. The synthesis was performed in a three step process. The carrier liquid (tetradecane) and the aqueous reactant solutions (reducing agent, surfactant, silver nitrate, colloidal Au nanoparticle solution ("seeds")) were fed into the mixer system by syringe pumps. Mixing was performed by T-type injectors and knot structures. The segmented flow principle supports a narrow distribution of residence times between the T-junctions and contributes to mixing by a shear force-induced segment-internal convection. The addition of surfactants reduces both the rate of nucleation and the particle growth rate. The effect of surfactant is dependent on the order of mixing of reactants and of the volume flow rate. In some cases, a bimodal distribution of colloidal products was observed. An optimization of flow rates leads to an improvement of nanoparticle size distribution. The obtained colloidal solutions were characterized by centrifugal sedimentation spectroscopy and by UV/VIS spectrophotometry. The size spectra and the ultramicroscopic images confirme the different quality of produced nanoparticles and show, that the optical properties of binary metal nanoparticles prepared by micro reaction technology can be modulated in a large frame in dependence on surfactant addition and micro fluidic conditions.

Keywords

nanoparticles, colloidal solution, micro reaction technology, slug-flow, fluid segments, flow rate

1. Introduction

Fast mass and heat transfer are two key advantages of micro reactors. This is of particular importance for synthesis procedures with fast partial processes and competitions between different reactions. Micro reactors can be used for homogenous as well as for heterogenous reactions. The implementation of continuous-flow processes and the precise control of reaction conditions enhance the attractivity of application of micro reactors for the synthesis of nanoparticles. So, dielectric inorganic [1-3], polymer [4-6], semiconductor [7-11] and metal nanoparticles [12-14] has been synthesized by micro reaction technology. Micro flow-through technique was applied for tuning semiconductor properties [15] and for controlling the size of gold and silver nanoparticles [16]. Microreaction technology was used for preparation of uniform gold nanorods and could be used for the generation of Au/Ag core/shell nanoparticles [17]. The absolute flow rate influences strongly the properties of colloidal product solutions in case of generation of Au/Ag nanoparticles by reduction of tetrachloroaurate and silver nitrate by ascorbic acid in a two step continuous-flow process [18].

Composed metal nanoparticles are of particular interest for different applications due to their specific surface states and optical and electronical as well as chemical properties. The microfluidic conditions influence the formation of AuAg nanoparticles due to the different rates of nucleation and particle growth for the different metals.

The quality of mixing and the rates of nucleation and particle growth are of particular importance for the nanoparticle diameter, the size distribution and the core/shell structure of nanoparticles obtained in two-step processes. It is well known that adsorbing polymer molecules and surfactants influences strongly the formation and growth rates of nanoparticles [19]. Surfactants influence also the formation and coalescence of micro fluidic slugs. Therefore, here the effect of surfactants on the two-step mixing process in formation of Au/Ag nanoparticles under micro flow-through conditions was studied.

2. Experimental

The experiments were carried out as two or three step micro flow-through processes. The segmented flow technique was applied for realization of a narrow distribution of residence times.between the first and the second and the second and the third mixing step. Tetradecane was used as inert, water-immiscible carrier liquid. Sodiumdodecylsulphate (SDS) was applied as an example of a frequently used surfactant. The formation of segments by injecting the first reactant solution and the addition of further reactants occured by T-injection elements. The mixing and residence loops were realized by knot structures of PTFE tube (inner diameter 0.5 mm).

Syringe pumps were applied for fluid actuation. They allowed flow rates up to 5 mL/min. The micro fluid arrangement is shown in fig. 1. The obtained colloidal solutions were characterized by UV/VIS spectrophotometry (Specord 200, Analytik Jena) and by centrifugal sedimentation spectrometry (CPS). The sedimentation spectrometry supplies a particle size distribution between about 5 and 500 nm diameter. Selected product samples have been aditionally investigated by SEM.

3. Results and discussions

Under the applied conditions, a lot of samples showed bimodal or polymodal product characteristics in the centrifuge spectra. It can be supposed that the appearance of bimodal or trimodal particle distributions is due to different populations of segments formed in the microfluidic system. The conentration of SDS is of significant importance for the ratio of different particle classes. In addition, the effect of enhanced surfactant concentration depends on the absolute flow rate, the flow rate ratios and the order of reactant addition in the micro flow through process.

In case of high flow rates a strong shift in the CPS peaks was found. A broad maximum at a particle diameter of 24 nm and two high shoulders reflect three main particle types at the low SDS concentration of 0.2 %.: one population with diameter above 30 nm and one population below 20 nm beside the 24-nm-population (fig. 2a). At high SDS concentration (5%) the main peak in CPS spectra is shifted to a particle diameter of 18 nm. A significant shoulder is only observed at the high particle diameter of about 50 nm (fig. 2b).

The particle yield of the 18-nm population (5% SDS) is considerable higher as the yield of 24-nm-population at low SDS concentration. Considering the fact that large particles seems to be overpresented in the mass-related CPS spectra, the experiment with 5% SDS supplied a significant improved particle distribution in the product solution.

The surfactant effect is still enforced under changed conditions (further slight enhanced flow rates). A maximum in the particle distribution of about 24 nm and shoulders below 20 nm and above 30 nm arise in the mass-related particle distribution spectra if 0.2% SDS solution was applied as in the case discussed above. The shoulder became reduced, the main maximum drifts down to 19 nm diameter and the yield of this particle population increases significantly if a SDS solution of 0.5 % was applied . A still smaller peak and a further increased yield of main particle population is the result of applicating a solution containing 2% SDS (fig. 3). With respect to the mass relation in the CPS spectra the colloidal product solution obtained under these conditions can be reagarded as approaching a monomodal distribution. The increase of concentration of surfactant up to 2% lead to a drastical improvement of the product quality under the applied flow rates and educt concentrations.

The both discussed examples show that SDS addition can lead to a significant change in the nanoparticle formation. But, it seems to be not possible to postulate a general improvement by the surfactant concentration enhancement. Micro flow-through synthesis experiments using the same experimental arrangement; but moderately changed flow rates of reactants show an other behaviour if the SDS concentration was varied. In case of reduced flow rates the maximum of particle diameter distribution remains in the range between 26 and 33 nm and the yield of particles and the band width are only moderately varied. A negative influence of higher concentrations of the surfactant SDS was found in same experiments if the order of application of reactants was changed.

4. Conclusions

The experiments show clearly that the addition of the surfactant SDS can influence drastically the quality of nanoparticles produced in a micro flow-through process. Obviously, the adsorption of surfactant molecules at the surface of gold particles can reduce the rate of silver deposition at gold cores significantly. So, the reaction behaviour shifts the size distribution and the optical properties of the obtained colloidal product solutions. An improvement of product quality (size, yield, homogeneity and size distribution) was found in most cases. But, only low influence or negative effects were found if specific other reactant ratios (modified flow rates) were applied. So, it can be concluded, that the addition of a surfactant can contribute decidingly to the improvement of nanoparticle quality in a miniaturized continuous flow synthesis, but the application must be optimized with respect of the order of reactant addition and the applied flow rates.

Acknowledgement

We thank A. Gross for discussions. The technical support of F. Möller and S. Schneider is gratefully acknowledged.

References

[1] Schenk, R.; Donnet, M.; Hessel, V.; Hofmann, Ch.; Jongen, N.; Löwe,H.: IMRET 5, Strasbourg, France, May 2001, 489

[2] V. Hessel, H. Löwe, Chemie Ingenieur Technik 2002, 74, 17 und 185

[3] R. Schenk, V. Hessel, N. Jongen, V. Buscaglia, S. Guillemet-Fritsch, A.-G. Jones, Nanopowders Produced using microreactors, in: Encyclopedia of Nanoscience and Nanotechnology (ed. by H.S. Nalwa) 2004, Vol 10

[4] P.M. Günther, J. Wagner, G.A. Groß, J.M. Köhler, 9th Internatl. Conf. On Miniaturized Systems : -TAS (Boston, October 9-13) 2005, 918-920

[5] G.A. Groß, C. Hamann, M. Günther, J.M. Köhler: Formation of polymer and nanoparticles doped polymer minirods by use of microsegmented flow principle, Chem. Eng. Technol. 30, 3 (2007), 341-346

[6] P.M. Günther, G.A. Groß, J. Wagner, F. Jahn, J.M. Köhler: Introduction of surface-modied Au-nanoparticles into micro flow-through polymerization of styrene, Chem. Eng. J. 135 S (2008), S 126-130

[7] H. Nakamura, A. Tashiro, Y. Yamaguchi, M. Miyazaki, T. Watari, H. Shimizu, H. Maeda, Application of a microfluidic reaction system for CdSe nanocrystal preparation: their growth kinetics and photoluminescence analysis. Lab on a Chip 2004, 4: 237-240

[8] E.M. Chan, R.A. Mathies, A.P. Alivisatos,Size-Controlled growth of CdSe nanocrystals in microfluidic reactors. Nano Letters 2003, 3: 199-201

[9] B.K.H. Yen., N.E. Stott, K.F. Jensen, M.G. Bawendi, Adv. Mater. 2003,15,1858-1862

[10] B.K.H. Yen, A. Gunther, M.A. Schmidt, K.F. Jensen, M.G. Bawendi, Angew. Chem. Internatl. Edit. 2005, 44, 5447-5451

[11] Link, S.; Wang, Z.L.; El-Sayed, M.A. J. Phys. Chem. 1999, 103, 3529-3533
[12] C. Sönnichsen, A.P. Alivisatos: Biophys. J. 2005, 88, 364 A - 365 A
[13] J. Wagner, T.R. Tshikhudo, J.M. Köhler, 9th Internatl. Conf. on Microreactor Technology IMRET 9 (Potsdam, 6.-8.Sept. 2006), 128
[14] J.M. Köhler, H. Romanus, U. Hübner, J. Wagner: Formation of star-like and core-shell nanoparticles during two- and three-step preparation in batch and in microfluidic systems, J. of Nanomaterials (2007), Article ID 98134
[15] J. DeMello, A DeMello, Lab Chip 2004,4,11N-15N
[16] J.Wagner, J. M. Köhler, Nano Letters 2005, 5, 685
[17] Sönnichsen, C.; Alivisatos, A.P.: Nanoletters 2005, 5, 301-304
[18] J.M. Köhler, M. Held, U. Hübner, J. Wagner: Formation of Au/Ag nanoparticles in a two step micro flow-through process, Chem. Eng. Technol. 30, 3 (2007), 347-354
[19] J. Wagner, T.R. Tshikhudo, J.M. Köhler: Microfluidic generation of metal nanoparticles by borohydride reduction", Chem. Eng. J. 135 S (2008), S 104- S 109

Figures

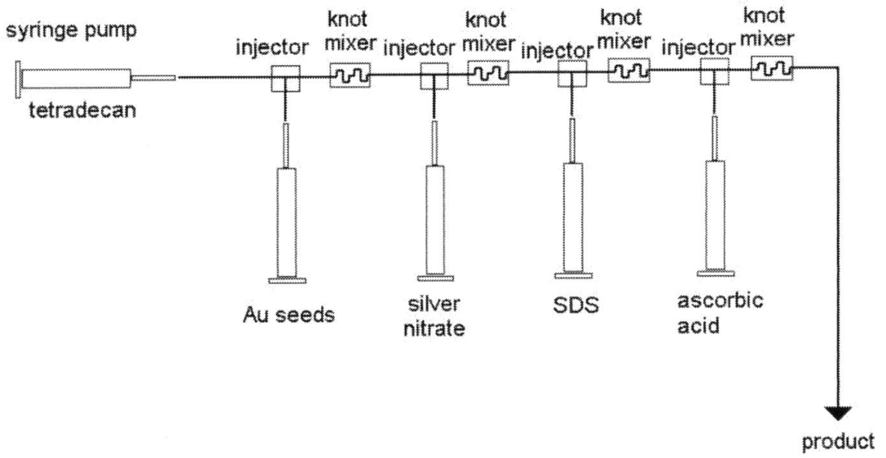

Fig. 1
Experimental set-up for micro flow-through synthesis of AuAg nanoparticles in the presence of surfactants

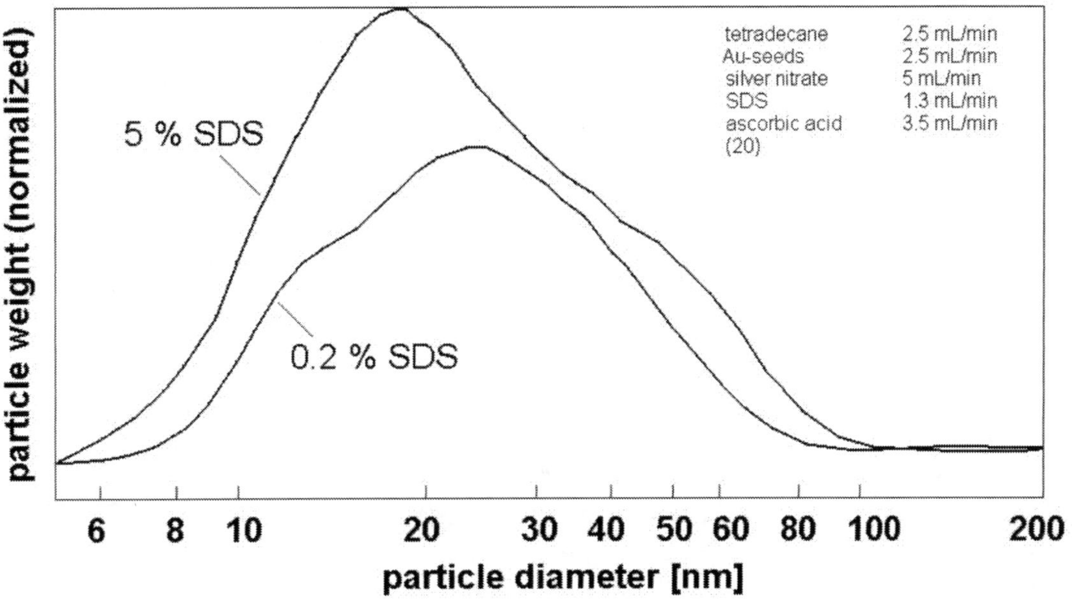

Fig. 2
Particle size distribution (CPS sedimentation spectrum) under following fluid mixing order and flow rates: tetradedcane 2.5 mL/min, Au-seeds (colloidal solution)2.5 mL/min, silver nitrate 5 : L/min, SDS (different concentrations) 1.3 mL/min, ascorbic acid 3.5 mL/min

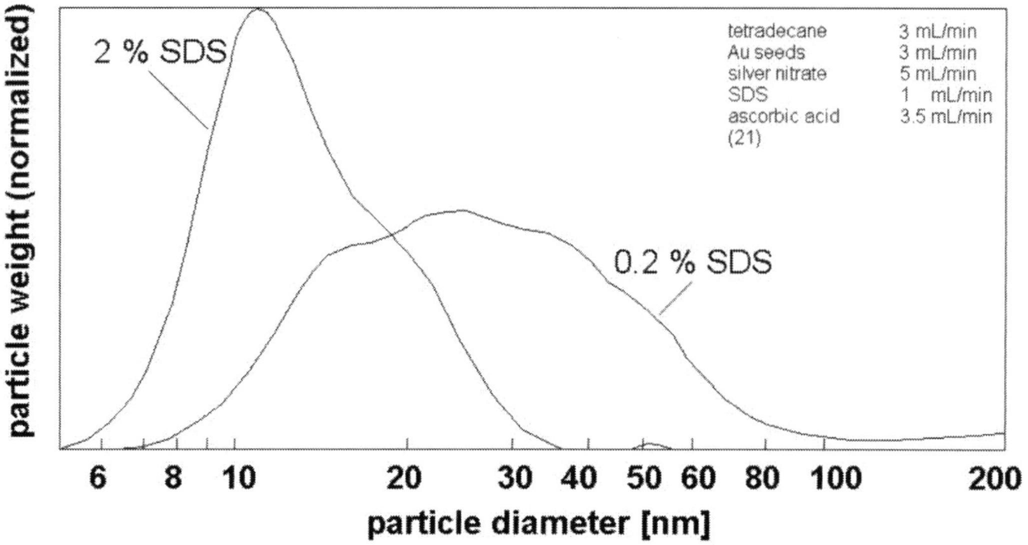

Fig. 3
Particle size distribution (CPS sedimentation spectrum) under following fluid mixing order and flow rates: tetradedcane 3 mL/min, Au-seeds (colloidal solution) 3 mL/min, silver nitrate 5 : L/min, SDS (different concentrations) 1 mL/min, ascorbic acid 3.5 mL/min

236d Controllable Preparation of Monodispersed Calcium Alginate Microcapsules in a Novel Microfluidic System

Jianhong Xu, Shangwei Li, Jing Tan, and Guangsheng Luo. Department of Chemical Engineering, The State Key Laboratory of Chemical Engineering, Beijing, China

In this work, we describe a novel method of prepare monodispersed calcium alginate microcapsules in a microfluidic device. The novel microfluidic device comprised of two cross-junction channels and a synthesizing channel. At the first junction, the side two streams, sodium alginate aqueous solution and calcium chloride aqueous solution, were separated to contact by the injection of de-ionized water in the middle channel. Then the aqueous mixture solution was induced into monodispersed droplets by the crossflow shear of octyl alcohol at the second junction. The formed droplets were in-situ solidified in the synthesizing channel with the rapid mixing of reactants in the droplets. The size of Ca-alginate microcapsules can be controlled from 50 ƒÝm to 500 ƒÝm in diameter with a variation of less than 5%, simply by changing two-phase flow rates. The results show that for a given fixed dispersed phase flow, the emulsion size decreases as the increase of continuous phase flow rates. The proposed microfluidic platform is capable of generating uniform emulsions and has the advantages of in-situ solidification and well control of size.

Key words: calcium alginate; microcapsule; monodispersed; microfluidic device

241e Nanoparticle Precipitation by Drops Flow in a Microchannel Device and Its Simulation

Shaowei Li, Jianhong Xu, Yujun Wang, and Guangsheng Luo. Department of Chemical Engineering, The State Key Lab of Chemical Engineering, Beijing, China

Well controlled two-liquid phase flow in a T-junction micro-channel device has been realized, and BaSO4 nanoparticles with good dispersibility were produced through micro-drops flow. Consistent with the well know results, the drop size decreases with the decrease of the dispersed phase flow rate or the increase of the continuous phase flow rate. The drop size has a prominent effect on the particle size. The increase of drop size results in the decrease of the particle size. This is might due to particularity of the mass transfer and mixing conditions in the drop formation stage. To prove this conclusion, the drop formation process as well as the mass transfer in the micro-device was simulation with CFD method. The results were compared with our previous drops flow mass transfer experiments and showed good accordance. Based on this result, a particle precipitation model was introduced to calculate the particle size.

Key words: nanoparticle, micro-channel, two-phase flow, drops flow, CFD, simulation

83ad An Effective Tool to Measure Reaction Heat with Micro-Pore Dispersion Reactor

Kai Wang[1], Yangcheng Lu[2], and Guangsheng Luo[2]. (1) The State Key Laboratory of Chemical Engineering, Beijing, China, (2) Department of Chemical Engineering, The State Key Laboratory of Chemical Engineering, Beijing, China

In this work, a new micro-dispersion reactor named micro-pore dispersion capillary reactor was designed for the first time to measure the reaction heat. The novel micro-dispersion reactor device comprised of a micro-pore dispersion channel and a capillary followed. With the advantages of microstructed reactor, such as safety, less reactant consumption and high selectivity, the measurement could be easily conducted. A data acquisition system was integrated with the reactor to collect the temperatures online, and the temperature difference between the reaction inlet and outlet at thermal isolation conditions was collected. The pre-mixing reaction, SINA Viscosa process was selected as the model system. Oleum was used as one reactant and cyclohexanecarboxylic acid diluted by n-hexane was used as the other. The reaction was started in the dispersion channel in which the reactants were pumped and achieved in the capillary. As a fast reaction with sulfonated consecutive by-product, high selectivity has been obtained in this microstructured reactor, thus the reliable results for reaction heat was obtained.

Key words: pre-mixing reaction; micro-dispersion; reaction heat

83ai Anionic Polymerization in Microreactor

Julia Getze and Andreas Greiner. Chemistry and Scientific Center of Materials Science, Philipps University of Marburg, Hans-Meerwein-Strasse, Marburg, D-35032, Germany

The intention of the work is develop novel polymeric materials by application of microreaction technology. A state-of-the-art pseudo anionic styrene polymerization was selected to test microreaction techniques. s-BuLi was used as initiator, cyclohexane as solvent and the reaction was carried out to 50 ° C. The high heat and mass transfer allow a living polymerization without additives at high concentration and temperature. Thus the efficiency of the polymerization was increased significantly.

103e Using Microreactors for Liquid-Liquid Extraction of Vanillin

Donata M. Fries, Tobias Voitl, and Philipp Rudolf von Rohr. Institute of Process Engineering, ETH Zurich, Sonneggstr. 3, Zurich, Switzerland

Within the development to "lab on a chip", all unit operations should be performed in a microreactor to get a powerful tool with all important processes available. In chemical engineering, extraction is an important process for cleaning carrier fluids or to separate the products from educts. As an example, we investigated the extraction of vanillin from the aqueous phase with toluene, which is part of an industrial process for the production of vanillin from lignosulfonates [1].

In order to extract vanillin out of the aqueous phase in a microreactor, three aspects had to be optimized in this project:

1) creating a large interfacial area and a stable flow pattern

2) achieving a sufficient long residence time

3) separating both phases effectively.

The used microreactor was designed in a way to fulfill all these requirements. The width was 300 µm, the height 200 µm. The two-phase channel length was 3.4 cm. Based on our experiences in gas-liquid flow, we designed the inlet in a way to achieve a stable flow without oscillations and a large interfacial area. An important aspect in extraction is the phase separation on the chip. Guenther et al. [2] demonstrated gas-liquid phase separation based on the capillary pressure difference: a rake-like shaped separator consisting of 20µm-wide capillaries sucks the continuous liquid phase out of the main channel. Based on this phenomenon, we designed a liquid-liquid separator. As PDMS is a very hydrophobic material, the continuous organic phase preferably wets the enhanced surface at the separator.

The flow has been characterized via Laser Induced Fluorescence (LIF), micro-Particle Image Velocimetry (µ-PIV) and standard microscopy. For the determination of the separation grade we used volumetric measurements. The concentrations of vanillin were determined by UV/Vis spectroscopy of the aqueous phase and by gas chromatography of the organic phase.

The separation of the organic and aqueous phase with the rake-like separator depended on the velocity: At higher velocities ($u>0.04$ m/s) separation degrees of more than 0.95 were observed. Whereas the separation grade was defined as the ratio of the organic phase volume to the total volume collected at the separator outlet.

As we used microreactors made of PDMS, the swelling behaviour of different solvents had to be investigated. The sorption of toluene into PDMS led to a mass increase. Saturation was achieved after 3 hours with a mass uptake of 100%. As the volume of the PDMS is large compared to the microchannel, geometrical changes of the structure can be neglected.

Using the microreactor described above gave for very small velocities ($u<0.03$ m/s) different flow patterns depending on the initial wetting of the reactor. For Taylor flow the extraction was more effective than for stratified flow. At higher velocities we obtained only Taylor flow. For a certain flow pattern the amount of vanillin left in the aqueous phase depended strongly on the residence time in the reactor. For residence times >2s in Taylor flow with optimized mass transfer the distribution ratio (concentration of vanillin in the organic phase versus the concentration of vanillin in the aqueous phase after the extraction) correlated with the thermodynamic equilibrium.

Based on these experiments we propose a microreactor to perform a multi-step continuous extraction. Volumetric flow rates are in the range of 30-500 µl/min. Possible applications are the continuous operation mode for the extraction in "lab on a chip" application and, as we are not mass transfer limited in the microscale [3], an analysis tool for kinetic measurements of extraction.

References

1. Hocking, M.B., Vanillin: Synthetic Flavoring from Spent Sulfite Liquor Journal of Chemical Education, 1997. 74(9): p. 1055-1059.

2. Guenther, A., Khan, S.A., Thalmann, M., Trachsel, F., Jensen, K.F., Transport and reaction in microscale segmented gas-liquid flow. Lab Chip, 2004(4): p. 278-286.

3. Fries, D.M., S. Waelchli, and P. Rudolf von Rohr, Gas-liquid two-phase flow in meandering microchannels. Chemical Engineering Journal. In Press, Corrected Proof.

Gas-liquid Taylor flow characteristics in straight and meandering rectangular microchannels

Norbert Völkel, Aloisiyus Yuli Widianto, Joelle Aubin and Catherine Xuereb
Laboratoire de Génie Chimique UMR 5503 CNRS, Toulouse, France
E-mail : Norbert.Volkel@ensiacet.fr
Joelle.AubinCano@ensiacet.fr

Taylor flow is a gas-liquid two-phase flow pattern which appears in microchannels and capillaries over a wide range of flow rates. It is characterized by elongated bubbles separated form each other by liquid slugs (Figure 1). The bubbles themselves are separated from the channel wall by a liquid film. A particular feature of Taylor flow is the velocity field within the liquid slugs; two symmetrical vortices are created at both sides of the channel centre line. This behavior is considered as advantageous for mass transfer and turns the Taylor flow regime into an interesting operating condition for gas-liquid or gas-liquid-solid microreactors (solid catalyst deposited on the channel walls) [1].

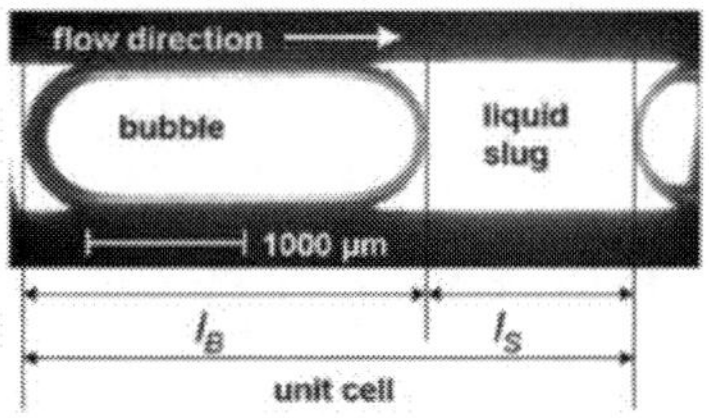

Figure 1: Example image of Taylor flow through a rectangular microchannel used in the frame of the present study (l_B = bubble length, l_S = slug length).

In order to provide sufficient residence time for a reactive system, typically long microchannels are needed. Since one of the design requirements of miniaturized devices is that they are compact, an obvious solution is to employ meandering channel geometries. In this case however, the channel bends induce a noticeable distortion of the recirculation motion within the liquid slugs [2]. The present work is part of a study, whose aim is to investigate if (and under which conditions) gas-liquid mass transfer in Taylor flow is enhanced by perturbations of the liquid velocity field. For this purpose three different rectangular microchannel configurations are used; one straight channel and two meandering channels (turning angle 180°), the latter two differing from each other in their bend-design (Figure 2). The microchannels were manufactured by etching a silicon wafer from the top to the bottom and enclosing it within two pyrex glass plates by anodic bonding (Figure 2(d)).

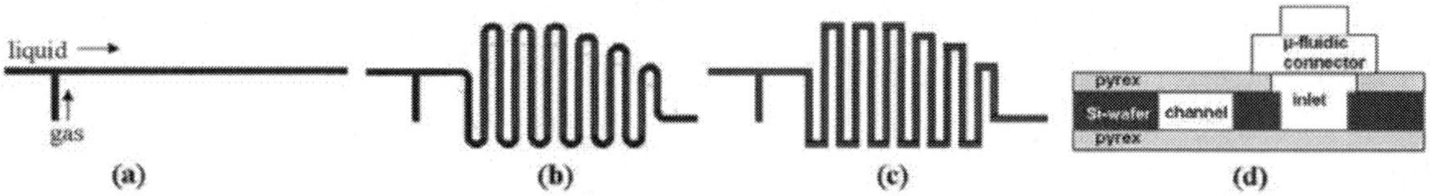

Figure 2: Microchannel geometries used in this study with w = 1000 µm, h = 500 µm; gas-inlet channel with w = 525 µm, h = 500µm. (a) straight channel l = 70 mm, (b) meandering channel with curved bends, (c) meandering channel with sharp corner bends, (b) and (c) $l \approx 300$ mm, (d) cross-sectional sketch of the microchannel plate.

For Taylor flow, the contact time between the bubble and the liquid film plays an important role on the gas-liquid mass transfer. It depends on bubble length, slug length, bubble velocity and liquid film flow [3, 4]. For this reason, a preliminary study of Taylor flow hydrodynamics was carried out for the three microchannels mentioned above. Air and ethanol were employed as working fluids at a temperature of 25°C (Bo = 0.16). The ranges of gas and liquid flow rates, as well as the characteristic values of the studied systems are given in Table 1. Movies of Taylor flow were recorded at the fixed distance from the inlet section (centre point at $x/d_{Ch} \sim 52$) using a high speed camera.

$\dot{V}_{air}$	0,3 – 1 ml/min	$\dot{V}_{ethanol}$	0,2 – 1 ml/min
U_{GS}	0,01 – 0,033 m/s	U_{LS}	0,007 – 0,033 m/s
Re_{GS}	0,45 – 1,49	Re_{LS}	3,27 – 16,33
We_{GS}	$4\cdot10^{-6} - 4\cdot10^{-5}$	We_{LS}	0,001 – 0,027
		$We_{LS}{\cdot}Eu_{LS}$	2,43 – 45,63

Table 1: Ranges of volumetric flow rates with resulting superficial velocities used in this investigation and their corresponding superficial dimensionless numbers (working fluids are air and ethanol).

In Figure 3, the superficial two phase velocity (U_{TP}) is plotted versus the bubble velocity (U_B) data. The experimental results are compared with correlations given by former Taylor flow studies in the literature. Firstly, it can be seen that the values measured in the straight channel are noticeably different from those found for the two meandering channels. This suggests that the bubble and slug lengths, as well as the bubble velocities are also a function of the two phase pressure drop. Based on a mass balance relationship derived for the Taylor flow regime, it can be stated that this graphical domain is divided into two regions by a parity line, where either the relative velocity between the bubble and the liquid film is zero or the liquid film does not exist. In the region above the parity line, the slope is greater than 1 and the liquid film flows with a zero velocity or a velocity which is smaller than the bubble velocity. In the lower region of the domain, the liquid film velocity is greater than the bubble velocity and the slope of the linear plots is less than 1.

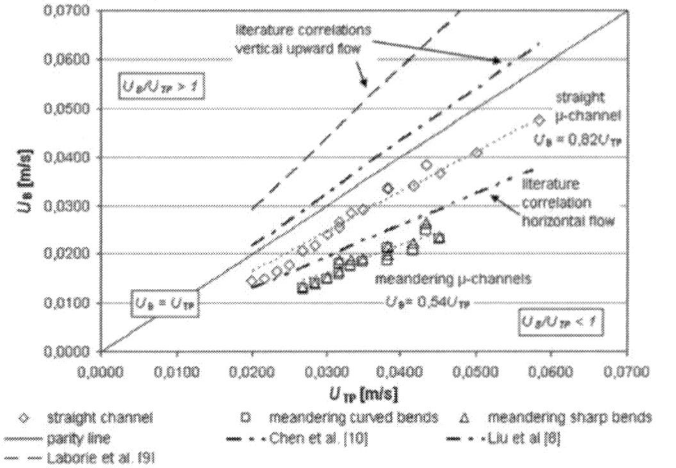

Figure 3: Superficial two-phase velocity (U_{TP} = U_{GS}+U_{LS}) versus bubble velocities (the lines represent correlations proposed in the literature; the resulting range of Capillary numbers is $2.33 \cdot 10^{-3} \leq Ca \leq 6.34 \cdot 10^{-3}$).

Generally for horizontal Taylor flow [5, 6, 7], the liquid film is assumed to be static, which suggests that the data should fall in the upper region of the domain. From Figure 3 it can be seen that the experimental results, as well as the literature correlation [10] for horizontal Taylor flow fall in the lower region of the plot. Hence, this suggests that the liquid film appears not to be static.

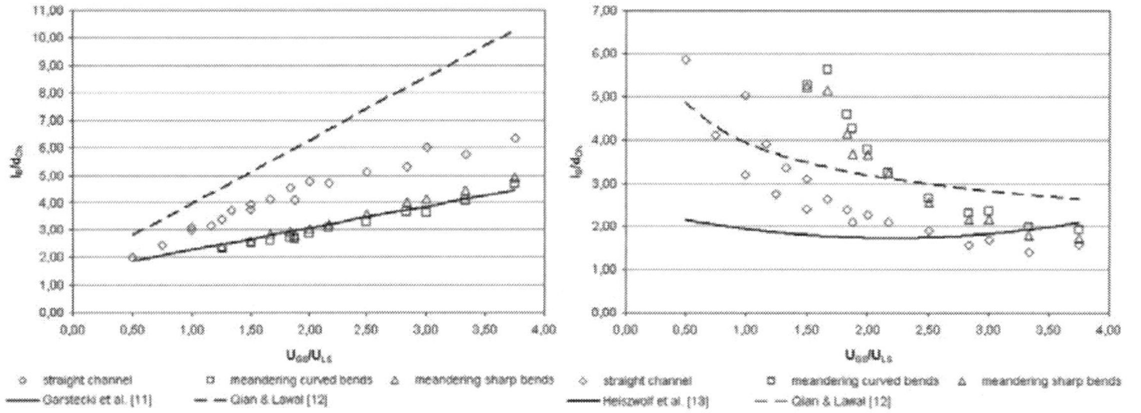

Figure 4: Superficial phase velocity ratio (U_{GS}/U_{LS}) versus **(a)** dimensionless bubble length (l_B/d_{CH}), **(b)** superficial phase velocity ratio (U_{GS}/U_{LS}) versus dimensionless slug length (l_S/d_{CH}), the drawn lines represent prediction criteria proposed by the literature, results obtained within a range of $2.33 \cdot 10^{-3} \leq Ca \leq 6.34 \cdot 10^{-3}$.

In Figure 4(a) it can be seen that the bubble length data for both meandering channels can be well predicted by the correlation given by Garstecki et al. [11], whilst the correlation proposed by Qian and Lawal [12] always over predicts the measured values. With respect to the slug length (Figure 4(b)), it can be seen that the experimental results generally do not follow the correlations [12, 13]. Furthermore, with an increasing superficial phase velocity ratio, the bubble length increases linearly, whereas the corresponding slug length decreases and tends towards an asymptotic limit. Since the cross-sectional bubble profile is assumed to remain fixed within the range of Capillary numbers studied here, a static behavior of the liquid film would also result in a linear relationship regarding the trend of the liquid slug lengths. Thus, this further suggests the possible existence of a liquid film flow.

In this paper we will use the findings of this hydrodynamic study to improve the existing models for the evaluation and estimation of gas-liquid mass transfer of Taylor flow in horizontal microchannels. In particular, a non-static liquid film will be accounted for. The sensitivity of the gas-liquid mass transfer to the characteristics of the liquid film will be evaluated.

Acknowledgements: This research is funded by the Integrated Project IMPULSE (project no. NMP2-CT-2005-011816), http://www.impulse-project.net/, within the 6th Framework Program of the European Commission. The authors also acknowledge the support of Professor Ryszard Pohorecki, Karolina Kula and Pawel Sobieszuk.

References:

[1] Kreutzer, M. T., Kapteijn, F., Moulijn, J. A., Heiszwolf, J.J., *Chem. Eng. Sci.*, 60, 5895-5916, **2005**.
[2] Günther, A., Khan, S. A., Thalmann, M., Trachsel, F., Jensen, K. F., *Lab Chip*, 4, 278-286, **2004**.
[3] Vandu, C. O., Liu, H., Krishna, R., *Chem. Eng. Sci.*, 60, 6430-6437, **2005**.
[4] van Baten, J. M., Krishna, R., *Chem. Eng. Sci.*, 59, 2535-2545, **2004**.
[5] Warnier, M. J. F., Rebrov, E. V., de Croon, M. H. J. M., Hessel, V., Schouten, J. C., *Chem. Eng. J.*, 135, Suppl. 1, S153-S158, **2008**.
[6] Suo, M., Griffith, P., *ASME J. Basic Eng.*, 86, 576-582, **1964.**
[7] Thulasidas, T.C., Abraham, M. A., Cerro, R. L., *Chem. Eng. Sci.*, 50, 183-199, **1995**.
[8] Liu, H., Vandu, C. O., Krishna, R., *Ind. Eng. Chem. Res.*, 44, 4884-4897, **2005**.
[9] Laborie, S., Cabassud, C., Durand-Bourlier, L., Laine, J. M., *Chem. Eng. Sci.*, 54, 5723-5735, **1999**.
[10] Chen, W. L., Twu, M. C., Pan, C., *Int. J. Multiphase Flow*, 28, 1235-1247, **2002**.
[11] Garstecki, P., Fuerstman, M. J., Stone H. A., Whitesides, G. M., *Lab Chip*, 6, 437-446, **2006.**
[12] Qian, D., Lawal, A, *Chem. Eng. Sci.*, 61, 7609-7625, **2006**.
[13] Heiszwolf, J. J., Kreutzer, M. T., van den Eijnden, M. G., Kapteijn, F., Moulijn, J. A., *Catal. Today*, 69, 51-55, **2001**.

65a Microchannel Based Integrated Desorber-Rectifier-Recuperator

Paul H. Humble, Ward E. Tegrotenhuis, Dustin D. Caldwell, and Dale A. King. Chemical & Biological Process Development Group, Energy and Efficiency, Pacific Northwest National Laboratory, 902 Battelle Boulevard, P.O. Box 999, Richland, WA 99352

Pacific Northwest National Laboratory has developed a family of microchannel based heat and mass transfer devices. Unit operations that have been developed include desorption, distillation, absorption, phase separation, and partial condensation. One of the promising attributes of microchannel based devices is the ability to combine several unit operations into a single device. This decreases the number of system components, eliminates piping, and reduces system size, weight and cost. The ability to combine unit operations is of particular interest in applications that require a minimal size and weight. One such application is portable cooling, where we have been developing microchannel based absorption cycle heat pump components for the U.S. Army. This ability to combine unit operations in a single device has been demonstrated with an integrated desorber-rectifier-recuperator for use in absorption heat pump cycles. This integrated device was developed using correlation based heat transfer models and finite element heat and mass transfer models. To better understand the desorption process, characterization experiments were conducted with a simplified single channel desorber, and this experimental data was used as design input for the integrated device. The 300 W scale desorber-rectifier-recuperator developed uses combustion heated air to generate refrigerant vapor; however, other heat sources such as exhaust from an internal combustion engine could be used. The performance of the desorber-rectifier-recuperator was tested in a breadboard system that includes a microchannel condenser, evaporator and absorber. The integrated desorber-rectifier-recuperator not only reduces system size but allows for more intensive heat integration and improved performance.

DEVELOPMENT AND OPTIMIZATION OF A MICROSTRUCTURED REACTOR FOR THE PARTIAL OXIDIZATION OF PROPANE TO PROPENE

Oliver Schwarz, Technical University of Berlin, Berlin, Germany
Phuong-Quyen Duong, Atotech Deutschland GmbH, Berlin, Germany
Gerd Schäfer, Atotech Deutschland GmbH, Berlin, Germany
Reinhard Schomäcker, Technical University of Berlin, Berlin, Germany

Abstract

The partial oxidization of propane to propene was used as a model reaction to develop and optimize a microstructured reactor for heterogeneously catalyzed gas phase reactions. Several methods for coating the reactor channels with catalytically active material were tested. The microstructured reactors were compared to conventional tubular fixed bed reactors (inner diameter 6 mm and 15 mm) in order to demonstrate their advantages. It was shown that the application of a microstructured reactor for partial oxidizations can be beneficial because of an efficient temperature control, narrow residence time distributions and enhanced safety.

Introduction

Although propene is one of the most important base chemicals in the world, it is primarily produced as a byproduct from steam cracking and other refinery processes. In addition, there are some on-purpose processes (such as the non-oxidative catalytic dehydrogenation of propane) commercially available [1]. However, these conventional production routes suffer from thermodynamic limitations and heavy coking. In addition, the ratio of desired products to undesired byproducts is difficult to influence. Since the demand for propene is forecasted to grow faster than the demand for corresponding lead products (e.g., ethane), a new process which avoids most of these problems appears to be attractive. The oxidative dehydrogenation of propane (ODP) over supported transition metal oxides (e.g., VO_x/Al_2O_3) has long been suggested as an alternative way to produce propene [2]. Figure 1 shows a simplified reaction network of the ODP. The key challenge is to minimize the formation of carbon oxides (CO_x), which occurs mainly via consecutive total oxidization of propene. Low propene selectivity has been the major barrier for an industrial application so far. In the future, propene yields have to increase substantially through improved catalyst or reactor performance or a combination of both to make the oxidative dehydrogenation of propane to propene economically feasible.

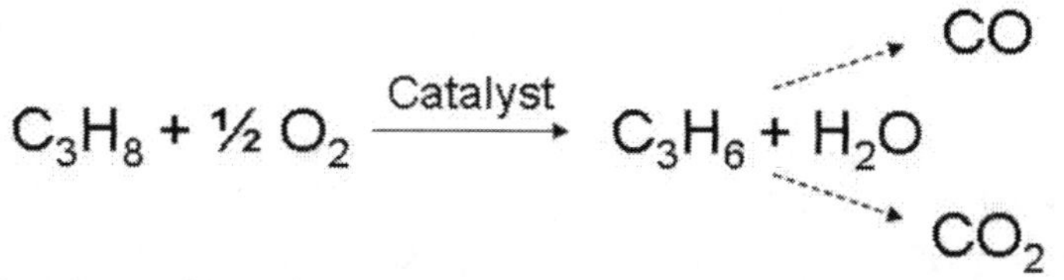

Figure 1. Simplified reaction network of the oxidative dehydrogenation of propane over a supported VO_x/Al_2O_3 catalyst [3]

Since the ODP is a strongly exothermic reaction that is sensitively influenced by mass and heat transport phenomena [3], microstructured reactor systems appear to be well suited for performing this reaction [4]. Their efficient heat transfer properties and narrow residence time distributions allow for a good control of reaction parameters, which is much more difficult in conventional reactor systems. In addition, enhanced process safety opens the field for less diluted feedstocks increasing achievable space time yields. As part of our studies on the ODP, a new concept for a microstructured reactor was developed and optimized to investigate this reaction in more detail and to use the obtained information for an evaluation of the potential of microstructured reactors for heterogeneously catalyzed gas phase reactions in the near future.

Experimental

Reactor Fabrication

The microstructured reactors were manufactured from stainless steel platelets by a new soldering technique under vacuum conditions at 400 °C. Different microstructure designs (single channel platelets, multi channel platelets, platelets for distributed reactant dosage) were combined with special platelets for inserting four thermocouples into the reactor as can be seen in Figure 2.

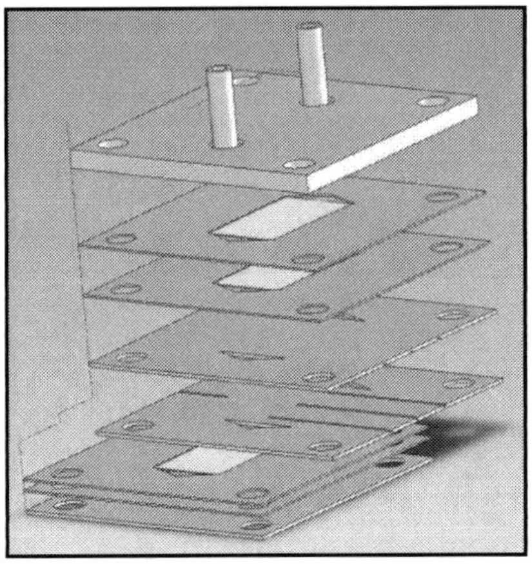

Figure 2. Scheme of a microstructured reactor with catalyst coated channels and special platelets for inserting thermocouples into the reactor

Three of them were used for temperature measurement, whereas the fourth was used for temperature control. The single channel platelets that are shown in Figure 2 can also be replaced by different channel designs for maximal flexibility.

Catalytic Coatings

For coating the reactor channels with catalytically active material, several methods were tested and optimized. Especially spray coating was used for applying the catalysts since it can easily be transferred to an industrial scale. For stabilizing the coatings, different binder systems (organic and inorganic) were used. In addition to the coating techniques, a new catalyst preparation method had to be developed. The obtained materials were thoroughly characterized by BET, XRD, Raman and UV/Vis spectroscopy and tested for their catalytic behavior. This was necessary to clearly separate catalyst effects from effects related to a change of the reactor geometry. Powders obtained from coatings were also characterized and tested for activity and selectivity in fixed bed reactors (inner diameter 6 mm) to elucidate the influence of the coating method on catalytic performance.

Residence Time Behavior

Residence time distributions of the different microstructure designs in comparison to a conventional tubular reactor (inner diameter 15 mm) were recorded using O_2/N_2 tracer experiments with volume flows between 30 and 120 ml min^{-1}. The signals were detected by a mass spectrometer. In addition, a simple theoretical model (axial dispersion model) was applied to fit experimental data and calculate corresponding Bodenstein numbers.

Thermal Behavior

The temperature of the catalytic bed of the conventional tubular reactor (inner diameter 15 mm) was measured using a movable thermocouple that was placed axially in a capillary in the center of the reactor. In comparison, the temperature of the catalytic bed of the microstructured reactors was measured using three fixed thermocouples that were placed at equal distances along the reactor channel (see Figure 2).

Catalytic Tests

Catalytic tests were carried out under the following standard reaction conditions: The composition of the reactant flow was 30 vol% C_3H_8, 15 vol% O_2 and 55 vol% N_2 (nitrogen being used for dilution of the reaction mixture). The catalyst loading of the reactors varied between 20 and 200 mg. Total volume flows were kept between 30 and 240 ml min^{-1} in a temperature range from 400 to 600 °C. As reaction products C_3H_6, H_2O, CO and CO_2 were detected and analyzed by gas chromatography. Only minor amounts of other alkenes and methane were identified.

Results and Discussion

Microstructured Reactors

Figure 3 shows a photograph of a microstructured reactor and the corresponding heating blocks. This special type of reactor is equipped with four inlet/ outlet tubes since it is designed for distributing one of the reactants over the channels coated with catalyst. Various performance tests resulted in sufficient temperature and pressure stability of the modules up to 600 °C, making them safe tools for performing strongly exothermic reactions at high temperatures in a large range of feed compositions.

Figure 3. Photograph of a microstructured reactor (designed for distributed reactant dosage) with corresponding heating blocks

For a variation of reactor volume or catalyst mass, additional channel platelets can be used for manufacturing the reactors. However, due to the fabrication process, the modules cannot be opened after a performance test without destroying them. Compared to the enhanced heat transfer properties through a solid material connection, this aspect seems to be only a minor drawback.

Catalytic Coatings

In order to choose a catalytic material with reasonable properties for an application in a microstructured reactor, four VO_x/Al_2O_3 catalyst samples with increasing vanadium content (1.4 to 7.8 wt%) were prepared and thoroughly characterized. The particle size of the Al_2O_3 starting material was chosen to be 3 μm in order to produce thin catalytic layers of about 20 μm. BET surface area of the catalysts decreases with increasing vanadium content probably due to clogging of the support material pores. XRD experiments revealed that the catalyst with the highest loading contains crystalline vanadia species that are known to be less active and less selective in the ODP. Therefore, this sample was ruled out to be used for further coating experiments. As can be seen from Figure 4, Raman spectroscopy confirmed results from XRD experiments (crystalline V_2O_5 in the catalyst with 7.8 wt% vanadium) and also revealed that all other samples contain only dispersed VO_x species that are more active and more selective in the ODP than crystalline vanadia. After characterization, the obtained samples were also tested for their catalytic performance. Activity for propane conversion increased with vanadium content for all samples but selectivity to propene at similar conversions constantly decreased. However, the catalyst with 5.3 wt% vanadium was chosen for further coating experiments, mainly because it was the most active (in terms of turn over frequencies) though not the most selective one. Detailed results of the catalyst characterization were reported in an earlier paper [5].

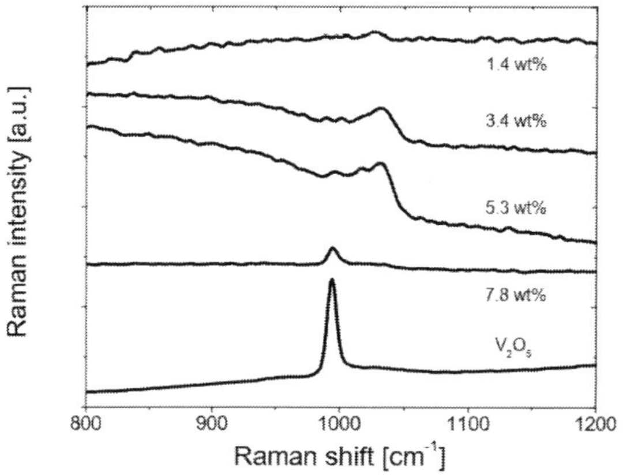

Figure 4. Raman spectra at room temperature after dehydration at 350 °C of the VO_x/Al_2O_3 catalysts available for coating experiments

The catalytic coatings produced from the catalyst described above were optimized with regard to a good adhesion of the ceramic support material on the metallic reactor surface. Alumina and silica containing binders produced coatings with a similar stability as formulations containing organic binders.

Residence Time Behavior

Residence time distribution is one of the major reactor parameters in continuous processes. Therefore, different microstructure designs were characterized by O_2/N_2 tracer experiments (see Figure 5) in order to compare them to the residence time distribution of a conventional tubular fixed bed reactor.

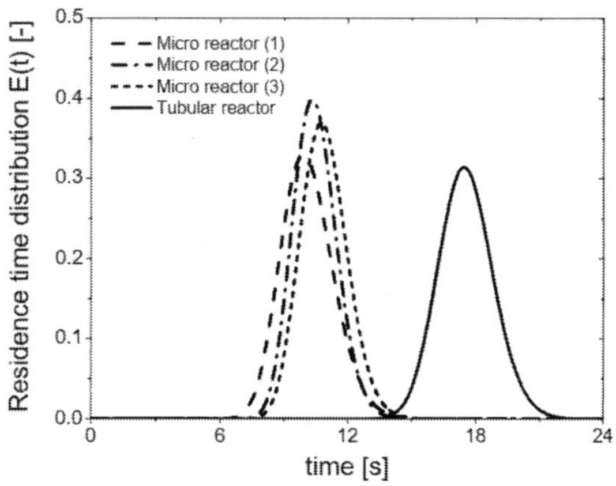

Figure 5. Residence time distributions compared for tubular reactor and different microstructure designs

Different reactor designs (assembly order, single channel or multi channel platelets) behave similar, resulting in almost identical residence time distributions. The tubular reactor shows a comparable residence time behavior. Bodenstein numbers were calculated to be well beyond 100, indicating only minor back mixing, which is favorable for suppressing undesired consecutive reactions of the initially formed product.

Thermal Behavior

Looking at the thermal behavior of both reactor types, the conventional tubular fixed bed reactor shows a pronounced temperature profile with a peak of more than 100 K under similar reaction conditions (temperature at reactor inlet 500 °C, propane conversion about 20 %, full oxygen conversion, see Figure 6).

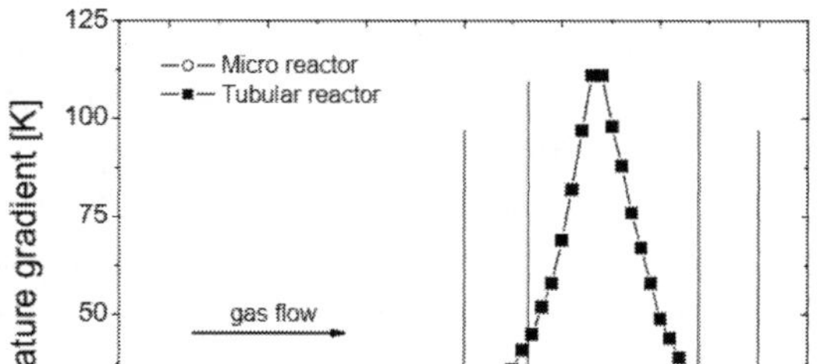

Figure 6. Temperature gradients compared for tubular and microstructured reactor (temperature at reactor inlet 500°C)

In both cases, it was not possible to measure the "true" temperature of the catalytic material. However, the thermocouples were placed very close to the catalyst bed giving a reasonable approximation of the "true" temperature. As can be seen from Figure 6, a good temperature control can be obtained only in the microstructured reactor. It is due to the high surface-to-volume ratio of about 8700 m^{-1} resulting in Stanton numbers of 1000 to 3000.

Catalytic tests

As described above, different methods for coating the reactor channels were developed. In particular, various binder systems (organic and inorganic) were applied in order to stabilize the catalytic coatings on the metallic reactor surface. These binders were found to greatly influence catalytic performance. Especially alumina and silica containing binders substantially decreased activity for propane conversion. This might be due to a change of the morphology of the catalysts, in particular the relationship between support material (Al_2O_3) and the active VO_x surface species. In contrast, organic binders made from cellulose derivates were found to have only minor influence on catalytic activity. Figure 7 shows propane conversion vs. residence time for the catalyst reference material (measured in a tubular rector, inner diameter 6 mm) and coatings made from different binders.

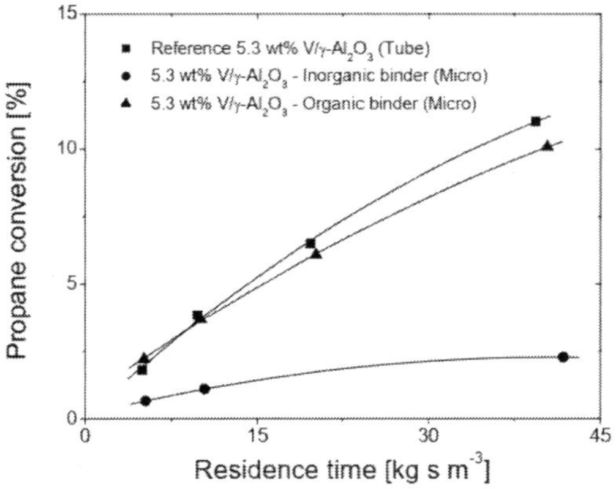

Figure 7. Activity of catalytic coatings compared to reference material (450 °C)

Similarly, the type of binder also influences selectivity towards propene as can be seen in Figure 8. Inorganic binders decrease propene selectivity, whereas organic binders do not show a major influence. Since all organic binders are assumed to be removed during preparation of the coatings by temperature treatment, the properties of the catalytic material appear to be unchanged. This is not the case when inorganic binders are applied. They stay within the coatings where they show major influence on catalytic performance. The origin of this influence has not been finally resolved yet. Since activity and selectivity are decreased, propene yields for these coatings were among the worst.

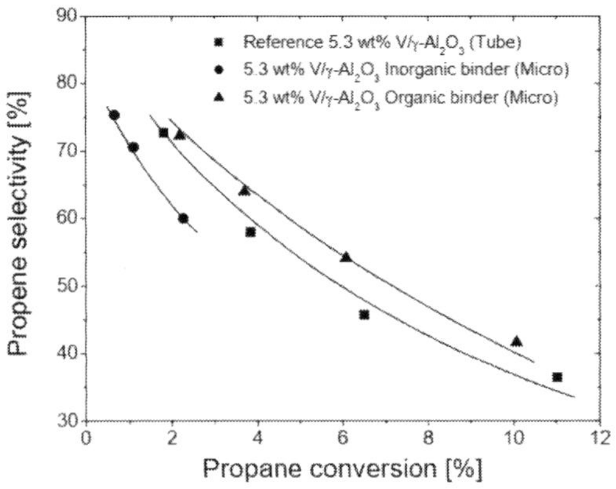

Figure 8. Selectivity of catalytic coatings compared to reference material (450 °C)

Using standard reaction conditions (450 °C), the conventional tubular fixed bed reactor and the microstructured reactor show similar catalytic behavior. However, the binder system that was used to stabilize the active coating in the microstructured reactors greatly influences the performance of the catalyst. Distinct differences between the investigated types of reactors can only be observed when standard reaction conditions are changed; in particular when reaction temperature is increased up to 600 °C. Since propene selectivity increases with temperature, these reaction conditions are favorable for high propene yields. However, near isothermal reactor behavior can only be obtained in the microstructured reactor whereas unselective gas phase reactions are enhanced in the tubular fixed bed reactor because of its pronounced temperature profile.

Conclusions

Taking the oxidative dehydrogenation of propane (ODP) to propene as a model reaction, a concept for a microstructured reactor was developed and optimized. It was shown that coating the reactor channels with catalytically active material is a major challenge in the development process. When comparing a tubular reactor (inner diameter 15 mm) to a microstructured reactor, no significant variations in residence time behavior were found, whereas thermal behavior showed pronounced differences. Especially, near isothermal reaction conditions can only be obtained in the microstructured reactor. Further research on the transferability of heterogeneously catalyzed gas phase reactions into microstructured reactors appears to be reasonable.

Acknowledgements

This work was funded by the Deutsche Forschungsgemeinschaft (DFG) as part of the Sonderforschungsbereich 546 and by the Fonds der Chemischen Industrie (FCI). Particularly, O. Schwarz would like to thank the State of Berlin for providing a NaFöG scholarship and Atotech Deutschland GmbH for supplying the microstructured reactors. Furthermore the authors are grateful to E.V. Kondratenko, C. Hess, and D. Habel for complementary experiments and fruitful discussions.

References

1. Eisele, P., and Killpack, R. (2000), "Propene", in *Ullman's Encyclopedia of Industrial Chemistry*, Electronic Release, Wiley-VCH Verlag GmbH & Co. KGaA.
2. Mamedov, E.A., and Cortés Corberán, V., "Oxidative dehydrogenation of lower alkanes on vanadium oxide-based catalysts. The present state of the art and outlooks", Appl. Catal. A: Gen. 127 (1995) 1-40.
3. Frank, B., Dinse, A., Ovsitser, O., Kondratenko, E.V., and Schomäcker, R., "Mass and heat transfer effects on the oxidative dehydrogenation of propane (ODP) over a low loaded VO_x/Al_2O_3 catalyst", Appl. Catal. A: Gen. 323 (2007) 66-76.
4. Steinfeldt, N., Dropka, N., Wolf, D., and Baerns, M., "Application Of Multichannel Microreactors For Studying Heterogeneous Catalysed Gas Phase Reactions", Trans IChemE, Vol. 81 Part A, August 2003, 735-743.
5. Schwarz, O., Frank, B., Hess, C., and Schomäcker, R., "Characterisation and catalytic testing of VO_x/Al_2O_3 catalysts for microstructured reactors", Cat. Comm. 9 (2008) 229-233.

On the Scalability of Microstructured Multilamination Mixing Devices

Andreas Kölbl, Manfred Kraut, and Klaus Schubert
Forschungszentrum Karlsruhe GmbH, Institute for Micro Process Engineering (IMVT),
P.O. Box 3640, DE-76021 Karlsruhe, Germany

Andreas.Koelbl@imvt.fzk.de
phone: + 49 7247 / 82-6653
fax: + 49 7247 / 82-3186

Introduction

Microstructured devices possess numerous advantages on the laboratory scale. The development of microstructured devices for industrially relevant throughputs has been rarely reported, and the corresponding upscale is still a challenging task.

The use of microstrucured V-type mixing devices has been previously reported in scientific literature for gases and liquids[1-5]. In this contribution the scalability of this type of devices is discussed with respect to 'mixing quality' and pressure drop.

Description of the Micromixing Devices

We have examined multilamination type mixers, which consist of a stack of micromachined foils with ten geometrically identical channels per foil. Figure 1 shows the construction principle of the examined V-type mixer inlays. The microchannels were machined in stainless steel foils (Figure 1a). Those micromachined foils were stacked, and the stack was diffusion bonded (Figure 1b) to a mixer inlay (Figure 2a). Mixing of liquids or gases occurs in the exit of the microchannel system.

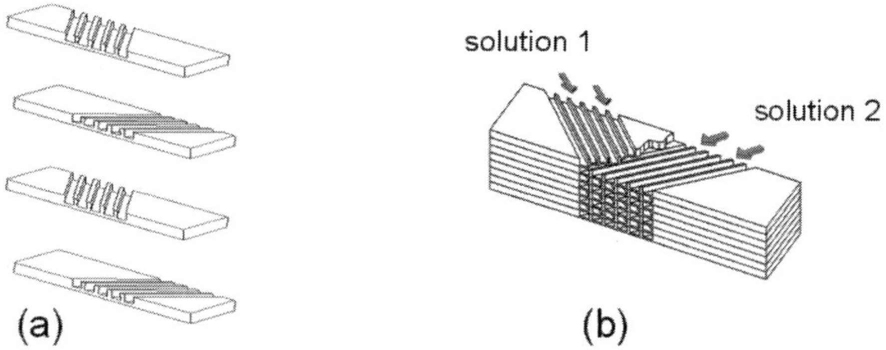

Figure 1: Construction principle of microstructured V-type mixers: (a) stainless steel foils with microchannels and (b) stack of stainless steel foils.

Mixing occurs by diffusion of the species into adjacent fluid layers. The theoretical thickness of those layers is as large as the thickness of the foils used (200 µm for the case of the mixing inlays used for this study yielding a diffusion path of 100 µm for a given molecule). The top and bottom layers are different regarding the theoretical diffusion path of the species.

Figure 2a shows a photo of the diffusion bonded foil stack, which serves as an inlay for a flange adapter (Figure 2b). This flange adapter can be conveniently attached to an experimental setup. Additionally, the outlet of the flange adapter serves as a mixing chamber at the exit of the microchannel system (8 mm diameter).

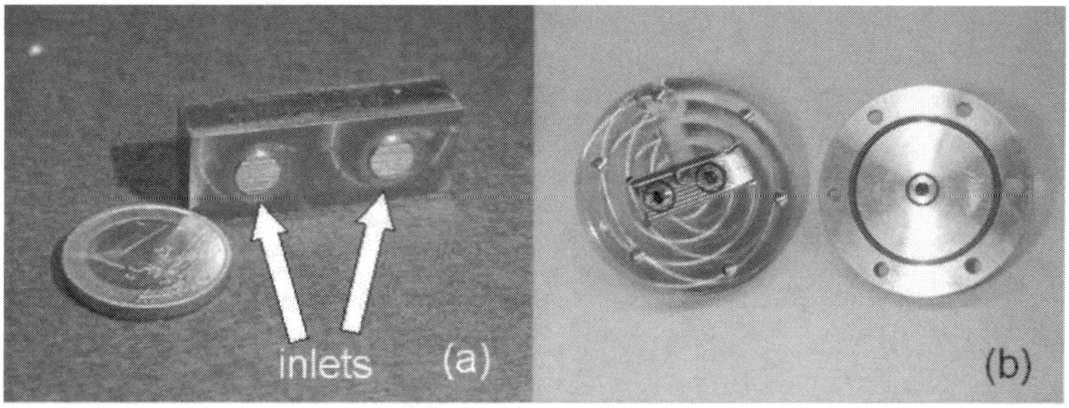

Figure 2: V-type mixer inlay (a) and flange adapter (b).

The examined V-type mixer inlays differ in the geometric parameters of the microchannel system. The parameters given in Table 1 are: the channel width, b; the channel height, h; the number of channels per foil, z_k; the fin width separating the channels, s; the bottom thickness, d; and the number of foils per inlay stack, z_f.

	b / µm	h / µm	z_k	s / µm	d / µm	z_f
s100-2foils	100	70	10	100	130	2
s100-4foils	100	70	10	100	130	4
s100-6foils	100	70	10	100	130	6
s100-12foils	100	70	10	100	130	12
s100-24foils	100	70	10	100	130	24

Table 1: Geometric parameters of the examined V-type mixer inlays.

Figure 3 shows a SEM of the exit of a mixer inlay with critical geometric parameters indicated.

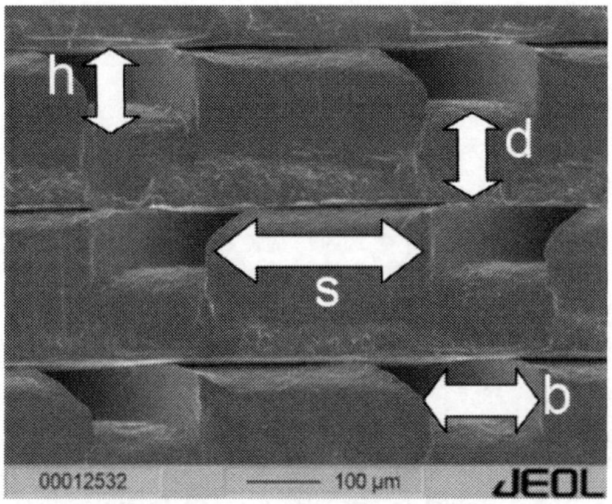

Figure 3: SEM of the exit side of a mixer inlay.

Description of the Test Method

The 'mixing quality' upscale assumptions are validated by means of the 'Iodide Iodate Method', a parallel-competitive reaction system, which produces more iodine the worse the mixing performance is. The amount of iodine is reflected by the absorbance value in the UV spectrum centered at 353 nm. The Iodide Iodate Method was developed for macroscopic stirred tanks[6-9] and later adapted to the needs of static microstructured mixing devices [10]. The concentrations of the reactants used to test the mixing devices are listed in Table 2.

	concentrations
solution 1	
sulfuric acid	0.03 mol/L
solution 2	
potassium iodide KI	0.016 mol/L
potassium iodate KIO_3	0.003 mol/L
sodium hydroxide NaOH	0.045 mol/L
hydroboric acid H_3BO_3	0.045 mol/L

Table 2: Concentrations of the reactants in the feed streams used for applying the Iodide Iodate Method on mixing devices.

Further details of the test procedure used for this examination can be found in a recent publication[5].

Variation of the Number of Microstructured Foils

Figure 3 compares the experimentally obtained absorption values of mixer inlays, which possess the same geometrical channel parameters (b, h, s, d, z_k; see Table 1) but differ in number of microstructured foils, i.e. the total number of channels.

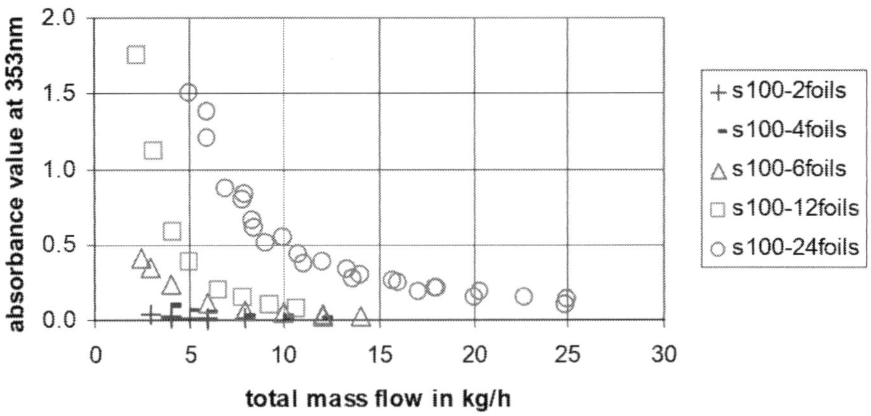

Figure 3: Experimentally obtained absorbance values over total mass flow for stacks of micromachined foils.

When the absorbance values are depicted in terms of linear velocity of the fluid flow at the exit of the microchannel system, all examined stacks show the same mixing performance (Figure 4).

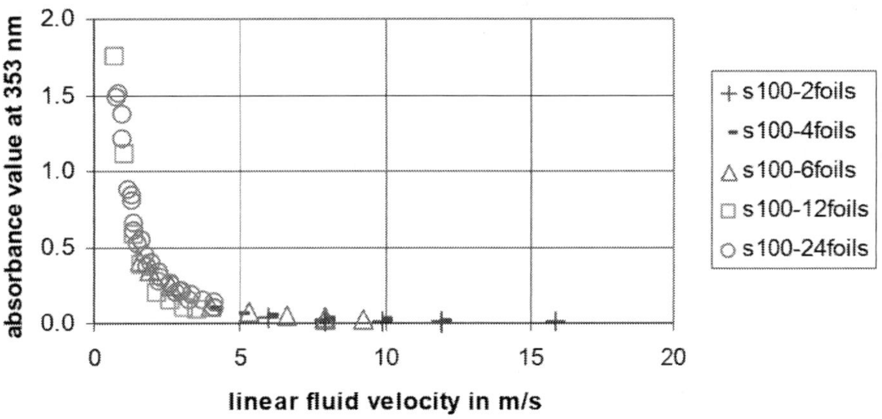

Figure 4: Experimentally obtained absorbance values over linear fluid velocity for stacks of micromachined foils.

The experimentally obtained absorbance values (OD) can be described within an error of ± 50% as a function of the total mass flow (dm/dt in kg/h) and the channel number ($z_k \cdot z_f$) by the empirical equation (I).

$$OD = 1.24\left(\frac{dm/dt}{3.6 \cdot z_k \cdot z_f}\right)^{-1.88} \quad (I)$$

The mixing devices were examined in a throughput range between 2.4 and 30 kg/h. The corresponding parity plot is given in Figure 5.

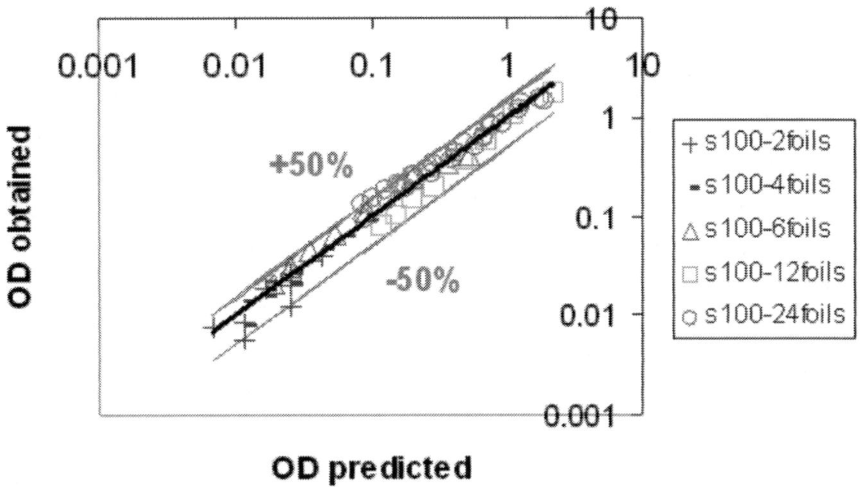

Figure 5: Parity plot of absorbance values measured and absorbance values predicted by equation I.

Since the fluid elements have the same linear momentum, and the same energy dissipation rate is achieved, it can be claimed that the mixers can easily be scaled by adding foils. This is an important point regarding the portability of laboratory experiments into industrial application.

Pressure Drop

Assuming that the pressure drop is mainly achieved in the microchannel system and the fluid flow is laminar, the pressure drop can be calculated from the linear fluid velocity,c; the Reynolds number, Re; the fluid density, ρ; the lengths of the micro channel, l; and the hydraulic diameter, d_h. Equation II[11] additionally contains corrections for the rectangular cross section of the microchannels (φ) and the additional pressure drop in the inlet and the outlet of the microchannel system (K).

$$\Delta p = \frac{\rho}{2} c^2 \left(\varphi \frac{64}{Re} \frac{1}{d_h^2} + K \right) \quad (II)$$

The experimentally determined pressure drops are 20% higher than the pressure drops predicted by equation II. This discrepancy could be explained with the shrinking of channel dimensions in the course of diffusion bonding. Shrinking of the hydraulic diameter of less than 9% results in an increase of the pressure drop of 20% (Equation II). Multiplying Equation II with a factor of 1.2, the pressure drops can be predicted within an accuracy of ± 5%.

Conclusions

Various microstructured V-type mixing inlays, possessing single channels of the same geometry but differing in the number of those channels were examined for pressure drop and 'mixing quality' (utilizing the 'Iodide Iodate Method'). The experimental results were compared in terms of scalability for this type of microstructured mixer. The experimental results show that the devices can be scaled by 'numbering up' the microstructured channels.

Literature

[1] S.A. Ehlers: "Einsatzpotentiale von Mikrostrukturmischern" PhD Thesis. University of Cologne, Germany, 1998.

[2] St. Ehlers, K. Elgeti, T. Menzel, G. Wießmeier: "Mixing in the offstream of a microchannel system" Chemical Engineering and Processing 39(2000)291-298.

[3] P. Pfeifer, L. Bohn, O. Görke, K. Haas-Santo, U. Schygulla, K. Schubert: "Microstructured Mixers for Gas-Phase Processes – Manufacture, Characterization and Applications" Chemical Engineering and Technology 28(2005)439-445.

[4] R.A. Rakoczy, M. Kraut, E. Anurjew, E. Hansjosten, U. Schygulla, K. Schubert: "Comparative study of micro mixing devices" Proceedings of the 7th World Congress of Chemical Engineering, Glasgow, Great Britain, 2005.

[5] A. Kölbl, M. Kraut, K. Schubert: "The Iodide Iodate Method to Characterize Microstructured Mixing Devices" AIChE Journal 54(2008)639-645.

[6] M.C. Fournier, L. Falk, J. Villermaux "A new parallel competing reaction system for assessing micromixing efficiency – experimental approach" Chemical Engineering Science 51(1996)5053-5064.

[7] M.C. Fournier, L. Falk, J. Villermaux: "A new parallel competing reaction system for assessing micromixing efficiency – determination of micromixing by a simple mixing model" Chemical Engineering Science 51(1996)5187-5192.

[8] P. Guichardon, L. Falk: "Characterisation of micromixing efficiency by the iodide-iodate reaction system. Part I: experimental procedure" Chemical Engineering Science 55(2000)4233-4243.

[9] P. Guichardon, L. Falk, J. Villermaux: "Characterisation of micromixing efficiency by the iodide-iodate reaction system. Part II: kinetic study" Chemical Engineering Science 55(2000)4245-4253.

[10] S. Panić , S. Loebbecke, T. Tuercke, J. Antes, D. Bošković: "Experimental approaches to a better understanding of mixing performance of microfluidic devices" Chemical Engineering Science 101(2004)409-419.

[11] VDI-Gesellschaft Verfahrenstechnik und Chemieingenieurwesen eds. "VDI-Wärmeatlas", 7. edition, Springer Verlag, Berlin a.o., 1994.

64f Development and Experimental Characterization of a Pilot-Scale Falling Film Microreactor and Outlook on Production Scale

Patrick Löb[1], Christian Hofmann[1], Bhanu Kiran Vankayala[1], Ulrich Krtschil[1], Volker Hessel[1], and Robert Franke[2]. (1) Mixing and Fine Chemistry Department, Institut für Mikrotechnik Mainz GmbH, Carl-Zeiss-Straße 18-20, Mainz, D-55129, Germany, (2) Evonik Oxeno GmbH, Paul-Baumann-Str. 1, D-45772 Marl, Germany

The standard Falling Film Microreactor (FFMR-STANDARD) developed by IMM in which liquid films of a few tens of micrometer thickness and interfacial areas up to 20,000 m^2/m^3 combined with an effective heat exchange can be obtained represents a broadly used microstructured lab tool in the area of gas/liquid contacting (see e.g. [1-3]). In the context of the German public funded project μ.Pro.Chem (Förderkennzeichen 16SV1991) with the partners BASF AG, Evonik Degussa GmbH and IMM one topic is to transfer this lab-scale device which is successfully applied for the Evonik Degussa GmbH process of an ozonolysis towards pilot scale and production scale. Targeted pilot throughput is thereby for the liquid about of 10 l/h which represents a one-hundred fold increase compared to the lab-scale device FFMR-STANDARD. To reach production scale a further increase of throughput by a factor of 10 – 100 is required.

The followed approach by IMM to reach pilot scale is first to design a larger base unit with an about tenfold increase of structured surface and then in a next step to design a pilot reactor based on (tenfold) internal numbering-up. Therefore in a first step a falling film microreactor with a larger reactor plate has been developed and realized. The microchannel width and depth has been kept constant compared to FFMR-STANDARD while both the number of channels and the length of the channels have been increased by a factor of $10^{0.5}$. The principal design in this new falling film microreactor (FFMR-LARGE) – removable reaction plate in a housing bearing the feeding lines – is similar to the one of the FFMR-STANDARD. Some modifications have been incorporated to improve or ensure equal distribution of gas and liquid in the reactor. The next step then was to come to a pure plate design for this larger base unit as basis for the later internal numbering-up. While the reaction plate structure for the gas/liquid contacting was kept, major changes have been necessary to realize the functionality of the housing with regard to stability and feed distribution by only structuring plane reactor plates by wet-chemical etching. Furthermore, the reactor plates design has been adjusted to the selected sealing process by brazing. Corresponding prototypes (called STACK-1x-FFMR-LARGE) have been successfully realized including a version with an open gas chamber which allows visual observation of the reaction plate. Finally, a pilot-scale reactor consisting of a stack of ten of these plate-based functional units has been realized (called STACK-10x-FFMR-LARGE).

The reactor development work has been accompanied and supplemented by experimental works dealing with the characterization of the reactor performance and also with studying equal distribution of gas and liquid in the reactor for all developed reactor types. The ozonolysis reaction has been performed by Evonik Degussa GmbH itself and can not be reported here. The reactor performance has additionally been studied by IMM using the CO_2 absorption in an aqueous sodium hydroxide solution and by an oxidation reaction of an organic compound by pure oxygen. These investigations have confirmed that the performance of the larger base unit FFMR-LARGE can be operated at a tenfold throughput compared to FFMR-STANDARD while keeping similar reactor performance (see also [4]). Via the CO_2 experimentation it was shown also that the step from FFMR-LARGE to STACK-1x-FFMR-LARGE has not changed reactor performance and thereby confirmed the design. Furthermore, just recently the experimentation with the pilot-scale reactor STACK-10x-FFMR-LARGE has started. So far already a reactor performance of about 70% at an again tenfold increased throughput has been reached. The works here are continuing to confirm 100% performance. Gas and liquid equal distribution in different falling film microreactors has been studied by different methods: evolution of the liquid film via dosing of a dyed liquid, liquid flow rate measurements per channel and discoloring of a with phenolphthalein dyed sodium hydroxide solution during CO_2 absorption. Thereby, the evaluation of the different reactor designs has been possible.

Concluding an outlook on the approach to reach production scale by a combination of internal and external numbering-up will be given.

[1] Ehrich, K., Linke, D., Morgenschweis, K., Baerns, M., Jähnisch, K., Chimia 56 (2002) 647-653.

[2] Yeong, K.K., Gavriilidis, A., Zapf, R., Kost, H.-J., Hessel, V., Boyde, A., Exerpimental Thermal and Fluid Science, 30 (2006) 463-472.

[3] Löb, P., Löwe, H., Hessel, V., Journal of Fluorine Chemistry 125 (2004) 1677-1694.

[4] Vankayala, B.K., Loeb, P., Hessel, V., Menges, G., Hofmann, C., Metzke, D., Krtschil, U., Kost, H.-J., International Journal of Chemical Reactor Engineering, accepted for publication.

83ab Fluid Flow in Deep Microchannel Reactor (FMR) for Pilot Scale Production

Ken-Ichiro Sotowa, Kaname Takagi, Shigeru Sugiyama, and Keizo Nakagawa. Department of Chemical Science and Technology, The University of Tokushima, 2-1 Minami-Josanjimacho, Tokushima, 770-8506, Japan

Fluid flow in a deep microchannel reactor (FMR) was investigated using computational fluid dynamics. As the name implies, FMR consists of deep microchannels which is typicaly in the order of ten centimeters. Theoretically the throughput of FMR is proportional to the channel depth, and thus a 1000 fold increase in the production rate can be easily achieved. The pressure drop can also be reduced by employing a deep microchannel. However, since FMR employs very deep channels, a care must be taken to suppress flow maldistribution that drastically decreases the performance of FMR as a microchannel reactor. We have already reported on the preliminary results on the channel structure in which flow maldistribution can be suppressed. The results were validated in an experiment using a FMR with a depth of 2.6 cm

In this work, we aim at desiging a pilot scale production plant using a FMR. The target production rate is 1000 tpa. The "depth" of FMR is more than 30cm. A computational fluid dynamics simulation was conducted to investigate the fluid flow in FMR with a depth of 30cm, and the effect of fabrication error on the flow distrubtion was studied.

83ac Scale-out Concepts in Chemical Micro Process Engineering Exemplified on Production Oriented Case Studies

Patrick Löb and Volker Hessel. Mixing and Fine Chemistry Department, Institut für Mikrotechnik Mainz GmbH, Carl-Zeiss-Straße 18-20, Mainz, D-55129, Germany

A major focus of IMMxs current work in the area of Chemical Micro Process Engineering is the transfer of (micro)structured components and microreactor based processes from lab to pilot and production scale. This topic is especially addressed in the two German public funded BMBF projects µ.Pro.Chem (Förderkennzeichen 16SV1991) and POKOMI (Förderkennzeichen 16SV1981) and the integrated EU-project IMPULSE. Within these projects different application areas are considered:

· Gas/liquid process (ozonolysis) in falling film microreators with an aimed pilot scale of 10 l/h liquid throughput and a production scale of about 100 – 1000 l/h (project µ.Pro.Chem).[1]

· Liquid/liquid process (Suzuki-polycondensation for OLED material synthesis) in microsturctured reactors or reactors with microporous metallic foams for production throughputs in the range of a few to several liters per hour (project POKOMI).[2]

· Surfactant dispersion in the context of liquid fabric enhancer production with a lab scale of about 10 l/h and an aimed pilot scale of about 1000 l/h (project IMPULSE Task 3.3).[3]

· High exothermic single phase ionic liquid synthesis with an aimed pilot scale of about 8 l/h (project IMPULSE Task 2.1).[4]

The approaches to reach larger throughputs in these application examples will be outlined supplemented by additional processes from IMM working areas but also by examples from literature (e.g. [5]). Furthermore, it is aimed to derive there from generic learning on the topic of scale-out approaches in Micro Chemical Process Engineering.

[1] Vankayala, B.K., Loeb, P., Hessel, V., Menges, G., Hofmann, C., Metzke, D., Krtschil, U., Kost, H.-J., International Journal of Chemical Reactor Engineering, accepted for publication.

[2] Hessel, V., Rothstock, S., Agar, D.W., Jadhavrao, P., Löb, P., Werner, B., in: Conference proceedings, 2007, AIChE spring national meeting: April 22-27, 2007, Houston. – New York: AIChE, 2007, 74d.

[3] Löb, P., Hessel, V., Hensel, A., Simoncelli, A., Chimica Oggi (Chemistry Today) 25, 3 (2007) 26-29.

[4] Löb, P., Hessel, V., Krtschil, U., Löwe, H., Chimica Oggi (Chemistry Today) 24, 2 (2006) 46-50.

[5] Iwasaki, T., Kawano, N., Yoshida, J., Org. Process Res. Dev. 10 (2006) 1126-1131.

Catalyzed Hydrogenation by Coated Catalysts in Microstructured Reactors

A microstructured high pressure hydrogenation reactor (µ-HPHR) relying on metal foams was developed and tested with regard to performance in solid catalyzed gas-liquid hydrogenation by the Institut für Mikrotechnik Mainz GmbH (IMM) and the University of Mainz. Extremely porous RECEMAT® nickel metal foams (MF) were used as catalyst carriers. The MFs were coated with nanocrystalline platinum or palladium stemming from electrochemical deposition. Corresponding aqueous electrolytes or ionic liquid based electrolytes were employed during deposition. The reactor design allowed easy exchange of the catalyst carrier. Therefore a quick and simple catalyst screening was possible. Hydrogenation of nitrobenzene to aniline in ethanol was chosen as model reaction. The reactor µ-HPHR was operated in a continuous way at temperatures in the range of 30-170 °C and at hydrogen pressures up to 35 bar. Mean residence time was in the range 0.5 - 85 s. The reactor was operated between the kinetic and mass transfer controlled regime. The impact of temperature, hydrogen pressure and catalyst on conversion and selectivity was investigated. Additionally the potential of the microreactor processing for process intensification was evaluated by comparison with a conventional batch procedure.

The metal foam bearing platinum as catalyst (6 mg/g metal foam, from aqueous solutions) and the nickel metal foam without additional catalyst hardly exhibited catalytic activity. Metal foam coated with palladium showed a high catalytic activity. The best result was obtained for the metal foam with Pd (30 mg/g metal foam) deposited from the corresponding ionic liquid electrolyte. At a nitrobenzene concentration of 1 mol/l, a flow rate of 1 ml/min, a temperature of 170 °C and a hydrogen pressure of 35 bar a selectivity of 98 % was reached. The space-time yield for the µ-HPHR reached here 24.1 kmol aniline per m^3 reactor volume and per hour. As comparison the following batch process described in literature can be taken: reactor of 3 m^3 volume, 20 bar, 125 °C, 5 % Pd/C catalyst, initial nitrobenzene concentration of 0.4 mol/l. Here the space-time yield - assuming no downtime between batches – amounts only to about 8.5 kmol/(m^3 * h).

Recently, these works were supplemented by corresponding experimentation at IMM using a newly developed Falling Film Microreactor version. Similar experimentation using the former Falling Film Microreactor have been reported by K.K. Yeong *et al.*

[1]. In comparison to the former version the new version contains a larger reaction plate (increase of the interfacial area by a factor of ten). The reaction plate was coated with aluminium oxide impregnated with Pd (5 %). The hydrogenation was performed at a temperature of 60°C, a 0.1 molar nitrobenzene solution in ethanol. At a liquid flow rate of 5 ml/min and a static hydrogen pressure of 1 bar conversion laid steadily around 80 % with an apparent 100 % selectivity in aniline. More details and discussion will be given in the presentation. The latter worked have been performed in the context of the EU-project IMPULSE.

[1] Yeong, K. K., Gavriilidis, A., Zapf, R., Hessel, V., Catalyst preparation and deactivation issues for nitrobenzene hydrogenation in a microstructured falling film microreactor, Catalysis Today 81 (2003) 641-651.

Poly-*L*-leucine catalysed chalcone epoxidation reaction in a microstructured PEEK reactor

*Suet-Ping Kee, Asterios Gavriilidis**
Department of Chemical Engineering, University College London, London, UK

Abstract

The poly-*L*-leucine catalysed epoxidation of chalcone reaction which allows access to highly enantioselective chalcone epoxides was investigated. A suitable continuous reaction protocol was previously established for this reaction system, achieving a conversion of 88.4 % and enantioselectivity of 88.8 %. Mixing was found to be critical due to the low diffusivity of the polymeric catalyst. In this work, a microstructured polyetheretherketone (PEEK) reactor with a footprint of 110 mm x 85 mm and production rate of ~ 0.5 g/day was designed. The reactor consists of two micromixer-reactor configurations in series for the deprotonation and epoxidation reactions, with 30 minutes and 16 minutes residence times respectively. A staggered herringbone mixer design was selected for enhancing the mixing in the microstructured reactor. The effects of varying reactant concentrations on reaction performance in the fabricated microstructured reactor were investigated. The base case condition (13.47 g/l PLL, 0.132 mol/l H_2O_2, 0.0802 mol/l chalcone, 0.22 mol/l DBU) was found to be optimal, achieving a conversion of 86.7 % and enantioselectivity of 87.6 %.

Introduction

Recent technological advances in the field of microreaction technology offer the opportunity to combine the benefits of continuous processing with the flexibility and versatility desired in the pharmaceutical and fine chemicals industry (Roberge et al. 2005; Schwalbe, et al 2004). The high surface to volume ratios enable highly efficient heat and mass transfer; these allow precise control of reaction parameters and access to new reaction regimes with improved yield and selectivity which are cost-effective. Production rates in microreactors can be increased by scaling out rather than scaling up, to supply increasing amounts of material for clinical trials as the drug progresses through the regulatory process. Scaling-out allows the process optimization from lab to pilot scales to be bypassed, saving both time and R&D costs. The suitability of the the poly-*L*-leucine (PLL) catalysed asymmetric epoxidation of chalcone for continuous processing has been established previously (Kee & Gavriiiliidis 2007). This work aims to demonstrate the design of a microchemical system for this reaction. This system allows highly enantioselective synthesis of chiral epoxides, which are useful intermediates for the synthesis of active ingredients and hence represent a commercially important synthesis for the pharmaceutical industry.

Chalcone is epoxidised to (2R,3S)-chalcone epoxide, in the presence of homogeneous polyethylene glycol-poly-*L*-leucine (PEG-PLL) catalyst, hydrogen peroxide extracted from urea-hydrogen peroxide adducts as oxidant and 1,8-diazabicyclo[5.4.0]undec-7-ene(DBU) as base. The solvent is a mixed Tetrahydrofuran (THF) and Acetonitrile (ACN) solvent, with a THF: ACN ratio of 2.15. The competing background epoxidation occurs in the absence of PEG-PLL catalyst, with chalcone epoxidised to both (2R,3S)-chalcone epoxide and (2S,3R)-chalcone epoxide, resulting in a racemic product. The rate equations for both the main and competing reactions at 23.1°C are shown in [1] and [2], with the values of the kinetic constants indicated in Table 1 (Mathew 2003):

* Corresponding author, Tel +44 (0)20 7679 3811, Fax: +44 (0)20 73832348
email:a.gavriilidis@ucl.ac.uk

$$Rate_{PLL-Catalysed} = \frac{k\,[Cat][Chalcone][H_2O_2]}{\left(1 + K_{Chalcone}[Chalcone] + K_{H_2O_2}[H_2O_2]\right)} \quad [1]$$

$$Rate_{Background} = k_{BG}\,[Chalcone][H_2O_2] \quad [2]$$

Table 1: Values for kinetic constants used in the rate equations for both catalysed and background chalcone epoxidation

k (l^2/g.mol.min)	$K_{Chalcone}$ (l/mol)	K_{H2O2} (l/mol)	k_{BG} (l/mol.min)
0.143	2.172	1.092	0.132

In our previous work (Kee & Gavriiiliidis 2007), the batch process was improved to prevent formation of precipitate and allow the reaction to be transferred to a continuous flow process. Following that, a continuous reaction protocol was established and several design issues were identified, notably insufficient mixing due to the low diffusivity of PEG-PLL. A single continuous reactor comprising of two mixer-reactors in series for the deprotonation and epoxidation reactions was therefore designed with the following requirements:

i. Enhanced mixing using a staggered herringbone mixer prior to reaction
ii. Deprotonation of peroxide by DBU in the presence of PEG-PLL in 15 – 30 minutes

Reactor Design/Fabrication and Experimental Set-up

Channels fabricated onto a planar surface do not have a circular cross section but a rectangular cross-sectional area which is better represented by slits. The flows in microchannels are typically laminar, with a parabolic flow profile. The velocity gradient across the rectangular channel causes a gradient in composition across the depth of the channel. Molecular diffusion tends to alleviate this gradient, hence the channels can be dimensioned appropriately to minimise the composition gradient. In a tubular reactor, for plug flow to be applicable the following criteria must be met (Raja et al. 2000):

$$\frac{d}{L} << \mathrm{Re}_d\, Sc << \frac{L}{d}, \quad \text{where } \mathrm{Re}_d\, Sc = \frac{\rho u d}{\mu}\frac{\mu}{\rho D_n} = \frac{ud}{D_n} \quad [3]$$

The lower bound ensures that the axial diffusive transport over the length scale L is negligible compared to axial convective transport, while the upper bound ensures that there are no radial variations. The reaction channel was sized for 16 minutes residence time at a volume of 0.32 ml. The selected channel dimensions for both deprotonation and epoxidation reactors are shown in Table 2.

Table 2: Reaction channel dimensions

	Deprotonation	Epoxidation
Channel Width (μm)	2000	2000
Channel Height (μm)	330	330
Channel Length (mm)	450	480
Residence time (min)	30	16

A laminar slit flow reactor was formulated, utilising kinetics [1] and [2] and was compared with experimental results. Both conversion, X and enantioselectivity, S values were computed using mixing cup average outlet concentrations.

PEEK is chemically compatible with THF and Acetonitrile below 70°C and was selected as it is relatively inexpensive and amenable to mass replication. The reactors were fabricated by Epigem (UK) using photolithographic methods for microstructures (< 150 μm deep and 300 μm wide) and machining methods for larger mili scale structures. The fabrication process uses an SU-8 layer spin coated on a substrate (which can be PEEK or

Acrylic), where the micro structures are patterned on the SU-8 layer and are therefore completely surrounded by SU-8. The mili scale structures, on the other hand consist of both substrate and SU-8. The reactors are bonded via diffusion bonding using SU-8. Figure 1 shows an assembled acrylic (not PEEK) reactor for visualization of the channels.

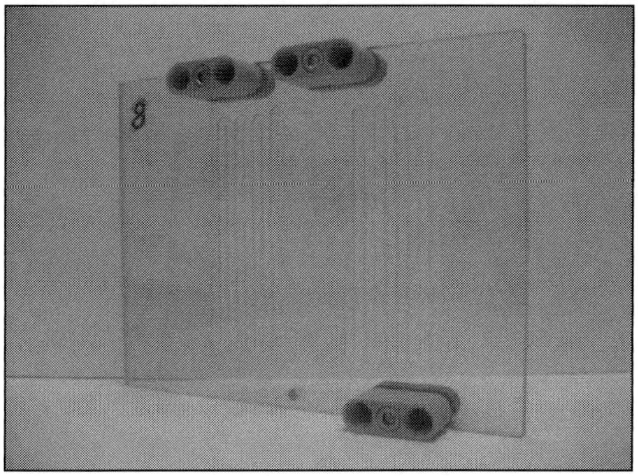

Figure 1: Assembled acrylic micro-structured reactor (85 mm x 110mm)

An XP 3000 Modular Digital Pump (Cavro) with three 50-μl syringes was used to pump solutions of chalcone, DBU and catalyst with peroxide.Three inline 2 μm stainless steel filters were connected at all three reactant inlets to prevent clogging of the reactor. DBU and PLL/peroxide solutions were pumped at a flowrate of 5 μl/min each, to the first staggered herringbone mixer on the PEEK reactor and entered a 0.3 ml delay loop with 30 minutes residence time. Subsequently, a 10 μl/min chalcone flow joined the combined streams in the second staggered herringbone mixer on the PEEK reactor in a 0.32 ml delay loop with a total residence time of 16 minutes. The PEEK reactor was maintained at the desired temperature by placing it in a water bath. The reaction was quenched at the reactor outlet by collecting the outlet flow in a stirred vial containing sodium sulphite.

Results and Discussion

Base Case

The PEEK reactor was tested with the PLL-catalysed asymmetric epoxidation of chalcone at base case conditions (PLL: 13.47 g/l, peroxide: 0132 mol/l, chalcone: 0.0802 mol/l, DBU: 0.22 mol/l). The results are shown in Table 3.

Table 3: Comparison of base case chalcone epoxidation at 23.1°C and 16 minutes residence time with other reactors and models under similar operating conditions.

	Conversion (%)	Enantioselectivity (%)
PEEK reactor	86.7	87.6
Slit flow reactor model	89.6	92.4
Continuous tubular reactor	88.4	88.8

There is good agreement with the values predicted using the slit flow reactor model and with previous experimental results obtained using a continuous tubular reactor at the same reaction conditions (Kee & Gavriiiliidis 2007). The predicted enantioselectivity value was larger than the experimental values although the experimental values were still within 5 % of the predicted value.

Effect of Chalcone Concentration

The changes in conversion and enantioselectivity with chalcone concentration are shown in Figures 2 and 3 respectively. The predicted values are also shown in dashed lines. The conversion gradually decreased from 87.9% to 64.5 % as the chalcone concentration was increased from 0.0401 to 0.1604 mol/l, displaying the same trend predicted by the model. The enantioselectivity gradually increased from 82.5 % to 88.0 % as the chalcone concentration was increased, while the model predicted a somewhat constant enantioselectivity.

Figure 2: Effect of chalcone concentration on conversion

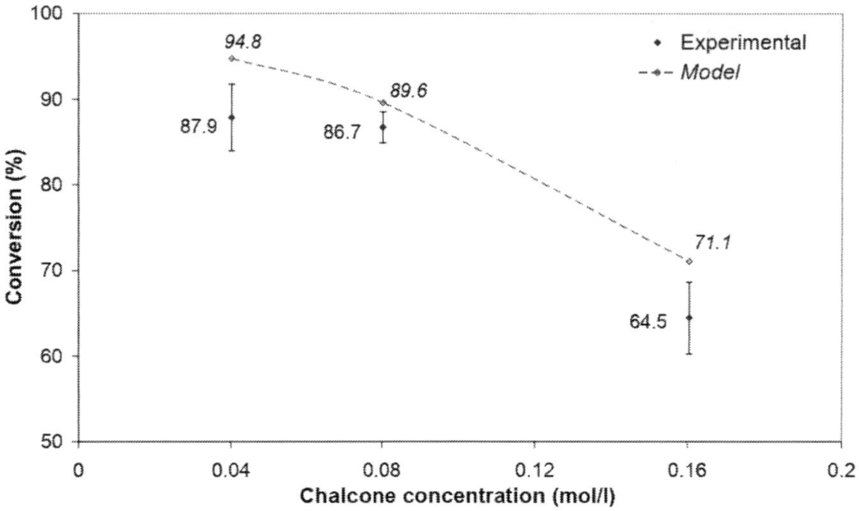

Figure 3: Effect of chalcone concentration on enantioselectivity

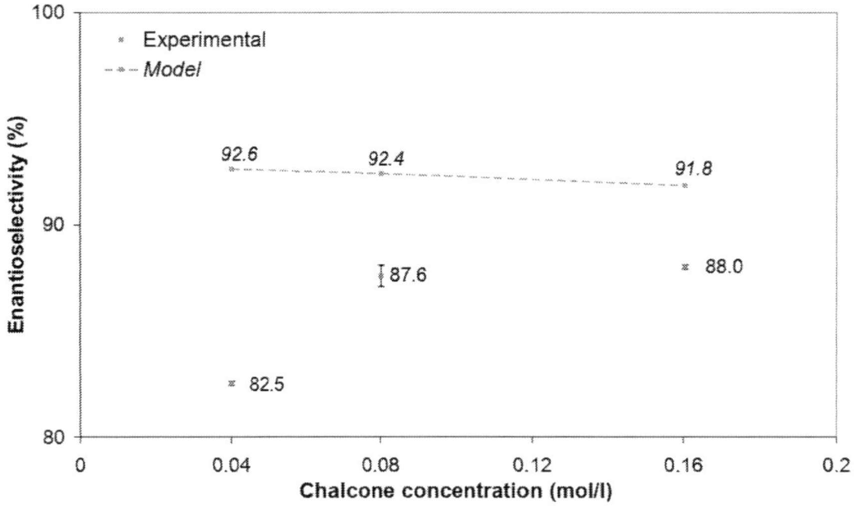

The higher predicted conversion and enantioselectivity values at all chalcone concentrations is likely due to higher catalytic activity in the batch of catalyst used to determine the rate equations, as catalytic activity has been observed to differ slightly from catalyst batch to batch. The difference between the predicted trend in enantioselectivity compared to experimentally observed trend can be attributed to the fact that the rate equation for the background reaction was obtained in the absence of the catalyst while in the presence of the catalyst this reaction is slowed down considerably, possibly due to the lack of free perhydroxyl ions, consistent with the view that the perhydroxyl ions are sequestered by the PLL catalyst (Lopez-Pedrosa et al. 2004). As the chalcone concentration is increased, the model predicts a decrease in the relative rate of catalysed to background reaction and hence a decrease in enantioselectivity. However, due to perhydroxyl ion sequestration by the catalyst, the background reaction is hindered and only the catalysed reaction rate will increase with an increase in chalcone concentration, resulting in improved enantioselectivity.

<u>Effect of Peroxide Concentration</u>

The changes in conversion and enantioselectivity with a change in peroxide concentration are shown in Figures 4 and 5 respectively. The conversion increased from 71.8 % to 86.7 % and then decreased to 69.7 % as the peroxide concentration was increased from 0.066 mol/l to 0.246 mol/l, in contrast with a predicted increase from 61 % to 99.2 % with increasing peroxide concentration. The enantioselectivity however showed a slight decrease from 91 % to 87.6 % and then decreased sharply to 63.8 % at 0.246 mol/l, in contrast with the predicted decrease from 92.6 % at 0.066 mol/l peroxide to 91.8 % at 0.246 mol/l peroxide.

Figure 4: Effect of peroxide concentration on conversion

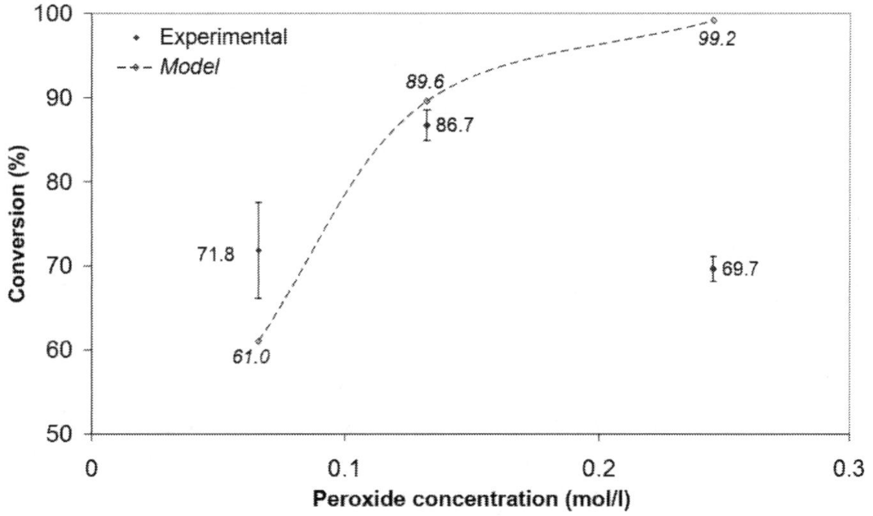

Figure 5: Effect of peroxide concentration on enantioselectivity

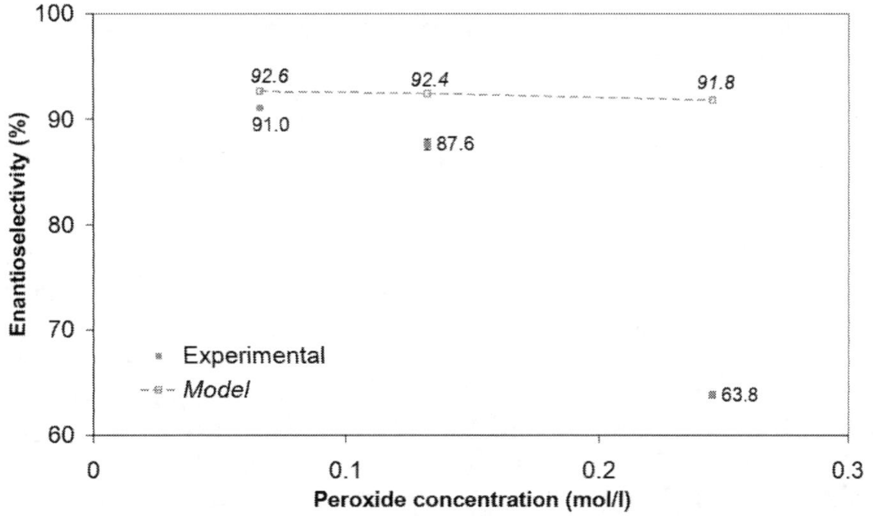

The deviation from expected performance can be explained by the fact that the reactive species is not the peroxide itself but rather the perhydroxyl ion (OOH^-) and the change in reactive species concentration with a change in peroxide concentration was not captured by the model. The concentration of this reactive species at the start of the reaction is dependent on the intial concentrations of both the peroxide and DBU and the residence time in the deprotonation reactor. The deprotonation of peroxide increases the concentration of the reactive species while the base catalysed decomposition of peroxide (Evans & Upton 1985) decreases the concentration of the reactive species. The base catalysed decomposition of peroxide is given by [4] and the corresponding rate equation has been shown to be of type [5].

$$H_2O_2 + OOH^- \rightarrow H_2O + OH^- + O_2 \qquad [4]$$

$$Rate_{decomposition} = k\,[H_2O_2][OOH^-] \qquad [5]$$

These factors suggest that an optimum condition may exist where the concentration of the reactive species is maximised. Other possible explanations for the drop in conversion at 0.246 mol/l peroxide concentration include incomplete deprotonation (longer time required to reach equilibrium concentration) and the base catalysed decomposition of peroxide (Evans & Upton 1985), which increases with increasing perhydroxyl ion concentration and may result in increased formation of oxygen bubbles which can reduce the residence time in the reactor.

The higher predicted enantioselectivity values at all peroxide concentrations can be explained in part by the higher catalytic activity of the catalyst batch used for the kinetic studies. The big decrease in experimental enantioselectivity at peroxide concentration of 0.246 mol/l is due to the increased availability of free peroxide, because of the lower ratio of perhydroxyl ion concentration to catalyst concentration. The higher rate of background reaction to rate of catalysed reaction implies that substrate saturation of the catalyst has been reached and therefore any further increase in peroxide concentration would only increase the background reaction.

Conclusions

The results of this work demonstrate the complexity of the epoxidation reaction of chalcone and the many factors which influence the system performance. The system is essentially a two-step reaction process, with the concentration of the perhydroxyl ions for the epoxidation influenced by the outcome of the deprotonation step. Methods to optimise the reaction conditions for PLL catalysed asymmetric epoxidation of chalcone must therefore take into account the effects of varying parameters on the deprotonation reaction since this affects the initial concentration of the peroxy anion. The results also suggest that the base case condition was the optimal condition.

Acknowledgements

Financial support from the Foresight LINK programme is gratefully acknowledged.

References

Evans, D. F. & Upton, M. W. 1985, "Studies on Singlet Oxygen in Aqueous-Solution .4. the Spontaneous and Catalyzed Decomposition of Hydrogen-Peroxide", Journal of the Chemical Society-Dalton Transactions no. 12, pp. 2525-2529.

Kee, S. P. & Gavriiiliidis, A. 2007, "Batch versus continuous mg-scale synthesis of chalcone epoxide with soluble polyethylene glycol poly-L-leucine catalyst", Journal of Molecular Catalysis A-Chemical, vol. 263, no. 1-2, pp. 156-162.

Lopez-Pedrosa, J. M., Pitts, M. R., Roberts, S. M., Saminathan, S., & Whittall, J. 2004, "Asymmetric epoxidation of some arylalkenyl sulfones using a modified Julia-Colonna procedure", Tetrahedron Letters, vol. 45, no. 26, pp. 5073-5075.

Mathew, S. 2003, From Micrograms to Multikilos Report: Kinetics of epoxidation of chalcone to chalcone epoxide using soluble poly-L-leucine catalyst, University of Hull, Hull, U.K..

Raja, L. L., Kee, R. J., Deutschmann, O., Warnatz, J., & Schmidt, L. D. 2000, "A critical evaluation of Navier-Stokes, boundary-layer, and plug-flow models of the flow and chemistry in a catalytic-combustion monolith", Catalysis Today, vol. 59, no. 1-2, pp. 47-60.

Roberge, D. M., Ducry, L., Bieler, N., Cretton, P., & Zimmermann, B. 2005, "Microreactor technology: A revolution for the fine chemical and pharmaceutical industries?", Chemical Engineering & Technology, vol. 28, no. 3, pp. 318-323.

Schwalbe, T., Autze, V., Hohmann, M., & Stirner, W. 2004, "Novel innovation systems for a cellular approach to continuous process chemistry from discovery to market", Organic Process Research & Development, vol. 8, no. 3, pp. 440-454.

101d Application of Encapsulated Enzymes in a Permselective Microscale Reactor

E. Ekkehard Barth, Sebastian Lohse, and David W. Agar. Biochemical and Chemical Engineering, Laboratory of Technical Chemistry B, Technische Universität Dortmund, Emil-Figge-Str. 66, D-44227 Dortmund, Germany

The immobilisation, or more generally heterogenisation, of homogeneous catalysts is a well-established technique for facilitating the separation or retention of an active catalytic component. For conventional supports the resulting mass transfer resistance often becomes a limiting factor, with only partial utilisation of the active internal surface area as a consequence. Such unfavourable concentration profiles within the support can reduce both conversions and selectivities. Such phenomena arise in most enzyme immobilisation techniques. One of these – microencapsulation – localises the resistance to mass transfer in the enclosing membrane, whilst free diffusion takes place within the liquid-filled capsule. This structured architecture can lead to higher effectiveness factors than for comparable entrapment immobilisation procedures. Furthermore, by bestowing permselective properties on the membrane shell, one can transform the curse of mass transfer resistance into the blessing of controlled catalyst access. By integration of the 'unit operation' of permselective diffusion at the microlevel, one achieves both process intensification through a good harmonisation between the rates of reaction and separation [1] as well as a powerful tool for manipulating the behaviour of non-selective catalytic sites, in analogy to the size-exclusion behaviour in zeolitic catalysis.

In the work presented, the non-selective hydrolytic enzyme dextranase and a permselective alginate membrane material were chosen as a test system for this advanced microencapsulation concept. Various types of hollow and solid spherical microcapsules containing dextranase were prepared using a specially designed nozzle with a superimposed peripheral air jet. The composition of the alginate was varied with respect to the two constituent monomers: ß-D-mannuronate and α-L-guluronate, the composition of the gelation solution of alkali earth metal ions [2] and using additives such as silica. The microbeads and microcapsules generated were characterised with respect to their physical structure, their permeability behaviour, enzyme leakage and catalytic performance. Variation of the experimental parameters enabled the systematic manipulation of both microcapsule size (1.8 mm) and the membrane thickness (> 150 μm). Initial work [3] demonstrated that the activity of the encapsulated biocatalyst in both microbeads and microcapsules is inferior to that of free dextranase, but also, as expected, that hollow microcapsules are superior to solid microbeads [4]. Enzyme leaching could be cut to a tolerable rate using low levels of colloidal silica to fine tune the diffusive properties of the alginate membrane. Subsequent studies revealed that the encapsulation of catalysts in microcapsules can be exploited, as envisaged, to improve their selectivity. In a mixed substrate containing two distinct dextran fractions with different molecular weight ranges (1.1 vs. 15-20 kDa) only the shorter oligosaccharide chains were hydrolysed to glucose. Although the permselectivity of the alginate membranes is based the exclusion of larger molecules above a given molecular weight threshold, the principle could be extended to alternative or multiple separation criteria to further enhance substrate selectivity and thus has considerable potential for conducting multistep reactions in a single vessel.

[1] Chai Y. Gelation Conditions and Transport Properties of Hollow Calcium Alginate Capsules. Biotechnology and Bioengineering. 87 (2). 228 233. 2004

[2] Yamagiwa K. Effects of alginate compositions and gelling conditions on diffusional and mechanical properties of calcium alginate gel beads. J. Chem. Eng. Japan. 28 (4). 463 467.1995

[3] Barth E E, Ufer A, Agar D W. Selectivity enhancement of microencapsulated enzymes with permselective shells. European Congress of Chemical Engineering (ECCE-6). Copenhagen, Denmark. 2007

[4] Konsoula Z, Liakopoulou-Kyriakides M. Starch hydrolysis by the action of an entrapped in alginate capsules α-amylase from Bacillus subtilis. Process Biochemistry. 41. 343 349. 2006

101f Development of Continuous Culture Microbioreactors

Daniel Schäpper[1], Rita L. Fernandes[1], Krist V. Gernaey[1], Anna Eliasson Lantz[2], Stuart Stocks[3], and Nicolas Szita[4]. (1) Department of Chemical Engineering, Technical University of Denmark, DTU - Building 227/250, Søltofts Plads, 2800 Kgs. Lyngby, Denmark, (2) BioCentrum-DTU, Technical University of Denmark, DTU - Building 227/250, Søltofts Plads, 2800 Kgs. Lyngby, Denmark, (3) Novozymes A/S, Krogshoejvej 36, 2880 Bagsværd, Denmark, (4) Department of Biochemical Engineering, University College London, Torrington Place, London WC1E 7JE, England

In industrial fermentation processes, starting up a new production is usually preceded by a tremendous research effort in which for example the productivity of different candidate production strains is compared (=screening). Ranking of strains according to productivity resulting from the screening phase is crucial in selecting strains for full-scale production. Initially, screening is done in microtiter plates or by using shake flask cultures. Experiments are rather easy to set up, but only batch experiments are possible. The information gained per experiment is limited, and typically only end-point measurements are performed. The ranking of strains resulting from these experiments does sometimes not reflect results at industrial scale, for example because these batch type processes are characterized by more extreme forms of nutrient excess/limitation compared to the fed-batch processes used at larger scales. In a later stage of production process development, experiments are performed in bench scale reactors to investigate the influence of process conditions on productivity. Bench scale reactors (typically with a volume of 1 to 10 L) have the advantage that they allow on-line measurements, and they are flexible since they can be operated in batch or fed-batch, but also as a continuous culture. However, the effort needed to prepare, operate and subsequently clean bench scale reactors is vast.

Microbioreactors offer the possibility to circumvent many of the above-mentioned problems: (1) Scaling out microbioreactors to systems with many parallel reactors allows for high-throughput screening; (2) The working volumes are very small (μL to mL range), keeping costs for culture media low; (3) On-line measurements are possible for the most important culture variables (optical density (OD), dissolved oxygen (DO), pH); (4) The reactors can be fabricated from polymers, thus making them disposable after use which greatly reduces assembly, cleaning and sterilisation efforts, thereby further reducing cost; (5) Finally, for batch type microbioreactors, the fermentation variables compare favorably with bench scale reactors, which indicates that the right culture physiology can be maintained at small vs. larger scale. Therefore, the ranking of any strains tested can be preserved across scales, making the technology commercially viable.

This work reports on the development of a continuous culture microbioreactor platform that can perform experiments with yeast (Saccharomyces cerevisiae). Compared to a batch experiment, the continuous culture has the advantage that steady-state conditions can be achieved. Additionally, it should be possible to induce step changes in the dilution rate, forcing the culture from one steady-state to the other with continuous measurement of the important culture variables, thus leading to dynamic information on the behavior of the culture under well-controlled experimental conditions.

The microbioreactor – with a volume of 100 μL – is fabricated out of the polymers poly (methylmethacrylate) (PMMA) and poly(dimethylsiloxane) (PDMS). Dissolved oxygen (DO) and pH are both measured with fluorescent sensor spots. Optical density (OD) is measured in the reactor itself. Contrary to conventional bench-scale reactors, microbioreactors are designed to work bubble-free: aeration is done through a semi-permeable PDMS membrane. A simpler membrane fabrication methodology was developed, allowing for cheaper manufacturing. Proper mixing is essential for good cultivation results in the microbioreactors, since substrate gradients might lead to a varying (location dependent) metabolic state of the culture. In the projected volume range, convective mixing is difficult to achieve due to the small Reynolds numbers. On the other hand, the volume is too large to be able to rely on diffusion alone. Different methods for mixing the reactor contents were investigated. Finally, results of S. cerevisiae cultivations performed in the microbioreactor will be reported and compared to literature data.

102a Residence Time Distribution Studies in Microchannels with Staggered Herringbone Structures

Alberto Cantu-Perez and Asterios Gavriilidis. Dept of Chemical Engineering, University College London, Torrington Place, London, WC1E 7JE, United Kingdom

Characterisation of residence time distribution (RTD) in microchannels is important since RTD has an impact on conversion and selectivity of reactions. This work focuses on numerical simulations and experiments to obtain the RTD for a rectangular microchannel. It also investigates the impact of using staggered herringbone structures on the bottom of the channel in the RTD [1]. The Navier Stokes equations are solved using a commercial finite element method (FEM) software. The RTD is obtained with a particle tracking algorithm in which a random-walk model is incorporated to simulate the effect of diffusion. Experiments are carried out by injecting a small quantity of a tracer dye and the outlet concentration is recorded as a function of time by means of a UV detector. The results indicate that the structured channel has a narrower RTD, which may be approximated by a model of axial dispersion exchanging mass with a stagnant zone [2]. Both simulations and experiments show that the difference between the RTD of a structured and unstructured channel is not substantial at short channel lengths, however this difference becomes more pronounced at longer lengths. The improved features of the structured microchannels open the possibility of using larger structured channels that behave in similar way to smaller unstructured channels. This can potentially eliminate some of the problems associated with smaller channels like higher pressure drop and susceptibility to clogging.

References

1. Stroock, A. D.; Dertinger, S. K.; Ajdari, A.; Mezic, I.; Stone, H. A.; Whitesides, G. M. Chaotic mixer for microchannels. Science 2002, 295 (5555), 647-651.

2. Castelain, C.; Berger, D.; Legentilhomme, P.; Mokrani, A.; Peerhossaini, H. Experimental and numerical characterisation of mixing in a steady spatially chaotic flow by means of residence time distribution measurements. International Journal of Heat and Mass Transfer 2000, 43 (19), 3687-3700.

A multi-objective analysis and optimization methodology for the design of passive micromixers based on their own topology

Cesar Augusto Cortes-Quiroz[1], Mehrdad Zangeneh[1] and Akira Goto[2]
(1) University College London, Department of Mechanical Engineering
Torrington Place, London
WC1E 7JE, UK
(2) Ebara Research Co. Ltd., Japan

The development of bio-MEMS and lab-on-a-chip devices requires efficient mixing of fluids. With this aim, a variety of passive and active micromixers have been developed. Computational Fluid Dynamics (CFD) performs now an important role in the development of microfluidics components, basically for the analysis of flow fields to adjust the design parameters and for the evaluation of the final design. But intuition and experience of the designer is usually behind the application of CFD for the design improvement. Therefore, the process has basically an empirical approach and it is still a build-and-test design procedure.

According to the strategy employed, the approaches to the optimization of microfluidics components can be grouped as follows: a) the modification of some geometry elements to improve the device performance, as it is made with the shape of a channel to minimize dispersion in an electrokinetically driven flow (Molho *et al* 2001, Mohammadi *et al* 2003), b) the definition of a set of geometric parameters in a component to examine by CFD the corresponding effects, as in the case of passive grooved micromixers (Stroock *et al* 2002a, Stroock and McGraw 2004, Aubin *et al* 2005, Yang *et al* 2005, Lynn and Dandy 2007), c) the combination of components to form more complex systems with different mixing effects as a result, e.g. Wang *et al* (2004, 2005) use analytical models in electrokinetic flows to evaluate the individual effects on transverse diffusion and convective dispersion generated by elements like straight or tapered channels, 90° bends and T-junctions, and predict the combined effect when these elements form composite geometries; Mott et al (2006) present a computational 'toolbox' where the flow field generated by one isolated feature, such as an specific groove, is determined by solving numerically the governing flow equations, the result is transposed to an advection map that projects a conserved scalar field across the feature to predict the lateral transport of fluid; the sequential application of the maps can predict the outflow distribution in designs that combine the basic features and helps to select the optimal combinations, and d) the use of numerical optimization techniques to improve the performance of microdevices, e.g. genetic algorithms have been used to define the optimum frequencies of the transverse injected flows controlled actively in a cross-channel micromixer (Müller *et al* 2004), as well as to minimize dispersion and reduce mixing time by optimizing the shape of microchannels (Ivorra et al 2006); very recently, Ansari and Kim have used three-dimensional Navier-Stokes analysis and the Radial Basis Neural Network (RBNN) method (2007a) as well as the Response Surface Method (RSM) technique (2007b) to optimize the shape of the grooves of a Staggered Herringbone Mixer, SHM (Stroock *et al* 2002), for enhancing mixing.

Therefore, there have been very few attempts in which automatic optimization techniques are exploited in the design of microfluidics components. The numerical optimization techniques applied so far do not seem to integrate yet a systematic and automatic optimization methodology that can be used in the design of optimum micromixers. Even more, they have dealt only with one performance criterion, the degree of mixing achieved at the outlet of the micromixer, and other outputs as it is the pressure loss are not included in the optimization of the designs. For this to happen, a multi-objective optimization methodology has to be implemented.

In this paper, a design and optimization methodology is presented which systematically integrates CFD with an optimization strategy based on the use of Design of Experiments (DOE), Function Approximation (FA) and Multi-Objective Genetic Algorithm (MOGA) techniques. The DOE method is a statistical approach to explore the design space for a given number of design parameters; the DOE creates an "experimental table" of design points which are then evaluated by CFD to get the values of the performance parameters. The design and performance parameters are used by the FA method to create the response surfaces or objective functions that correlate these parameters and can be used to carry out sensitivity analysis. Finally the MOGA is run on the objective functions to determine the Pareto Front (PF) which represents the optimum values of design parameters and the best compromise between the performance parameters can be chosen to fill the design requirements.

Passive micromixers are attractive to MEMS engineers since no extra forcing mechanism is required for mixing. In particular, the SHM is of great interest due to its effectiveness for flows at low Reynolds numbers (Re < 10) and it has been used for applying the proposed methodology. Geometric parameters (design factors) of the SHM (Figure 1) have been examined by choosing a set of discrete values in appropriate ranges (Table 1) and combining them in arrays defined by the DOE. These arrays give the set of simulations which are solved by CFD.

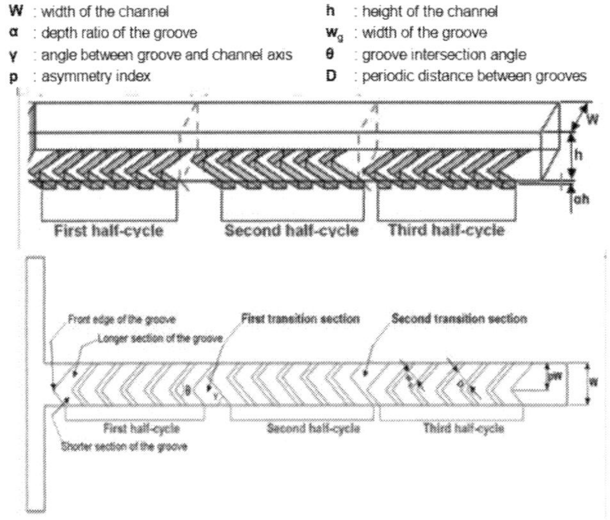

Figure 1. Geometric parameters of the SHM model.

Table 1. Design factors and levels for the DOE

Levels \ Factors	A. w/h	B. α	C. w_g/D	D. p
1	1.6	0.25	0.25	0.55
2	2.4	0.5	0.5	0.7
3	3.2	0.75	0.75	0.85

Mixing efficiency and Pressure drop in the mixer are the performance parameters in evaluation. The set of design factors and performance parameters define the input data for the Function Approximation method which approximates the objective functions with response surfaces that can be visualized as the example in figure 2. The MOGA is applied on these approximated functions to give a set of design points from which the PF of optimum designs can be determined (Figure 3). To validate the designs on the PF, CFD simulations are run for comparison of results. The improvement in mixing can be seen in figure 4 by comparing the contours of species concentration on cross-sections of the mixing channel of the original and the optimum models. The results show comparatively the improvement in mixing for the required design specifications.

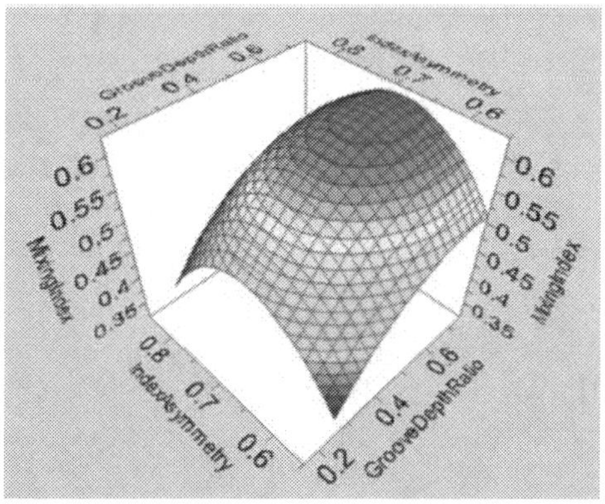

Figure 2. Response surface generated with the FA.

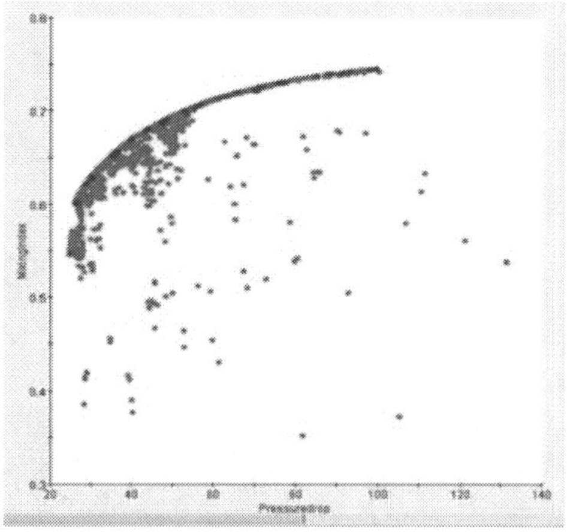

Figure 3. Iteration points showing the PF from the application of MOGA on Pressure drop vs. Mixing Index optimization.

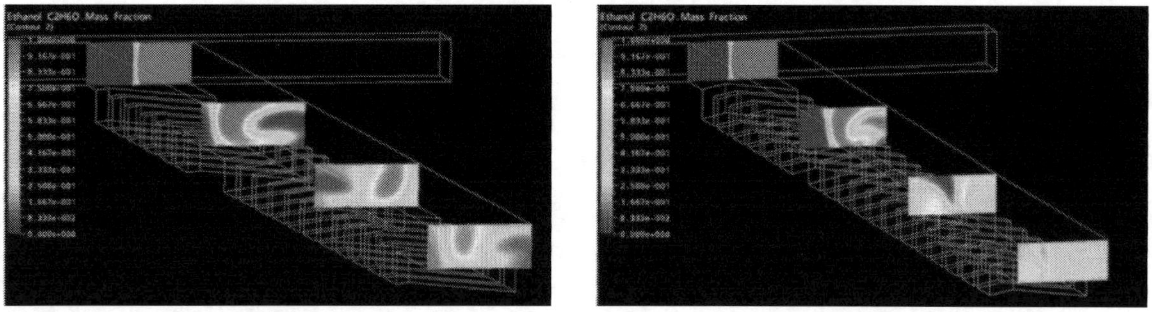

(a) (b)

Figure 4. Mass fraction (concentration) contours on different cross-section planes of the mixing channel: (a) Original mixer. (b) Optimum mixer.

The effects on mixing of selected geometric parameters of the SHM have been evaluated and the parameters optimized accordingly. On the performance parameters, the optimization gives an efficient mixing with the highest mixing index and lowest pressure drop for any combination of the optimum design parameters. The methodology allows the simultaneous investigation of the effect of geometric features on the mixing performance of micromixers whose design strategy is based fundamentally on the generation of chaotic advection.

References

Ansari M. A. and Kim K. –Y. (2007). Application of the Radial Basis Neural Network to Optimization of a Micromixer. *Chemical Engineering Technology* **30** (7), 962-966.

Ansari M. A. and Kim K. –Y. (2007). Shape optimization of a micromixer with staggered herringbone groove. *Chemical Engineering Science* **62**: 6687-6695.

Aubin J., Fletcher D. F. & Xuereb C. (2005). Design of micromixers using CFD modelling. *Chemical Engineering Science* **60** (8-9), 2503-2516.

Ivorra B., Hertzog D. E., Mohammadi B. & Santiago J. G. (2006). Semi-deterministic and genetic algorithms for global optimization of microfluidic protein-folding devices. *Int. Journal for Numerical Methods in Engineering* **66**: 319-333.

Lynn N. S. and Dandy D. S. (2007). Geometrical optimization of helical flows in grooved micromixers. *Lab on a Chip* **7**: 580-587.

Mohammadi B., Molho J. I. & Santiago J. G. (2003). Incomplete sensitivities for the design of minimal dispersion fluidic channels. *Comput. Methods Appl. Mech. Eng.* **192** (37-38), 4131-4145.

Molho J. I., Herr A. E., Mosier B. P., Santiago J. G., Kenny T. W., Brennen R. A., Gordon G. B. & Mohammadi B. (2001). Optimization of turn geometries for microchip electrophoresis. *Anal. Chem.* **73** (6), 1350–1360.

Mott D. R., Howell Jr. P. B., Golden J. P., Kaplan C. R., Ligler F. S. & Oran E. S. (2006). Toolbox for the design of optimized microfluidic components. *Lab on a Chip* **6**: 540-549.

Müller S.D., Mezić I., Walther J. H. & Koumoutsakos P. (2004). Transverse momentum micromixer optimization with evolution strategies. *Computers & Fluids* **33**: 521-531.

Stroock A. D., Dertinger S. K. W., Ajdari A., Mezić I., Stone H. A. & Whitesides G. M. (2002). Chaotic mixer for microchannels. *Science* **295**: 647-651.

Stroock A. D. and McGraw G. J. (2004). Investigation of the staggered herringbone mixer with a simple analytical model. *Philos. Trans. Royal Soc. London, Ser. A* **362** (1818), 971-986.

Wang Y., Lin Q. A. & Mukherjee T. (2004). System-oriented dispersion models of general-shaped electrophoresis microchannels, *Lab on a Chip* **4** (5), 453-463.

Wang Y., Lin Q. A. & Mukherjee T. (2005). A model for laminar diffusion based complex electrokinetic passive micromixers. *Lab on a Chip* **5** (8), 877-887.

Yang J. -T., Huang K. -J. & Lin Y. -C. (2005). Geometric effects on fluid mixing in passive grooved micromixers. *Lab on a Chip* **5** (10), 1140-1147.

65e Continuous Synthesis of Environmentally Friendly Lubricants in Scalable Reactors

Daniela H. Mueller, **Manuela Wielgosch**, Denise Herzog, Florian Pontzen, Marcel A. Liauw, and Lasse Greiner. Itmc, RWTH Aachen, Worringerweg 1, Aachen, 52074, Germany

The continuous synthesis of environmentally friendly lubricants based on oleic acid methyl ester is studied as part of the collaborative research center 442 (SFB 442). Overall aim is to develop environmentally friendly high performance tribosystems. Therefore, engineers, material scientists, chemists, and biologists are working closely together. Our research is focused on the development of new environmentally friendly lubricants. We use oleic acid methyl ester from renewable sources as starting material. This is processed to give a lubricant within the two steps epoxidation and successive alcoholysis.

The reaction is currently carried out routinely as a batch process on a 50L scale to satisfy the needs of the collaborators for the use in machining and property testing. For the development of an environmentally friendly product, evaluation of toxicity and biodegradability has to be taken into account which could be demonstrated successfully for the batch process. Thus, besides yield and selectivity these factors are taken into account directly as the process is developed and are used to judge alterations and process intensification directly

We will present results concerning the transfer to a continuous process and the impact on product properties and performance. Part of the strategy for process intensification is the use of microreaction engineering.

102b Evaluation of Numerical Methods for Describing Mass Transfer in Microstructured Devices Using Experimentally Determined Residence Time Distributions

Sebastian Lohse[1], Jens F. Acker[2], Zhu Fan[1], E. Ekkehard Barth[1], Joachim Franzke[3], Stefan Turek[2], and David W. Agar[1]. (1) Biochemical and Chemical Engineering, Laboratory of Technical Chemistry B, Technische Universität Dortmund, Emil-Figge-Str. 66, D-44227 Dortmund, Germany, (2) Institute of Applied Mathematics, University of Dortmund, Vogelpothsweg 87, Dortmund, 44227, Germany, (3) Miniaturisation, ISAS - Institute for Analytical Sciences, Bunsen-Kirchhoff-Str. 11, D-44139 Dortmund

The benefits of miniaturisation in chemical reactors for the intensification of chemical processes have been demonstrated by numerous examples over the last few years. Enhanced mass and heat transfer are the most important phenomena contributing to improved reactor performance. Despite experimental proofs-of–concepts for various microreactors, there remains a considerable need for accurate reactor models describing the actual reactor behaviour. In traditional chemical engineering three main modelling levels are distinguished: (i) ideal reactors, (ii) globally characterised real reactors and (iii) locally resolved rigorous reactor models.

The work presented here is focused on an evaluation of the precision and reliability of different numerical approaches to describe the liquid phase mass transfer within microreactors (i.e. level (iii)). Due to the enormous aspect ratios in microreactors, i.e. the ratio of reactor length to the microscaled dimension, an accurate treatment of the mass transfer equations is a major challenge for reliable simulations using computational fluid dynamics (CFD) of an acceptable complexity. In particular, the propagation of steep fronts can be distorted by virtual numerical diffusion effects caused by the discretisation of the second order derivatives in the mathematical model.

An experimental technique generating such steep concentration gradients is the determination of the residence time distribution by measuring the transient response to a Dirac- or an ideal step stimulus signal applied to the reactor inlet at the outlet of the reactor. For this reason data obtained using a novel non-intrusive method for the determination of residence time distributions in microreactors [1] are used to evaluate the simulation results at the global modelling level (ii). An estimate of the error introduced by numerically superimposed diffusion can thus be obtained.

Numerical concepts for simulating residence time distribution in microreactors

Two alternative approaches are conceivable. On the one hand, an Euler-Lagrange approach using virtual (mass-free) particles can be applied to a steady-state solved flow field (Navier-Stokes equation together with continuum condition) as a post-processing tool [2]. In this case, the movement of the individual particles is along the streamlines of the flow field, i.e. no diffusive terms are included in the transport equations for the tracer phase. For the comparison between this kind of simulation and experimental data, a given number of particles is introduced into the inlet cross-section of the microreactor at time zero, weighted according to the actual local flow rate, and the exit time for the individual particles is subsequently recorded at the outlet. Summing the particles obtained over a particular time interval in a histogram then yields the residence time distribution.

In the second technique, the tracer phase for the determination of the residence time distribution is treated as a second continuous phase according to an Euler-Euler approach. Unlike the virtual particles, the continuous tracer phase is affected by two transport mechanisms, namely both convection and diffusion. The resulting system of partial differential equations considers the physically diffusion actually occurring, together with a certain amount of additional numerical diffusion introduced when solving with standard CFD-methods. For comparison of the simulated concentration field, the exit concentration is monitored and integrated over the cross-section, weighted according to the actual flow velocity in each section of the exit plane.

Finally, the Euler-Euler approach was refined by implementing advanced mathematical methods within the solving algorithm for the partial differential equations to diminish the contribution of numerical noise to the diffusive mass transfer (Total variance diminishing (TVD) method) [3].

Experimental technique for determining residence time distribution in microreactors

The experimental determination of residence time distribution is conducted using a novel non-intrusive technique developed by the authors [1]. Basically, a caged fluorescent dye dissolved in a carrier fluid is used and an ideal stimulus step signal is generated by exposing the reactor inlet under steady-state flow conditions to UV-light emitted by a laser diode with a Gaussian-shaped intensity profile, to ensure compatibility of the optical tracer 'injection' process with the prevailing laminar flow conditions. The determination of the residence time distribution is then easily conducted by detecting the two dimensionally resolved fluorescence signal at the reactor outlet and integrating the detected concentrations, weighted according to the laminar flow profile at each position, over the whole cross-section.

Results and discussion

Two different geometries were analysed using the different concepts described: a simple straight rectangular channel (150 μm x 70 μm) with a length of 20 mm was investigated for five different mean residence times and the behaviour of an intricately structured reactor with several parallel channels, as an example of a device designed for internal parallelisation, was examined at the same mean residence times.

In the first case, the analytical solution is well-known for both the Euler-Lagrange as well as the Euler-Euler approach. By virtue of the purely laminar flow behaviour, the residence time distribution obtained using the virtual tracer particles matches the theoretically calculated analytical solution for laminar flow in pipes, first derived by Levenspiel [4], exactly. As would be expected from the enhanced radial mass transfer in microstructured reactors, both the Euler-Euler approach and the experimental results show a significantly stronger influence of diffusive mass transfer and thus deviations from the ideal solution form. Nevertheless, the standard Euler-Euler approach exhibits an even stronger deviation on the same grid, probably caused by additional numerical diffusion effects. First results demonstrate that this effect occurring in particular at high flow rates can be suppressed successfully by applying TVD-methods, for example, when simulating the tracer as a continuous second phase. These results can also be used to estimate the amount of numerical diffusion taking place in microstructured devices. Using the second intricately structured device, the consequences of such effects are elucidated for a system without an analytical solution.

Conclusion

The comparison between various numerical approaches and experimental results for microreactor modelling, using the example of residence time distribution determination is an appropriate test system for the evaluation of mass transfer in a system with sharp concentration gradients, providing a useful indicator of the potential reliability of simulation results using the currently available methods for describing heat and mass transfer with chemical reaction. An accurate determination of diffusive mass transfer is of particular interest in such instances e.g. for the simulation of fast reactions in catalytic wall reactors, where steep gradients over the miniaturised dimension might arise.

[1] S. Lohse, I. Gerlach, D. Janasek, P.S. Dittrich and D.W. Agar, IMRET 9 (2006)

[2] J. F. Acker and S. Turek, Ergebnisbericht 193, Universität Dortmund (2000)

[3] D. Kuzmin and S. Turek, J. Comput. Phys., 198, p. 131 (2004)

[4] O. Levenspiel and W.K. Smith, Chem. Eng. Sci., 6, p. 227 (1957)

46a Microreactor Technology in the Fine Chemical and Pharmaceutical Industry: Is the Revolution Underway?

Dominique M. Roberge[1], Bertin Zimmermann[2], Fabio Rainone[2], and Norbert Kockmann[2]. (1) Chemical Engineering, University of Ottawa, 161 Louis Pasteur (A-329), Ottawa, ON K1N 6N5, Canada, (2) Process Optimization, Lonza, Lonzastr., Visp, 3930, Switzerland

Few years ago we published results of a reaction analysis and performed an initial economical analysis for microreactor technology applied in the fine chemical and pharmaceutical industry (Chem. Eng. Technol. 28, 318-323, 2005). This publication was well received by the international community and a general echo was perceived to further develop the subject. Thus, this presentation will introduce additional economical outcomes from the use of microreactor technology in an industrial context. The analysis is based on the last two years of manufacturing experience.

One of the first comments in this new perspective is that the published reaction analysis should not be used for forecast, namely for the prediction of the number of reaction candidates that can be performed using microreactor technology. That analysis is a factual picture at a given time in space based on batch reactor technology. The future development of various reaction types using a new technology is not foreseen.

The analysis has the major advantage of encouraging the development of new reactor technologies in terms of unit operation. Different modules were developed to perform critical tasks such heating/ cooling, tempering, manifolding and reaction to enable a complete toolbox. Thus, Lonza, together with its partners Corning and Siemens, was able to construct reactor system to operate a large number of reactions in the fine chemical and pharmaceutical industry.

Finally, it is also worth noticing after industrial experience that among the new factors playing a crucial role in the economical benefit of this technology were the additional advantages such as reduction in labor costs, quick change-over (high flexibility) in system setup, and fast kg-scale productions. The goal of this presentation is also to portray a real case study with the calculated economical advantage.

Ionic liquids in continuous flow micro reactors

Johannes Klos, Daniel Wilms, Andreas F. M. Kilbinger, Holger Löwe, and Holger Frey**

Institut für Organische Chemie, Johannes Gutenberg-Universität Mainz, Duesbergweg 10-14, 55099 Mainz, Germany

Abstract

Over the last years, ionic liquids (ILs) have attracted increasing attention due to their thermal stability, non-volatility and almost negligible vapor-pressure. Numerous reports on various applications since have been published. Most of them describe the employment of ILs as alternatives to highly polar solvents for synthetic applications. However, the preparation and subsequent work-up of these compounds is tedious and time-consuming, mostly because purification cannot be achieved by simple distillation methods like for conventional organic solvents. Here we report on the development of an alternative approach for the facile synthesis of various novel imidazolium-based ILs that affords highly pure products without the necessity of subsequent purification steps. The highly exothermic quarternization reaction involving the attack of an alkylating agent (typically an alkyl bromide) on an imidazolium species frequently leads to the formation of undesired side products when the reaction is carried out in a standard batch reactor. Microreactors offer a high degree of control over such reactions because of their high surface-to-volume ratio and superior heat- and mass transport. 1-hexyl-3-methylimidazolium bromide was used as a model compound and prepared in a microreactor under continuous flow conditions in order to evaluate the ideal reaction conditions. Based on the afforded results, we prepared several novel ILs by reacting imidazoles with different alkyl bromides. Within the scope of these studies, we focused on ILs containing differently substituted phenyl rings and characterized these compounds with respect to further use for direct application or subsequent reaction sequences.

Introduction

Ionic liquids (ILs) are fluids consisting of ions, which are differ from conventional molten salts in a variety of properties.[1] While ILs are usually liquid below 100 °C, non-volatile and posses an almost neglible vapor pressure, conventional molten salts have high melting points, viscosity and corrosiveness. The first ILs synthesized were inherently sensitive to hydrolysis.[2-5] Noteworthy potential for practical applications was not given until the first hydrolysis-stable ILs where introduced in the early 1990ies.[6] Within the last 15 years, a wide variety of ILs has been prepared and characterized with respect to their physical and chemical properties. Since then, these species have become a fascinating alternative to classic organic solvents, mainly because of their high dissolving power for otherwise sparingly soluble compounds.[1,7,8] The key step in the synthesis of ionic liquids is the quaternization of an amine, phosphane or oxygen. The nature of the alkylating agent determines the anion. In most cases halide ions are chosen. Due to their inherent non-volatility, ionic liquids cannot be conveniently distilled prior to use as solvents or reactants, which leads to the major challenge in the respective synthesis procedures that have to result in compounds of very high purity, which is essential for the practical use of ILs. This prerequisite commonly requires tedious purification steps subsequent to the synthesis, leading to high cost and limited potential for industrial application.

One of the reasons why microreactors have recently attracted increasing interest is due to their significant benefits originating in very short diffusion pathways and large interfacial contact areas per unit volume (10,000 – 50,000 m^2/m^3)[9] that result in superior heat and mass transfer compared to conventional lab reactors. Therefore, higher yields and selectivities as well as improved product qualities have been achieved by transferring chemical reactions, especially those that proceed highly exothermic or endothermic, from classical batch reactors to microstructured devices.[9-14] Parallel set-ups of several microreactors can provide convenient access to large scales of products.

Synthesis

Herein we present the synthesis of novel ionic liquids in a microstructured reactor, which can be obtained in high purity without any additional purification steps.

All syntheses presented were carried out in a reactor equipped with a stainless steel slit-interdigital mixer provided by the *Institut für Mikrotechnik Mainz* (IMM). The layout of the micromixer allows mixing processes to take place within milliseconds due to the combination of the regular flow pattern created by multi-lamination with geometric focusing (Figure 1). Residence times are very short as the inner volume of the mixer is extremely small (approximately 8 µL). High pressure stability of the set-up (up to 100 bar) permits continuous flow at viscosities of up to 10,000 mPas.

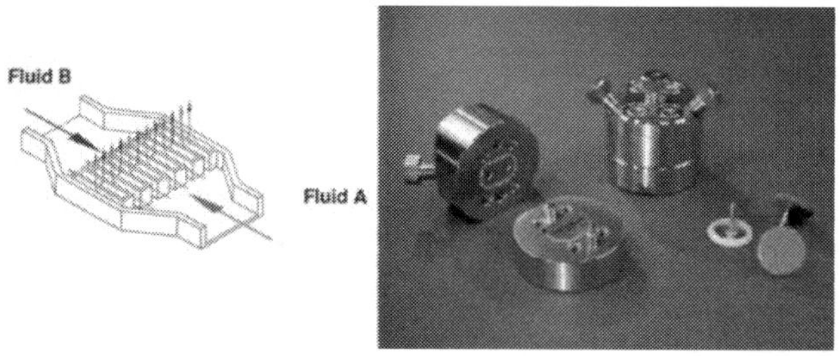

Figure 1. Mixing principle and set-up of the slit-interdigital mixer.

In order to determine the applicability of the microreactor set-up for the intended continuous synthesis of ionic liquids, a well known compound (1-hexyl-3-

methylimidazolium bromide **IL-1**) was targeted. It can be obtained from alkylation of 1-methylimidazole with 1-bromohexane and exhibits very low viscosity at elevated temperatures. Therefore, this product was chosen for the initial test-runs of the applied set-up, which included two HPLC pumps, two loops for preheating the liquid reactants, the micromixer and a heated delay loop. By varying the flow rate and/or the length of the delay loop the reaction time can be easily adjusted.

Compound **IL-1** was obtained as a slightly yellow liquid. The highly efficient mixing process permits working at a temperature of 150 °C without noticeable hotspot formation in reaction times of 5 – 10 min depending on the flow rates. The ^{1}H-NMR spectrum of **IL-1** (Figure 2) shows no impurities. As this method is not very practicable for the determination of small quantities of impurities, HPLC was performed. The chromatograms only show a very little amount of 1-bromohexane remaining in the ionic liquid (less than 1%) which can be easily removed in vacuum.

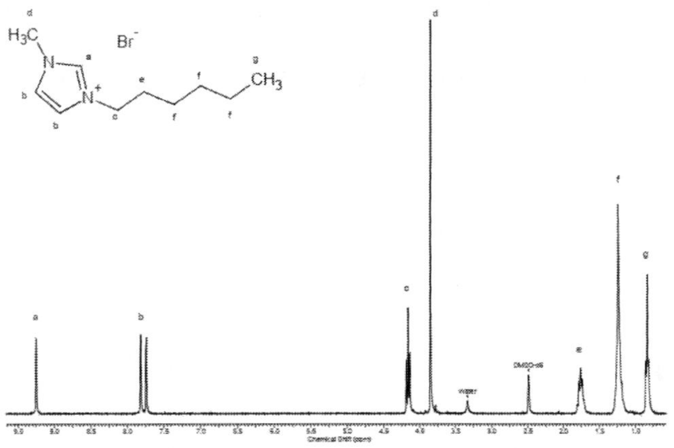

Figure 2. ^{1}H-NMR spectrum of 1-hexyl-3-methylimidazolium bromide (IL-1).

Encouraged by these promising results we directed our efforts towards the synthesis of entirely new ionic liquids that have not been reported previously. These novel compounds were synthesized in the microstructured reactor by alkylation of

1-methylimidazole with different phenylalkyl bormides resulting in **IL-2** to **IL-5** (Scheme 1).

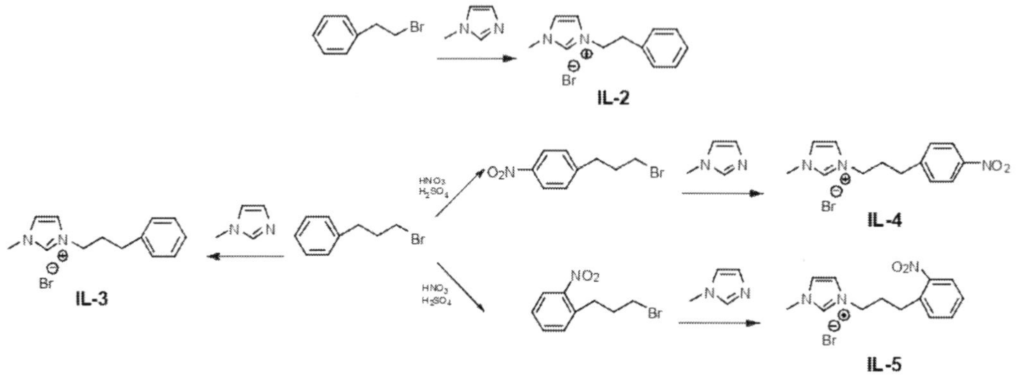

.***Scheme 1.*** Synthesis of phenylalkyl imidazolium bromides.

We considered these compounds to represent an interesting class of ILs for probing and establishing our new versatile continuous flow process. All compounds were obtained as slightly yellow or orange liquids at high throughputs between 100 and 200 ml/h at analogous conditions to the preparation of **IL-1**. The ^{1}H-NMR spectra of these compounds were measured directly after recovery from the reactor and indicate complete conversion as well as the absence of any undesired side product.

IL-2 differs from **IL-3** in only one methylene group, however, it does crystallize after several hours and can be recrystallized from acetone to form colorless needles, whereas **IL-3** remains liquid at room temperature. The additional methylene group incorporated into compound **IL-3** is sufficient to prevent formation of a highly ordered crystal lattice at room temperature. The ionic liquids possessing a nitro group on the phenyl substituent (**IL-4** and **IL-5**) remained liquid at room temperature for prolonged periods, but started crystallizing after several weeks.

Another interesting ionic liquid was synthesized by reacting vinyl imidazole with 1-bromohexane (Scheme 2). In order to avoid thermal poylmerization of the vinyl function, the reaction was carried out at 100 °C with an increased delay loop of an inner volume if 30 mL: The obtained 1-hexyl-3-vinylimidazolium bromide **IL-6** offers great potential for structural diversification as the compound can easily be used in different reactions such as olefin cross metathesis, radical polymerization, hydrosilylation and other olefin addition reactions.

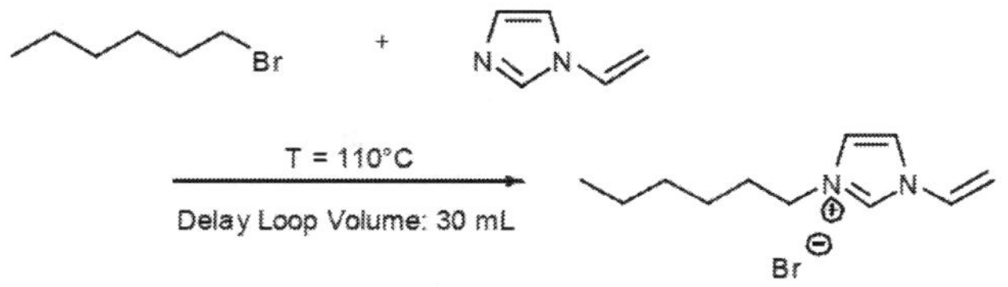

Scheme 2. Synthesis of 1-hexyl-3-vinylimidazolium bromide.

Conclusion

The newly developed flow process opens a quick and facile synthetic route to ionic liquids in a microstructured reactor. This remarkably versatile approach permitted the first preparation of imidazolium based ILs bearing an additional aromatic unit in the cation. The presented approach represents a highly versatile alternative to conventional synthesis methods and can be easily adopted to a large variety of ionic liquids.

Experimental Part

An organic synthesis bench scale plant (OSCB) reactor device provided by the IMM was used for all syntheses. In a typical experiment, alkyl bromide and imidazole were separately pumped into the mixing chamber (inner volume: 8 µL) by HPLC pumps. Behind the mixing device, the reaction mixture was directed through a delay loop with an inner volume of 10 mL (30 mL for **IL-6**) at constant flow rates between 0.6 and 2 ml/min (molar ratio of the reactants) and recovered at the outlet of the plant.

^{1}H-NMR spectra were recorded at 300 MHz on a Bruker AC and were referenced internally to residual proton signals of the deuterated solvent.

DSC curves were recorded on a Perkin Elmer DSC 7 and a Perkin Elmer Thermal Analysis Controller TAC 7/DX.

X-Ray structure determination of **IL-2** was collected on a Bruker AXS Smart CCD diffractometer with graphite monochromated Mo-K_α radiation (λ = 0.71069 Å).

Acknowledgement

We thank the Institut für Mikrotechnik Mainz for continuous technical support. Dr. Dieter Schollmeyer is acknowledged for X-ray single crystal measurements. H. F. acknowledges continuous support by the Fonds der Chemischen Industrie. D. W. is grateful for a fellowship from the Fonds der Chemischen Industrie. A. F. M. K. and J. K. thank the International Research Training Group (IRTG) Mainz for financial support.

Literature

[1] Wasserscheid, P.; Keim, W. *Angew. Chem.* **2000,** *39*, 3773-3789.

[2] Wasserscheid, P.; Welton, T., *Ionic Liquids in Synthesis.* Wiley-VCH: Weinheim, 2003.

[3] Sugden, S.; Wilkins, H. *J. Chem. Soc.* **1929**, 1291.

[4] Chum, H. L.; Koch, V. R.; Miller, L. L.; Osteryoung, R. A. *J. Am. Chem. Soc.* **1975,** *97*, 3264.

[5] Wilkes, J. S.; Levisky, J. A.; Wilson, R. A.; Hussey, C. L. *Inorg. Chem.* **1982**, 1263.

[6] Wilkes, J. S.; Zaworotko, M. J. *J. Chem. Soc. Chem. Commun.* **1992**, 965.

[7] Holbrey, J. D.; Seddon, K. R. *Clean Products and Processes* **1999,** *1*, 223.

[8] Welton, T. *Chem. Rev.* **1999,** *99*, 2071-2083.

[9] Hessel, V.; Löwe, H.; Müller, A.; Kolb, G., *Chemical Micro Process Engineering 1+2. Fundamentals, Modelling and Reactions/Processes and Plants*. Wiley-VCH: Weinheim, 2005.

[10] Ehrfeld, W.; Hessel, V.; Löwe, H., *Microreactors: New Technology for Modern Chemistry*. Wiley-VCH: Weinheim, 2000.

[11] Thayer, A. M. *Chem. Eng. News* **2005,** *83*, 43.

[12] Geyer, K.; Codée, J. D. C.; Seeberger, P. H. *Chem. Eur. J.* **2006,** *12*, 8434-8442.

[13] Jähnisch, K.; Hessel, V.; Löwe, H.; Baerns, M. *Angew. Chem. Int. Ed.* **2004,** *43*, 406-446.

[14] Kockmann, N.; Brand, O.; Fedder, G. K., *Micro Process Engineering*. Wiley-VCH: Weinheim, 2006.

Investigation of Condition for Bubble Detachment from Microchannels

Tadashi Sano and Shigenori Togashi
Hitachi, Ltd. Mechanical Engineering Research Laboratory,
832-2, Horiguchi, Hitachinaka, Ibaraki 312-0034, Japan

1. Introduction

Micromachining technologies have recently been used to make miniature devices for synthetic applications. These devices are called microreactors[1]. Several successful microreactors have been developed[2], but additional work is required before using in industrial processes.

In a microreactor, two liquids make contact inside minute channels in order to mix due to molecular diffusion. By definition, a minute mixing channel is narrow; therefore, the throughput of such a mixing channel is inevitably low. To increase the throughput, alternate and parallel flow of two liquids is formed in one channel, which is called multilayer flow. Each layer of liquid has a width of about 100 μm in a single channel.

The microreactor, which was developed by Hitachi Ltd., forms 48 pairs of multilayer flow and has a throughput of 20 ml/min. Multilayer flow microreactors use only 1 pair of pumps, which means that if one channel is blocked by a bubble, the subsequent change in flow effects the other channels. This means it is impossible to accurately calculate a microreactor's performance.

It is important for microreactor performance to remove bubbles; therefore we developed the calculation method of removing bubbles. We first developed our method as a model and then confirmed its effectiveness by experiment.

2. Bubble blockade in microchannels

A photograph of the microreactor is shown in Fig. 1. Two liquids are pumped into the mirocreactor. In the reactor, the each liquid is divided into the plural channels, and forming multilayer flow. An enlarged internal view of the microreactor is shown in Fig. 2. One liquid A flows from the 6 microchannels; the other liquid B flows from the 6 nozzles between the microchannels. Then these two liquids form a multilayer flow of 6 pairs.

There are three bubbles attaching to the nozzles in Fig. 2. These bubbles blocked liquid B flow. In this state, multilayer flow forms only 3 pairs; therefore, it is impossible to form uniform flow, and the local mixture ratio of the two liquids is unequal, meaning that the product yield decreases.

The bubbles attaching to the nozzles are able to remove by differential pressure dependent on the flow rate. Therefore we developed a calculation method of removing bubbles that uses the bubble detachment model.

3. Bubble detachment model

The surface tension of a liquid determines the conditions that a bubble detaches from a nozzle; therefore, it is possible to calculate the bubble removing conditions using surface tension as a parameter. We speculated the conditions detailed below when a bubble detached due to a differential pressure.

The process of a bubble detachment is shown in Fig.3. First, a bubble which is floating in a liquid container or tube, attaches to a nozzle(Fig.3-(i)-a or (i)-b). When the bubble stops moving, it is possible to use Eq. (1) because the forces on both sides of the bubble are in equilibrium.

(P_1: downstream pressure, P_2: upstream pressure, γ: surface tension, θ: contact angle of upstream, φ: contact angle of downstream, r: nozzle radius)

$$P_2 + \frac{2\gamma}{r}\cos\varphi = P_1 + \frac{2\gamma}{r}\cos\theta \quad \cdots\cdots\cdots (1)$$

When the upstream pressure is higher than downstream one, the bubble moves downstream (Fig.3-(ii)). When the bubble moves to end of the nozzle, the bubble stops and downstream side of surface expands. When the bubble radius is equal to the nozzle radius, the bubble radius becomes a minimum and the surface tension becomes a maximum. Therefore, as the contact angle of the downstream side of the bubble reaches 0 degrees, it is possible to derive Eq. (2).

$$P_1 - P_2 = \frac{2\gamma}{r}(1-\cos\theta) = \Delta P \quad \cdots\cdots\cdots (2)$$

According to Eq. (2), when a larger differential pressure than ΔP is generated, the bubble will detach (Fig.3-(iii)).

We used the ethanol aqueous solution to confirm Eq. (2). It was adjusted the surface tension of the solution by varying ethanol concentration. The several concentrations were used as a parameter in our calculations and experiments.

4. Experimental result and discussion

An ethanol solution was pumped into the microreactor, and the flow rate was gradually increased. The flow rate was measured when all the bubbles were detached. Then, the detaching differential pressure was calculated by use of the Hagen-Poiseuille formula.

Comparison of the differential pressure for all the bubbles detachment between the experimental result and the calculation data is shown in Fig. 4. As the calculated data and the experimental result agreed with a margin of error of plus or minus 9%, it found that the bubble detachment model was practicable.

5. Conclusion

We constructed the bubble detachment model in order to investigate the conditions, which a bubble detached from a nozzle in the microreactor. The differential pressure that a bubble detached from nozzles was calculated by use of this model. As the calculation data of the differential pressure and the experimental results were in good agreement, it is concluded that this model is practicable method to calculate the condition for a bubble detachment.

REFERENCES

1. Benson, R. S. and J.W. Ponton, "Process Miniaturisation - A Route to Total Environmental Acceptability", Trans. IChemE, Vol. 71, Part A, (1993) 160–168.

2. Ehrfeld, W., Golbig, K., Hessel, V., Lowe, H. and Richter, T., "Characterization of Mixing in Micromixers by a Test Reaction: Single Mixing Units and Mixer Arrays", Ind. Eng. Chem. Res. 38, 3 (1999) 1075–1082.

Fig. 1 Microreactor

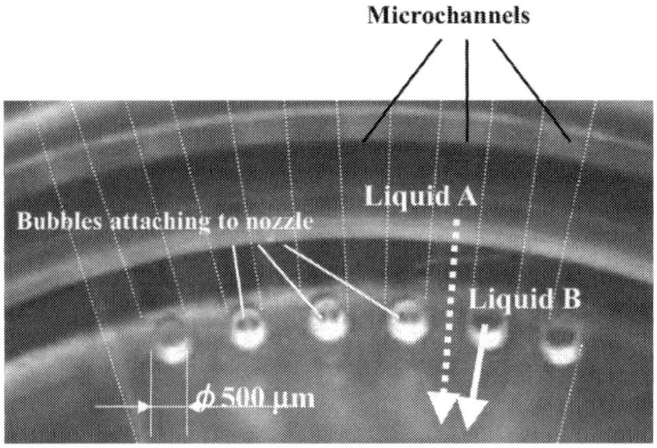

Fig. 2 Microreactor internal structure

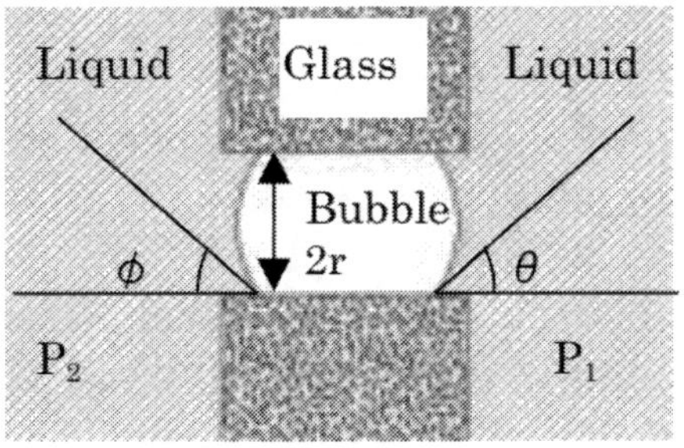

i) Bubble attaching

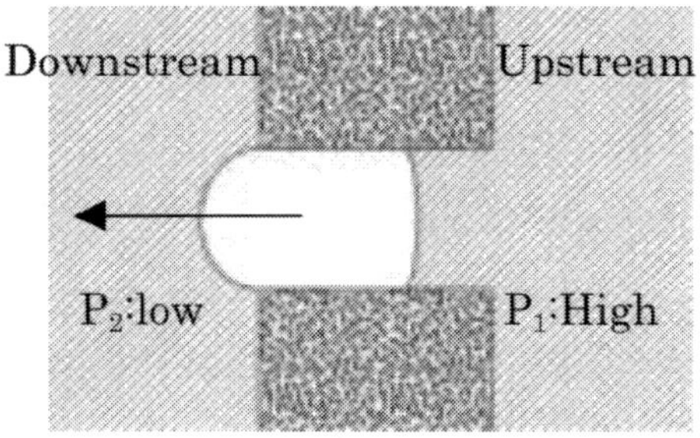

ii) Bubble transfer

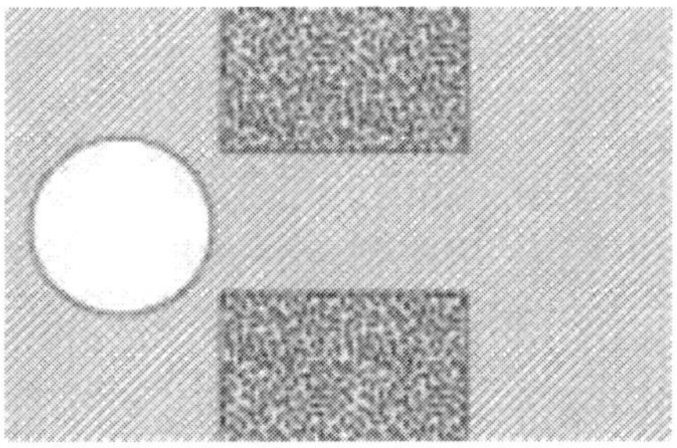

iii) Bubble detachment

Fig.3 Process of a bubble detachment at a nozzle

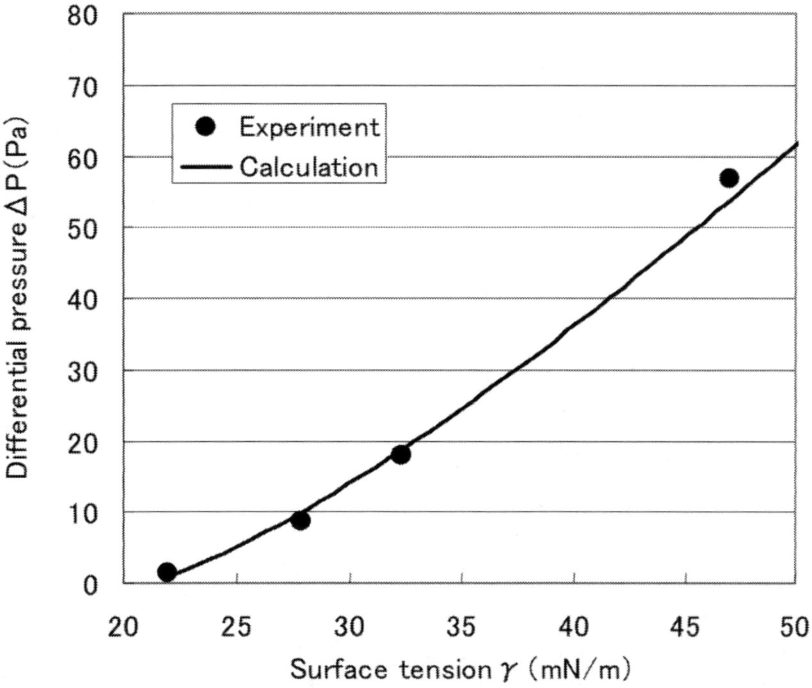

Fig. 4 Comparison of the calculated data and experimental result for the differential pressures that bubble detaches

240a Characterization of Porous Silicon Layers on Ordered Pillar Arrays Located in Flow Channels of Microfluidic Devices

Roald M. Tiggelaar, Vincent Verdoold, and Han J.G.E. Gardeniers. Mesoscale Chemical Systems, MESA+ Institute for Nanotechnology, University of Twente, P.O. Box 217, 7500 AE, Enschede, Netherlands

The high surface-to-volume ratio makes porous silicon attractive for microfluidic applications, like carrier matrix for proteins in enzymatic microreactors [1], catalyst support in microreactors for heterogeneous multiphase reactions [2] or in microfluidic devices for liquid chromatography separation [3]. A method to obtain porous silicon is via the anodization of highly doped silicon in hydrofluoric acid solutions [4]. The characteristics of porous silicon layers realized with this electrochemical process – such as thickness, porosity, pore-size distribution and effective surface-area – can be determined using gravimetric measurements [5], SEM/TEM inspection and nitrogen adsorption isotherm measurements [6]. Main disadvantages of these conventional techniques are, that they require a lot of porous material (in the order of grams) to obtain reliable data, which can only be comprised by anodization of large surfaces, i.e. wafers with a diameter of 3" or more. This aspect makes these techniques less suitable for the characterization of porous silicon layers in microfluidic devices, where porous silicon is usually locally fabricated in flow channels (i.e. small areas) and onto non-planar shapes and structures (3D instead of 2D). Here we report on fast and easy methods for the characterization of porous silicon films created on the surface of micropillars located in ordered arrays in flow channels. These ordered porous pillar arrays will be functionalized in chips for liquid chromatography separation [3] and in microreactors for studying three-phase (G-L-S) catalytic reactions, both fabricated using micromachining (cleanroom) technologies in silicon and vitreous materials.

Channels containing solid silicon pillars were made using Reactive Ion Etching with the Bosch process (gases: SF_6 and C_4F_8) in highly p-doped silicon (0.01 Ω/cm). Pillars with various shapes and heights were fabricated, with a minimal radius of 3 µm (standard lithography) and heights up to 50 µm (Fig. 1). The exterior of these pillars was anodized using the galvanostatic operational mode, for various conditions (HF-concentration and anodization current). All anodization conditions applied to solid pillars were also applied to flat substrates, in order to compare the characteristics of porous layers on pillars (3D-structures) with the properties of porous silicon realized on flat surfaces subjected to identical experimental conditions.

The characteristics of porous silicon films on flat substrates are investigated with High Resolution SEM imaging (thickness, uniformity), BET/BJH N_2 adsorption measurements (average pore-size distribution, effective surface area) and refractive index measurements (porosity). In figure 2 a typical picture of porous silicon is shown, in table 1 the properties of porous silicon films are summarized. For all investigated conditions mesoporous silicon is formed [7], of which the pore-size and thickness depends on HF-concentration and current density: the higher the HF concentration and the lower the current density, the smaller the pore- size and the porosity. Furthermore, for a fixed HF-concentration the thickness of the porous layer becomes larger for higher current densities, and for a constant current density its thickness increases with increasing HF-concentrations. The porous layers are homogeneous in depth and have a rather narrow pore-size distribution. According to literature the porosity of layers can be measured with ellipsometry [8,9]. The porosity of a porous silicon layer is closely related to its pore-size distribution [5], and indeed different refractive indices are measured for various electrolyte concentrations. However, for fixed HF-concentrations no clear difference in refractive index was found for increasing current densities, despite growing average pore sizes. In figure 3 HR-SEM images are shown of silicon pillars subjected to the electrochemical process, resulting in silicon pillars with a porous cladding. The thickness of the porous cladding is uniform across the height of the pillars, but seems less then on flat substrates, which can be attributed to differences in the E-field density across a pillar and a flat surface. Estimated average pore sizes on pillars yield roughly identical values as for flat substrates (e.g. 5-6 nm for 5 % HF, based on HR-SEM pictures). Currently the pore-size distribution of porous pillar arrays is investigated using HR-SEM imaging in combination with Focused Ion Beam Etching, as well as how to relate the average pore-size and porosity to ellipsometric data obtained from porous pillars arrays.

In conclusion, the characteristics of porous silicon layers are investigated with HR-SEM images, adsorption measurements and refractive index measurements. Since for the investigated anodization times (≤ 10 minutes) no significant differences are found between porous films created on flat surfaces and micropillars, data obtained from flat substrates can be used to quantify the properties of porous micropillar arrays, thereby avoiding destructive methods and/or time-consuming techniques. Channels filled with ordered porous pillar arrays will be implemented in microfluidic devices for liquid chromatography purposes (to enhance the available surface) and in microreactors for studying three-phase (G-L-S) catalytic reactions (as a structured support for the catalytic active phase).

References

[1] J. Drott, K. Lindström, L. Rosengren and T. Laurell, J. Micromech. Microeng., 7, 14-23, 1997.

[2] M.W. Losey, R.J. Jackman, S.L. Firebaugh, M.A. Schmidt and K.F. Jensen, J. Microelectromech. Syst., 11, 709-717, 2002.

[3] W. de Malsche, D. Clicq, V. Verdoold, P. Gzil, G. Desmet and J.G.E. Gardeniers, Lab. Chip, DOI: 10.1039/B710507J, 2007.

[4] R.W. Tjerkstra, M.J. de Boer, J.W. Berenschot, J.G.E. Gardeniers, A. van den Berg and M.C. Elwenspoek, Electrochim. Acta, 1997, 42, 3399-3406.

[5] R. Herino, G. Bomchil, K. Barla, C. Bertrand and J.L. Ginoux, J. Electrochem. Soc., 134, 1994-2000, 1987.

[6] S. Brunauer, P.H. Emmett and E. Teller, J. Am. Chem. Soc., 60, 309-319, 1938.

[7] V. Lehmann, J. Electrochem. Soc., 140, 2836-2846, 1993.

[8] S. Uehara, K. Taira, T. Hashimoto, H. Sasabu and T. Matsubara, Phys. Stat. Sol. (a), 182, 443-446, 2000.

[9] V. Torres-Costa, F. Pászti, A. Climent-Font, R.J. Martín-Palma and J.M. Martínez-Duart, Phys. Stat. Sol. (c), 9, 3208-3212, 2005.

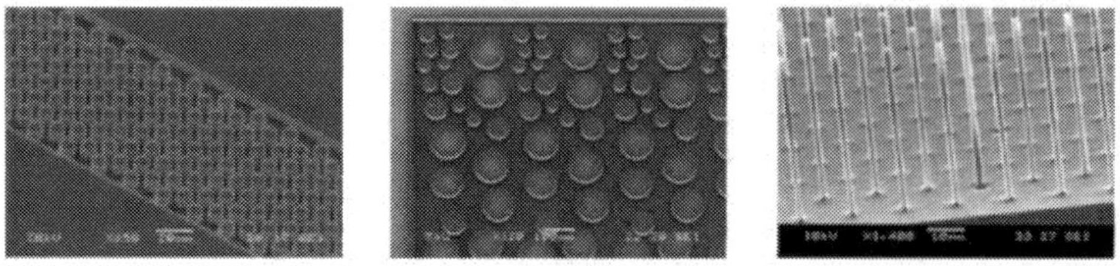

Figure 1: SEM pictures of solid silicon pillars

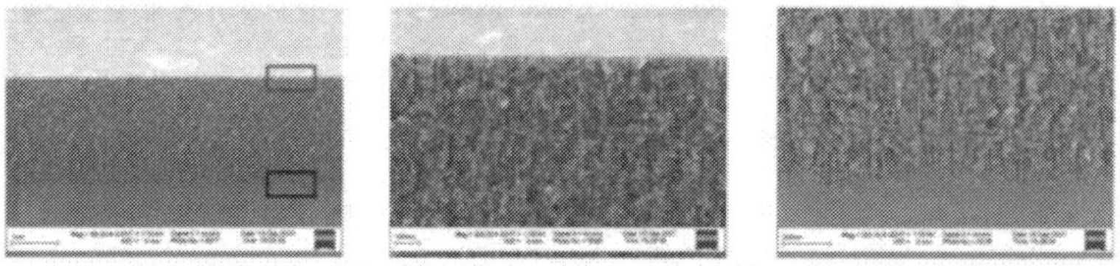

Figure 2: High Resolution SEM pictures of porous silicon formed on a flat substrate (settings: 15% HF, 200 mA, 300 sec)

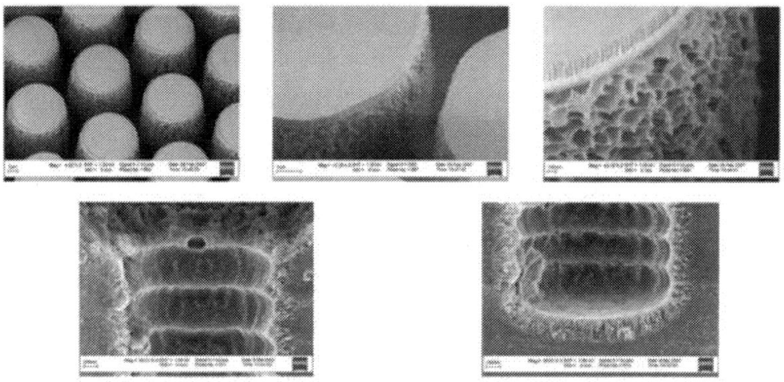

Figure 3: High Resolution SEM pictures of micropillar array with porous silicon exterior (bottom row: top view and bottom view of porous pillar)

Exp. settings: **- HF concentration** **- current** **- time**	**Thickness PS-layer [HRSEM; nm]**	**Average pore-size [BJH; nm]**	**Effective surface [BET; m^2/g]**	**Refractive index [n]**
5 %, 50 mA, 300 sec	612 ± 4	4.2 ± 0.2	151	2.48 ± 0.08
5 %, 100 mA, 300 sec	745 ± 7	4.9 ± 0.1	251	2.47 ± 0.14
5 %, 200 mA, 300 sec	1645 ± 11	5.4 ± 0.1	477	2.49 ± 0.31
15 %, 50 mA, 300 sec	892 ± 20	-	-	3.37 ± 0.28
15 %, 100 mA, 300 sec	1632 ± 22	-	-	3.29 ± 0.22
15 %, 200 mA, 300 sec	2266 ± 17	4.4 ± 0.5	534	3.32 ± 0.61
20 %, 50 mA, 300 sec	950 ± 9	2.1 ± 0.2	112	3.62 ± 0.57
20 %, 100 mA, 300 sec	2164 ± 5	2.5 ± 0.2	231	3.87 ± 0.93
20 %, 200 mA, 300 sec	3201 ± 4	3.5 ± 0.4	511	3.64 ± 0.62

Table 1: Characteristics of porous silicon layers for various hydrofluoric acid concentrations and anodization currents

198f Reaction Engineering Investigations of the Uncatalysed Cyclohexane Oxidation in A Microstructured Reactor

Thomas Lange, Johannes Fischer, and **Elias Klemm**. Professorship Chemical Technology, Chemnitz University of Technology, Strasse der Nationen 62, Chemnitz, 09111, Germany

A flow apparatus with a microstructured reactor was developed for reaction engineering investigations of the uncatalysed cyclohexane oxidation. Uncatalysed cyclohexane oxidation is a highly exothermic radical partial oxidation reaction. It is known to be autocatalytic and hazardous. Due to an intensification of heat and mass transport and an enhanced safety the microstructured reactor allows elevated operation conditions ("new process window"), e.g. at higher pressures and temperatures than in conventional industrial bubble columns. Thus, with the microstructured reactor working at increased reaction rate a higher space time yield could be expected. (see Tab. 1).

parameter	**bubble column**	**microstructured reactor**
T	140 – 160 °C	200°C
p	10 – 20 bar	80 bar
t	60 – 100 mins	app. 2 mins
X	app. 7%	app. 7%
S	> 90%	app. 80%

Tab. SEQ Tab. * ARABIC 1: Comparison of the performance of a bubble column and a microstructured reactor [1, 2].

The reason for the decrease in selectivity observed may be with the temperature dependence of the intrinsic kinetics or by the increased specific surface of the microstructured reactor. This effect was investigated by varying the diameter of the microchannel.

A significant influence of the specific channel surface on the dependence of cyclohexyl hydroperoxide concentration upon residence time was observed (Fig. 1). With reduction of the channel diameter and increase of the specific channel surface the maximum of the cyclohexyl hydroperoxide concentration decreased and shifted towards shorter residence times. Simultaneously an earlier increase of the product concentrations of cyclohexanol and cyclohexanone as well as carboxylic acids was detected (not shown as diagram). It can be assumed that cyclohexyl hydroperoxide is decomposed on the channel walls, most likely into hydroxyl and cyclohexyloxy radicals. Hydroxyl radicals will start new radical chains (through reaction with surrounding cyclohexane) and, thus, the consumption rate of cyclohexane and the formation rate of products increases. Cyclohexyloxy radicals underlie ring opening and are further oxidized to carboxylic acids.

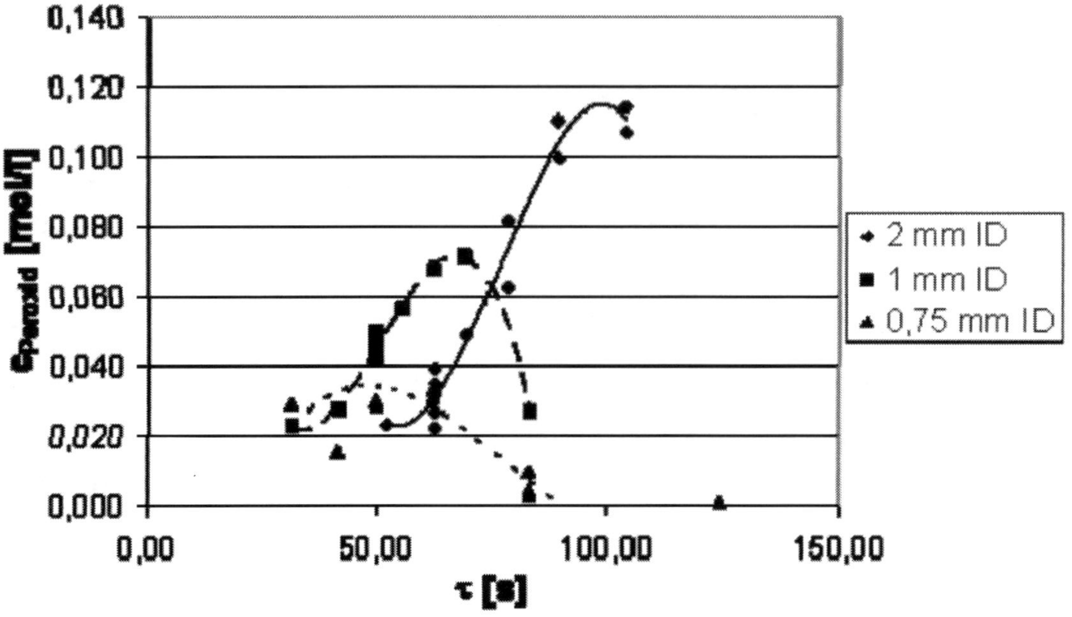

Fig. SEQ Abb. * ARABIC 1: Concentration of the peroxide versus residence time for different diameters of the channel, T = 200°C, p = 80 bar, 90 wt% cyclohexane (rest toluene).

Kinetic measurements are in progress in order to develop a kinetic model which properly describes cyclohexane oxidation in a microstructured reactor. This model must consider surface reactions of the cyclohexyl hydroperoxide and is based on the reaction mechanism of Jacobs et al.[3, 4, 5].

This project was financially supported by the German Federal Ministry of Education and Research (16SV1992) and the BASF.

[1] W. B. Fisher, J. F. van Reppen, Cyclohexanone Cyclohexanol, Kirk-Othmer Encyclopedia of Chemical Technology, Wiley-VCH, **2000**.

[2] M. T. Musser, Cyclohexanol and Cyclohexanone, Ullmann's Encyclopedia of Industrial Chemistry, Wiley-VCH, **2000**.

[3] I. Hermans, T. L. Nguyen, P. A. Jacobs, J. Peeters, Autoxidation of Cyclohexane: Conventional Views Challenged by Theory and Experiment, ChemPhysChem, 6, 637 – 645, **2005**.

[4] I. Hermans, P. A. Jacobs, J. Peeters, To the Core of Autocatalysis in Cyclohexane Autoxidation, Chem. Eur. J., 12, 4229 – 4240, **2006**.

[5] I. Hermans, P. A. Jacobs, J. Peeters, The Formation of Byproducts in the Autoxidation of Cyclohexane, Chem. Eur. J., 13, 754 – 761, **2007**.

240c Preparation of Polymer Colloidal Particles Using Spinning Disk Reactor

Zhen Lai, Xerox Corporation, 800 Phillips Rd, Bldg 143-02S, Webster, NY 14580

Product quality consistency is the major challenge faced by Process Research and Development groups in chemical industries. Recently the approach of process intensification has attracted more and more attention. It offers many attractive features, not least dramatically reduced capital investment required for commercial-scale operation, but also easier scaling out rather than scaling up, which has the potential for reduced lead times to commercialization of processes. Many approaches have been made to intensifying processes, and one is continuous processing, where there is an additional benefit of offering better control of product quality.

Recently, the spinning disk reactor (SDR) was brought forward as an alternative to traditional stirred tank processing technology, claiming to offer distinct advantages with respect to mixing characteristics, heat transfer, and residence time distribution, through creating a more intense fluid dynamic environment. Compared with classical stirred tank reactors, significant enhancements in polymerization rate have been obtained using SDR for a wide range of homogenous phase polymerization reactions that involve condensation, free radical and ionic mechanisms. And the continuous production of nano- and micro-size particles via reactive crystallization using SDR was also reported. There are also reports predicting a thousand-fold reduction in reaction, hundred-fold reduction inventory, and a 93% reduction in impurity level by using SDR for a process.

However, no studies have been reported that using SDR in heterogeneous-phase polymerization and no disclosed process involved using multiple SDRs.

In this paper, a method to produce submicron sized polymer particles with tunable particle composition and particle morphology using continuous SDR (single or multiple stages) for heterogeneous-phase polymerization (copolymerization) (i.e. emulsion, dispersion, and miniemulsion, microemulsion polymerization, etc.) will be discussed.

EFFICIENT GAS-LIQUID-SOLID REDUCTION AND OXIDATION SYSTEMS BY MICRO-PACKED BED REACTOR

Nungruethai Yoswathananont, Naoki Sasagawa, Yoji Fukuda, and Masaaki Sato

Faculty of Science, Osaka Prefecture University, Sakai, Osaka, Japan

Prepared for presentation at AIChE Spring National Meeting/ April 10, 2008/ IMRET 10/ 235 Multiphase Reactions, Dispersions and Foams – Part 1

Abstract

An efficient gas-liquid-solid microflow system was applied to reduction and oxidation reactions by using micro-packed bed reactor. The reactor consisted of a column, 1 mm in diameter and 25 cm in length, in which heterogeneous catalysts containing palladium were packed firmly. The substrate solution was mixed with either hydrogen gas or oxygen gas at the T-shaped mixer just before the inlet of the column. The mixture of gas and liquid were traveled through the tiny pathways in the column under an elevated pressure up to 2.5 MPa, giving rise to efficient reduction or oxidation reactions of organic molecules. In the case of hydrogenation reaction, acetophenone derivatives and hydrogen gas were passed through the flow reactor, where efficient hydrogenation took place giving rise to the corresponding alcohol or fully reduced ethylbenzene derivatives. On the other hand, efficient oxidation reaction took place when benzylalcohol and oxygen gas were passed through the flow reactor. Here, either benzaldehyde or benzoic acid was obtained selectively depending on the flow conditions.

Keywords: microreactor, microflow, hydrogenation reaction, oxidation reaction, heterogeneous catalyst

Introduction

Catalytic reactions are widely utilized for making chemical products in many chemical industries including food, textiles, pharmaceutical, plastics, agriculture, etc. Therefore, new technologies for the catalytic reactions to get higher efficiency and selectivity are desired to meet recent requirement of less waste and lower energy consumption.

In the last decade, there have been increasing interests in carrying out catalytic reactions in a flow regime through microstructured channels, since it is expected to open a new field in producing chemical substances. For example, efficient hydrogenation[1-3], oxidation[4], Suzuki-Miyaura coupling[5], Mizoroki-Heck reaction[6], Sonogashira coupling[7], and Kumada-Corriu reaction[8] have been reported with the use of the microreactors. The advantages of the microreactors relative to conventional batch reactors have been well documented in the literatures.[9] In general, it is accepted that high surface-to-volume ratio, fast heat and mass transfer properties, and minimizing risk of explosive reactions are the key factors of the efficient microreactors.

In our previous report, we have shown that our continuous microflow reactor possessed high activity toward hydrogenation reaction of 4-cyanobenzaldehyde and various nitrile derivatives.[10] The schematic draw of the microflow reactor is shown in Figure 1. Efficient hydrogenation reactions were taken place by mixing substrate solution and hydrogen gas at T-shaped mixer and letting the gas-liquid mixture go through a stainless steel tube column or a Teflon tube column, where palladium on carbon catalyst (Pd/C) was packed firmly. The high catalytic activity was thought to be derived from an increase in the mass transfers between different phases in the Pd/C packed column. Therefore, our microflow reactor was expected to be effective to the other palladium catalyzed multiphase reactions.

In the present work, the microflow reactor containing heterogeneous palladium catalyst has been applied to hydrogenation reaction of ketone derivatives and also to oxidation reaction of alcohols to prove the effectiveness of our microflow system. High yield and selectivity were achieved by simply changing the flow condition such as reaction temperature, residence time, and solvent.

Experimental

All reagents were of analytical grade and were used without further purification. Pd/C (AER type, 5%wt Pd) was purchased from N. E. CHEMCAT Co.

The micro-packed bed reactor consisted of the stainless steel tube or Teflon tube, 1mm in inner diameter with 5-25 cm in length. The Pd/C was packed into the tube by filling a slurry solution of Pd/C in methanol at a flow rate of 0.18 mL/min for 12 hours. At both ends of the tube, two filters were connected as to hold the catalyst inside.

The experiments were performed in microflow system and the results were compared with those of the conventional batch reactor system. The substrate solution was fed to one inlet of the T-shaped mixer by a HPLC pump. Hydrogen gas or oxygen gas was brought to the other inlet of the T-shaped mixer by adjusting the volume of the gas at 0.5-2.5 MPa with a mass flow controller. The mixture of liquid and gas was introduced to the inlet of the tube. The tube was heated by a ribbon heater, and the temperature was monitored by a digital thermometer. The residence time in the tube was only 2 minutes. The resulting solution coming out from the outlet of the tube was collected. After removing the solvent by evaporation under reduced pressure, the residue was dissolved in $CDCl_3$ and the yield of the product was determined by ^{1}H-NMR.

In the case of batch system, the Pd/C was suspended in the substrate solution, where the amounts of catalyst and substrate solution were the same as in the microflow experiment. The hydrogenation reaction and oxidation reaction were conducted in an autoclave (50 mL capacity) after purging the hydrogen gas or oxygen gas up to around 0.5-2.5 MPa. The reaction time was settled to 2 minutes. Then, Pd/C was filtered off and the solvent was evaporated under reduced pressure to isolate the product. The product yield was determined as shown in the microflow system.

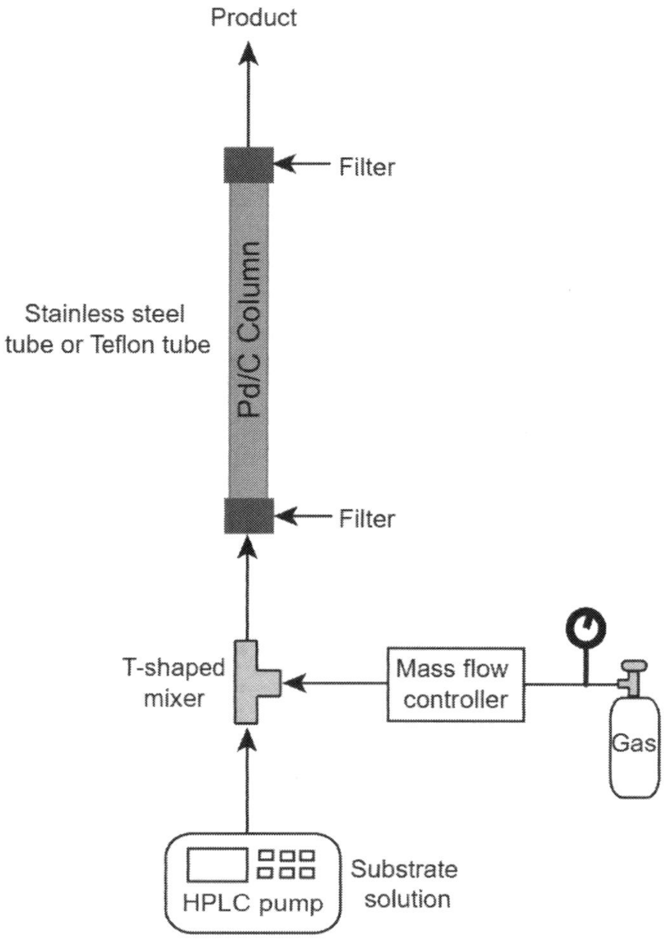

Figure 1 Schematic representation of the gas-liquid-solid microflow system.

Results and Discussion

In this work, the microflow system was applied to the hydrogenation reaction of acetophenone derivatives and to the oxidation reaction of benzylalcohol. The efficiency of the microflow system was studied by comparing the product yield with that of conventional batch system, autoclave.

Hydrogenation reaction of acetophenone derivatives

Hydrogenation reaction of various acetophenone derivatives was investigated in both microflow and batch systems under the corresponding conditions. Reduction of acetophenone derivatives (**I**) gave ethylbenzene (**III**) derivatives through an intermediate, benzylalcohol (**II**) derivatives, as shown in scheme 1. Yields of the hydrogenated products are summarized in Table 1.

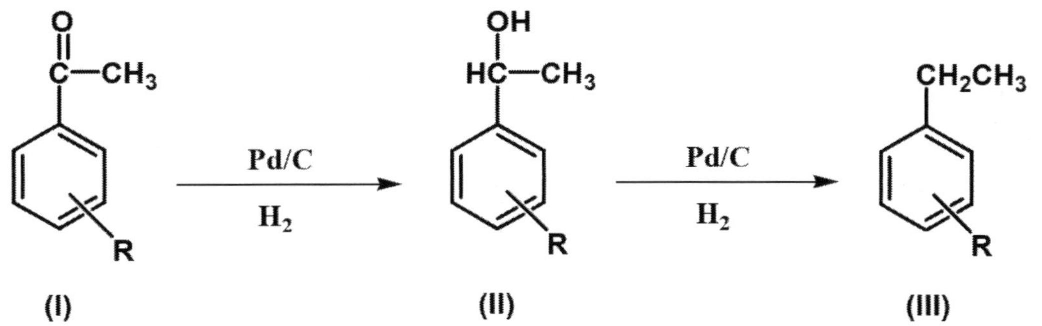

Scheme 1. Hydrogenation reaction of acetophenone derivatives.

Table 1. Hydrogenation reactions of acetophenone derivatives (**I**) in methanol solution using the microflow and the batch systems at 25°C under the hydrogen gas pressure at 2.5 MPa for 2 minutes.

R	Experimental system	Compound (%) (I)	(II)	(III)
p-COOH	microflow	0	0	100
	batch	24	76	0
p-H	microflow	0	0	100
	batch	30	70	0
p-CH_3	microflow	0	0	100
	batch	47	53	0
m-NH_2	microflow	0	100	0
	batch	96	4	0

Using the microflow reactor at 25 °C, *p*-acetylbenzoic acid, acetophenone and *4*-methylacetophenone were converted completely to the fully reduced product ethylbenzene derivatives (compound **III**) even though the residence time of the microflow system was only 2 minutes. The hydrogenation reaction using the batch reactor, however, benzylalcohol derivatives (compound **II**) were the main product and the fully reduced product (compound **III**) were not obtained at all. In the case of *m*-aminoacetophenone, the alcohol product (compound **II**) was obtained quantitatively when the reaction was performed in the microflow system, while in the batch system, the product yield was only 4% and 96% of starting material remained. These results indicated that efficient hydrogenation reaction was taken place in the microflow system compared with that in the batch system.

Oxidation reaction of benzylalcohol

To evaluate the efficiency of our microflow system to the other multiphase reactions, we have chosen an oxidation reaction of benzylalcohol. The microflow reactor used in this oxidation reaction was similar to that applied to the hydrogenation reaction except for the use of oxygen gas instead of hydrogen gas. Mass flow controller was set to 0.3-0.5 mL/min under around 0.5 MPa. Oxidation reaction of benzylalcohol (compound **IV**) gave benzaldehyde (compound **V**) and benzoic acid (compound **VI**) as shown in scheme 2. The oxidation reactions were carried out at 60-120 °C in both microflow and batch systems and the results are summarized in Table 2.

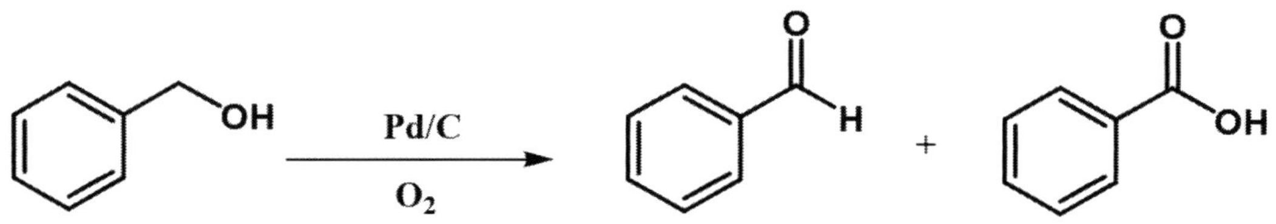

(IV) (V) (VI)

Scheme 2. Oxidation reaction of benzylalcohol (**IV**)

Oxidation reaction of benzylalcohol in the microflow system at 60°C gave benzaldehyde (43%) and benzoic acid (4%), while about half of the starting material remained. Increasing the reaction temperature to 80°C, almost all the benzylalcohol was oxidized, where benzaldehyde (89%) was obtained as a main product and benzoic acid (9%) as a minor product. At 100°C, benzylalcohol was converted to benzaldehyde (39%) and further to benzoic acid (61%). When the reaction was performed at 120 °C, benzylalcohol was oxidized to benzoic acid selectively (97%). Using the batch system at 120°C, on the contrary, the yield of benzaldehyde was as low as 2%. So, it was obvious that the microflow system was also efficient to the oxidation reactions. These results indicated that our microflow system had large advantages over the batch system in term of efficient hydrogenation and oxidation reactions.

Table 2. Oxidation reactions of benzylalcohol in the microflow and batch systems for 2 minutes.

Temperature (^{o}C)	Experimental system	Compound (%)		
		(IV)	**(V)**	**(VI)**
60	microflow	53	43	4
	batch	100	0	0
80	microflow	2	89	9
	batch	100	0	0
100	microflow	0	39	61
	batch	100	0	0
120	microflow	0	3	97
	batch	98	2	0

Conclusion

In summary, we have demonstrated an efficient micro-packed bed reactor for heterogeneous palladium catalyzed hydrogenation reaction of acetophenone derivatives and oxidation reaction of benzylalcohol. As the structure of our microflow reactor is simple, the heterogeneous palladium catalyst can be easily replaced and renewed after losing its catalytic activity. An interesting point of the micro-packed bed reactor is that either hydrogenation reaction or oxidation reaction can be conducted simply by introducing hydrogen gas or oxygen gas to the microflow reactor. These results suggest that our microflow system has a large potential to various types of multiphases reactions.

Acknowledgement

The authors wish to thank the Research Association of Micro Chemical Process Technology (MCPT), Japan and New Energy and Industrial Technology Development Organization (NEDO), Japan for financial support.

References

1. Wieβmeier and Hönicke, D., *Ind. Eng. Chem. Res.* 1996, **35**, 4412; Losey, M. W., Jackman, R. J., Firebaugh, S. L., Schmidt, M. A. and Jensen, K. F., *J. Microelectromech. Syst.*, 2002, **11**, 709; Kobayashi, J., Mori, Y., Okamoto, K.,Akiyama, R., Ueno, M., Kitamori, T. and Kobayashi, S., *Science*, 2004, **304**, 1305.

2. de Bellefon, C., Tanchoux, N., Caravieilhes, S., Grenouillet, P. and Hessel, V., *Angew.*

Chem. Int. Ed., 2000, **39**, 3442; de Bellefon, C., Abdallah, R., Lamouille, T., Caravieilhes, S. and Grenouillet, P., *Chimia*, 2002, **56**, 621; de Bellefon, C., Lamouille, T., Pestre, N., Bornette, F., Pennemann, H., Neumann, F. and Hessel, V., *Catalysis Today,* 2005, **110(1-2)**, 179; Radwan, A., Bruno, F., Valeria, M. and de Bellefon, C., *Catalysis Today,* 2007, **125(1-2)**, 34.

3. Surangalikar, H., Ouyang, X. and Besser, R. S., *Chem. Eng. J.*, 2003, **93**, 217; Liu, W., Roy, S. and Fu, X., *AIChE J.*, 2005, **51 (8)**, 2285; Saaby, S., Knudsen, K. R., Ladlow, M. and Ley, S. V., *Chem. Commun.*, 2005, 2909; McGovern, S., Harish, G., Pai, C. S., Mansfield, W., Taylor, J. A., Pau, S. and Besser, R. S., *Chem. Eng. J.*, 2007, **135 (SUPPL. 1)**, S229; Iles, A., Habgood, M., De Mello, A. J. and Wootton, R. C. R., *Catalysis Letters*, 2007, **114 (1-2)**, 71; Tadepalli, S., Halder, R. and Lawal, A., *Chem. Eng. Sci.*, 2007, **62 (10)**, 2663; Enache, D. I., Hutchings, G. J., Taylor, S. H., Raymahasay, S., Winterbottom, J. M., Mantle, M. D., Sederman, A. J., Gladden, L. F., Chatwin, C., Symonds, K. T. and Stitt, E. H., *Catalysis Today*, 2007, **128 (1-2 SPEC. ISS.)**, 26.

4. Kawaguchi, T., Miyata, H., Ataka, K., Mae, K. and Yoshida, J., *Angew. Chem. Int. Ed.*, 2005, **44**, 2413; Jachuch, R. J. J., Selvaraj, D. K. and Varma, R. S., *Green Chem.*, 2006, **8**, 29; Yi, J. P., Fan, Z. G., Jiang, Z. W., Li, W. S. and Zhou, X. P., *J. Comb. Chem.*, 2007, **9 (6)**, 1053; Yube, K., Furuta, M., Mae. K., *Catalysis Today*, 2007, **125 (1-2)**, 56.

5. Greenway, G. M., Haswell, S. J., Morgan, D. O., Skelton, V. and Styring, P., *Sens. Actuators B*, 2000, **63**, 153; Kunz, U., Schönfeld, H., Kirschning, A. and Solodenko, W., *J. Chrom. A.*, 2003, **1006 (1-2)**, 241; Basheer, C., Jahir Hussain, F. S., Lee, H. K. and Valiyaveettil, S., *Tetrahedron Lett.*, 2004, **45 (39)**, 7297; Comer, E. and Organ, M. G., *J. Am. Chem. Soc.*, 2005, **127 (22)**, 8160.

6. Liu, S., Fukuyama, T., Sato, M. and Ryu, I., *Org. Process Res. Dev.*, 2004, **8**, 477; Styring, P., *Chem. Eng.*, 2005, **765**, 40.

7. Fukuyama, T., Shinmen, M., Nishitani, S., Sato, M. and Ryu, I., *Org. Lett.*, 2002, **4**, 1691; Hillerich, J. and Plenio, H., *Chem. Commun.*, 2003, 3024; Solodenko, W., Wen, H., Leue, S., Stuhlmann, F., Sourkouni-Argirusi, G., Jas, G., Schönfeld, H., Kunz, U. and Kirschning, A., *Eur. J. Org. Chem.*, 2004, 3601.

8. Haswell, S. J., O'Sullivan, B. and Styring, P., *Lab Chip*, 2001, **1**, 164; Phan, N. T. S., Brown, D. H. and Styring, P., *Green Chem.*, 2004, **6 (10)**, 526.

9. for example see Ehrfeld, W., Hessel, V. and Löwe, H., *Microreactor: New Technology for Modern Chemistry*; Wiley-VCH: Weinheim, 2000; Jas, G. and Kirschning, A., *Chem. Eur. J.*, 2003, **9**, 5708.

10. Yoswathananont, N., Nitta, K., Nishiuchi, Y. and Sato, M., *Chem. Commun.*, 2005, 40; Yoswathananont, N., Nitta, K., Nishiuchi, Y. and Sato, M., *2005 AIChE Spring National Meeting Conference Proceedings*, 2891.

MICROPLASMA REACTORS WITH INTEGRATED CARBON NANOFIBERS AND TUNGSTEN OXIDE NANOWIRES ELECTRODES

Anil Agiral[1], K. Seshan[2], Leon Lefferts[2], J. G. E. (Han) Gardeniers[1]
Mesoscale Chemical Systems[1] and Catalytic Processes & Materials[2], MESA+ Institute for Nanotechnology, University of Twente, The Netherlands

Introduction

Miniaturized plasma sources have generated considerable interest due to a number of important applications [1]. Atmospheric pressure cold plasma processing techniques have been widely investigated in the field of energy, fuel and environmental issues. Dielectric barrier discharge (DBD) has been widely used to generate microplasma at atmospheric pressure. DBD consists of filamentary microdischarges of nanosecond duration. It is characterized as weakly ionized and highly reactive non-equilibrium plasma at atmospheric pressure. Performing the plasma process in a microreactor leads to precise control over the reactants to selectively produce desirable products [2]. Inelastic collisions with processing gas can excite the molecules and atoms giving radicals, electronically and vibrationally excited species. Extremely small volume and large surface area of microplasma reactors can result in the change of pathways of plasma assisted conversion.

Nanostructured electrodes such as transition metal oxide nanowires and carbon nanofibers (CNFs) show good sensing and field emission properties with low turn-on field and stability in poor vacuum conditions [3]. Field electron emission can supply primary electrons during the initiation of microdischarge. Local field enhancement on the apex of the nanotips due to high aspect ratio can lead to field emission at negative voltages and field ionization at positive voltages. These primary charged species can contribute to pre-breakdown current during initiation of discharge and can lower the breakdown voltage for microplasma reactors.

CO_2 decomposition in plasma reactors has generated considerable interest because of its high energy efficiency [4]. Electron collisions have been found to be very important in the decomposition of CO_2 to atomic oxygen and carbon monoxide [5].

In this study, CNFs and $W_{18}O_{49}$ nanowires were incorporated into a silicon chip electrode which was placed in a glass microreactor. Characteristics of dielectric barrier discharge with nanowires and CO_2 decomposition reactions with CNFs and their comparison with plane-to-plane electrodes without nanostructures were investigated. We simulated dielectric barrier discharge with pure CO_2 gas in order to understand the importance of electron impact and radical reactions in CO_2 plasma.

Experimental apparatus and description of model

Tungsten oxide nanowires and CNFs were grown in a quartz reactor. Hydrogen and nitrogen gases with high purity (99.995 % Hoek Loos, The Netherlands) were used.

Si substrates were heated to 973 K (6 K/min) in N_2 (80 sccm) at atmospheric pressure and when the 973 K was reached, ethylene (27 sccm) was introduced into the quartz chamber for 20 mins. Finally, the samples were cooled down in N_2 flow (80 sccm) to room temperature. N-type Si (100) wafers (1-10 Ohm cm) were used as substrates. Tungsten oxide nanowires were grown on tungsten film which was sputtered with dc sputtering from a 99.999 % pure tungsten target. For CNFs, silicon wafers were patterned in a standard photolithography process finishing with 20 min 393 K postbake. 10 nm nickel catalyst and 10 nm tantalum adhesion layer were deposited by electron beam evaporation. Samples with regular arrays of metal dots with 4 μm in diameter were fabricated by removing excess metal. Structural and compositional characterization of the nanostructures were carried out using scanning electron microscopy (SEM) (LEO 1550 FEG), high resolution transmission electron microscopy (HRTEM) equipped with EDX, selected area diffraction (SAD), X-Ray diffraction and X-ray Photoelectron Spectroscopy (PHI Quantera Scanning ESCA Microprobe System).

Field electron emission from tungsten oxide nanowires and CNFs were tested in diode structure with 2-μm electrode distance.

Pyrex type microreactor is shown in Figure 1a. Pyrex microreactors were used to generate microplasma by DBD at atmospheric pressure. Rectangular cross section of microreactor is 50 mm in length and 15 mm in width. Copper tape electrode is attached to Pyrex plate and Si chip electrode (with or without nanostructures) in the microchannel as shown in Figure 1b. Microchannels in the chips were made by means of chemical etching and thermal bonding. Microchannels were realized with hydrofluoric acid etching using a chromium0gold mask layer. Gas outlet and inlets were created by powderblasting. Teflon tubes were glued to the chips via steel ferrules with epoxy. Before gluing, the surface of the tubes were activated by a fluoro-etchant and acetic acid so that active acetyl and hydroxyl groups on the surface can allow good adhesion of Teflon to the Pyrex glass. We Chromium-gold layer was stripped and the bonding surfaces were cleaned with nitric acid and potassium hydroxide. Channel depth is 500 μm. Completed microplasma reactor with electrode and gas connections is shown in Figure 1c. The power supply unit consists of Trek 610 E amplifier and function generator and generates output voltage variable from 1 to 15 kV and frequency from 100 Hz to 15 kHz. Power can be varied by changing the applied voltage and the frequency. Power absorbed by the microplasma was evaluated on the basis of Lissajous figure measured with an oscilloscope. Power consumption by the microplasma reactor is between 0.05 and 1 W. Voltage drops across a capacitor and a resistor were used to record transferred charge and current, respectively. Light emission from microplasma reactor was collected by inserting an optical fiber to the microchannel from side and optical emission spectrometer (HR 4000, Ocean Optics) was used obtaining spectra in the range of 200-800 nm. Microplasma reactor during operation is shown in Figure 1c. The reaction mixture was processed in the discharge region. The reactor temperature was at 300 K and a tiny thermocouple was inserted in the post discharge region of microreactor to record the gas temperature. Maximum 320 K was measured at the highest power applied. The experiment was conducted under atmospheric pressure varying voltage. The gas residence time was set at 10 ml/min. When external ac or pulsed voltage was applied to Si chip electrode with nanostructures, dielectric Pyrex layer was polarized and intense electric field was formed, resulting in filamentary discharges.

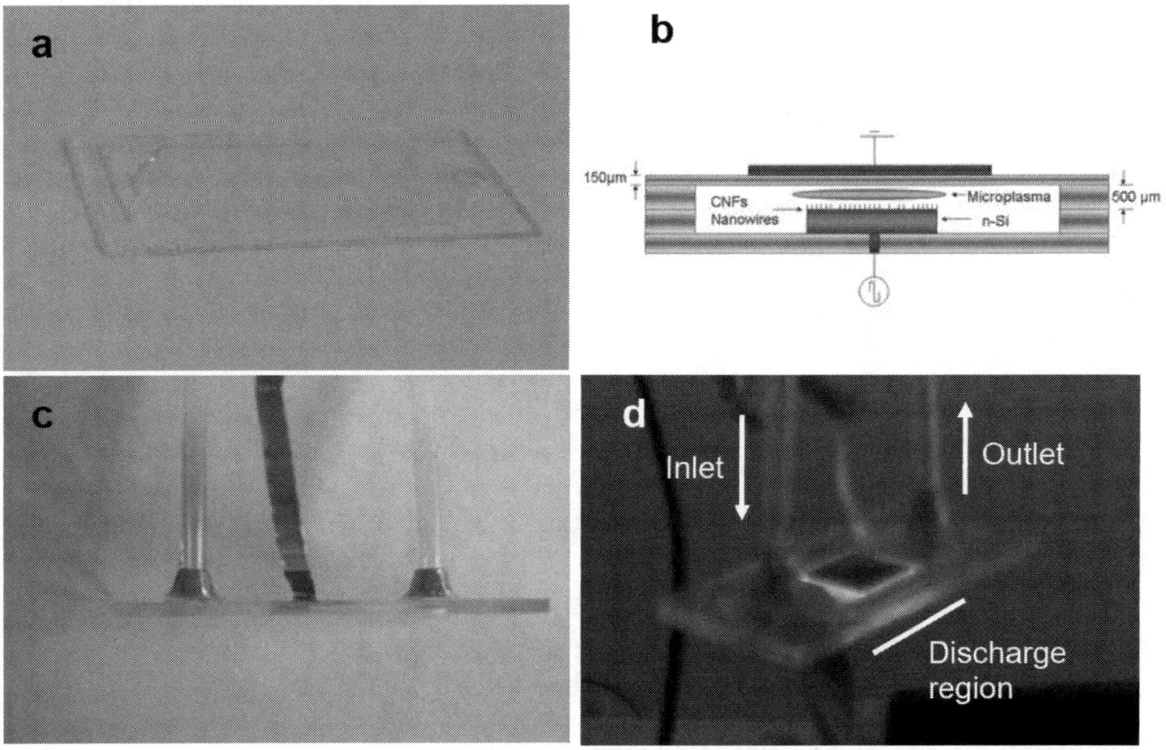

Figure 1. Microplasma reactor fabricated from Pyrex glass (a), general diagram of the device with integrated nanostructures on Si chip as electrodes (b), Copper electrodes and Si chip electrode attached to glass microreactor (c), microplasma device during operation (d).

Experimental apparatus for testing CO_2 conversion using CNFs as electrodes is shown in Figure 2. A quadrupole ion-trap mass spectrometer (Pfeiffer QMS 422) with residual gas analyzer was used to analyze reaction mixture. Mass flow controllers were used to set constant flow rate. Helium was introduced to reactor with 10 sccm flow rate and CO_2 signal of the m/e = 44 peak was recorded to calculate the conversion. CO_2 signal before and after the microplasma treatment was used to calculate CO_2 flow after discharge treatment using the equation:

CO_2 flow = He flow * (Fraction of CO_2)/(Fraction of He) (1)

We simulated dielectric barrier discharge with pure CO_2 as precursor gas. Simulation employs well mixed plasma reactor model, as available in Chemkin 4.1 [6]. The electron energy balance equates the time rate of change of electron's swarm internal energy to the net flow of electron enthalpy into and out of the reactor, accounting for net chemical production rates, surface losses, collisional losses and power deposition from the externally applied electromagnetic field. A modified residence time was introduced to solve the non-homogeneity of the barrier discharge. Assumption was done to match percent conversion obtained by experimental data to the simulation. The reduced electric field was calculated considering the capacitances of dielectric wall and applied voltage. BOLSIG+ software was used to solve 0-D Boltzmann equation [7].

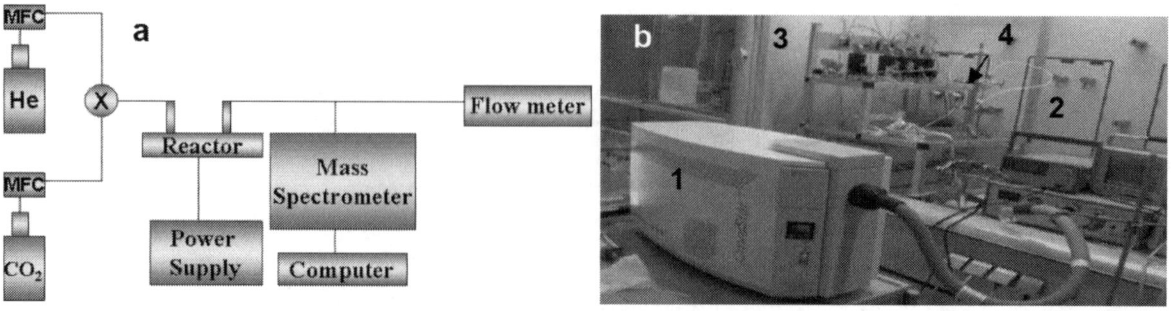

Figure 2. Experimental apparatus for testing CO_2 conversion using CNFs as electrodes integrated in DBD type microplasma reactor: Schematic diagram (a), photograph of the setup; 1: mass spectrometer, 2: power supply unit, 3: mass flow controllers, 4: microplasma reactor (b).

Results and discussion

Characterization of CNFs and tungsten oxide nanowires

Figure 3a and 3b show SEM micrographs of dense structure of CNFs grown on the arrays of dots of nickel catalyst. Nanofibers reaches the maximum height around 2-3 μm. The fibers contain metal particles in the top of fiber and diameter of the fibers varies between 10 and 30 nm.

Figure 3c and 3d show SEM images of tungsten oxide nanowires what were grown on tungsten thin films exposed to ethylene at 700°C at atmospheric pressure. The diameter and length of the nanowires were in the range of 15-20 nm and 50-250 nm, respectively. SAD and HRTEM analyses (not shown) indicated that d-spacing along the long side of the nanowire is 3.78 Å, corresponding to the (010) plane of monoclinic $W_{18}O_{49}$ (JCPDS no. 36-0101). [010] is the growth direction of nanowire. EDX analyses confirmed that oxygen is in the order of 2.73 times the amount of tungsten as indicated by HRTEM and SAD. XPS investigation revealed that oxygen source of nanowires comes from the native oxide layer chemisorbed on the tungsten thin film. XRD analyses on the nanowire sample showed characteristic peaks for WO_{3-x}, α-W_2C, β-W and Si. SAD analyses performed in the interfacial structure between nanowires and thin film as shown in the cross section SEM image in Figure 4b indicated two d-spacings 2.36 Å and 1.36 Å corresponds to the phase of α-$W_2C(002)$ and α-$W_2C(103)$, respectively. Tungsten carbide formation enhances the nucleation of crystalline nanowires. It may be due to creation of an interfacial strain that drives nucleation points of nanowires.

Atmospheric pressure field electron emission from CNFs and tungsten oxide nanowires

Atmospheric pressure field emission properties of CNFs and nanowires showed a turn on field around 3.3 V/μm with a current density of 10 $\mu A/cm^2$. Exponential dependence of current density (J) – electric field (E) (not shown) suggested a field emission process [8]. Similar experiment performed in a diode structure without nanostructure did not show any rectification behavior. Fowler-Nordheim plots (ln I/V^2 versus 1/V) of both nanostructures indicated that electron emission from nanowires and nanofibers follows Fowler Nordheim electron tunneling theory. Field enhancement at the apex of the tips of nanostructures resulted in electron emission at atmospheric pressure in air.

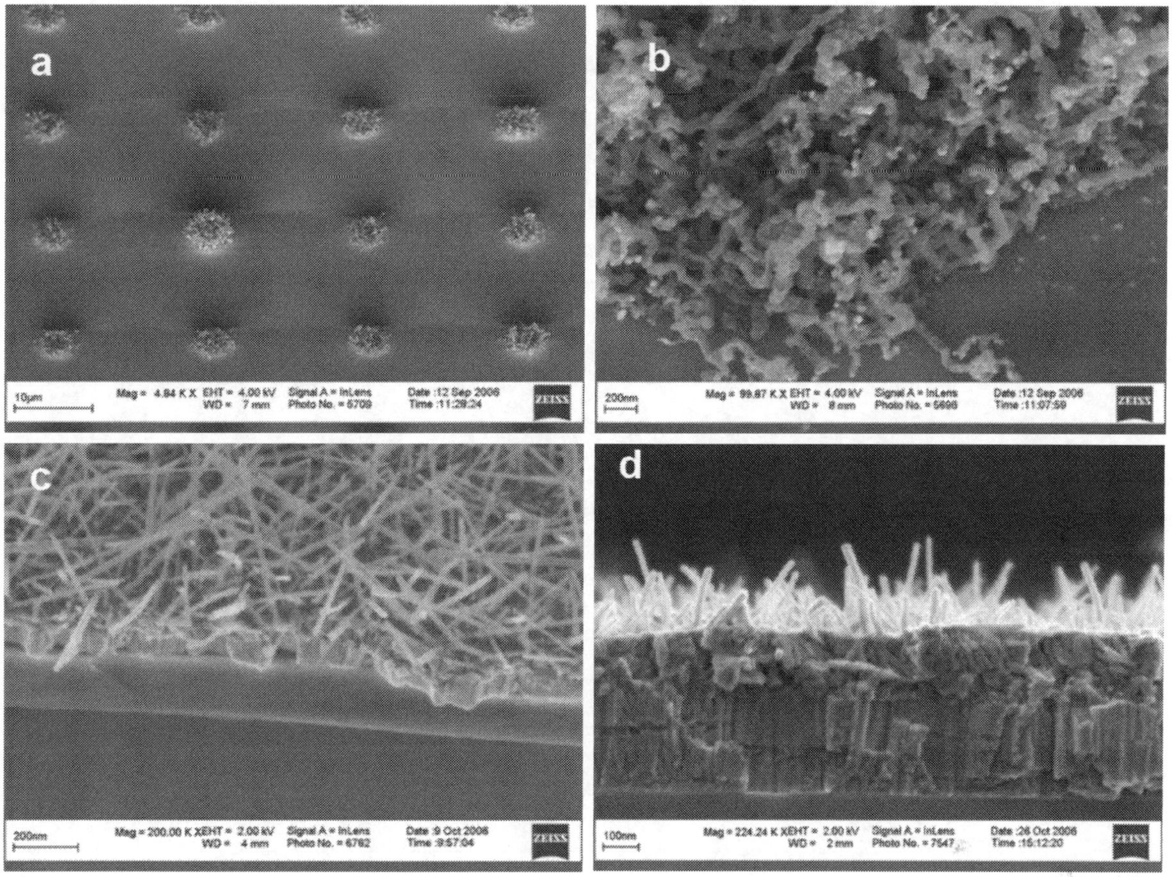

Figure 3. Arrays of nanofibers formed on nickel thin film catalyst exposed to ethylene and nitrogen at 973 K at atmospheric pressure (a), close look to the nanofibers showing individual fibers with Ni nanoparticles on top (b), tungsten oxide nanowires formed on the tungsten thin film (c), cross section of the thin film with nanowires.

Electrical characteristics of microplasma generated in the microreactor with CNFs electrode

The silicon chip with nanowires was incorporated into the glass microchannel and 1 kHz square wave voltage was applied to generate microplasma while the copper tape electrode was grounded. Breakdown voltage of the gas gap was determined from light emission using optical emission spectrometer and transferred current using oscilloscope. Breakdown voltage reduced from 4800 V to 3720 V when tungsten oxide nanowires and plane tungsten film electrodes were used, respectively. Number of microdischarges, which was deduced from the number of current spikes, and the amount of transferred current are much higher in the case of nanowires electrode comparing with plane electrode. Reduction of breakdown voltage and increase in transferred current in the microplasma reactor with nanowire electrode is suggesting electric field amplification and emission of electrons. Comparison of deposited power into the barrier discharge from nanowires and plane electrodes calculated from Lissajous figures indicated that higher energy density microplasma was obtained with nanowires electrode at the same voltage. Electric field enhancement due to high aspect ratio of nanowires resulted in generation of higher number of streamers comparing to plane tungsten film.

CO_2 conversion in microplasma reactor with CNFs electrode

CO_2 conversion to CO and O_2 is an endothermic reaction with an enthalpy of 283 kJ/mol. Electrical characterization of the microplasma with CNFs showed that energy density deposited into the discharge is much higher with respect to plane cathode. CO and O_2 were detected as main reaction products by mass spectrometer. Figure 4 shows the conversion enhancement using CNFs electrode at the same measured voltage levels with plane electrodes due to increase in energy deposition and reactive microdischarges.

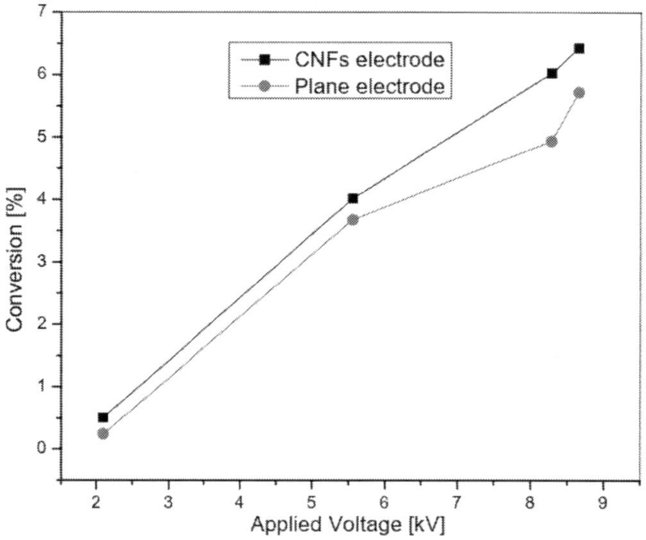

Figure 4. CO_2 conversions a function of applied voltage on CNFs and plane electrodes at 1kHz, square wave voltage.

Modeling results

Simulation of CO_2 conversion in the microplasma reactor can help us to clarify the role of energy deposition on electron impact and neutral reactions. We neglected all spatial effects and surface kinetics in the plasma. The reaction mechanism includes neutral-neutral chemistry, electron induced dissociation and ionization and surface recombination of ions at the walls. Reaction mechanism starts from pure CO_2. Main electron impact reactions are tabulated in the Table 1 with referring numbers to individual reaction channels. To determine the reaction rate coefficients of electron induced reactions, 0-D solution of Boltzmann equation was used by using BOLSIG+. Electron energy distribution function was obtained using the two-term spherical harmonic expansion. Cross sections of partial excitation cross sections as tabulated in Table I, were obtained from BOLSIG+ database. Electron impact reactions of O_2 and CO molecules were also included in the model. Rate coefficients of reactions of these molecules were obtained from calculated electron energy distribution function of CO_2 plasma.

Table 1. Electron impact reaction channels considered in CO_2 microplasma

Reaction Description	Reaction	Threshold Energy [eV]	Reaction Number
Momentum transfer	$CO_2 + e^- \rightarrow CO_2 + e^-$	-	R #1
Excitation	$CO_2 + e^- \rightarrow CO_2^* + e^-$	0.083	R #2
Excitation	$CO_2 + e^- \rightarrow CO_2^* + e^-$	0.167	R #3
Excitation	$CO_2 + e^- \rightarrow CO_2^* + e^-$	0.291	R #4
Excitation	$CO_2 + e^- \rightarrow CO_2^* + e^-$	0.339	R #5
Excitation	$CO_2 + e^- \rightarrow CO_2^* + e^-$	0.252	R #6
Excitation	$CO_2 + e^- \rightarrow CO_2^* + e^-$	0.422	R #7
Excitation	$CO_2 + e^- \rightarrow CO_2^* + e^-$	0.505	R #8
Excitation	$CO_2 + e^- \rightarrow CO_2^* + e^-$	2.5	R #9
Dissociative Excitation	$CO_2 + e^- \rightarrow CO + ½O_2 + e^-$	7	R #10
Dissociative Excitation	$CO_2 + e^- \rightarrow CO + ½O_2 + e^-$	10.5	R #11
Total Ionization	$CO_2 + e^- \rightarrow CO_2^+ + e^-$	13.3	R #12
Electron attachment	$CO_2 + e^- \rightarrow CO_2^-$	3.85	R #13

The relationship between cross section data of reaction channels (Table I) and average electron energy is shown in Figure 5a. We considered the momentum transfer, vibrational and dissociative electronic excitations, ionization and attachment reactions. Figure 5b shows the relations between electric field and average electron energy or electron temperature. We estimated the electron temperature from experimentally determined reduced electric field. Estimated electron temperature is around 4 eV. Electron temperature was used to use as an input to Chemkin to get a steady state solution. Thermo chemical data were used to calculate reverse reaction rate of reversible reactions. For metastables like ions and radicals, thermo chemical data were generated by using temperature dependent specific heat and entropy data for the ground state atom and adding excitation energy to the heat of formation of the corresponding species. We suggested two processes occurring on the surface; recombination of positive ions and electrons on the wall and de-excitation of metastable species on the surface. Surface of microreactor channel was considered as 300 K.

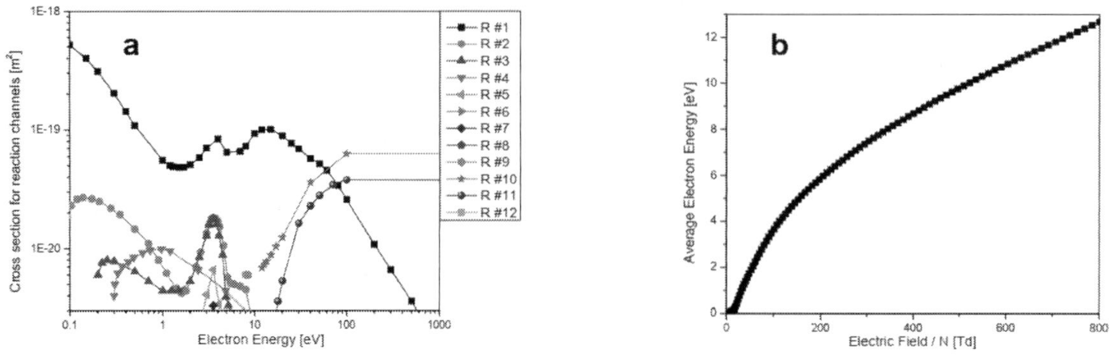

Figure 5. Cross section of electron impact reactions tabulated in Table I as a function of electron energy (a) and calculated electron energy as a function of electric field (b).

Figure 6a shows the rate coefficients calculated from the cross sections that obtained from correspondent references by BOLSIG+ code. At low electron energies (0-3 eV) rate coefficients for vibrational excitations (R #2 and R #3) are reaching maximum. Further increase in electron energies results in considerable increase in dissociative electronic excitation (R #10 and #11) and ionization (R #12). Relationship between rate coefficients and electron energy was described as three-parameter Arrhenius fit and employed as input to the Chemkin. Figure 6b shows the energy loss coefficients of reaction channels as a function of average electron energy. Energy spent on dissociative excitation and ionization reactions considerably increased starting from 3 eV. This indicated that as the electron energy is increasing, energy is mostly spent on major chemical reactions which are dissociation and ionization. Rate-of-production analysis in the Chemkin model determined that R #10 and #11 are the dominant reaction paths. When nanofibers were used as electrodes, increase in energy deposition and number of reactive streamers with respect to plane cathodes, increase the number of dissociative electronic excitations. CNFs can also affect the heterogeneous processes and open new reaction routes on the surface due to its high surface area and potential active functional groups.

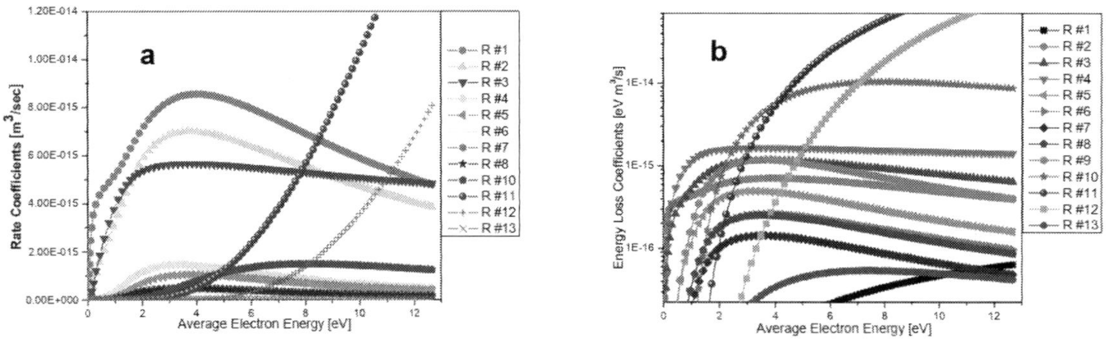

Figure 6. Rate coefficients (a) and energy loss coefficients (b) calculated as a function of average electron energy.

Conclusion

CNFs and $W_{18}O_{49}$ nanowires were incorporated into silicon chip electrodes which were placed in glass microplasma reactors. Characteristics of dielectric barrier discharge generation with nanowires and CO_2 decomposition reaction with CNFs and their comparison with plane-to-plane cathodes without nanostructures were investigated. Reactivity of microplasma was increased as a result of higher energy deposition when electrodes with nanostructures used. Chemical kinetic modeling of CO_2 conversion in the microplasma reactor showed that dominant pathways for the consumption of CO_2 are dissociative excitation channels. In the case of CNFs electrode, reactivity of microdischarges towards CO_2 conversion increased as a result of enhancement of dissociative excitations.

References

1. K.H. Becker, K.H. Schoenbach, J.G. Eden, J.Phys.D:Appl.Phys., 39, (2006), 55.
2. T. Nozaki, A. Hattori, K. Okazaki., Catal. Today, 98, (2004), 607.
3. R. Seelaboyina, J. Huang, J. Park, D. H. Kang, W. B. Choi, Nanotechnology 17 (2006) 4840-4844.
4. V.D. Rusanov, A.A. Fridman, G.V. Sholin, Sov. Phys. Usp., 24, 447.
5. T. Mikoviny, M. Kocan, S. Matejcik, N. J. Mason, J. D. Skalny, J.Phys.D:Appl.Phys., 37, (2004), 64.
6. R. J. Kee, F. M. Rupley, J. A. Miller, M. E. Coltrin, J. F. Grcar, E. Meeks,H. K. Moffat, A. E. Lutz, G. Dixon-Lewis, M. D. Smooke, J. Warnatz, G. H. Evans,R. S. Larson, R. E. Mitchell, L. R. Petzold, W. C. Reynolds, M. Caracotsios, W. E. Stewart, P. Glarborg, C. Wang, C. L. McLellan, O. Adigun, W. G. Houf, C. P. Chou, S. F. Miller, P. Ho, P. Young, D. J. Young, CHEMKIN Release 4.0, Reaction Design, San Diego, CA (2004).
7. G.J.M. Hagelaar, L.C. Pitchford, Plasma Sources Sci. Technol., 14, (2005), 722-733.
8. A. Agiral, J. G. E. Gardeniers, Synthesis and atmospheric pressure field emission operation of $W_{18}O_{49}$ nanowires, unpublished data.

83q Micro Structured Reactors for Heterogenously Catalyzed Gas Phase Reactions – a Concept Study

Elias Klemm[1], Helke Döring[1], Steffen Schirrmeister[2], and Andreas Geisselmann[3]. (1) Department of Chemical Technology, Chemnitz University of Technology, Strasse der Nationen 62, Chemnitz, 09111, Germany, (2) Uhde GmbH, Dortmund, Germany, (3) Degussa AG, Hanau, Germany

Introduction The use of micro structured reactors in large scale processes like the production of vinyl acetate from ethylene, acetic acid and oxygen using a Pd/Au catalyst poses several challenges. Processes performed in conventional reactors like multi tubular reactors are well known and in case of the vinyl acetate production high selectivities can already be achieved. Nevertheless the increase of conversion can lead to reduced recycle flow rates and thus to energy cost savings. In the following the already known concept of the DEMiS® reactor shall be evaluated in terms of productivity and heat respectively mass transport limitations. **Results** The space-time-yield related to the reactor volume is directly dependent on the thickness of the catalyst coating. As shown in Fig. 1 the productivity achieved in a conventional fixed-bed reactor [1] can be obtained when using a catalyst thickness of about 500 µm. This is valid for a catalyst with the same activity and selectivity as the one in the conventional reactor [2]. The rate constant for an improved catalyst is shown as line a) in Fig. 1 [3]. Here the activity is increased by a factor of 3. However due to design effects this leads to an increased productivity of only 2.3 for a wall coated micro structured reactor with a catalyst thickness of 500 µm. Further increases in productivity can be achieved by further improvements of catalyst activities or by increasing the catalyst coating thickness. First, the focus will be on increasing catalyst thickness: although the maximum thickness obtainable is dependent on both, mechanical stability and beginning mass transfer limitations. From the mechanical point of view a catalyst coating with a thickness of 1000 µm is still stable [4]. The mass transport limitation derived by a degree of efficiency for the catalyst of 95% using the Weisz-Prater criterion starts for the given reaction conditions with a rate constant of approximately 0.5 $m^3_{VAM}/kg_{cat}h$. The rate constants actually achieved allows an increase in productivity up to 260 $kg_{VAM}/m^3_{reactor}h$. Compared to data from conventional reactors this means 4.7 times higher productivity. The other option for increasing productivity is the increase of catalyst activity. Due to high heat transfer abilities of micro structured reactors the catalyst used as a wall catalysts is able to have a higher activity while the reactor maintains a high selectivity towards the wanted product. This was shown in [3] for a wall catalyst with 2.5 wt% Pd and 1.1 wt% Au. **Summary** An evaluation of the DEMiS® reactor concept for the acetoxylation of ethylene to form vinyl acetate was performed. Mass transport limitations only take place for higher activities than the ones being achieved until now and permit the use of catalyst thicknesses up to 1000 µm for achieving increased reactor productivities. Higher reactor productivities enable the reduction of reactor sizes for the production of same amounts of vinyl acetate as in multi tubular reactors. **Literature** [1] G. Roscher in: Ullmann's Encyclopedia of Industrial Chemistry, Wiley-VCH, Weinheim, 2005. [2] E. Klemm, H. Doering, A. Geisselmann, S. Schirrmeister, Chemie Ingenieur Technik 79 (2007) 697-706. [3] WO 2006/042659 A1, 2006. [4] DE 10 2005 019 000 A1, 2006. Fig. 1: Space-time-yield for a wall catalyst as a function of reaction rate for the formation of vinyl acetate

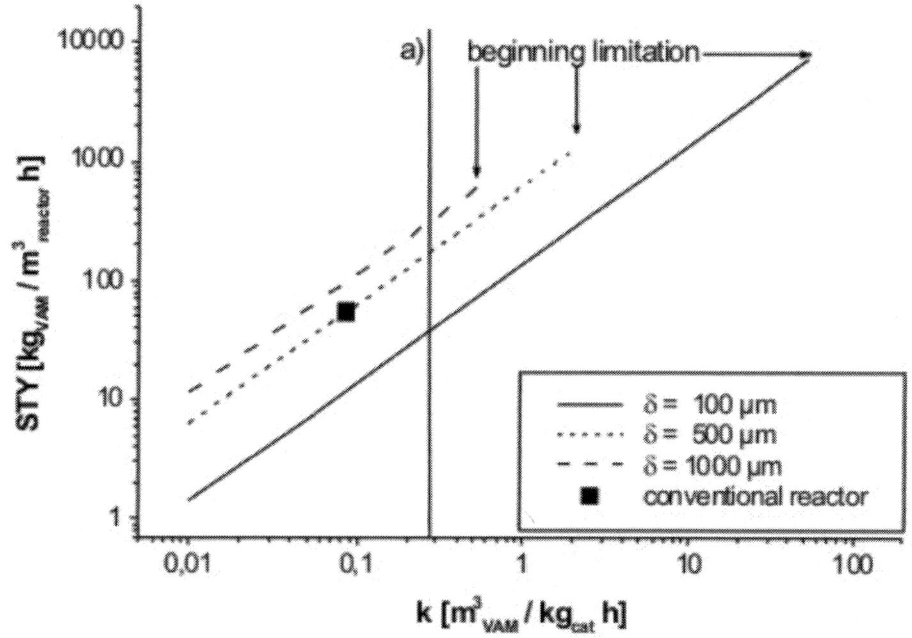

a)
beginning limitation
δ = 100 μm
δ = 500 μm
δ = 1000 μm
conventional reactor
STY [kg_VAM / m³_reactor h]
k [m³_VAM / kg_cat h]
10000
1000
100
10
1
0,01
0,1
1
10
100

Micro-reaction technology (MRT) for the safe industrial application of diazoalkane-chemistry: New chances for Process Intensification

Michael Struempel; Annegret Stark; Bernd Ondruschka

Institute for Technical Chemistry and Environmental Chemistry, Friedrich-Schiller University of Jena, Lessingstrasse 12, 07743 Jena, Germany
Annegret.stark@uni-jena.de

Diazomethane can be safely delivered on demand using micro-reaction technology (MRT) both on R&D and the industrially relevant scale, to allow for in situ conversion to the desired product while avoiding the cumbersome handling and storage of explosive diazomethane solutions. The first part of this contribution thus shows that MRT is a potent technology to make diazomethane accessible for the pharmaceutical and fine chemical sector.[1] However, the inherent formation of the by-product p-toluenesulfonic acid (p-TsOH) from the precursor Diazald® is not acceptable from an ecological and economical point of view. Thus, following the concept of process intensification, the goal of our ongoing research is the recycling of the p-TsOH and the efficient regeneration of the precursor Diazald® using MRT.

Many very reactive organic compounds are known to be theoretically applicable to the selective derivatization to yield fine chemicals or pharmaceutical products under mild conditions. However, since most processes in these areas are operated semi-continuously or even with standard laboratory batch equipment, the use of highly toxic, reactive, and therefore often explosive reagents is in most cases not possible from an occupational health and safety point of view.

In general, the key points for the intensification of a process include a) the increase of the concentrations of educts in a (if possible, single) solvent system throughout all process steps, b) minimizing the use of excess chemicals, solvents and auxiliaries to improve the atom efficiency, and c) to work under energy efficient conditions (temperature, pressure) to enhance yield, selectivity and conversion while maintaining short residence times. Regarding the Green Chemistry aspect as part of Process Intensification, other advantages of MRT are that an inherently safe process can be achieved at an early R&D stage, thus reducing both investment costs and time-to-market.

The industrial relevance of diazomethane, as an example of a highly reactive organic compound, is severely limited by its hazardous properties (explosiveness, toxicity, carcinogenicity[2-4]), although it would offer a vast chemistry as mild methylating agent, 1,3-dipolar compound and carbene source. Hence, its industrial relevance, for example in the derivatization of labile biogenic substances and complex pharmaceutical compounds,[5,6] is nil. Thus, a technology allowing for its save and sustainable use would open new pathways to these technically interesting compounds for the fine chemical and pharmaceutical sector.

We and others have recently shown that MRT[1,7-10] can serve as a consummate tool to overcome the above-mentioned safety issues: due to the extremely small internal volumes of micro-reactors, uncritically low spatiotemporal concentrations of diazomethane result. The peripheral equipment and concise arrangement of the reactor

allows for the installation of extra safety equipment, such as temperature and pressure sensors and a surrounding acid bath.
Another benefit arises in the context of effective process development in addition to the advantages relating to process safety: since the continuous production process is optimized already at an early R&D stage, the scale-up to the required product quantities is achieved by numbering-up to multiple manufacturing lines running in parallel.
From a chemical point of view, diazomethane is generated from the precursor Diazald® by treatment with a strong base followed by a rapid, direct conversion with the substrate.[11] In the present study, benzoic acid yielding methyl benzoate (Figure 1) was used as a model substrate,[5] thus a continuous two-stage micro-reaction process was developed and optimized by varying important parameters in combination with comparative batch experiments. Since the liberation of diazomethane from the precursor Diazald® and successive reaction with an carboxylic acid is known to be quantitative,[12] the yield of methyl benzoate is directly proportional to the yield of diazomethane. A commercially available micro-reactor (Little Things Factory GmbH, Ilmenau, Germany, 0.12 ml internal volume, type "ST MI018") was used as reaction device. It consists of four layers providing 32 mixing elements distributed over the length of the reactors, which twist the product stream by 90° each.

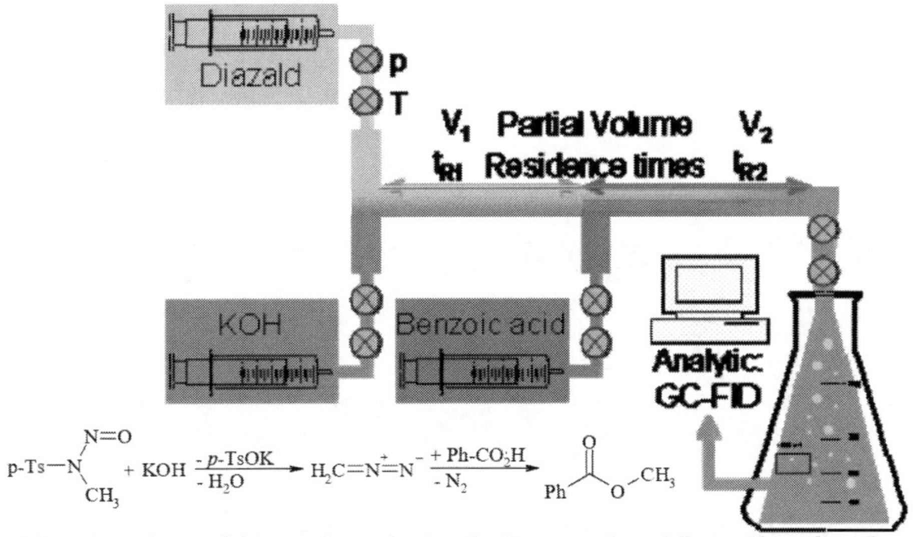

Figure 1 Reaction scheme of the experimental set-up for the generation of diazomethane from the precursor Diazald® with KOH, and its successive reaction with benzoic acid to methyl benzoate using MRT.

Since it is of crucial importance in a MRT process to ensure particle-free conditions to avoid clogging due to precipitation from feeds with high saline load, a thorough solvent study was carried out in batch experiments. Using 2-propanol as solvent for the base KOH, and diethyldiglycol or carbitol as solvents for the organic feeds, high yields of 75 and 60%, respectively, in combination with optimal solubility and minimal formation of particles were achieved. In comparison, other solvents for the organic feeds, such as THF, diethylether or toluene, show precipitation. Also, the use of an aqueous base solution yielded methyl benzoate < 30%. Hence, the solvent combination diethyldiglycol/2-propanol was chosen for the transfer of the reaction into continuous mode, as it promised the highest yields. Surprisingly, as a result of locally exceeding the maximum solubility, precipitation did occur in the micro-reactor unlike in batch mode. Therefore, carbitol/2-propanol was investigated in the micro-reaction set-up (Figure 1) as the second favorite alternative, giving a yield of 60%, which corresponds well with the result obtained in batch.

The influence of the parameters residence time, flow rate and reaction temperature were investigated in order to further optimize the process conditions. Using the optimized solvent system (2-propanol for the base and carbitol for the organic feeds), the effect of the residence time for the generation of diazomethane from the precursor by treatment with the base (t_{R1}) was studied, and compared with a capillary reactor which did not contain mixing elements (PTFE, Upchurch Scientific, internal diameter 1.55 mm).

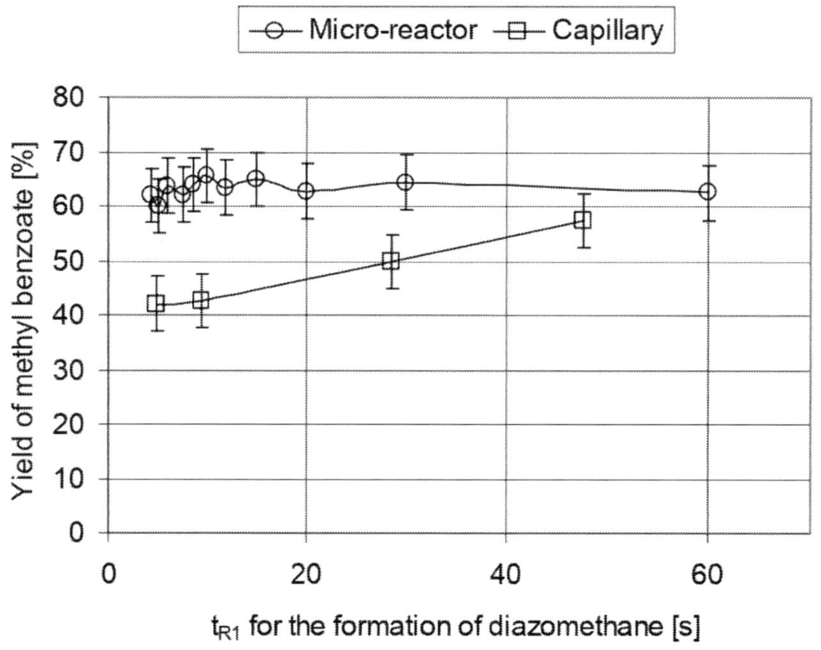

Figure 2 Dependence of the yield of methyl benzoate (relative to Diazald®) on the residence time t_{R1}; formation of diazomethane, in the micro-reactor and a capillary.

The different residence times shown in Figure 2 were achieved by adjustment of the feed flows. There is a clear dependency of the yield of methyl benzoate (relative to Diazald®) on t_{R1} if a capillary reactor is used as reaction device: For maximum yields, residence times > 50 s are required, whereas this is not the case for the ST MI018 micro-reactor with t_{R1} > 5 s. These findings are likely due to the presence of mixing elements in the latter reactor.

For the reaction of the thus generated diazomethane with benzoic acid, the variation of the second residence time (10 s < t_{R2} < 70 s), in contrast to t_{R1}, did not exhibit any effect on the yield in both the micro-reactor and capillary reactor. Additionally, as long as the residence time t_{R1} was sufficient (micro-reactor > 5 s, capillary > 50 s; Figure 2), neither of the reactors showed an effect on the total flow rate.

Due to the inherent safety exhibited by the MRT set-up, it was possible to investigate the dependence of the temperature on diazomethane formation as well. A constant yield of 65 % of methyl benzoate was obtained at temperatures between 0 and 50°C.

At higher temperatures (50–85°C), a decrease in yield became apparent, possibly due to starting diazomethane decomposition.[13,2,4] The set-up thus demonstrates its applicability to reactions at higher temperatures (up to 85 °C), although for the reaction investigated, an increase of temperature does not improve the performance. Hence, the set-up should prove to be applicable to both, reactions of diazomethane with other, less reactive substrates, and to other highly hazardous reactants with potential synthetic relevance, for example phosgene.

Finally, a continuous stability test was performed to demonstrate that the method and technique are suitable for industrial production. Under optimized conditions, a constant yield of 75 % of methyl benzoate was generated over a period of > 1 hour, and longer production periods appear feasible without the risk of clogging. Hence, at a flow rate of 840 ml h^{-1}, it should be possible to produce and directly convert about 2.5 mol d^{-1} of diazomethane with a single micro-reactor. Albeit based on extrapolation, the latter number shows that through its continuity and the inherent safety, the process makes previously hazardous chemistries available for the pharmaceutical, fine chemical and specialty sectors, while providing yields which compare well with the classic batch synthesis.

A critical point of this method for industrial application is the high amount of *p*-toluenesulfonic acid (*p*-TsOH) which is generated as side-product from Diazald®, resulting in a low atom economy with respect to diazomethane (14% at a yield of 70%). In our presently ongoing project, a recycling strategy of this acid to the precursor Diazald® is developed (Figure 3), and preliminary results are presented in the second part of this contribution.

The precursor is synthesized from *p*-toluenesulfonic acid in multiple steps, which themselves involve highly reactive compounds, e.g. chlorinating agents, sodium nitrite or intensively odorous and volatile compounds such as alkyl amines. Hence, MRT offers itself as a tool to carry out the envisaged complete synthesis.

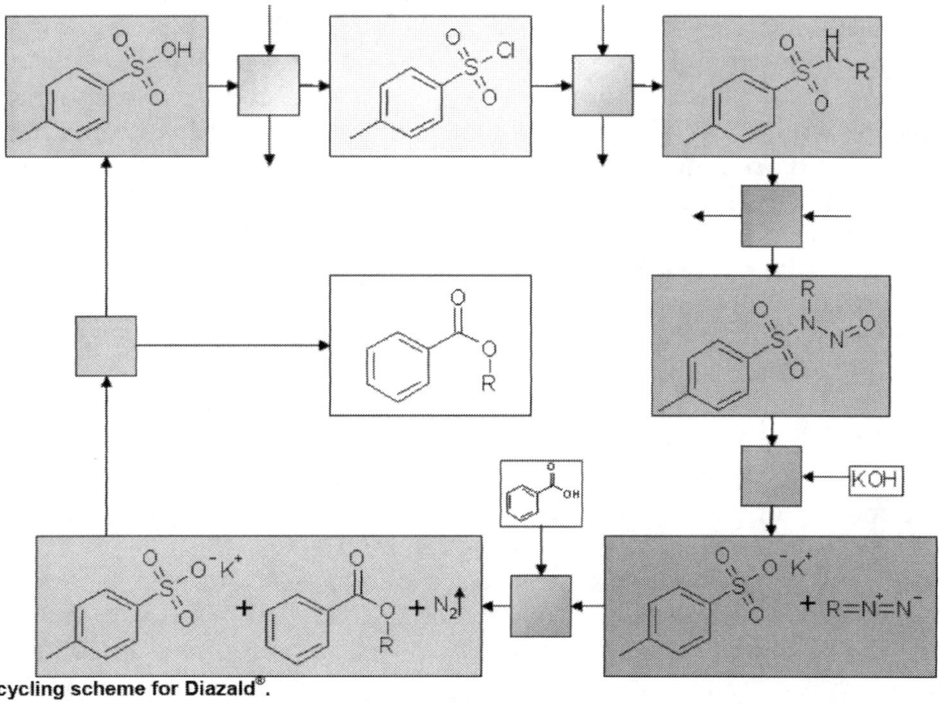

Figure 3 Recycling scheme for Diazald®.

For the complete synthesis (Figure 4), we have chosen to start with the chlorination of *p*-TsOH or its salt, a reaction which is typically carried out with thionyl chloride. The *p*-toluenesulfonyl chloride (*p*-TsCl) thus formed is then converted to *N*-methyl-*p*-toluenesulfonamide (*p*-TsNHMe) using methylamine. In the last step, *p*-TsNHMe is converted to Diazald® using sodium nitrite.

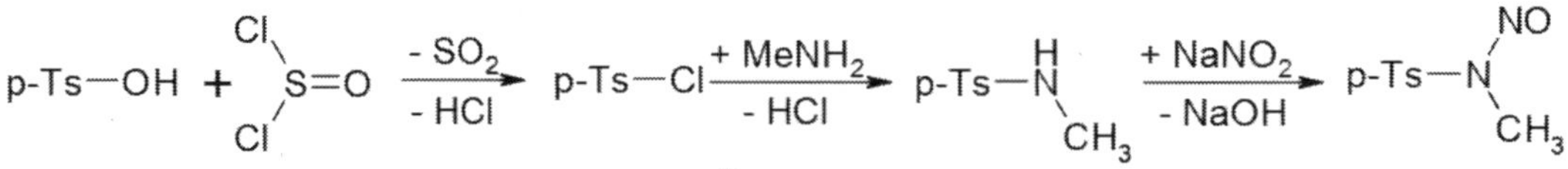

Figure 4 Example of reaction steps to synthesize Diazald® from *p*-TsOH.

Our preliminary results indicate that the chlorination is the first critical step, since the typical batch conditions[14] are difficult to transpose to a continuous set-up: In batch using thionyl chloride, the product is isolated by precipitation on ice. To avoid the precipitation in the continuous MRT set-up, *p*-TsOH was dissolved in toluene in an attempt to extract the product *p*-TsCl directly in the work-up step with water. However, this approach proved to be not practical, as hydrolysis of the product quickly occured. The chlorination of *p*-TsONa using thionyl chloride gave very satisfactory yields (80 %) in batch modus. Unfortunately, no solvent was yet identified which is capable of dissolving *p*-TsONa, thionyl chloride, and the by-product NaCl.

Besides the hydrolysis of the product and frequently occurring precipitation under certain conditions, another problem encountered is the acidic pH after the chlorination reaction, making a neutralization step prior to the following aminolysis necessary. Due to the low atom efficiency of this reaction type, the excess of chlorination agent for the quantitative conversion has to be taken into account. Hence, phosphorous pentachloride as alternative chlorination agent is even more prohibitive than thionyl chloride. Our present efforts focus on the optimization of this reaction, and on the use of catalysts to reduce the environmental burden. Comparative energy balances will show whether the efficiency of the reaction can be improved with MRT, or if a batch or semi-batch approach in regular sized apparatus is more advantageous.

In contrast to the chlorination, the aminolysis is easily applicable in the continuous MRT process giving quantitative yields. The methylamine is added as aqueous solution (40%) to *p*-TsCl in an organic solvent and in three-fold excess in order to also allow for the neutralization of HCl. For the transfer to MRT, the choice of solvent for *p*-TsCl is important: It was found that in water-immiscible solvents, such as toluene or diethylether, temporary precipitations occurred. Surprisingly, a biphasic system was obtained when acetonitrile was used as solvent at certain ratios of methyl amine to *p*-TsCl. This finding can be explained by the separation of the aqueous phase driven by the insolubility of *p*-TsNHMe in water.

Additionally, MRT shows some advantages for this reactions step: The concentrations of *p*-TsCl can be increased to 2M while allowing for isothermal conditions, whereas in batch-mode, the reaction is highly exothermic (T>80 °C).

The third reaction step to form Diazald® by reaction of *p*-TsNHMe with sodium nitrite has not yet been investigated to be transferred into MRT.

In summary, this contribution details the safe and efficient use of MRT for the generation of diazomethane from Diazald®. The production of Diazald® itself with MRT is the focus of our present study, and preliminary results thereof are presented.

Experimental Notes: Diazomethane is highly toxic, explosive and should therefore only be handled according to recommended procedures.[15,5]
The total flow rate is defined as the sum of the three inlet flow rates. The relative ratio of feeds and the reactant ratios were kept constant.

Acknowledgement: M.S. gratefully thanks the German Federal Environmental Foundation (DBU) for a PhD scholarship.

References:

1 M. Struempel, B. Ondruschka, R. Daute and A. Stark; Green Chem., 2008, 10, 41.
2 OSHA, www.osha.gov/SLTC/healthguidelines/diazomethane/recognition.html, 17.02.00; www.gifte.de/chemikalien/diazomethan.htm, 17.08.05; ILO-CIS, ICSC 1256, www.ilo.org/public/english/protection/safework/cis/products/icsc/dtasht/_icsc12/icsc1256.htm, 03.95.
3 C. D. Gutsche, Org. Reactions, 1954, 8, 391.
4 E. B. LeWinn, Am. J. Med. Sci., 1949, 218, 556.
5 T.H.Black, Aldrichim. Acta, 1983, 16, 3.
6 T. Archibald, Diazomethane, www.thomasarchibald.com/adobe/diazomethane.pdf, 28.03.2007.
7 E. Klemm, M. Rudek, G. Markowz and R. Schuette, 'Mikroverfahrenstechnik', in Chemische Technik - Prozesse und Produkte, eds. Winnacker, Kuechler, Wiley-VCH, Weinheim, 2004.
8 K. Jaehnisch, V. Hessel, H. Loewe and M. Baerns, Angew. Chem., Int. Ed., 2004, 43, 406.
9 W. Ehrfeld, V. Hessel and H. Loewe, *Microreactor*s - New Technology for Modern Chemistry, Wiley-VCH, Weinheim, 2005.
10 Micro Process Engineering - Fundamentals, Devices, Fabrication, and Applications, ed. N. Kockmann, Wiley-VCH, Weinheim, 2006.
11 W. F. Ferstl, M. S. Schwarzer and S. L. Loebbecke, Chem.-Ing.-Tech., 2004, 76, 1326.
12 E. K. Marshall and S. F. Acree, Ber. Chem. Ges., 1910, 10, 2323.
13 J. A. Moore and D. E. Reed, Org. Synth., 1961, 41, 16.
14 O. G. Lutsenko, G. P. Shaposhnikov, and V. P. Kulinich, Russ. J. Gen. Chem., 2004, 74, 300.
15 Th. J. de Boer and H. J. Backer, Org. Synth., 1963, Coll. Vol. 4, 250.

236e Electrospray Emulsification of Wax

Andres F. Mejia[1], Dawei Luo[2], Manuel Marquez[3], and **Zhengdong Cheng**[2]. (1) Department of Chemical Engineering, Universidad Industrial de Santander, Bucaramanga, Colombia, (2) Artie McFerrin Department of Chemical Engineering, Texas A&M University, College Station, TX 77843, (3) Arizona State University; National Institute of Standards and Technology; Los Alamos National Laboratory

Monodisperse wax emulsions were prepared by electrospraying of melt wax. The size of the emulsions was tailored by varying experimental parameters of the electrospray, such as the flow rate of the discontinuous phase, its electric conductivity, and the applied voltage. The monodispersity of the emulsions was greatly influenced by the surface tension and density of the collection liquid. We also demonstrated the fabrication of uniform wax microdisks using the phase transition of wax molecules inside these emulsions.

Web Page: che.tamu.edu/orgs/groups/Cheng/

236c Microfluidic Synthesis of Macroporous Copolymer Particles

Stanislav Dubinsky[1], Hong Zhang[1], Zhihong Nie[1], Ilya Gourevich[1], Dan Voicu[1], Eugenia Kumacheva[1], and Martin Deetz[2]. (1) Chemistry, University of Toronto, 80 St. George Street, Toronto, ON M5S 3H6, Canada, (2) Advanced Biosciences, Rohm and Haas Chemicals, LLC, 727 Norristown Road, Spring House, PA 19477-0904

Macroporous polymer resins are a class of materials with a permanent well-defined porous structure. A broad range of applications of these polymers(ion exchange, catalysis,biomedical applications)requires a "tailor-made" particle/pore size and particle/pore size distribution. A method that yields monodispersed macroporous particles with dimensions in the range from 50 to 100 micrometers and controlled distribution of the size of pores is highly desirable.The ability to generate porous particles by means of microfluidic synthesis has been demonstrated,however, currently no thorough study exists on the formation of macroporous particles with pre-determined dimensions and controlled pore size distribution. Here,we report semicontinuous photoinitiated microfluidic synthesis of macroporous polymer particles with a narrow size distribution in the size range from tens to hundred microns. The microbeads were synthesized by (i) emulsification of the mixture of glycidyl methacrylate (GMA), crosslinking agent and porogen in the microfluidic droplet generator; (ii) on-chip photopolymerization of GMA and (iii) off-chip post-photopolymerization of GMA. The use of GMA was motivated by the ability to carry out subsequent surface modification of the resulting microbeads, due to the reactivity of the pendant epoxide groups. Combination of the advantages of conventional suspension polymerization and microfluidic synthesis allowed us to produce polymer particles with a narrow size distribution and a controlled size of pores.

198b Synthesis of 5-Hydroxymethylfurfural from Renewable Resources Using Microreaction Technology

Stefan Loebbecke[1], Tobias Tuercke[1], Srinivas Yada[1], Slobodan Panic[1], Detlef Schmiedl[1], Birgit Kamm[2], and Thomas Frank[3]. (1) Fraunhofer Institute for Chemical Technology ICT, Joseph-von-Fraunhofer-Strasse 7, Pfinztal, D-76327, Germany, (2) Biorefinery.de GmbH, Potsdam, 14471, Germany, (3) Little Things Factory LTF GmbH, Ehrenbergstrasse 1, Ilmenau, 98693, Germany

Introduction

The interest in renewable resources as chemical feedstocks is growing considerably fast. For example, furan derivatives obtained from renewable carbohydrates such as fructose have a huge potential to become sustainable substitutes for petroleum-derived chemical components [1, 2]. For example, 5-hydroxymethylfurfural (5-HMF) is a key intermediate for the synthesis of 2,5-furandicarboxylic acid, furandiamine and other potential building blocks for polymers [1-5].

Today, the synthesis of 5-HMF from renewable resources is usually based on the catalytic dehydration of mono- and polysaccharides using organic acids (e.g. oxalic acid), inorganic acids (e.g. H_3PO_4, HCl), zeolithes or salts ($MgCl_2$). However, all these batch processes show considerable limitations in performance. When dehydration is conducted in pure water at high temperatures only low selectivities and a wide variety of by-products are obtained. On the other hand, processing at low temperature give certainly fair selectivities (>80%) but only poor conversion of < 50% can be achieved. To overcome restrictions in conversion and selectivity dehydration processes have also been conducted in biphasic systems based on water and an organic co-solvent such as methyl-isobutyl-ketone (MIBK), 2-butanol or dimethylsulfoxide (DMSO) [1], but also in sub- and supercritical mixtures of acetone/water, in sub- and supercritical methanol as well as in subcritical acetic acid [3]. Although conversion and HMF selectivity could be improved in comparison to conventional HMF syntheses the major drawback of these processes is their negative impact on the sustainability of the process since organic solvents are used that are partly hazardous, harmful to the environment and raise process costs significantly.

In recent years microfluidic structures have received an increasing interest as reactors that allow applying new process windows to chemical reactions as a result of improved mass and heat transport characteristics. Here, we report on the microreactor-based green synthesis of 5-hydroxymethylfurfural in water avoiding the use of any organic co-solvent. The process conditions were deliberately shifted towards high temperature and pressure regimes in combination with short residence times to significantly increase space/time yields and achieve satisfying selectivities.

Experimental

10% aqueous solutions of fructose were processed with diluted hydrochloric acid in various microreactors providing different mixing efficienies. Syringe pumps (Sykam S1610) were used for continuous pulsation-free flow of the reactants and back pressure valves were installed to avoid inhomogeneous flow behaviour. At elevated temperatures cartridge check valves were used to keep pressure above the vapour pressure of water (15-20 bars). The reaction was thermally quenched at the reactor outlet by rapid cooling in capillaries (800 µm diameter). Quenched reaction mixtures were analyzed quantitatively by HPLC (Agilent 1200) including potential by-products such as levulinic acid and formic acid.

The dehydration process was carried out under systematic variation of process temperature (100°C - 200°C) and residence time (1 min - 3 min) applying different flow rate ratios (fructose $_{aq}$: HCl $_{aq}$ = 1:1 – 1:5) and hydrochloric acid concentrations (0.1 mol/L and 0.5 mol/L).

Results and Discussion

Table 1 shows some exemplary results obtained for the dehydration of fructose in microrectors at a residence time of one minute. The best selectivity for 5-HMF (75%) was obtained at 185°C employing an HCl concentration of 0.1 mol/L. The conversion of fructose can be increased, as expected, by raising the temperature. However, at temperatures above 190°C insoluble humins and polymers are formed causing a blockage of the reactors. Higher concentrations and flow rates of hydrochloric acid lead to higher conversion but diminish 5-HMF selectivity due to an increase in byproduct formation.

Tab. 1: Exemplary results for fructose dehydration in a microreactor within one minute residence time.

No.	temperature	HCl concentration	fructose conversion	5-HMF selectivity
1	135°C	0.1mol/L	4%	60%
2	170°C	0.1mol/L	40%	60%
3	**185°C**	**0.1mol/L**	**71%**	**75%**
4	185°C	0.1mol/L*	75%	46%
5	185°C	0.5 mol/L	78%	43%

* flow rate ratio fructose:HCl = 1:5

Usually, the batch synthesis of 5-HMF in pure aqueous solution leads to conversion and selectivity values in the range of 50% or less [1]. Due to the intensified mass and heat transport attainable in continuously operated microreactors the dehydration process can be performed under conditions of significantly increased temperature and reduced residence time. Consequently, the dehydration of fructose could be optimized providing 20% higher conversion and 25% higher 5-HMF selectivity (Fig. 1). Moreover, the space/time yield could be significantly improved in comparison to the batch process.

To improve conversion and selectivity in batch processes organic co-solvents like MIBK and 2-butanol have been used to extract HMF from the aqueous phase equilibrium [1]. In addition, DMSO has been used to stabilize the more reactive furanose form of fructose [1]. In Fig. 1 the values of fructose conversion and 5-HMF selectivity obtained in batch processes using different combinations of co-solvents are compared with the results obtained from the microreaction process. The data show that 5-HMF selectivity obtained in the microreaction process is significantly higher than in most of the biphasic batch processes. Only if a combination of three different organic co-solvents (namely DMSO, MIBK and 2-butanole) is used a slightly better conversion (75% vs. 71%) and 5-HMF selectivity (80% vs. 75%) were obtained than achievable in the microprocess. However, the microreaction process offers a much more faster, cheaper and environmental friendly way to synthesize 5-HMF from fructose.

Conclusion

The green synthesis of 5-Hydroxymethylfurfural (5-HMF) from aqueous solutions of fructose was successfully conducted in a continuous microreaction process without using any organic solvent. By shifting reaction parameters such as temperature, pressure and residence time towards new process windows 5-HMF was obtained with increased selectivity of 75% at high space/time yields.

The subsequent catalytic oxidation of 5-HMF to 2,5-furandicarboxylic acid (FDCA) in microreactors is currently under investigation.

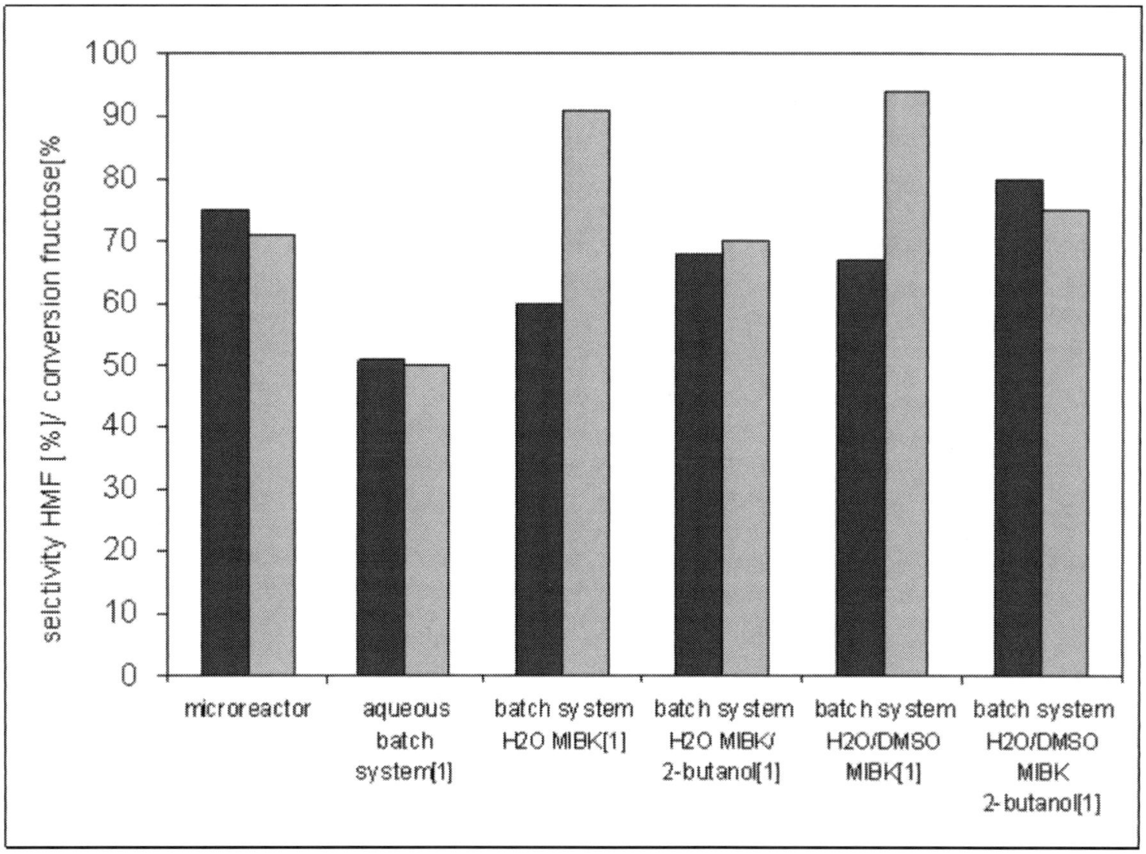

Fig. 1: Comparison of 5-HMF selectivity (black bars) and fructose conversion (grey bars) obtainend in continuous microreactor and in batch [1] processes.

Acknowledgement

Financial support by the Deutsche Bundesstiftung Umwelt (German Environmental Foundation) is gratefully acknowledged.

References

[1] a) Y. Román-Leshkov, J. N. Chheda, J. A. Dumesic, *Science* **2006**, *312*, 1933;

b) J. N. Chheda, Y. Román-Leshkov, J. A. Dumesic, *Green Chem.* **2007**, *9*, 342

[2] B. Kamm, *Angew. Chem. Int. Ed.* **2007**, *46*, 5056

[3] a) M. Bicker, J. Hirth, H. Vogel, *Green Chem.* **2003**, *5*, 280

b) M. Bicker, D. Kaiser, L.Ott, H. Vogel, *J. Supercritical Fluids* **2005**, *36*, 118

[4] C. Carlini, P. Patrano, A.M. Galetti, G. Sbrana, *Appl. Catal. A* **2004**, *275*, 111

[5] B. F. M. Kuster, *Starch/Staerke* **1990**, *42*, 314

102d Characterization of Flow Regimes and Transport Phenomena in Complex Microchannels

Norbert Kockmann, Process Optimization, Lonza, Lonzastr., Visp, 3930, Switzerland

Enhanced transport phenomena are essential for chemical reactions in continuous flow microstructured equipment. Flow regimes and heat transfer are governed by laminar conditions, especially in liquid phase systems. To achieve high mass flow rates, typical dimensions of channels in microstructured equipment are in the range from 50 to 1000 μm, while flow velocities are ranging from cm/s to few m/s. In channel curves and meandering channels, secondary flow structures appear, starting from Dean flow with a vortex pair to more chaotic flow structures at higher Re numbers. In complex channel arrangements, the narrowest cross section mainly determines the pressure loss and energy dissipation of the entire system. Nevertheless, a detailed analysis of the flow regimes is essential for the understanding of transport phenomena in these channel structures.

Heat and mass transfer is greatly influenced by the convective flow regime, which leads to further enhancement conjointly with miniaturization. Typical examples of curved flow and mixing in T-shaped micromixers are given. It is shown that the pressure loss and energy dissipation can be used to determine heat transfer coefficients and mixing characteristics.

Chemical reactions are characterized by their kinetics and stoichiometry. Rapid mixing, short residence time within the device, and narrow residence time distribution are important parameters for to design microstructured devices for process development and chemical production. Dimensionless numbers assist the appropriate and successful design of microreactors concerning mixing, residence time, and heat transfer.

83w Optimizing the Deprotection of the Amine Protecting P-Methoxyphenyl (PMP) Group in An Automated Microreactor Platform

Kaspar Koch[1], B.J.A. Van Weerdenburg[1], J.M.M. Verkade[1], P.J. Nieuwland[1], F.P.J.T. Rutjes[1], and J.C. M. Van Hest[2]. (1) Synthetic Organic Chemistry, Radboud University Nijmegen, Toernooiveld 1, Nijmegen, Netherlands, (2) Bio-Organic Chemistry, Radboud University Nijmegen, Toernooiveld 1, Nijmegen, Netherlands

We have previously shown that small microreactors with volumes in the range of a few microliters can be successfully applied to screen and optimize biosynthesis reactions consuming only minute amounts of reaction fluids[1,2]. Although our in-house developed microreactor platform is capable of producing such a large amount of data points in a single day, the generation of the amount of data required for optimizing three or more parameters would be very time-consuming if all possible reaction conditions would be screened in a grid with regular intervals. Therefore we applied the principles of Design of Experiments (DOE), more specifically D-optimal design[3] to optimize an industrially relevant reaction: the removal of a p-methoxyphenyl (PMP)[3] amine-protecting group. The applied DOE resulted in a very efficient screening of this reaction. It could be concluded that the PMP group could readily be removed within five minutes using the optimal reaction conditions. These optimal conditions were also applied to synthesize gram quantities of the deprotected amine in the continuous flow reactor.

[1] K. Koch, R.J.F. van den Berg, P.J. Nieuwland, R. Wijtmans, H.E. Schoemaker, J.C.M. van Hest, F.P. J.T. Rutjes, Biotechnol. Bioeng, in press.

[2] P.J. Nieuwland, K. Koch, J.C.M. van Hest and F.P.J.T. Rutjes, Proceedings of µTAS 2006 Conference, Tokyo, 2006, 870-872.

[3] P. F. de Aguiar, B. Bourguignon et al., Chemom. Intell. Lab. Syst., 1995, 199-210.

[4] J.M.M. Verkade, L.J.C. van Hemert, P.J.L.M. Quaedflieg, P.L. Alsters, F.L. van Delft, F.P.J.T. Rutjes, Tetrahedron Let., 2006, 8109-8113.

64c Microreactors for the Basic Organic Teaching Lab Course

Swen Koersten[1], Dana Kralisch[1], Kristina Dubnack[1], Peter Kreitmeier[2], Burkhard Koenig[2], and Guenter Kreisel[1]. (1) Institute of Chemical Engineering and Environmental Chemistry, Friedrich Schiller University of Jena, Lessingstraße 12, Jena, Germany, (2) Institute of Organic Chemistry, University of Regensburg, Universitaetsstraße 31, Regensburg, Germany
Microreactor technology currently finds its way to applications in chemical industry and research. The method offers significant opportunities and advantages: Chemical syntheses may proceed more efficient with increased selectivity and the small volumes used in the technology make it inherent safe. However, the application of microreactors in production, development and research calls for trained personnel. Only skilled scientists, who are aware of opportunities and limitation of the technology, will gain its potential. Therefore, microreactor technology should become part of the educational curriculum at universities, in particular in chemistry and chemical engineering. To meet the needs of university education, microreactors to be used in a basic teaching laboratory must be affordable, reliable and robust, and allow a direct monitoring of the chemistry. We develop such systems based on glass microstructured components produced by a sand blasting method by the industrial project partner Little Things Factory GmbH (Ilmenau, Germany). The microreactors are made from borosilicate glass which is inert to most chemicals. To connect the microreactor a set of easy to use fittings and tubing was developed. The flow of the reactants is either realized by gravity using reservoirs placed at different height or by the use of a syringe pump. A flow rate without pulses up to 10 mL/min is feasible. To control the reaction temperature the whole microreactor is simply emerged into a cooling or heating bath. If a precise or liquid-free control of the temperature is desired, thermoelectric modules are used which allow a temperature range between -10 to +60 oC. On-line reaction monitoring is envisaged by the use of a UV flow cell based on LED technique and conductivity measurements inside the device. The teaching laboratory experiments developed for the microreactor devices span the full range of the organic chemistry at the Bachelor level. To be instructive and easy to monitor, reactions with visible colour change are selected. So far aldol condensations, redox reactions and the synthesis of an azo compound have been realized. Overall, a set of 15 reactions will be developed with full instructions for students and supervisors. All experimental set-ups, lab manuals, background material and instructions for syntheses will become part of the internet teaching material database NOP at www.oc-praktikum.de. The documents are available in several languages, e.g. in German and English. All material is available free of charge and supports docents and lab supervisors developing state of the art laboratory classes.

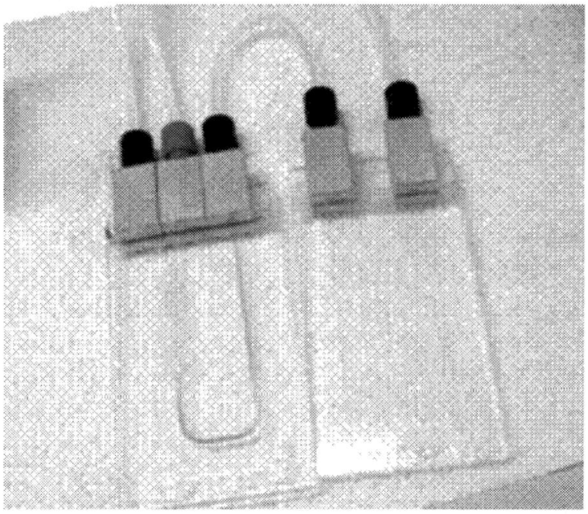

181e Alkane Activation at Ambient Temperature – Unusual Selectivities, C-C, C-H Bond Scission Vs C-C Bond Coupling

Cristiano Trionfetti[1], Anil Agiral[2], Han J.G.E. Gardeniers[3], L. Lefferts[1], and **K. Seshan**[4]. (1) Catalytic Processes and Materials, University of Twente, PO Box 217, Enschede, 7500 AE, Netherlands, (2) Mesoscale Chemical Systems, MESA+ Institue for Nanotechnology, University of Twente, P.O. Box 217, Enschede, 7500 AE, Netherlands, (3) Mesoscale Chemical Systems, University of Twente, Meander, Room 146, P. O Box 217, Enschede, Netherlands, (4) Catalytic Processes and Materials, MESA+ Institue for Nanotechnology, University of Twente, P.O. Box 217, Enschede, 7500 AE, Netherlands

Availability and low cost of light alkanes (C_1-C_4 range) make them interesting as feedstock for commercial fuels and chemicals. However, direct conversion of lower alkanes for such purposes is a challenging problem [1]. Two commonly attempted solutions to the problem are to use oxygen and/or selective catalysts for efficient conversions. However, in the case of light alkanes (C_1-C_3) higher temperatures, in the range 550-850°C, are required to get appreciable alkane conversions. Activation of C-C and C-H bonds at higher temperatures, even in the presence of heterogeneous catalyst systems, tend to be initiated by homogeneous splitting of bonds, creation of radicals and radical chain reactions leading to products. High temperature alkane conversions have inherent difficulties, *viz.* (i) they lead to extensive, endothermic C-C and C-H bond cleavage causing formation of cracking products or coke, respectively (ii) cause loss of catalyst activity due to sintering and (iii) favor non-selective combustion of alkanes.

Alkane activation at lower temperatures, even though challenging, is very interesting from a commercial point of view. Gaseous plasma, generated by dielectric barrier discharge and which consists of energetic electrons, is reported to be able to activate hydrocarbons [2]. It is possible to generate a cold plasma at ambient conditions (atm pressure) in a micro-reactor [3]. This presentation highlights the oxidative conversion of light alkanes, C_1-C_3 range, in the presence of cold plasma in a microreactor. Thus, C-C and C-H bond activation at lower temperatures and its influence on products selectivities are discussed. The employed plasma microreactor consists of a Pyrex rectangular chip with microchannels dimension of 30 mm length x 5 mm width and a channel depth of 500 μm (Fig. 1).

The temperature of the gas was firstly determined by the optical spectrum emission [4] recorded during alkanes activation and additionally monitored by inserting a thermocouple inside the micro channel outside the discharge area. Under the conditions used (1 atm, 3 W power input), the temperature was below 40°C.

In the case of C_1-C_3 range alkanes and under the same conditions conversion levels of 22 mol% for propane, 15 mol% for ethane and 10 mol% for methane was observed. Most remarkably, in all the experiments, high selectivity to products with higher molecular weight than the starting hydrocarbon (coupling products) were also observed (Fig. 2). In fact, in the presence of a cold plasma, alkane molecules can be directly activated/converted via collisions with energized electrons that produce radicals for eg., $C_3H_7^{\bullet}$ in the case of propane due to cleavage of C-H bonds ($C_3H_8 + e^- \rightarrow C_3H_7^{\bullet} + H^{\bullet} + e^-$) [2]. This can initiate radical chain reactions and their propagation and/or termination reactions take place at ambient temperature. The presence of products containing more C atoms than the starting alkane requires C-C bond formation/coupling under the conditions present in the microreactor. Typically formation of C-C bonds is an exothermic process and therefore favored at lower temperatures. Coupling reactions between radicals also imply a decrease in entropy because the number of molecules/radicals decreases. Thus, in our studies direct homologation of alkanes is observed.

This represents an interesting novelty to directly upgrade cheap low molecular weight alkanes to commercially useful fuels and/or feedstock materials for the chemical industry.

References

1. T.V. Choudary, E. Aksoylu, D. Wayne Goodman, Catalysis reviews 45 (1) (2003) 151-203

2. U. Kogelschatz, Plasma Chemistry and Plasma Processing 23 (1)(2003)1-46

3. C. Trionfetti, A. Agiral, J.G.E. Gradeniers, L. Lefferts, K. Seshan, J. Phys. Chem. C, *submitted*

4. S. Kado, K. Urasaki, Y. Sekine, K. Fujimoto, T. Nozaki, K. Okazaki, Fuel 82 (2003) 2291-2297

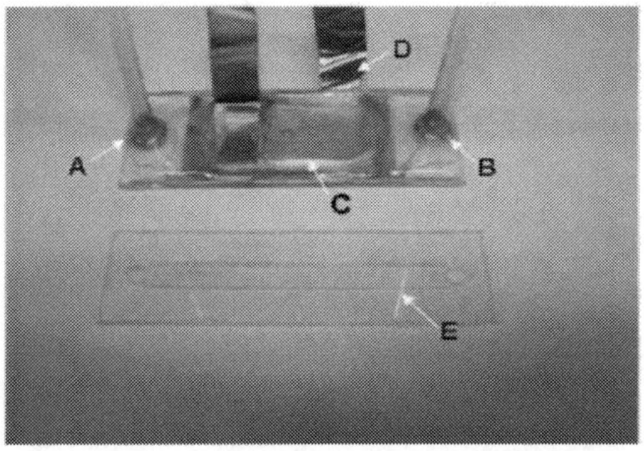

Fig.1 Employed plasma microreactor; top view, the inlet and out let (A, B), the copper plate (C) connected to a power supply using adhesive copper foils (D) and the microchannel (E) are shown in the picture.

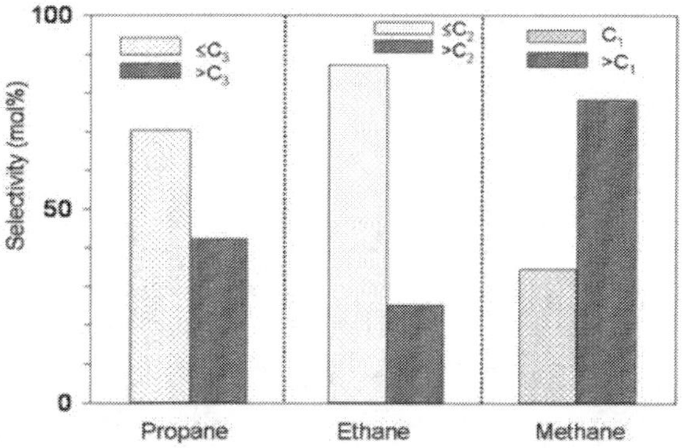

Fig.2 Oxidative conversion of C_1-C_3 range alkane. Selectivity toward products with, respectively, lower (cracking) and higher (coupling) molecular weight than the starting hydrocarbon. All the experiments were performed under plasma conditions at RT.

83ag Modular Microreaction System

Samrat Mukherjee[1], Matthias Kroschel[2], Carmine Raffa[2], Olaf Stange[2], Eric Boonstra[1], and Thomas Daszkowski[1]. (1) Bayer Corporate and Business Services, 8500 West Bay Road, Baytown, TX 77520, (2) Ehrfeld Mikrotechnik BTS GmbH, Mikroforum Ring 1, 55234 Wendelsheim, Germany

Micro process technology opens new pathways for the development of economical, innovative and intensified processes. In the chemical and pharmaceutical industry, this technology plays an increasingly important role in research, development and production because of its decisive advantages over conventional technology. For example: microstructured devices have up to three orders of magnitude higher surface to volume ratio compared to conventional apparatuses. This results in intensified heat transfer. Microreactors are an excellent solution for extremely fast and highly exothermic or endothermic reactions. These reactions can be much better controlled in a microreactor and the possibility of hot spot for homogenous or heterogeneous reaction will be minimized. In conventional reactors strong temperature profiles could influence the product spectrum, and in some case even the catalyst can be irreversibly damaged. Microreaction technology therefore can be used to perform hard to control reactions, which otherwise cannot be performed in a conventional reactor.

Due to very high surface to volume ratio, the mass transfer limited processes can also take advantage of microtechnology. In micromixers the mixing velocity is an order of magnitude higher then in conventional mixers, which results in smaller mixing sections, a few millimeters or even less. Another advantage of micro devices is the option to control and adjust the process parameters much better. This leads not only to a better yield, but also to a higher level of safety. Additionally costs for downstream processing can be reduced significantly: for example by a solvent free production, production of reduced waste material during start up or shut down, change of product or potential production interference because of the small hold ups in continuous operating microreaction plants etc.

The Modular Microreaction System, which will be presented in this poster, has been demonstrated for liquid, gas-liquid and gas phase reactions and also for heterogeneous reactions. In addition to the Modular Microreaction System can also be used to produce micro- and nanoparticles. For example, modules, which were developed for precipitation reactions, can be used to produce pigments and catalysts. The advantages of suspension handling and the production of solid particles with specified properties lead to process innovations, which have already been installed in small production plants. Additionally some microreactors, allow the handling of slurries. The operating parameters include a temperature range of –20 to 200 °C; operating pressures up to 100 bar (at room temperature); and rated throughputs of up to 25 L/h for low-viscosity media. Special modules available significantly extend this process range. Among the items available are special reactors for reaction temperatures from –100 °C to 500 °C and stand-alone, high-performance production modules with rated throughputs of several thousand liter per hour.

With minimal effort, a whole lab or pilot plant can be installed with the Modular Microreaction Systems.

In conjunction with process control system and peripheral equipments, these microreaction systems are powerful development and production tools, which allow innovations in all areas of the chemical industry. Sustained development of modules and new module materials will widen the field of application for the Modular Microreaction System. This will be shown in this poster.

Modular Micro Process Technology - New Devices and Applications

Olaf Stange[1], Frank Schael[1], Frank Herbstritt[1], Hans-Erich Gasche[2], Eric Boonstra[3] and Samrat Mukherjee[3]

[1]Ehrfeld Mikrotechnik BTS GmbH, Mikroforum Ring 1, 55234 Wendelsheim, Germany, [2]Bayer Technology Services GmbH, 51368 Leverkusen, Germany, [3]Bayer Corporate and Business Services, 8500 West Bay Road, Baytown, TX 77520

Micro process technology opens new pathways for the development of economical, innovative and intensified processes. In the chemical and pharmaceutical industry, this technology plays an increasingly important role in research, development and production because of its decisive advantages over conventional technology. For example: microstructured devices have up to three orders of magnitude higher surface to volume ratio compared to conventional apparatuses. This results in intensified heat transfer. Microreactors are an excellent solution for extremely fast and highly exothermic or endothermic reactions. These reactions can be much better controlled in a microreactor and the possibility of hot spot for homogenous or heterogeneous reaction will be minimized. In conventional reactors strong temperature profiles could influence the product spectrum, and in some case even the catalyst can be irreversibly damaged. Micro process technology therefore can be used to perform hard to control reactions, which otherwise cannot be performed in a conventional reactor.

Due to very high surface to volume ratio, the mass transfer limited processes can also take advantage of micro process technology. In micromixers the mixing velocity is an order of magnitude higher then in conventional mixers, which results in smaller mixing sections, a few millimeters or even less. Another advantage of micro devices is the option to control and adjust the process parameters much better. This leads not only to a better yield, but also to a higher level of safety. Additionally costs for downstream processing can be reduced significantly: for example by a solvent free production, production of reduced waste material during start up or shut down, change of product or potential production interference because of the small hold ups in continuous operating microreaction plants etc.

Using micro process technology, the development of products or continuous processes, e.g. transfer from batch to continuous production could be accelerated significantly. Additionally, new synthesis routes are possible, which can not be realized in a conventional apparatuses in a cost effective manner, such as highly exothermic reactions. For this kind of applications micro process technology is most suitable. Modular micro process systems consist of different types of modules, including mixers, heat exchangers, reactors, sensors, actuators and fluidic and electrical interfaces. All these modules are fully or partially equipped with microstructures. These modules cover essential unit operations within the reaction technology. Additionally the systems are equipped for measuring pressure, temperature and flow rates. Also analytic modules are available, which allow us to determine the quality of the product e.g. using NIR- or UV/VIS-spectroscopy. New developed modules (residence time, photo chemistry reactor, separation etc.), sensors (e.g. pH, coriolis flow meter), actors (temperature control unit, valves) and module materials will be shown in the presentation.

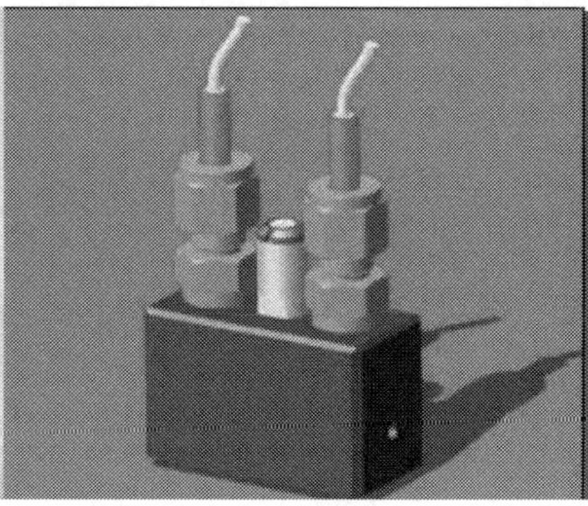

Fig. 1: Photo chemistry reactor Fig. 2: pH-sensor Fig. 3: 3-way-valve

With minimal effort, a whole lab or pilot plant can be installed with the modular microreaction systems. In conjunction with process control system and peripheral equipments, these microreaction systems are powerful development and production tools, which allow innovations in all areas of the chemical industry.

Fig. 4: Modular Microreaction System

A micro process system as a flexible pilot plant has been demonstrated for small production. Also, if the holdup of the unit is small, production rates up to several tons is possible. In addition to liquid, gas-liquid and gas phase reactions, microreaction technology can also be used to produce micro- and nanoparticles. For example, modules, which were developed for precipitation reactions, can be used to produce pigments and catalysts. The advantages of suspension handling and the production of solid particles with specified properties lead to process innovations, which have already been installed in small production plants. Additionally some microreactors, allow the handling of slurries.

The scale-up from lab to bulk production can be achieved by parallel connection of microreaction technology plants or by scale-up of the modules by multiplying the microstructures reaching production rates up several thousand tons per year. This indicates that microreaction technology could be an alternative to conventional production units.

For successful process development, a combined knowledge of micro process technology concepts along with conventional plant design process, reaction engineering and process chemistry is essential. This holistic approach is the key to innovative, micro process technology-based solutions for efficient processes and high-performance plants. This will be also shown in the presentation for selected applications.

181a Heat Recuperation Strategies for Microreactors: Thermal Management in Gas-Phase and Catalytic Propane/air Oxidation

Justin A. Federici[1], Eric D. Wetzel[2], Bruce R. Geil[2], and Dionisios G. Vlachos[1]. (1) Chemical Engineering, University of Delaware, 353, Colburn Laboratory, Newark, DE 19716, (2) Army Research Laboratory, AMSRD-ARL-WM-MA, Bldg. 4600, Aberdeen Proving Ground, MD 21005-5069

There has been increasing interest in the development of new portable power generation devices. Based on the relatively high gravimetric energy density of liquid fuels [1], an integrated thermoelectric/ microreactor system with relatively low chemical to electrical energy conversion efficiency has an overall energy density on par with conventional batteries. It has recently been suggested that effective thermal management of the released heat can improve the efficiency of these systems so that their energy density exceeds that of traditional batteries[2]. Based on the increased area to volume ratio, heat losses play in important role in the overall stability of these microsystems. In addition, the thermal efficiency of the microreactor must be optimized, as enthalpy is lost in the exiting gases. One thermal management strategy to overcome these loses is the utilization of "excess enthalpy" burners, suggested first by Lloyd and Weinberg [3], where the hot exiting products are used to exchange heat with the incoming cool reactants. Heat recirculation is an example of these "excess enthalpy" reactors and was used in this study of the enhancement of catalytic microcombustor stability with respect to single channel systems.

The heat loss stability of both gas-phase [4] and catalytic [5] microcombustion of propane/air mixtures was studied in single channel and heat recirculation reactors. Using the commercial, computational fluid dynamics (CFD) package Fluent, the effect of wall thermal conductivity on the critical heat loss coefficient was determined. It was observed in both gas-phase and catalytic systems that the heat recirculation benefit is only observed in the limit of lower thermal conductivity walls. This is based on the recirculated gases providing additional preheating of the reaction zone, when the axial conductive heat transfer via the walls is insufficient. The heat recirculation effect on the microsystem stability with respect to inlet velocity is also discussed. Comparison of the stability of homogeneous and catalytic microreactors is also done. Select simulation results were tested experimentally. Specifically, stainless steel, single channel and heat recirculation microreactors were fabricated and tested with Pt/anodic alumina catalysts. Consistent with CFD results, the conductive wall heat recirculation system was not observed to demonstrate a significant increase in stability.

References:

1. Sitzki, L., et al. Combustion in Microscale Heat-Recirculating Burners. in The Third Asia-Pacific Conference on Combustion. 2001. Seoul, Korea.

2. Federici, J.A., et al., Catalytic microcombustors with integrated thermoelectric elements for portable power production. J. Power Sources, 2006. 161(2): p. 1469-1478.

3. Lloyd, S.A. and F.J. Weinberg, A burner for mixtures of a very low heat content. Nature, 1974. 251: p. 47-49.

4. Federici, J.A. and D.G. Vlachos, A Computational Fluid Dynamics Study of Propane/Air Microflame Stability in a Heat Recirculation System. Comb. and Flame, 2007. Accepted: p. doi:10.1016/j. combustflame.2007.09.009.

5. Federici, J.A. and D.G. Vlachos, Catalytic Combustion of Propane/Air Mixtures in Single Channel and Heat Recirculation Microburners: An Experimental and Computational Fluid Dynamics Modeling Comparison of Stability. 32nd International Symposium of Combustion, 2007. Submitted.

235b Liquid-Liquid Two-Phase Flow Patterns and Mass Transfer Characteristics in Rectangular Glass Microreactors

Anne-Laure Dessimoz[1], Laurent Cavin[2], Albert Renken[1], and Lioubov Kiwi-Minsker[1]. (1) Ecole Polytechnique Fédérale de Lausanne, GGRC, ISIC-SB, station 6, Lausanne, 1015, Switzerland, (2) Givaudan Suisse SA, Vernier, 1214, Switzerland

Intensification of chemical processes aiming at the effective use of raw materials and energy implies miniaturization of chemical reactors.

In order to define the reaction conditions and to design a microstructured reactor for a specific multiphase chemical transformation, detailed knowledge of the hydrodynamics in the microchannels is necessary. The common modes of interface in the case of liquid-liquid two-phase flow are "slug flow" and "parallel flow". The main problem in the control of flow pattern is due to the flow sensibility to the experimental parameters like: linear velocity, ratio of the phase volume, fluid properties, channel geometries, and the construction material. Despite its importance, systematic experimental study of the effect of fluid properties on flow patterns has not been reported. The mass transfer performance is another key characteristic of microstructured devices applied for reacting systems. Only Burns and Ramshaw (2001) provided data on mass transfer performance for the slug flow regime during chemical reaction. The mass transfer characteristics for the parallel flow have not been reported so far.

In this context, the flow of two immiscible fluids was investigated in two rectangular glass microchannels with Y- and T- junctions and with equivalent diameters of 269 and 400 μm, respectively. The influence of the fluid properties on two-phase liquid-liquid flow patterns was studied using deionised water and toluene with volumetric flow rates between 1 - 6 ml/h and monitored by a photo-camera. Sodium hydroxide and/or trichloroacetic acid were added to both phases in order to change density, viscosity and interfacial tension of the fluid/fluid system. The mass transfer performance for the slug flow and parallel flow pattern was determined using the instantaneous neutralization reaction (controlled by mass transfer) between the trichloroacetic acid dissolved in the organic phase and sodium hydroxide present in the aqueous phase. The total volumetric flow rate was varied between 2 to 12 ml/h and the concentration of NaOH in the aqueous phase between 0.1- 0.3 M. The volume ratio of the organic/aqueous phases was kept constant 1:1.

The results showed that the shape of the interface between two immiscible liquids is controlled by a competition between the viscous forces and the interfacial tension. The viscous forces act to extend and drag the interface downstream (Squires and Quake, 2005) favouring the parallel flow. The increase of interfacial tension reduces the interfacial area resulting in a slug flow. The observed flow patterns were correlated with the mean Capillary and Reynolds numbers. The model allows prediction of the flow patterns depending on the values of the Reynolds and Capillary numbers for equal volumetric flow rates of the two phases.

For the conditions applied during the study, the slug flow and the parallel flow showed similar mass transfer performance with the global volumetric mass transfer coefficient (k_{gl}*a) in the range of 0.2 to 0.5 s^{-1}. The values of the mass transfer coefficients (k_{gl}) are strongly dependent on the way used to estimate the specific interfacial area. If the wall film of the continuous phase is taken into account in the case of slug flow, the global mass transfer coefficient was found to be in the range of 10^{-5}-10^{-4} m/s for both flow patterns. For the parallel flow, the mass transfer coefficient was mainly affected by the base concentration in water. In the case of slug flow, no influence of the NaOH concentration on k_{gl} was observed. The k_{gl} values were influenced by the linear velocity only. The results can be rationalized using a film model, however the presence of secondary phenomena, like interfacial instabilities must also be taken into consideration.

References

Burns, J. R. ,Ramshaw, C., (2001). The intensification of rapid reactions in multiphase systems using slug flow in capillaries. Lab on a Chip, 1, 10-15.

Squires, T. M. ,Quake, S. R., (2005). Microfluidics: Fluid physics at the nanoliter scale. Reviews of modern physics, 77, 977-1026.

MICRO-EXTRACTOR FOR LIQUID-LIQUID EXTRACTION, CONCENTRATION AND IN-SITU DETECTION OF LEAD

Van Man TRAN[1,2], Jean BERTHIER[1], Régis BLANC[1], Olivier CONSTANTIN[1], Claude VAUCHIER[1], Nicolas SARRUT[1]

[1]*Department of micro-Technologies for Biology and Healthcare*
LETI-Minatec, 17 rue des Martyrs, 38054 Grenoble Cedex 9, France
[2] *Grenoble Institute of Technology, INPG*
E-mail: van-man.tran@cea.fr and nicolas.sarrut@cea.fr

ABSTRACT

In this work, we present the design, fabrication and characterization of a novel micro-extractor which performs on-line extraction-concentration-detection (ECD) of target molecules flowing in a carrier liquid. The system comprises a primary microchannel containing a flowing aqueous carrier liquid and a secondary organic storage fluid circulating in an adjacent channel. The interfaces between the two immiscible fluids are stabilized by vertical micro-pillars. The system encompasses three functions: (i) extraction of the target molecules from the carrier fluid through the pillar-stabilized interfaces, (ii) concentration of the targets in the secondary organic solvent due to its very low—or zero—velocity, (iii) on-line detection via optical spectrometry.

We successively present the analysis of the physics of the system which has led us to a specific design, then the microfabrication of the chip, and finally demonstrate the extraction, concentration and detection of lead ions (Pb^{2+}) from a water flow.

***Keywords**: liquid-liquid extraction, microflow, interface stability, Monte-Carlo modeling, lead detection*

1. Introduction

Microfluidic systems have recently brought dramatic improvements for the monitoring of dangerous molecules and/or pathogens in solutions [1]. They allow for continuous and sensitive detection of these targets in liquids. However, integration of the three key functions required by such systems, namely efficient extraction, sufficient concentration and sensitive detection (ECD) remains a difficult task. In this paper, we present a new prototype of integrated microsystem based on liquid-liquid extraction.

Different principles can be used to realize an integrated system [2,3,4]. The solution chosen here is that of liquid-liquid extraction (LLE) currently used as sample preparation technique in environment analysis or bioanalysis. This method differentiate the compounds based on their affinity for two different immiscible liquids, usually water or water-based (aqueous) and an immiscible organic solvent or organic solution. More precisely the selective affinity is triggered by the binding on the interfaces of the targets with a hydrophobic ligand contained in the solvent. A solution sometimes found in the literature is a LLE based on the formation of an emulsion by mixation followed by a phase separation process [5]. This solution being difficult to integrate in a compact microsystem, we have preferred the approach pioneered by W. Ehrfeld [6] based on the stabilization of the interface between the two liquids. This stabilization is performed by the presence of micro-pillars upon which the interface is anchored (Fig. 1). Because the interfacial area must be large, many interfaces are needed; hence, the contact area between the two liquids is constituted by many pillar-stabilized interfaces. Extraction is realized by the migration of the targets through the interfaces. On the other hand, concentration is achieved by keeping the secondary liquid (solvent) at rest—or flowing at a low speed. And finally detection is made by optically illuminating the solvent channel with a UV-VIS detector.

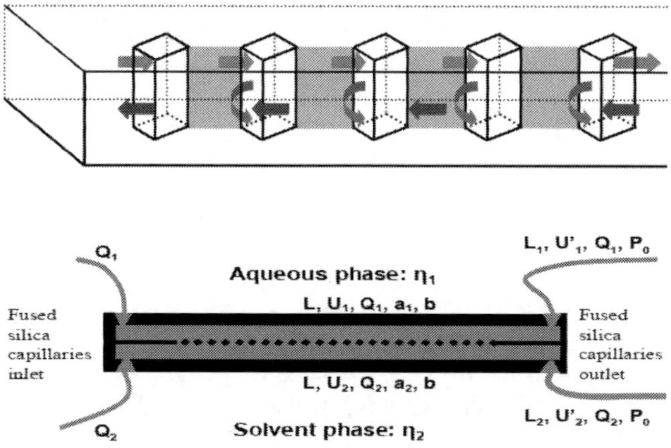

Fig.1. The interfaces are anchored on pillars (top); schematic of the flow channels (bottom).

This type of microsystem covers a wide range of applications, like the monitoring of the presence and concentration of chemical or biological molecules in water. Especially, an application of ECD device is aimed at the evaluation of the presence of the heavy metals in water. Some heavy metals, like lead, mercury and arsenic are extremely toxic even at very low concentration; others like copper and zinc are useful for the human metabolism at trace concentration, but dangerous at higher concentration. Macroscopic conventional systems used for the determination of heavy metal in water are Inductive Coupled Plasma – Mass Spectrometry (ICP-MS), Electrothermal Atomisation Atomic Absorption Spectrometry (EA-AAS) or polarography. However, the use of spectrophotometry or fluorescence associated with the micro scale offers many advantages in term of simplicity and low cost. Moreover these methods could be integrated easily in portable instrument.

In this text, we successively present the analysis of the physics of the system which has led us to a specific design, then the microfabrication process of the chip, and finally we demonstrate ECD of lead ions (Pb^{2+}) from water.

2. Physical analysis: stability and extraction

In this section, we focus on the physics of the system. Two constraints determine the physical behavior of the system. The first constraint is that the interfaces must be stable and attached to the pillars at all times. Remember that we do not want mixation of the two phases. The second constraint is that the interfacial area must be sufficient to provide an efficient mass transfer. We show here that the first constraint is associated with a maximum length L_{st} of the extraction channel. It is impossible to have interface stability if the length of the extraction channel is larger than L_{st}. Next, we show that the second constraint can also be translated in a characteristic length L_e. It is clear that the longer the extraction channel, the larger the efficiency of extraction. We note L_e the length sufficient to obtain the capture of 70% of the flowing targets. The other parameters for sensitivity refer to chemistry, i.e. a good association constant between targets and ligands diffusing in the solvent.

2.1. Stability length

It is observed first that the constraints on the interfaces are normal constraints, the tangential constraints being negligible [7, 8]. Hence, the force applied on the interfaces is related to the pressure difference across the interface. When the interfaces are stable, the two microflows evolve independently, like two independent channels. The pressure at location x depends on the pressure drop between x and the outlet of the channel (Fig. 1)

$$P(x)-P_0 = f(\eta,Q,a,b)\,(L-x)+g(\eta,Q,D_H)$$

The pressure difference across the interface at location x region is then given by

$$P_1(x)-P_2(x) = \left[f(\eta_1,Q_1,a_1,b_1)-f(\eta_2,Q_2,a_2,b_2)\right](L-x) + \left[g(\eta_1,Q_1,D_{H1})-g(\eta_2,Q_2,D_{H2})\right] \quad (1)$$

where Q_1 and Q_2 are the flow rates of aqueous phase and solvent. Taking into account the modification of the Washburn equation for the pressure drop for rectangular channels [9], the pressure difference across the interface is

$$P_1(x)-P_2(x) = 8(L-x)\left[\frac{\eta_1 Q_1}{w_1^3 d}\xi(\alpha_1)-\frac{\eta_2 Q_2}{w_2^3 d}\xi(\alpha_2)\right] + 128\left(\frac{\eta_1 Q_1 L_1}{\pi D_H^4}-\frac{\eta_2 Q_2 L_2}{\pi D_H^4}\right) \quad (2)$$

where α is the aspect ratio of the rectangular channel between the width w and the depth d; $\xi(\alpha)$ is defined by $2(1/3-64\alpha/\pi^5 \tanh \pi/2\alpha)$. The maximum pressure difference that an interface can sustain is derived from Laplace's law. In case of pillars with sharp edges, it can be shown that pinning conditions lead to [10]

$$\Delta P_{max} = \frac{\gamma}{R_{curv}} = \frac{2\gamma}{\delta} \quad (3)$$

Then, upon substitution of (3) in (2), the maximum stability length L_{st} is given by

$$L_{st} = \frac{\Delta P_{max} - 128\left(\frac{\eta_1 Q_1 L_1}{\pi D_H^4}-\frac{\eta_2 Q_2 L_2}{\pi D_H^4}\right)}{\frac{8\eta_1 Q_1}{w_1^3 d}\xi(\alpha_1)-\frac{8\eta_2 Q_2}{w_2^3 d}\xi(\alpha_2)} \quad (4)$$

Interpretation of equation (4) is shown in the Fig. 2.

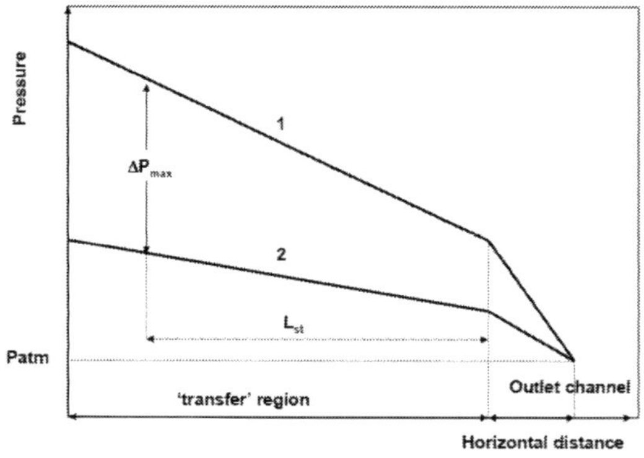

Fig.2. Pressure profiles and determination of the maximum length of interface stability (here, the outlet channels have the same length)

At the outlet, the pressure is the atmospheric pressure; going upstream, the pressure evolves differently in each channel, and the pressure difference in both channels increases linearly. At a distance equal to L_{st}, the pressure difference across the interface reaches ΔP_{max}, and the interface cannot be stable anymore. If flow rate conditions do not correspond to the stability length, break down of interfaces occurs at the inlet of the channel (Fig. 3).

The model also shows that sharp-edged pillars perform better than smooth, rounded pillars [10].

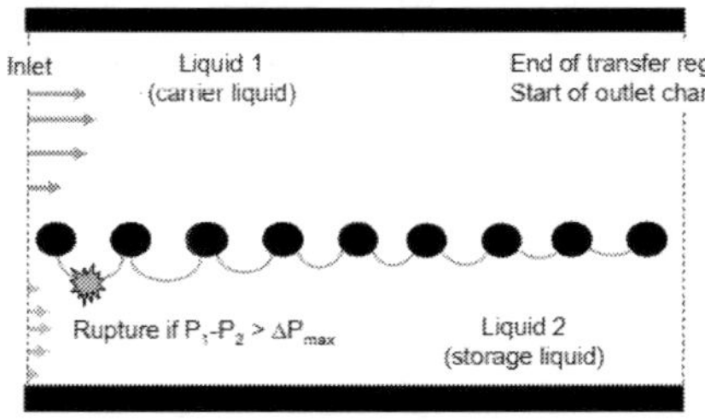

Fig.3. Schematic view of the interfaces shape and potential break down of the interface

2.2 Extraction efficiency

Extraction is realized by convective-diffusive transport of the targets to the interfaces followed by chemical complexing of these targets on the interfaces with the ligands in excess in the solvent. We suppose here that extraction is limited by the transport of targets and not by the chemical reaction at the interface, i.e. there is always an excess of ligands present in the vicinity of the interface and the kinetics of the reaction is much faster than mass transfer.

We have developed two different approaches to determine the extraction efficiency: the first method is numerical and we use a Monte-Carlo approach; the second is analytical based on a spatial Fourier series expansion. In the Monte-Carlo approach, the model is similar to that of the diffusive random walk of particles, except that a convective motion is superposed to the random-walk itself (Fig.4 and 5).

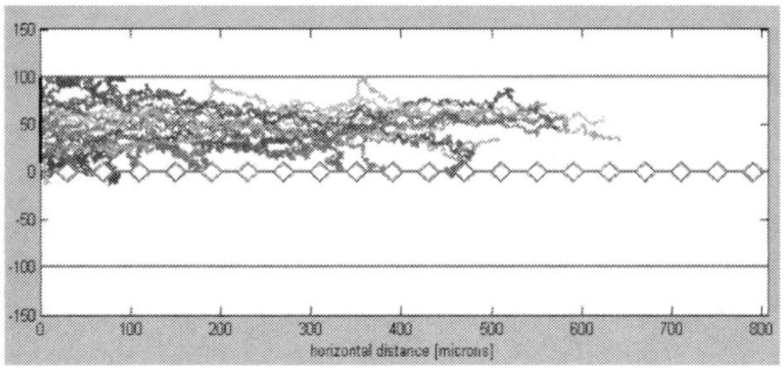

Fig.4. Random walk of particles in a microflow calculated by Monte-Carlo modeling. The pillars have a diamond shape and for simplicity the interfaces are linear segments between pillars.

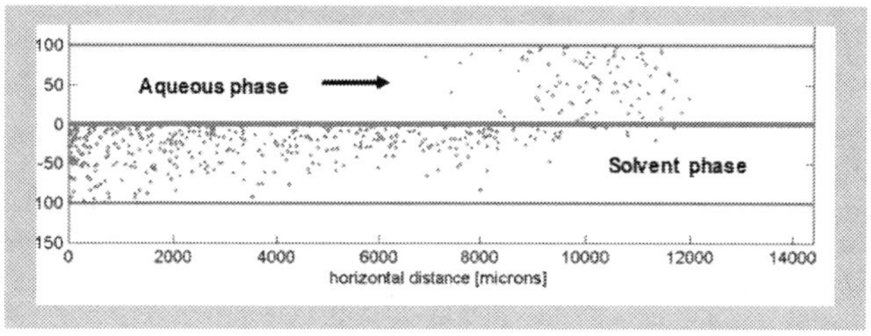

Fig.5. Extraction of targets from an aqueous carrier liquid to a solvent (Monte-Carlo simulation)

The analytical model requires some simplifications: first, we assume that the channel has a rectangular shape and that the bottom wall represents a non-interrupted, continuous, rectilinear interface. An order of magnitude of the mass transfer length is given by equating a diffusion time with a convective time [11, 12]. An approximate time for the diffusion of a particle initially located on the other side of the channel to reach the wall—or the interface—is $\tau_d \approx w^2/4D$. In this time interval τ_d, the particle has been transported by the flow along a horizontal distance $L \approx U\tau_d$. We obtain

$$\frac{L}{w} \approx \frac{1}{4}\frac{Uw}{D} = \frac{1}{4}Pe \tag{5}$$

where D is the diffusion constant and Pe is the Péclet number of the carrier flow. The ratio L/w is also known as the Graetz number [13]. However relation (5) is only a coarse approximation; it is not a usable model. In the following, we develop a more precise model for the mass transfer length. Let us introduce another simplification: along the x-axis, the convective transport is much larger than the diffusive transport, and we will neglect axial diffusion. Then by constructing a mirror image of the channel (Fig.6), the boundaries are identical, and the concentration can be written under a Fourier's series expansion in the Lagrangian system moving with the average velocity U

$$c(y,t) = 2c_0 \sum_{k=0}^{\infty} \frac{(-1)^k}{\pi\left(k+\frac{1}{2}\right)} \cos\left(\left(k+\frac{1}{2}\right)\frac{\pi y}{w}\right) \exp\left(-\left(k+\frac{1}{2}\right)^2 \frac{\pi^2}{w^2} D t\right) \tag{6}$$

Integration of (6) with respect to y produces the time dependence of the concentration

$$C(t) = 2c_0 \sum_{k=0}^{\infty} \frac{(-1)^k}{\pi^2\left(k+\frac{1}{2}\right)^2} \exp\left(-\left(k+\frac{1}{2}\right)^2 \frac{\pi^2}{w^2} D t\right) \tag{7}$$

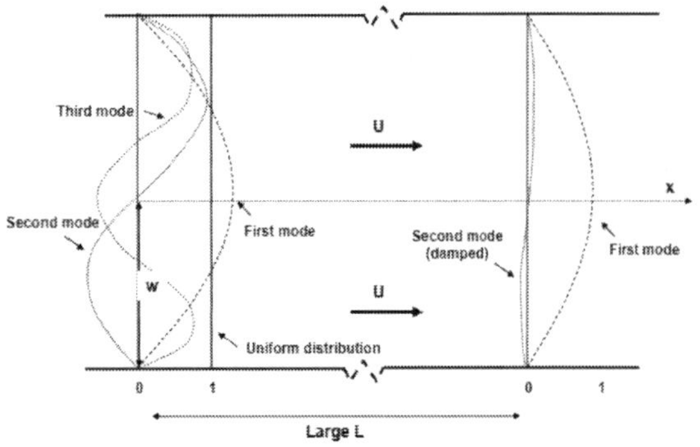

Fig.6. At inlet, the uniform distribution is decomposed in spatial Fourier series; after some translation length L, only the first mode remains, all the other being damped.

Substitution of t by x/U brings back to the fix Eulerian coordinate system. Numerical simulations have shown that, after a very short distance, all the modes are damped except the first mode corresponding to k=0 (Fig. 6). The attenuation in the x-direction is then

$$a = \frac{8}{\pi^2}\exp\left(-\frac{\pi^2 D x}{4 w^2 U}\right) \qquad (8)$$

The number a represents the fraction of targets still in suspension in the carrier liquid at the length x. The fraction of targets transferred to the solvent at the length x is then eff=1-a. A reduction of 70% (corresponding to $a = 8e^{-1}/\pi^2$) of the number particles continuing to flow in the aqueous channel is reached at the length L_e determined by

$$\frac{L_e}{w} = \frac{4}{\pi^2}\frac{U w}{D} = \frac{4}{\pi^2} Pe \qquad (9)$$

Equation (9) has exactly the same form than equation (5). However, the coefficient before the Péclet number has been explicitly determined.

The analytical development leading to equation (8) has been made using simplifying hypothesis. In reality, a comparison with MC modeling results (Fig.7) suggests the slightly modified expression

$$eff = 1 - \exp\left(-\frac{\pi^2 D\, R_{s,eq} L}{4 w^2 U}\right) \qquad (10)$$

which takes into account the ratio of the gap length (or the gap area) to the total length (or the total area) $R_s = S_{gap}/S_{tot}$ modified by a geometrical constant that depends on the shape of the pillars $R_{s,eq} = \kappa\ R_s$.

2.3. Discussion

The system must be designed so that the stability length L_{st} is maximum. It can be shown that reducing homothetically the dimension of the pillars and gaps increases the stability length without decreasing the efficiency. Besides, if we aim for an efficiency of 70%, then the dimensions must be chosen so that $L_e < L_{st}$. It can be shown that this inequality requires a limitation on the flow rate and on the dimensions of the channels. Especially an increase of the etching depth of the channels improves considerably the efficiency of the system. In conclusion, the physical constraints that we have determined translate into microfabrication constraints.

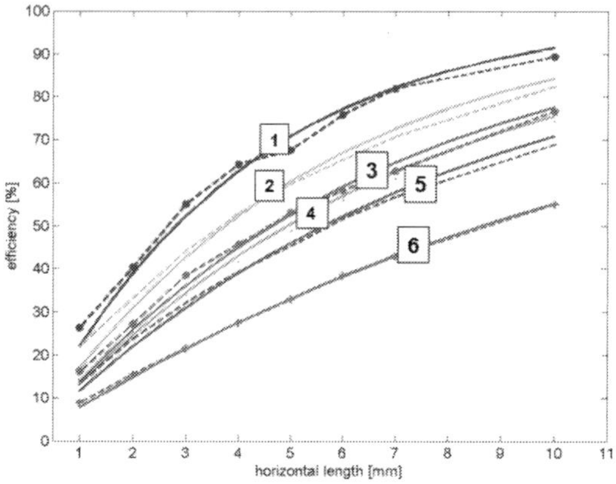

Fig. 7. Extraction efficiency vs. axial distance. The dots and dotted lines correspond to the Monte-Carlo simulation results and the continuous lines are best fit based on relation (10). 1- Uninterrupted interface R_s=1, 2- ideally thin membrane $R_{s,eq}$=0.75, 3- lozenges $R_{s,eq}$=0.52, 4- thick membrane $R_{s,eq}$=0.49, 5- circles $R_{s,eq}$=0.43, 6- squares $R_{s,eq}$=0.28. Note that R_s=0.22 from 2 to 6; Q_1=0.6 µl/min and $D=10^{-9}$ m^2/s.

In the following section, we present the microfabrication of the device. We will see that, for lead ions extraction, channels of depth d=150 µm, width w=100 µm, and a length L= 5 mm are for the moment the best solution that can be realized.

3. Fabrication of micro-extractor

3.1 Microfabrication

The extraction microsystem is designed and manufactured on silicon substrate (100 mm wafer) of 525 µm thickness using the traditional techniques for microtechnology on silicon, namely photolithography, DRIE (Deep Reactive Ion Etching) and thermal oxidation. Fig. 8 summaries the process which is completed by molecular direct bonding and surface treatment by silanisation. The microsystem has a dimension of 5 x 8 mm^2. The first 80 µm depth etched at the inlets and outlets allows for the insertion of capillary tubes. The second principal etching is used to obtain two adjacent channels of 100 and 150 µm width and 150 µm depth. This etching creates the micropillars too. A cycle of de-oxidations followed by thermal oxidations is used for the final cleaning. The Pyrex top lid is assembled by molecular direct bonding. Capillaries of 75-150 µm diameter and 30 mm length are connected using UV-polymerized Masterbond glue. Devices of different pillar shapes and dimensions are designed on a same 100 mm wafer.

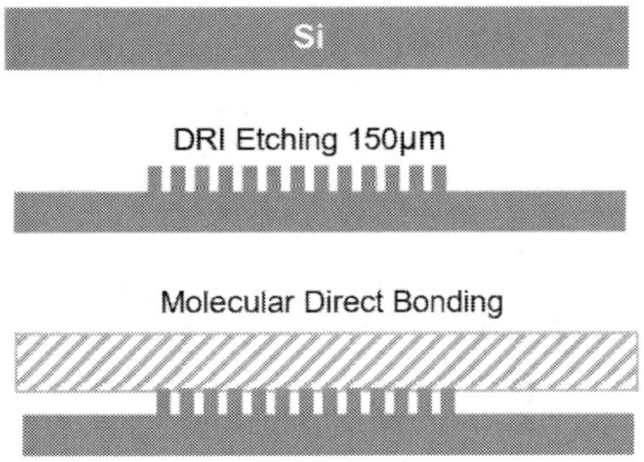

Fig.8. Fabrication process of the micro-extractor

The Deep Reactive Ion Etching (DRIE) process is realized according to Fig. 9. A 3 µm oxidation is first realized (9a); then the wafer undergoes the photolithography step. Deposition of resin is followed by a structuring step and DRI etching of the 3 µm layer of oxide (Fig. 9b) which constitutes the mask for the step (9e). The process is repeated for the bottom side of the substrate in order to obtain a 150 µm DRI etching (Fig. 9c) which help for easier cutting of the device from wafer. In both cases the resin is eliminated by treatment with oxygen plasma and a HNO_3 solution. Next, inlet and outlet ports are made by photolithography and DRIE (Fig. 9d). Finally, a 150 µm DRI etching step creates the pillars (Fig. 9e). A 1% HF solution is used in 1 minute to remove the silicon oxide waste on the vertical surface. A total de-oxidation of the plate followed by a thermal oxidation of 100 nm and the second total de-oxidation assure the cleaning of all the fluorinated contaminants brought by the DRIE process and produces the final device (Fig. 9f).

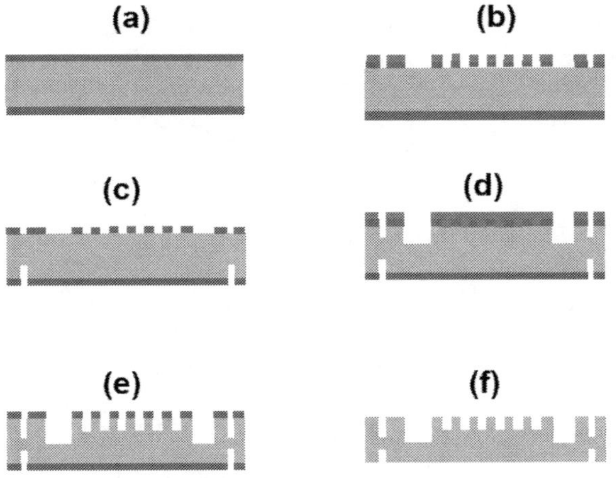

Fig. 9. Details of the DRIE process

Fig. 10 shows the SEM (Scanning Electron Microscopy) images of the device. A 550 µm thick Pyrex wafer is mounted on the silicon plate by plasma activated direct bonding. The Pyrex top lid is treated by rinsing with a solution of H_2SO_4:H_2O_2 2:1 and then by oxygene plasma. The two wafers are wetted and finally are aligned and then heat treated at 400°C. This process first creates the hydroxyl bridges linking the two surfaces at a distance of 11.54 Å. After that the bridges switch partially from Van der Waals links to siloxane bonding (Si-O-Si) at 3.18 Å [14] as shown in following equation

$$SiOH + HOSi \rightarrow Si\text{-}O\text{-}Si + H_2O$$

The thermal cycling accelerates this transformation. Fig. 11 shows the mechanism as presented by Q. Tong et al in our temperature condition [15].

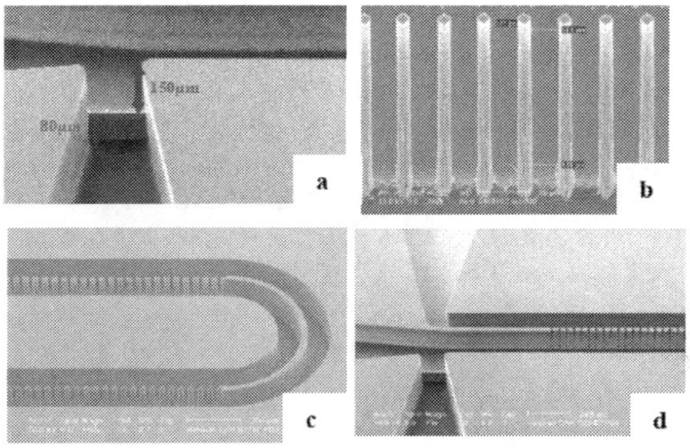

Fig.10. SEM images of micro-extractor showing the two levels of deep etching (a), view of the micropillars (b), a continuous wall can be fabricated so that the channels can make turns (c), view of the channels and pillars: the carrier fluid channel is at the top, the solvent channel at the bottom, and the space for the optical fiber in the middle (d).

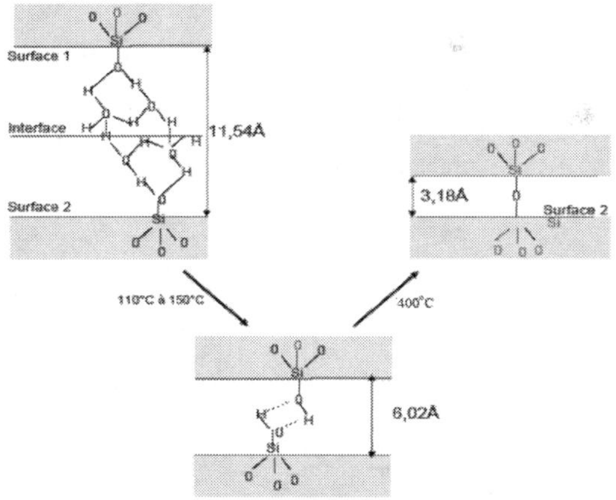

Fig.11. The molecular direct bonding depends on the formation of 6.02 Å hydrogen bond array between two surface by wetting process and then the heat transformation to 3.18 Å siloxane Si-O-Si bond.

The packaging is achieved by connecting the inlet/outlet fused silica capillaries (Polymicro 150 µm OD) to the device.

3.2 Surface treatment and microfluidic tests

Optimal stability conditions would be obtained if the walls of the aqueous-phase channel were hydrophilic and that of the solvent (organic phase) were hydrophobic. This solution however is not practical for MEMS microfabrication. We chose to render the whole system hydrophobic by grafting a suitable monolayer of silane, i.e. FDTS perfluorodecyltriclorosilane, $CF_3(CF_2)_7(CH_2)_2$-$SiCl_3$. The silanisation is realized at 35°C in a vacuum chamber (MVD-100). The first step consists in wetting the surface to produce Si-OH bonds and the second step creates Si-O-Si bonds. Using of trichlorosilane compound assures the rate of silanisation because kinetics of Si-Cl is faster than Si-OMe and Si-OEt respectively.

The microsystem is placed in the vacuum chamber at 23 mTorr. Vapor is injected to activate the surfaces of silicon and glass and a 500 mTorr pression of gazeous FDTS is maintained during 30 minutes. The linear alkyl chain of monolayer FDTS provides for the hydrophobicity of surface (Fig. 12). The measurements of contact angles and surface tension of the 1 μL water droplets on silicon or silicon oxide surface with a DIGIDROP BBX equipped with WinDrop software showed that the initial contact angle of water between 65° and 75° reaches 100-116° after the treatment.

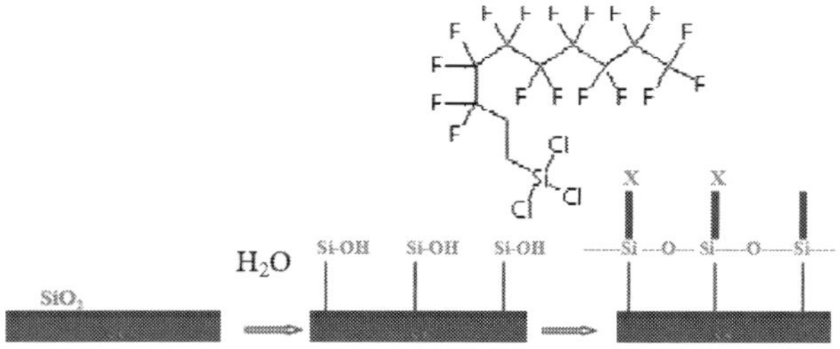

Fig. 12. Silanisation in gaseous phase by FDTS.

The silane layer is stable with temperature. The table 1 shows the constancy of the contact angle at different temperatures.

Table 1. Contact angle of a 1 μL water droplet on silane treatment at different levels of temperature (average values of five measurements). Contour and Manual are two methods of measurement (accuracy is about 5°).

Temperature °C	Measurement method of angle (°)	
	Contour	Manual
20 (RT)	113.4	109.9
80 (30min)	112.2	108.6
100 (30min)	101.8	104.1
120 (30min)	100.7	103.0
140 (30min)	103.6	104.7
160 (30min)	104.8	109.7
160 (90min)	104.3	110.8

The microfluidics tests have been performed by injecting liquids from two syringe pumps (precision 0.1 μL/min) and/or with an electronic nanopump (Dionex Ultimate 3000; precision 1 nL/min). The flow in the microdevice has been observed under microscope using a digital camera controlled by an acquisition system of Pinacle Studio Plus. The experimental results showed an excellent agreement with the calculation; the stability at the water-air interface can be obtained till 5 $\mu L.min^{-1}$ (Fig. 13) and we have reached a value of 10 for the water/solvent flow rate ratio.

Using the extractor shown in Fig.10.d where the solvent is at rest, the 112.5 nL solvent microchamber is used as a spectrophotometric cuve.

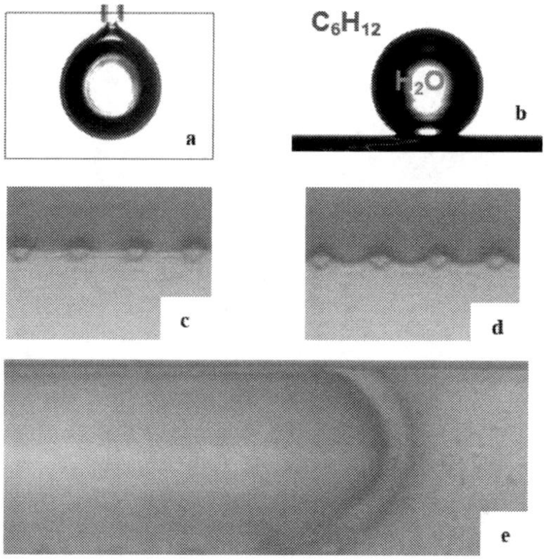

Fig.13. Water droplet hanging in solvent (a), 1 μL water droplet on silanized silicon substrate in cyclohexane (b), water-air interface in non-flowing water: $P_{air}=P_{atm}$ (c) and flow rate of water Q=0,4 μL/min (d), the experimental view advancing front of water in the channel, confirming the strongly hydrophobic properties of the walls (e).

4. Liquid-liquid extraction of lead in a microflow

Our first goal is the ECD of lead ions in water. In the primary channel, the water is flowing with a maximum flow rate of 2-3 μL/min whereas the solvent (chloroform) is at rest. Dithizone (H_2Dz) molecules are mixed to the solvent in order to extract the lead ions on the interface.

4.1 Experimental set-up

For liquid-liquid extraction, the lead solution (Pb^{2+}) is prepared from $Pb(NO_3)_2$ 99,99% Aldrich Sigma in deionized water (18.2 MOhm.cm). The dithizone solution (H_2Dz) in $CHCl_3$ is prepared and purified by the protocol presented by Marczenko [16]. In order to fix the pH to 7.5, an HEPES buffer is used at a concentration equal to 10 times that of the lead in water. The VARIAN Cary UV-Screw 300 spectrophotometer and an optical fiber UV-VIS detector of ZEISS were used to monitor the extraction in the wavelength range of 300-700nm.

The liquids are injected using two syringe pumps (precision 0.1 μL/min) and/or an electronic nanopump (Dionex Ultimate 3000; precision 1 nL/min) (Fig. 14).

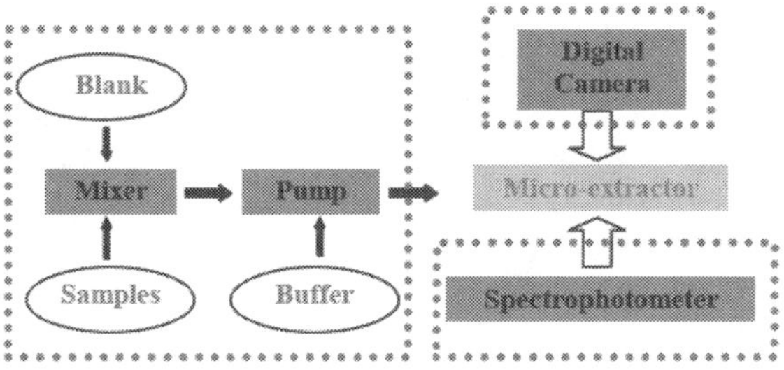

Fig.14. Schematic of experimental set-up with three modules: injection, observation and detection.

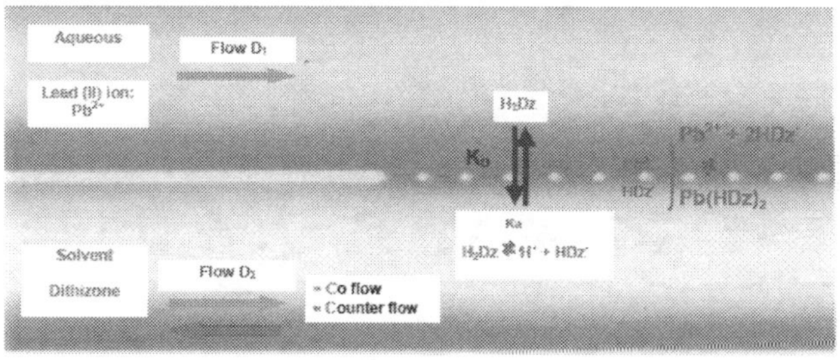

Fig.16. Lead extraction process in micro-extractor

Fig. 15 shows the inlet of the micro-extractor with the optical fiber aligned with the organic channel, used as spectrophotometric cuve. The optical fiber has a numerical aperture N_a of 0.22 rad, and its core radius is 55±1.5 μm. The solvent is injected first, filling both channels, with a high flow rate Q_1 = 5.0-10.0 μL/min in order to remove the air initially present in the micro-extractor. This step is followed by the injection of pure water followed by the water/lead sample at flow rate Q_2 = 0.1-1.0 μL/min. The measurement begins when the flow rate of the organic fluid is set to zero and the absorbance is recorded versus time.

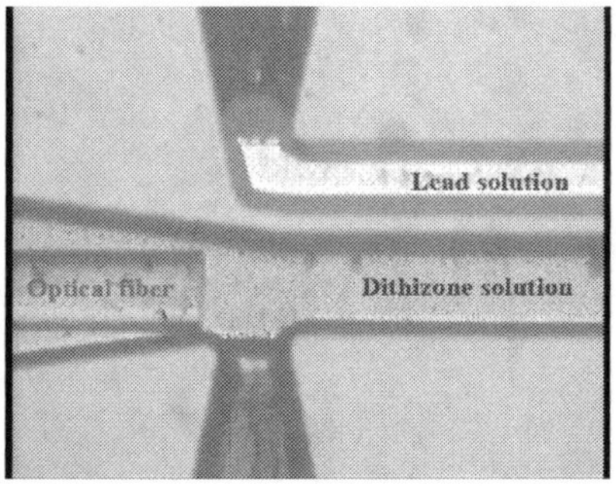

Fig.15. View of the inlet of the micro-extractor with the emission optical fiber

4.2 Extraction, concentration and in-situ detection of lead in the micro-extractor

The spectrophotometric determination of lead ions uses dithizone as an organic reagent. Dithizone (diphenylthiocarbazone) is a weak acid which dissociates in alkali media and forms a complex with lead. Dithizone gives a green taint to organic solvents like chloroform ($CHCl_3$) or cyclohexane (C_6H_{12}), due to the HDz^- ions

$$H_2Dz \rightleftarrows H^+ + HDz^-$$

The maximum absorbance of H_2Dz is located at the wavelength 605-606 nm and the absorption increases linearly with the concentration of H_2Dz. After reaction with Pb^{2+}, a pink color appears due to the formation of lead dithizonate complex which absorbs at wavelength of 520-525nm [16, 17]. Hence, the extraction of lead ions produces a new peak of absorption at 520-525nm. The reaction process on chip is shown in Fig. 16.

The concentration of the complex is determined by spectrophotometry. The Fig. 17 shows a typical spectrum of liquid-liquid extraction of lead.

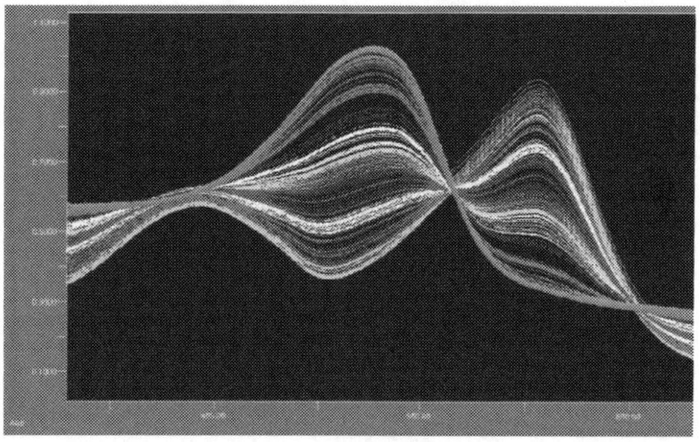

Fig.17. Absorbance spectrums of dithizone and lead dithizonate obtained from the extraction of 1 ppm Pb^{2+} solution by 10 ppm H_2Dz /$CHCl_3$ solution (250 scans, 12 scans/min) at a flow rate of 0.4 μL/min.

Fig. 18 shows the kinetics of absorbance of dithizone and dithizonate, and fig. 19 the corresponding concentration kinetics.

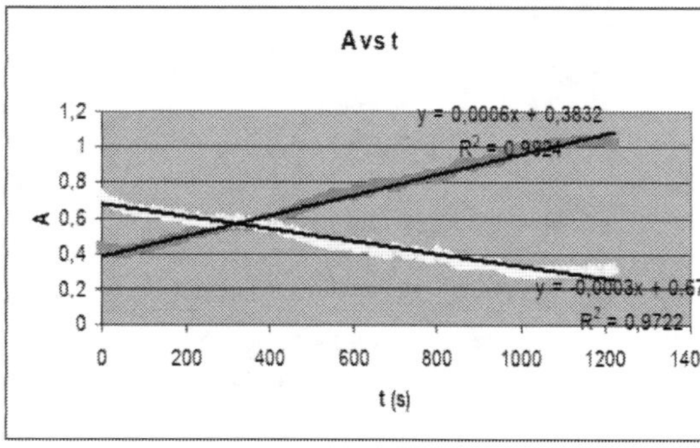

Fig.18. On-chip extraction of 1 ppm lead solution flowing at 0,6 μl/min. The curves are deduced from the analysis of the absorbance spectrums 606 nm (dithizone) and 525 nm (lead dithizonate).

The curves are linear until saturation is reached. The slopes indicate the dithizone consumption (-0.0006 AU/s) and the lead dithizonate production (+0.0003 AU/s). They satisfy the stoichiometric value of 2.

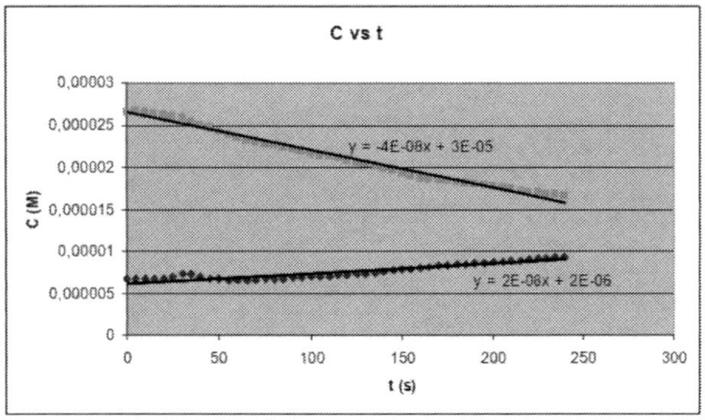

Fig.19. Kinetics of dithizone consumption and production of lead dithizonate in 1 ppm Pb^{2+} solution at 0.6 µL/min.

Liquid-liquid extraction of lead ions in the range of 1-5 ppm at a flow rate in the range 0.1-1.0 µL/min has been done with success. The duration of total extraction was about twenty minutes. Thus, at the relatively high concentration of Pb^{2+} of 1 ppm, we obtained a concentration factor of 4. In the future, working at a lower concentration (10-100 ppb), we expect a larger concentration factor since the ratio between dithizone and Pb^{2+} concentration will be higher.

Conclusions

A new microfluidics system aimed at liquid-liquid solvent extraction has been presented in this paper. This microsystem is based on the transfer of targets from a primary aqueous phase to a solvent through liquid interfaces stabilized by vertical micro-pillars. The conception of the system stems from the analysis of the stability of the interfaces and from the physics of transport and diffusion in the aqueous phase. The key of microfabrication is the use of DRIE to obtain a high aspect ratio for the micropillars, followed by a hydrophobic treatment by silanization.

It has been shown that the microchip achieves the extraction of lead ions at a concentration of 1-5 ppm using addition of dithizone in the solvent. Implementation of an in-situ detection by optical fiber UV-VIS detector in order to enhance portability has been done successfully.

Using this microfluidic platform, we plan to optimize the chemical process in order to increase the concentration factor and consequently to lower limit of detection.

Acknowledgments

The authors would like to thank Claude Chabrol, Nadine David, Emilie Dubard, Alain Troussier, Guillaume Costa and Hubert Jeanson for their collaboration and valuable discussions.

References

[1] A. Van den Berg, P. Berveld, *Labs-on-a-chip: origin, highlights and future perspectives*, Labs on chip Vol. 6 pp. 1266-1273 2006.

[2] M. Tokeshi, T. Kitamori, *Continuous flow chemical processing on a microchip using micro unit operations and a multiphase flow network*, Progress in Nuclear Energy, Vol. 47, No.1-4, pp. 434-438 2005.

[3] K. Sato, A. Hibara, M. Tokeshi, H. Hisamoto, T. Kitamori, *Microchip-based chemical and biochemical analysis systems*, *Advanced Drug Delivery Reviews*, Vol. 55, pp. 379-391, 2003.

[4] X. Wang, C. Saridara, S. Mitra, *Microfluidic supported liquid membrane extraction*, Analytica Chimica Acta Vol 543, pp. 92-98, 2005.

[5] Jason G. Kralj, Hermantkumar R. Sahoo, Klavs F. Jensen, *Integrated continous microfluidic liquid-liquid extraction*, Lab-on-a-chip, Vol 7, pp. 256-263, 2007.

[6] W. Ehrfeld, V. Hessel, H. Lowe, *Microreators*, Wiley-VCH, pp. 126-130, 2000.

[7] A. Günther, M. Jhunjhunwala, M. Thalmann, M.A. Schmidt, K.F. Jensen, "*Micromixing of miscible liquids in segmented gas-liquid flow*," Langmuir, Vol. 21, pp. 1547-1555, 2005.

[8] J. Berthier, *Microdrops and Digital Microfluidics*, William-Andrew Publishing, pp. 401-410, 2008.

[9] M. Bahrami, M.M. Yovanovich, and J.R. Culham, *Pressure drop of fully-developped, laminar flow in microchannels of arbitrary cross-section*, Journal of Fluid Engineering, Vol. 128, pp. 1036-1044, 2006.

[10] J. Berthier, Van Man Tran, H. Jeanson, N. Sarrut, *On the stability length of fluid/liquid interfaces sustained by microfabricated pillars*, Proceedings of the 2007 Nanotech Conference, Santa Clara, Ca., 20-24 May 2007.

[11] J. Berthier, P. Silberzan, *Microfluidics for Biotechnology*. Artech House, 2005.

[12] M. R. Bringer, C. J. Gerdts, H. Song, J. D. Tice, R. F. Ismagilov, *Microfluidic systems for chemical kinetics that rely on chaotic mixing in droplets*, Phil. Trans. R. Soc. Lond. A, Vol. 362, pp. 1087–1104, 2004.

[13] T. Gervais, K. F. Jensen, *Mass transport and surface reactions in microfluidic systems*, Chemical Engineering Science, Vol. 61, pp. 1102 – 1121, 2006.

[14] P. Amirfeiz, S. Bengtsson, M. Bergh, E. Zanghellini, L. Borjesson, *Formation of Silicon Structures by Plasma-Activated Wafer Bonding*, J. Electrochem. Soc., Vol. 147, pp. 2693-2698, 2000.

[15] Q. Tong, U. Gosele, *Semiconductor wafer bonding*, John Wiley and sons, p. 82, 1999.

[16] Z. Marczenko, *Separation and spectrophotometric determination of Elements*, Ellis Horwood Limited, 1986.

[17] B. Romberg, H. Muller, *Photometric screening-test for heavy metals under flow injection conditions using extractive determination with dithizone*, Analytica Chimica Acta Vol 353, pp. 165-172, 1997.

Horizontal Taylor flow hydrodynamics, pressure drop and overall mass transfer in a capillary reactor

M.J.F. Warnier, E.V. Rebrov, M.H.J.M. de Croon, V. Hessel, J.C. Schouten
Department of Chemical Engineering and Chemistry, Eindhoven University of Technology, P.O. Box 513, 5600 MB Eindhoven, the Netherlands
j.c.schouten@tue.nl; www.chem.tue.nl/scr

Introduction. The interest for applying micro structured reactors and monoliths for performing gas-liquid-solid (catalytic) reactions has increased considerably over the past years [1,2,3,4]. The hydrodynamics of the gas-liquid flow in the small channels (diameter < 1 mm) and the relation to pressure drop and gas-liquid and liquid-solid mass transfer has to be well understood for a proper design of these reactors. Taylor flow is the most studied regime in these systems in the literature. Pressure drop studies have been mostly performed in optically transparent channels without chemical reaction and without a catalytic coating on the channel walls. High speed imaging techniques were then used to obtain information about the hydrodynamic parameters. Three phase mass transfer studies have been performed mostly in devices which are not transparent. The hydrodynamic parameters were then obtained from visual observation of the flow patterns outside the actual reaction channel, assuming that the conditions in the reaction channel were similar to those in the viewing section. Mass transfer and pressure drop have therefore seldom been studied simultaneously in the same device and/or in the same experiment. We believe however that this combined measurement will contribute to a proper interpretation and understanding of mass transfer processes in micro channels.

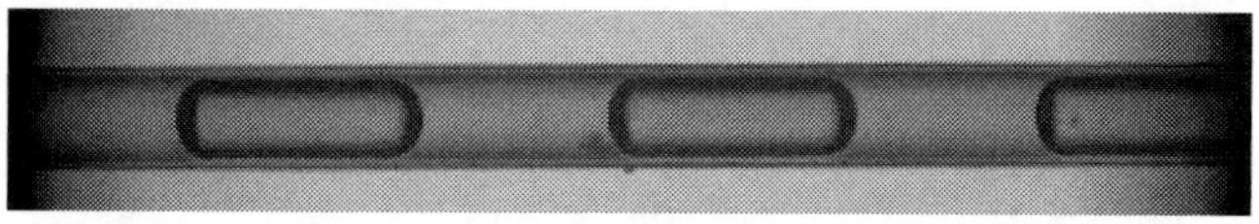

Figure 1: Image of water/nitrogen Taylor flow in a fused silica capillary internally coated with a 120 nm TiO_2 layer loaded with 1 wt.% Pd. The internal diameter is 250 μm. The superficial gas velocity is 0.13 m/s and the superficial liquid velocity is 0.14 m/s.

Objective of this work. Therefore, in this work, both overall mass transfer and pressure drop measurements have been simultaneously performed during Taylor flow at different operating conditions. High speed imaging techniques were used to obtain information on the hydrodynamic parameters (Figure 1). The catalytic hydrogenation of phenyl-acetylene to styrene and ethylbenzene was performed in a catalytically active fused silica capillary with an inner channel diameter of 250 μm and a length of 4.5 m. The capillary wall was coated with a 120 nm thick porous TiO_2 layer loaded with 1 wt.% Pd. Catalyst activity was demonstrated by hydrogenating a 10 vol.% phenylacetylene in methanol solution in the capillary. At a temperature of 30 ^{0}C the catalyst activity was in the order of 1 $mol_{H2}mol_{Pd}^{-1}s^{-1}$ [5]. In subsequent experiments isopropanol was used because of its higher boiling point compared to methanol. The catalyst activity could then be increased by increasing the reaction temperature without reaching the boiling point of the solvent. The hydrogen partial pressure was controlled by diluting the hydrogen flow with nitrogen.

Pressure drop. First, pressure drop experiments were performed with in a similar capillary with a length of 1 m. A water /nitrogen system was chosen instead of an isopropanol/ phenylacetylene/hydrogen/nitrogen system. More literature data are available for the first system allowing for a better comparison of the data and model described in this work to the work of other authors. Nitrogen was fed and the superficial gas velocity (U_g^0) was varied between 0.03 and 0.52 m/s (the velocities are given at a pressure of 1 bar and a temperature of 298 K). The superficial velocity (U_l) of water was varied between 0.11 and 0.36 m/s. The experiments were performed at a temperature of 292 K.

The liquid flow was split into two equal flows which were contacted with the gas flow in a cross-mixer (Figure 2). A differential pressure sensor (Honeywell 26PC 05 KD, response time of 1 ms) measuring the pressure difference with ambient pressure was connected close to the capillary inlet and 13 cm upstream from the cross-mixer to ensure a fully developed Taylor flow passing the sensor. All connectors and tubing had an inner diameter of 250 μm. The flow exited the capillary under ambient conditions.

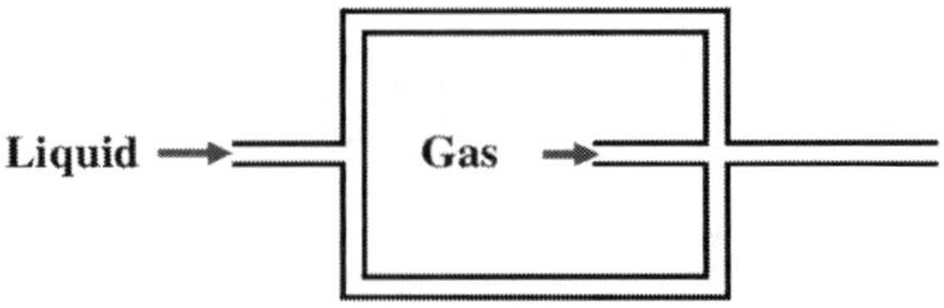

Figure 2: *Schematic drawing of the cross-mixer. The liquid flow is split into two equal flows and contacted with the gas flow at an angle of 90°. All channels in the cross-mixer have a diameter of 250 μm.*

The capillary is optically transparent, even with the catalytic coating present. Images of the Taylor flow were recorded using a high-speed camera connected to a microscope at a frame rate of 2500 frames per second for 2 seconds. Images were recorded for 58 combinations of gas and liquid velocities and analyzed in Matlab using scripts developed by the authors. The gas bubble velocity, the number of gas bubbles formed per unit time and the gas and liquid slug size distributions were estimated directly from the recorded images. A mass balance based model for Taylor flow was then applied to the data to obtain the gas hold-up and the film thickness of the liquid film surrounding the gas bubble [6]. The recordings of the pressure signals and the video images were synchronized.

The pressure drop over the capillary is caused by the liquid flow in the liquid slug, assuming that the liquid film surrounding the gas bubble is stagnant and that the pressure drop in the gas bubble is negligible. The gas bubbles influence the pressure drop because their presence disturbs the otherwise parabolic velocity profile in the liquid slug. The average velocity in the liquid slug is equal to the average superficial velocity in the channel (U_g+U_l) (Figure 3). Note that the superficial gas velocity (U_g) in equations 1-4 is the velocity at the conditions in the channel. Calculations show that the pressure drop caused by the acceleration of the flow due to expansion of the gas phase is in the order of 0.1% of the total pressure drop. Acceleration effects are therefore not taken into account. The pressure drop ($(dP/dz)_{exp}$) over the capillary can then be described in terms of an apparent friction factor f_{obs} times the Reynolds (Re) number, see Equation 1. Other parameters in equation 1 are: the liquid slug length (L_s), the gas bubble length (L_b), the hydrodynamic diameter (D_h) and the liquid viscosity (μ_l).

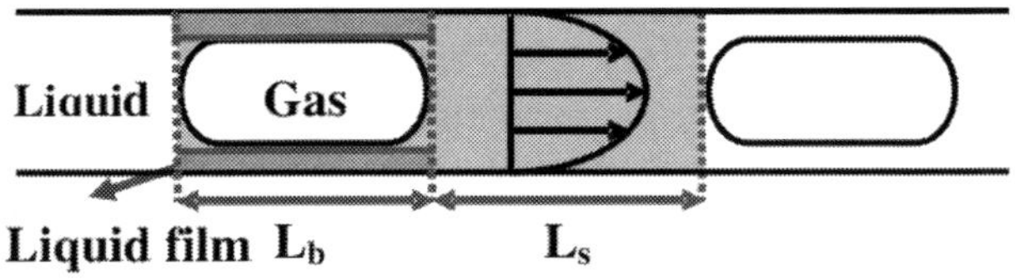

Figure 3: *Schematic drawing of a gas-liquid Taylor flow. The gas bubble length (L_b), the liquid slug length (L_s) and the stagnant liquid film are indicated. Only the liquid in the liquid slug contributes to the pressure drop. The average velocity in the liquid slug is equal to the average superficial velocity in the channel (U_g + U_l).*

$$f_{obs}\,\mathrm{Re} = -\left(\frac{dP}{dz}\right)_{\exp}\left(\frac{D_h^{\,2}}{2\mu_l\left(U_g+U_l\right)}\right)\left(\frac{L_s+L_b}{L_s}\right) \quad (1)$$

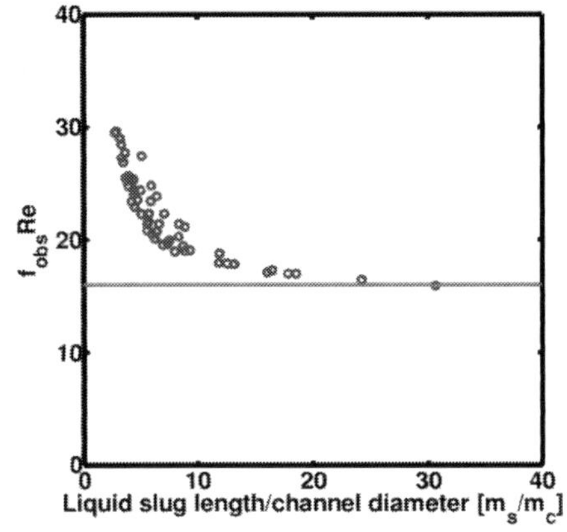

Figure 4: *Apparent friction factor multiplied with the Reynolds number as a function of the liquid slug length. The red line indicates the single phase limit for infinitely long liquid slugs ($f_{obs}Re=16$).*

Figure 4 shows that the apparent friction factor multiplied with the Reynolds number decreases with increasing liquid slug length. This value approaches the single phase limit value of 16 for infinitely long slugs. This indicates that the number of bubbles per unit channel length determines the excess pressure drop over single phase liquid flow. This was already reported earlier for channels with an approximately 10 times larger diameter and for various liquids [1].

Mass transfer. A model was derived describing the pressure (P) and the hydrogen molar flow rate (F_{H2}) as a function of axial position z in the capillary based on the following assumptions:

1. *The gas and liquid phases exhibit plug flow behavior.* Since both the gas bubbles and the liquid slugs move through the capillary as isolated packages with internal recirculation cells, both of them are considered to be well mixed without axial dispersion.
2. *No nitrogen dissolves into the liquid phase.* Since hydrogen is consumed from the liquid phase and nitrogen is not, the nitrogen flux from the gas into the liquid is considerably smaller than the hydrogen flux. The partial pressure of hydrogen will therefore hardly be influenced by nitrogen dissolving in the liquid phase.
3. *The reaction is fully mass transfer limited.* The reaction rate is then zero order in phenylacetate and first order in hydrogen.
4. *Only the liquid flow in the liquid slugs contributes to the pressure drop.* The same assumptions were made that lead to equation 1.

These assumptions lead to the following set of equations (2-4) from which the overall mass transfer coefficient k_{ov} and the observed friction factor f_{obs} were estimated. The model reduces to two independent equations with these two unknowns if equation 2 is substituted into equations 3 and 4:

$$U_g = \frac{F_{H_2} + F_{N_2}}{A_r P} RT \tag{2}$$

$$\frac{dF_{H_2}}{dz} = -A_r k_{ov} \left(1 - \frac{U_g}{u_b} \right) \frac{P}{H_{H_2}} \frac{F_{H_2}}{F_{H_2} + F_{N_2}} \tag{3}$$

$$-\frac{dP}{dz} = \left(\frac{2 f_{obs} \mathrm{Re} \mu_l \left(U_g + U_l \right)}{D_h^2} \right) \left(\frac{L_s}{L_s + L_b} \right) \tag{4}$$

The product concentrations at the exit of the capillary were determined by GC measurements. The total amount of reacted hydrogen was then calculated from the overall mass balance. Thus, both the inlet and outlet hydrogen molar flows (F_{H2}) and the pressure drop were known from the measurements. The bubble velocity (u_b), the gas bubble length (l_b) and the liquid slug length (l_s) were obtained from image analysis. The molar flow of nitrogen (F_{N2}), the cross-sectional channel area (A_r), the channel diameter (D_h), the liquid viscosity (μ_l), and the hydrogen Henry coefficient (H_{H2}) were constant. The mass transfer coefficient k_{ov} and the observed friction factor (f_{obs}) were obtained by simultaneously solving equations 2, 3 and 4 with the appropriate boundary conditions using a boundary problem solver routine in Matlab. The paper will report the dependences of these measured overall mass transfer coefficients and observed friction factors on the Taylor gas bubble velocity, the gas bubble and liquid slug size distributions, and the film thickness of the liquid film surrounding the gas bubble, at different hydrodynamic operating conditions.

References

[1] M.T. Kreutzer, F. Kapteijn, J.A. Moulijn, J.J. Heiszwolf, Multiphase monolith reactors: Chemical reaction engineering of segmented flow in microchannels, Chem.Eng.Sc., 60 (22) (2005) 5895-5916.

[2] M.W. Losey, M.A. Schmnidt, K.F. Jensen, Microfabricated multiphase packed-bed reactors: Characterization of mass transfer and reactions, Ind.Eng.Chem.Res., 40 (12) (2001) 2555-2562.

[3] V. Haverkamp, V. Hessel, M.A. Liauw, H. Lowe, M.G. Menges, Reactor Model for Fast Reactions in the Micro-Bubble Column and Validation, Ind.Eng.Chem.Res. (2007).

[4] V. Hessel, H. Lowe, P. Angeli, A. Gavriilidis, Gas-liquid and gas-liquid-solid microstructured reactors: Contacting principles and applications, Ind.Eng.Chem.Res., 44 (25) (2005) 9750-9769.

[5] E.V. Rebrov, A. Berenguer-Murcia, H.E. Skelton, B.F.G. Johnson, J.C. Schouten, Selective Hydrogenation of Phenylacetylene to Styrene on a Pd/TiO2 Coating in a Microreactor, *Conference proceedings IMRET 10, 2008, AIChE spring national meeting.*

[6] M.J.F. Warnier, E.V. Rebrov, M.H.J.M. De Croon, V. Hessel, J.C. Schouten, Gas hold-up and liquid film thickness in Taylor flow in rectangular microchannels, Chem.Eng,J., 135S(-), S153-S158

128c Ignition Strategies for Catalytic Microdevices

Niket S. Kaisare[1], **Georgios D. Stefanidis**[2], and **Dionisios G. Vlachos**[2]. (1) Department of Chemical Engineering, Indian Institute of Technology-Madras, Chennai, 600-036, India, (2) Department of Chemical Engineering and Center for Catalytic Science and Technology (CCST), University of Delaware, 150 Academy Street, Newark, DE 19716

Integrated microdevices are increasingly being explored for application in distributed and portable power generation (1-4) due to the high energy density of fuels. Aside from meeting safety and stability requirements, these devices should also be fast starting, i.e., the time required for device warm-up from cold-start conditions to the operating temperature should be short. This is also desirable from an environmental standpoint to minimize emissions during startup. Two commonly employed startup strategies are preheating of the feed stream to a temperature higher than the ignition point and resistive heating. The aim of this work is to compare these two startup strategies for microscale devices at various operating parameters using a hierarchy of models, including pseudo-homogeneous 2D and full 2D computational fluid dynamics (CFD) simulations. Furthermore, the role of reactor solid structure in transient light-off is investigated.

Our results show a remarkably different behavior of ignition of microdevices from that of large-scale counterparts. When inlet preheating is applied, microburners with insulating walls ignite earlier, i.e., with a lower inlet feed temperature. The reason for the earlier ignition of insulating walls can be traced to their lower heat losses. When electric resistive preheating is employed, highly conducting walls require lower electric power for ignition. This dependence of steady state ignition characteristics on wall conductivity is rather weak. However, it has a strong effect on the transient time required to attain steady state. Highly conducting walls have a shorter response time than insulating walls. One reason for this behavior is their higher axial heat recirculation through the conducting walls. The heat capacity of the solid structure also affects the startup time. Higher heat capacity requires, as expected, longer time to reach steady state. The higher the inlet velocity, the greater is the amount of electric power required to preheat the gas to a certain (ignition) temperature. A good start-up strategy is to preheat the reactor to a desired temperature using resistive heating, and then start the flow. Finally, it was found that faster steady state is achieved with only part of the reactor is preheated. In this strategy, the effect of preheating location on the power required for ignition is rather small, but is significant on the time to steady state. Preheating of the front-end of the reactor is preferable.

References

1. J.D. Holladay, E.O. Jones, M. Phelps, J. Hu, J. Power Sources 108 (2002) 21-27.

2. S.R. Deshmukh, D.G. Vlachos, Chem. Engng. Sci. 60 (2005) 5718-5728.

3. J.C. Ganley, E.G. Seebauer, R.I. Masel, AIChE J. 50 (2004) 829-834.

4. A.C. Fernandez-Pello, Proc. Combust. Inst. 29 (2003) 883-889.

128d Fluid Distribution and Blockage Diagnosis in Parallel Microchannel Configurations

Osamu Tonomura, Satoshi Nagahara, Manabu Kano, and Shinji Hasebe. Chemical Engineering, Kyoto University, Kyoto, 615-8510, Japan

It is clear from worldwide research that micro chemical process technologies offer a unique approach to the spatial and temporal control of chemical reactions. The numbering-up approach is effective at increasing the production capacity of micro chemical plants and opens up the possibility of a direct transfer from laboratory scale to production scale. However, some difficulties are left behind. One of the numbering-up problems is the flow maldistribution among microchannels and/or microdevices due to the inadequate design of micro chemical plants. The other problem is the blockage in microchannels, and it is necessary to develop a systematic method that can detect and diagnose the blockage without many miniaturized sensors. In the previous papers, several types of micro distributors have been investigated to realize the uniform flow distribution at a normal condition. However, such distributors are hardly suitable for fault detection and diagnosis, since many sensors have to be installed in the parallel microchannels to monitor changes in the process conditions. This work is intended to develop a new distributor which makes the fluid distribution uniform and realizes the fault diagnosis based on a limited number of sensors. In addition, a blockage diagnosis procedure that isolates the blocked microchannel from the parallel microchannels is proposed. Upon examination with both of experiments and numerical simulations, it is concluded that the proposed distributor and diagnosis procedure can realize the stable long-term operation of micro chemical plants with numbering-up structures.

129c Characterization of High Heat and Mass Transfer in T-Shaped Microreactors and Its Optimal Design

Masato Kubota, Osamu Tonomura, Manabu Kano, and Shinji Hasebe. Chemical Engineering, Kyoto University, Kyoto, 615-8510, Japan

Mixing is one of the key unit operations in micro chemical plants. The requirements for industrial micromixers are rapid mixing, rapid heat transfer, high throughput, low pressure drop, blockage prevention, and so on. To shorten mixing time, the interfacial area between the fluids should be increased, and the diffusion length should be decreased. Various types of passive and active micromixers have been presented in the literature, but such mixers mainly focus on how to achieve rapid mixing. For example, T-shaped micromixers have a simple structure and may offer the possibility of meeting the requirements for real production. Several papers say that mixing efficiency in T-shaped micromixers is drastically enhanced in the presence of convective flow generated at high flow rate. However, there are few papers investigating the effect of convective flow not only on mixing but on heat transfer, chemical reaction, and residence time distribution.

This work is intended to investigate both T-shaped micromixers and interdigital micromixers by using CFD simulations. In the first part of this work, the influences of design and operation parameters upon mixing, heat transfer, chemical reaction, and residence time distribution are investigated. The simulation results show that the convective flow in T-shaped micromixers is effective in increasing heat transfer to realize uniform temperature distribution. In addition, the effects of convective flow are highly dependent on design and operation parameters. In the second part of this work, a systematic design method is proposed. Namely, response surface method (RSM) is introduced into the CFD-based optimal design of micromixers. RSM is one of effective statistical analysis approaches and is often used in the product design works based on experimental data. The validity of the proposed design method is assessed through case studies. The combination of CFD and RMS promotes the design efficiency of various types of micromixer.

128a Manipulating Homogeneous Chemistry in Catalytic Microreactors: Operation Strategies and Design Principles

Georgios D. Stefanidis and **Dionisios G. Vlachos**. Department of Chemical Engineering and Center for Catalytic Science and Technology (CCST), University of Delaware, 150 Academy Street, Newark, DE 19716

Combined catalytic and homogeneous chemistry of light hydrocarbons is a key issue in the design and thermal management of several microchemical (sub-millimeter scale) devices, used either for energy production, e.g., heat, or the production of chemicals, e.g., oxidative dehydrogenation of ethane to ethene [1]. Depending on the application, homogeneous chemistry in these devices should either be purposely avoided or promoted. The aim of this work is to study how the interplay between homogeneous and catalytic chemistry is affected by different operating conditions, such as the inlet composition and inlet velocity and by different reactor design parameters, such as the channel gap size (characteristic dimension) and the wall material. To this end, fuel lean propane/air combustion in a microburner is simulated in 2D using Computational Fluid Dynamics. Single-step homogeneous and catalytic chemistry is employed [2,3]. The results show that if only catalytic chemistry is desired, the microburner should either be operated at very low inlet velocities, near the extinction limit, or at high inlet velocities beyond the flame blow-out limit. Within the coupled catalytic/homogeneous chemistry regime, the extent of homogeneous chemistry and consequently the desirable temperature level can be maintained through adjustment of the inlet composition. Our results also show that although external heat losses increase decreasing gap size, homogeneous chemistry is sustained for channel gap sizes as low as 200 microns. The latter finding is attributed to the enhanced gas-to-solid mass transport that assists catalytic chemistry followed by enhanced solid-to-gas heat transfer that sustains homogeneous chemistry as the gap size decreases. Finally, we find that catalytic chemistry without gas reactions is sustained over a substantially wider inlet velocity range when employing high-conductivity wall materials.

References

1. Caputo, F. D. T.; Russo, G.; Benedetto, A. D.; Pirone, R. Modeling ethane oxy-dehydrogenation over monolithic combustion catalysts. AIChE J. 2004, 50(9), 2233-2245.

2. Kaisare, N. S.; Vlachos, D. G. Extending the region of stable homogeneous micro-combustion through forced unsteady operation. Proceedings of the Combustion Institute 2007, 31, 3293-3300.

3. Kaisare, N. S.; Deshmukh Vlachos, D. G. A reduced mechanism for methane and one-step rate expressions for fuel-lean catalytic combustion of small alkanes on noble metals. Combustion and Flame 2007, 149(4), 366-383.

181b Design Principles of Multifunctional Microdevices for Hydrogen Production Via Short Contact Time Steam Reforming

Georgios D. Stefanidis[1], Niket S. Kaisare[2], Matteo Maestri[3], Enrico Tronconi[3], and **Dionisios G. Vlachos**[1]. (1) Department of Chemical Engineering and Center for Catalytic Science and Technology (CCST), University of Delaware, 150 Academy Street, Newark, DE 19716, (2) Department of Chemical Engineering, Indian Institute of Technology-Madras, Chennai, 600-036, India, (3) Dipartimento di Chimica, Materiali e Ingegneria Chimica "Giulio Natta", Politecnico di Milano, Piazza Leonardo da Vinci 32, Milano, 20133, Italy

Autothermal hydrogen generation in a catalytic reactor requires coupling of endothermic reforming reactions with exothermic combustion or partial oxidation reactions. The oxidation and reforming reactions can be carried out in separate reactors, as in the current industrial practice, or combined in a multifunctional device (1). Steam methane reforming is a major industrial process for hydrogen production in refineries and petrochemical plants and may become an important player in biorefineries. Therefore, investigation of its potential to provide hydrogen generation at the microscale is of great interest for portable and distributed power generation.

We use a pseudo two-dimensional (2D) model developed in (2) to model a microdevice, which consists of two parallel catalytically coated, closely spaced plates. Propane/air combustion on Pt takes place in one of them and steam methane reforming on Rh in the other. Fast heat exchange is achieved through the thin wall that separates them. This relatively simple configuration provides compactness, intimate coupling between the two processes, and the flexibility of different catalysts, fuels and flow configurations. We investigate whether we can make steam methane reforming workable at small scales at short contact times by capitalizing on process intensification while avoid microburner extinction. This requires not only fast reforming chemistry but also fast heat transfer. Aside from that, we aim at developing general design guidelines of integrated multifunctional microreactors. To this end, the effect of operating conditions, of the choice of the wall material and the reactor size are studied.

Our results show that the intrinsic steam methane reforming chemistry on Rh is very fast and only slightly slower than methane partial oxidation. Catalytic plate reactor operation in co-current mode enables compact steam methane reforming, fast transverse heat transfer and millisecond operation. Design maps for efficient operation have been generated; they are demarcated from the maximum power output point, the breakthrough point and the material stability limit (an upper temperature above which materials' stability is an issue). Low conductivity wall materials result in increased methane conversion at the expense of hot spot formation and steeper axial wall temperature gradients. Finally, it was found that decrease in the reforming channel gap size at constant inlet velocity drastically increases fuel conversion and system efficiency.

References

1. Deshmukh, S. R.; Vlachos, D. G. Chemical Engineering Science 2005, 60, 5718-5728.

2. Kaisare, N. S., Vlachos, D. G. Proceedings of the Combustion Institute 2007, 31, 3293-3300.

235f Reactor Design Optimization for Direct Synthesis of Hydrogen Peroxide

Tomoya Inoue[1], Yoshikuni Kikutani[2], Satoshi Hamakawa[1], Fujio Mizukami[1], and Takehiko Kitamori[3]. (1) Research Center for Compact Chemical Process, National Institute of Advanced Industrial Science and Technology (AIST), 4-2-1, Nigatake Miyagino-Ku, Sendai, Miyagi, 983-8551, Japan, (2) Microchemistry group, Kanagawa Academy of Science and Technology, 3-2-1 Sakado, Takatsu, Kawasaki, Kanagawa, Japan, (3) Department of Applied Chemistry, Shool of Engineering, The University of Tokyo, 7-3-1 Hongo, Bunkyo, Tokyo, 113-8656, Japan

Reactor design was investigated for the direct reaction between hydrogen and oxygen to produce hydrogen peroxide. Hydrogen peroxide is primarily produced by the anthraquinone process, in which 2-alkyl-9, 10-anthraquinones are hydrogenated with palladium catalyst in the first step, and the products are subsequently oxidized by air to produce hydrogen peroxide. This process is safe and easily scaled-up, but anthraquinone decomposes gradually during the redox cycle. That results in trace amounts of contamination, and further purification step is required for obtaining products.

Compared with the current anthraquinone process, direct production from hydrogen and oxygen can enjoy less energy consumption and environmental benignancy. This reaction is conducted as multiphase reaction among hydrogen - oxygen mixture (gas phase), aqueous medium (storing hydrogen peroxide produced, liquid phase) and supported palladium catalyst (solid). In spite of the advantage, widespread application of the process has been limited by the risks inherent in handling the explosive mixture. Moreover, high pressure from 2 MPa to 10 MPa are typically needed to promote mass transfer from gas to liquid phase.

Recently, microfabricated reactor technology has been successfully applied to selective hydrogen peroxide production by handling explosive mixture of hydrogen and oxygen, as shown by one of the authors. 1 Still, there is more room for optimizing the microfluidic design to promote the productivity of hydrogen peroxide. Scaling out technology should be also implemented so that the productivity meets the industrial requirement. As the reaction is conducted as three phase reaction, reactor design is crucial for the productivity.

We conducted the reaction in a glass fabricated microreactor. As it is transparent, flow regime inside the reactor can be easily monitored during the reaction, which is crucial for the reactor performance. 1 Also glass as the reactor material has been proved to be suitable for fine chemical processes, for example in Lonza's c-SSP process. 2 Microchannels were fabricated on a glass wafer (30mm width X 70mm length X 0.7mm depth) by the combination of chemical etching (Institute for Micro Technology (IMT), Japan) and mechanical machining. Chemical etching was suitable for fabrication of a microchannel with less than 100 micrometer width, while mechanical machining is suitable for that of larger channels with more than 200 micrometer width. Once microchannels as well as inlet/outlet holes were fabricated on a glass wafer with the same dimensions, the two wafers were thermally bonded (IMT). Catalyst was loaded into the channel (600 micrometer width X 300 micrometer depth) as slurry.

We found that the gap between the size of microchannels for gas and liquid flows is crucial for the reactor design, especially for validating proper gas - liquid distribution into the reactor. When gas channel was with more than 100 micrometer width, the flow regime was unstable because of the crosstalk between gas and liquid dispensing channels, causing pulsing flow into the catalyst bed. The crosstalk between the gas - liquid dispensing channel was significantly suppressed when the width of gas inlet channel was reduced to 50 micrometer (depth of the channel became 25 micrometer because of the geometrical characteristics of wet etching). As both of the gas and liquid flows were driven by , we regard that the difference of pressure drop by gas and liquid is responsible for this crosstalk issue. 3 Gas channels should be narrow enough in order to compensate the pressure drop difference between gas and liquid channel.

Proper gas - liquid distribution in the reactor is crucial for the reactor performance, which will be presented at the conference. Also another effort on scaling out is in progress, and reactor performance will be presented at the conference.

(1) Inoue, T.; Schmidt, M. A.; Jensen, K. F. Microfabricated Multiphase Reactors for the Direct Synthesis of Hydrogen Peroxide from Hydrogen and Oxygen. Ind. Eng. Chem. Res. 2007, 46, 1153.

(2) Roberge, D. M.; Bieler, N.; Thalmann, M. Microreactor technology and continuous processes in the fine chemical and pharmaceutical industries. PharmaChem. 2006, June, 15.

(3) Wada, Y.; Schmidt, M. A.; Jensen, K. F. Flow Distribution and Ozonolysis in Gas - Liquid Multichannel Microreactors. Ind. Eng. Chem. Res. 2006, 45, 8036.

83b Mass Transfer during Taylor Flow in Microchannels with and without Chemical Reaction

Nan Shao, Asterios Gavriilidis, and **Panagiota Angeli**. Department of Chemical Engineering, University College London, Torrington Place, London, WC1E 7JE, United Kingdom

The characteristics of mass transfer from gas to liquid during Taylor flow in capillaries smaller than 1mm in diameter with and without chemical reaction were investigated using Computational Fluid Dynamics modelling. The effects on mass transfer of tube size, bubble velocity and bubble geometry, such as bubble diameter (or thickness of film between the bubble and the tube wall) and bubble/slug lengths, were studied. Separate gas and liquid domains were solved and the interfacial concentration profile was updated so that the concentration gradients caused by the gas mass transfer were better addressed than in the single (liquid) domain module [1]. The second order fast reaction of CO2 absorption in NaOH aqueous solution was modelled. It was found that mass transfer improves with decreasing channel size and increasing bubble velocity, as a result of decreasing diffusion path length and enhancement of liquid convection respectively. Mass transfer is also improved when liquid film at the wall thickens, because this increases the volume of the film region, i.e. increases its capacity to take-up CO2. In both cases with and without chemical reaction, increase of the liquid slug length reduces mass transfer because the liquid circulation frequency is decreased in long slugs. On the other hand, bubble length has an opposite effect on mass transfer in both cases. When chemical reaction takes place, longer bubbles contribute to larger contact areas along the film region thus improving mass transfer. In the absence of reaction, however, the increased amount of gas transferred species is not consumed and saturates the film region quickly, thus preventing any further mass transfer. The simulation results also reveal the existence of concentration gradients within the gas bubbles. This results in interfacial concentration lower than the average gas phase concentration which also affects the rate of mass transfer from gas to liquid.

Keywords: Taylor flow, mass transfer, microchannel, chemical reaction, CO2 absorption

[1] Vandu, C.O., Ellenberger, J., and Krishna, R., 2005. Hydrodynamics and mass transfer in an upflow monolith loop reactor. Chemical Engineering and Processing 44, 363-374.

235e A Transparent Micro-Structured Reactor for the Investigation of Gas-Liquid Reactions at Elevated Pressures and Temperatures

Claude De Bellefon[1], Arnaud Leclerc[1], Vivian Houzelot[1], Daniel Schweich[1], Patrick Pouteau[2], and Cyril Delattre[2]. (1) LGPC ESCPE Lyon, CNRS, 43, bd 11 Novembre 1918, Villeurbanne, 69616, France, (2) CEA-LETI, Grenoble, 38054, France

At the laboratory scale, micro-structured reactors can bring several advantages such as: to lower inventory of reacting material, to use dangerous/toxic reagents, to perform reactions under non-conventional operating windows, to decrease heat and mass transfer limitations, etc. Most of these capabilities have been demonstrated for monophasic systems, i.e. when only one fluid, either a gas or a liquid, is involved, a solid catalyst being generally coated on the microstructure wall. However, some of these potential applications have not been demonstrated with multiphase reactions since it is much more difficult to control gas-liquid flow while performing fast and exothermic reactions such as hydrogenations, oxidations, fluorinations and others. In fact, most of the micro-structured reactors designed for gas-liquid or gas-liquid-solid applications cannot meet the multiple challenge of high temperature, high pressure and controlled flow. In this report, a micro-structured reactor answering these challenges is reported and applied to the investigation of reactions under non-usual pressure and temperature ranges. It is well known that Taylor flow can be obtained easily under standard conditions and using components as simple as capillary tubes. Many reports actually deals with such simple and efficient reactors. While experimentation using capillaries are easy to set and quite inexpensive they cannot lead to compact systems to allow visual inspection of the all channel which precludes the study of the reaction conversion and solid formation and the mapping of the flow regime. Also, the arrangement of the capillary is “space demanding”, in particular when heating is required (oil bath, oven etc.). last, the design of the gas-liquid injector, which is determining for the perfect control of bubble size and liquid slug length, cannot be easily changed nor precisely machined. Silicon made micro-reactors offer the possibility to conduct Taylor flow in compact systems. The microreactor used here is made from silicon by well known lithography and etching techniques. The channels (300x300 µm, 1 m long) are covered by bonding a Pyrex glass wafer to the oxidized silicon device which provides visual access to the reaction medium. This is particularly useful for the monitoring of gas-liquid reactions. The test reaction was the exothermic (> 200 kJ.mol-1) oxidation of an alkane into a mixture of the corresponding ketone (K), alcohol (A) and hydroperoxide (HP). Working with pure oxygen is very tempting to increase the gas-liquid mass transfer rate, to speed-up the reaction rate and to decrease side products formation. Because of the safety issues raised by the alkane/oxygen mixture in the gas phase, such investigations are very scarce. In order to check for the possible intensification of the process, the reaction was investigated in a large range of pressure and temperature including pure oxygen. At 200°C and ca. 12 bar O2 pressure, a conversion of more than 4 % is measured at residence time as short as 2 min in the micro-reactor. The selectivity in the desired products HP, K and A is also very good (88%). The productivity in the desired products is increased fourfold compared to a similar experiment performed with air. Attempts to perform the reaction at higher temperature and/or residence time resulted in over-oxidation and by-product formation. Such limits are easily identified by visual observation using a microscope (See Figure). A very nice Taylor flow is generated at the injector (a) and no significant bubble shrinking due to oxygen consumption by the reaction is observed throughout ca. 60% of the channel (b). At some point however, a serious shrinking of the bubbles is evidenced (c) which ultimately leads to a liquid monophasic flow (d). Interestingly, new bubbles are formed close to the outlet of the channel which likely reveals over-oxidation driving to formic acid, ethylene and propylene and ultimately to CO2. Thus, the Pyrex capped silicon component allows the direct visualisation of the reaction progress under high pressure and temperature. However, further experiments have revealed that gas-liquid mass transfer still limits the rate of reaction despite the low dimensions of the channel

(300 µm). Further studies are in progress to design a more efficient tool able to provide both mass transfer free operations and oxygen refilling.

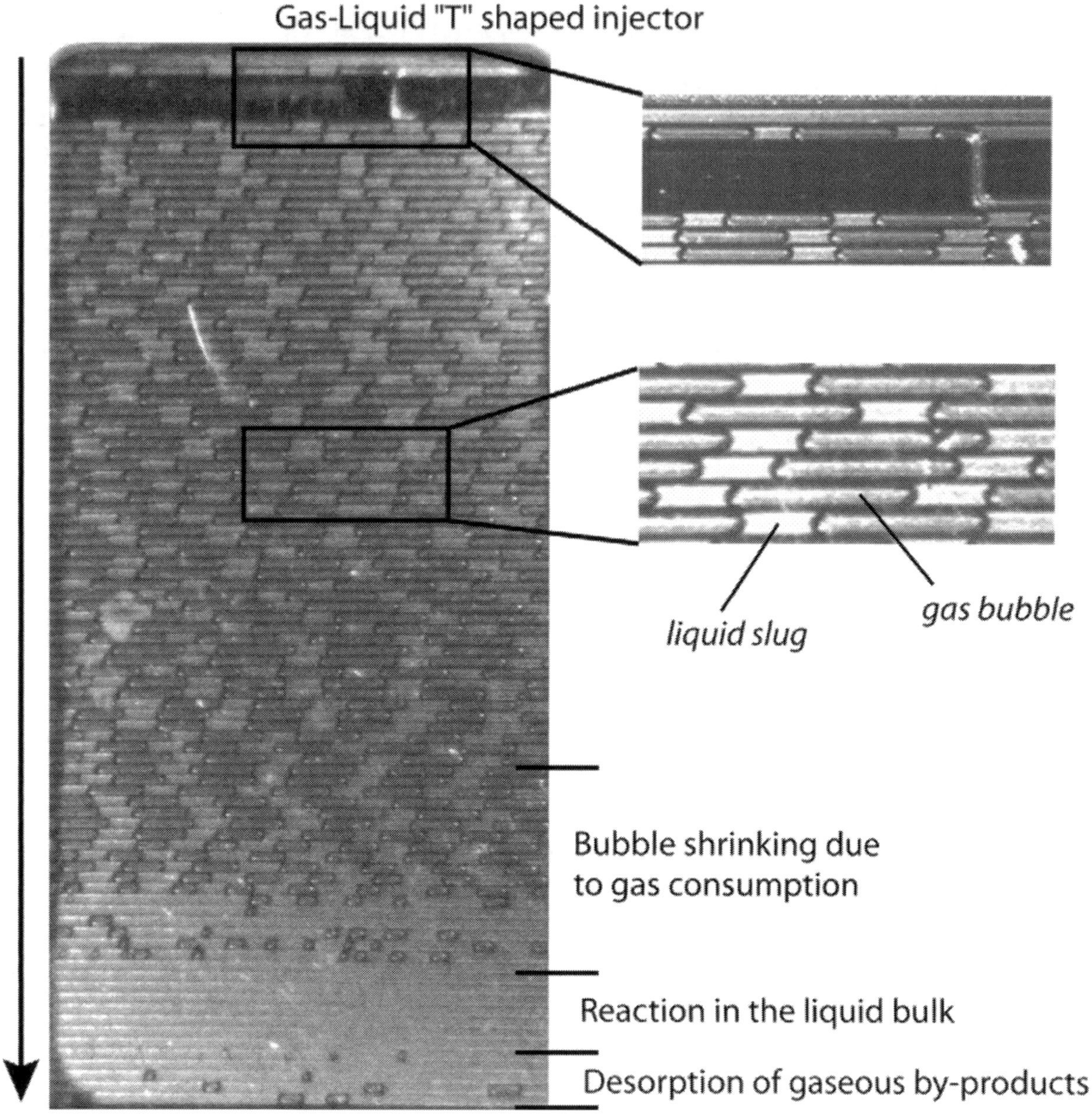

130a Screening and Optimization in a Microreactor Platform

Kaspar Koch[1], N.J.H. Van Harskamp[1], P.J. Nieuwland[1], J.C.M. Van Hest[2], and F.P.J.T. Rutjes[1]. (1) Synthetic Organic Chemistry, Radboud University Nijmegen, Toernooiveld 1, Nijmegen, Netherlands, (2) Bio-Organic Chemistry, Radboud University Nijmegen, Toernooiveld 1, Nijmegen, Netherlands

The variety of chemical reactions performed in continuous flow microreactors is steadily increasing. Important advantages over conventional batchwise reactions are a generally enhanced control over selectivity and significant suppression of side product formation as a result of improved mass and heat transfer processes. An additional valuable feature of microreactor technology is that due to the small dimensions reaction screening can be efficiently carried out without using large amounts of substrate or catalyst. These advantages have been substantiated over the past decade with a growing amount of examples[1-4]. We have developed a microreactor platform which is capable of acquiring large numbers of data points in a single day, consuming only small amounts of reaction fluid (typically 5 mL). In order to make the data acquisition process as efficient as possible, we have applied the principles of Design of Experiments (DOE), more specifically D-optimal design[5] to optimize industrially relevant reactions.

[1] G. N. Doku, S. J. Haswell, T. McCreedy, G. M. Greenway, Analyst 2001, 126, 14.

[2] K. Geyer, J. D. C. Codee, P. H. Seeberger, Chemistry-a European Journal 2006, 12, 8434.

[3] V. Hessel, P. Lob, H. Lowe, Current Organic Chemistry 2005, 9, 765.

[4] P. Watts, C. Wiles, Chemical Communications 2007, 443.

[5] P. F. de Aguiar, B. Bourguignon et al., Chemom. Intell. Lab. Syst., 1995, 199-210.

83aa On-Chip Parallelization of Microreactors: Up-Scaling Synthesis Microreactor Technology

Jurjen Emmelkamp[1], Ruud Olieslagers[1], Albert Prak[2], Hans H. Van den Vlekkert[2], Ronald J.J. Boot[1], and René M. De Zwart[3]. (1) Instrument Modelling & Control, TNO Science&Industry, Eindhoven, Netherlands, (2) LioniX BV, Enschede, Netherlands, (3) Precision Manufacturing, TNO Science&Industry, Eindhoven, Netherlands

Several microreactors exist, mainly single ones, though the first steps for up-scaling by parallelization have been taken [1, 2]. This presentation describes the design models one has to obey when up-scaling a common on-chip fluidic synthesis microreactor consisting of several inlets and one outlet, without loss of yield and conversion.

Up-scaling a fluidic synthesis microreactor, with remaining volume-to-surface ratio at the reaction area, can only be achieved by parallelization of the microreactors. An on-chip system has preferably a minimum number of inlets, one for each reactant and/or fluid catalyst, and one outlet. To achieve this, flow splitters are required, to split the inlet flows into n flows, where n is the number of parallel reactors, see Figure 1. The schematic in this figure shows necessity for crossing fluidic channels, acquiring a stacked wafer chip of at least three layers. At the start of the mixing area, or reaction area for instantaneous reactions, the different flow splitters are connected. Just connecting flow splitters to form a parallel microreactor chip without a well considered design easily lead to large errors in mixing ratio per micromixer, or in bubble encapsulation in one of the parallel channels. Both scenarios will lead to decrease of yield.

In microchannels small deflections and imperfections always exist, easily varying the hydraulic resistance over parallel channels in the range of a few percent, resulting in flow variations. We will present that flow variations can lead to mixing ratio errors in unbalanced flow splitters, and how balanced flow splitters can be achieved. In highly exothermic reactions flow splitter balancing becomes very important, due to the relation between temperature, viscosity and flow resistance. We will present a model to calculate the yield drop based on the deflections in channel cross-sectional area and shape.

The second objective in designing parallel microreactors is avoiding bubbles blocking channels. At encapsulated bubbles two forces are applied: capillary forces that tend to keep the bubble in position and pressure drops across the bubbles that try to push the bubble out of position. We will present the mechanism behind these forces, based on conventional microfluidics theory, and we will give a calculation model for minimizing the risk for bubble blocking to a well defined level.

Currently, flow splitter reactor chips with 8 parallel reactors are being fabricated by LioniX BV, Enschede, the Netherlands. At the conference, the fabricated chip and experimental results will be shown.

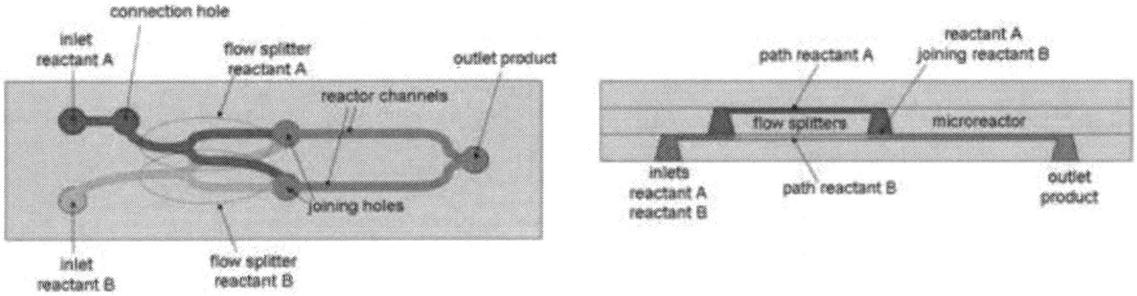

Figure 1. Schematic of a simple 2-reactor parallel microreactor chip including two flow splitters. Left: Top view; Right: Side view.

241c Development of a Novel Compact 3-Phase Hydrogenation Reactor

Martin Grasemann, Albert Renken, and Lioubov Kiwi-Minsker. Chemical Engineering, Ecole Polytechnique Federale de Lausanne, Lausanne, Switzerland

Selective catalytic three-phase hydrogenations are widely used for the manufacturing of fine chemicals, vitamins and pharmaceuticals. The reactor performance and catalyst selectivity and activity of these systems are considerably influenced by mass transfer. To minimize mass transfer limitations, production is often based on slurry reactors in batch or semi-batch mode using catalyst particle sizes of a few micrometers. The subsequent filtration of the product is a costly and time consuming additional process step. Another aspect of reactor design concerns the evacuation of reaction heat, since hydrogenations are highly exothermic (e.g. •HR ≈ -170kJ/mol for acetylenic alcohols). To increase heat transfer capacity and maximize viable catalyst loadings, many applications resort to recycle reactors with external heat exchangers. This, however, entails a large increase in reactor volume as well as complete backmixing of reactants and products, leading to limited reactor performance and losses in process selectivity.

Continuous catalytic multiphase processes based on structured catalysts are a promising alternative to slurry batch technology in terms of productivity, energy consumption, catalyst handling and process safety. Various structured catalyst systems have been reported for different multiphase reactions, often yielding performances comparable or superior to industrial slurry catalysts. However, to tap the full potential of these catalysts, product recycle and the resulting backmixing have to be avoided, making the removal of reaction heat a far more demanding task.

Herein we report the use of sintered metal filters (SMF) as micro-structured catalyst support for multi-phase hydrogenations. The material consists of metal fibers (d ≈ 20μm) forming highly porous sheets of ~ 0.3mm thickness. The SMF-sheets are self-supporting and can easily be shaped into a structured staged bubble column. The open structure with porosities >80% ensures a low pressure drop and acts as a static mixture for enhanced bubble redistribution.

The SMF catalyst layers are integrated into a micro-structured cross-flow heat exchanger and used as stages in a bubble column reactor. Due to the high heat transfer performance of the micro-heat exchanger element in combination with the high activity of the SMF-based catalyst, the system can be operated as a compact 3-phase reactor cascade with minimum hold-up on each stage. This results in the reduction of total reactor volume by several orders of magnitude as well as largely increased reactor performance and process safety as compared to conventional batch or recycle reactors.

The reactor is tested in the selective hydrogenation of 2-methy-3-butyn-2-ol over a Pd/ZnO catalyst supported on SMF.

83ah Universal Micro-Reactors Made of Glass for Applications to Study the Reaction Kinetics

Thomas Frank, Little Things Factory LTF GmbH, Ehrenbergstrasse 1, Ilmenau, 98693, Germany and Di Zhi, Faculty for Mechanical Engineering, University of Technology Ilmenau, Ehrenbergstraße 29, Ilmenau, 98693, Germany.
Universal Micro-reactors made of glass for applications to study

the reaction kinetics

Thomas Frank, Little Things Factory GmbH, 98693 Ilmenau/ Germany

Di Zhi, University of Technology Ilmenau/ Germany

Introduction

Micro-reactors for the chemical reactions in general have a volume of a few milliliters. The dwell correspond to the reaction time. In conventional terms the reaction time is for a variety of compounds in the area of fractions of a second to a few minutes. Although the reaction times in the second area are very suitable for micro-reactors, essentially bring us prolonged problems. Because of the long dwell , at acceptable flow rates, the micro-reactors must be refined with an extremly large innervolume. In order to avoid this, we should to try to accelerate the reaction speed. The following three variables affect the response rate particularly strong.

Concentration: The higher the concentration of the reactant, the more likely it is that particles mix with the appropriate energy.

Temperature: The higher the temperature, the more particles just require the minimum energy or more.

Surface: The larger the surface, the greater the reactive surface is mixing.

Process intensification in glass micro-reaktors

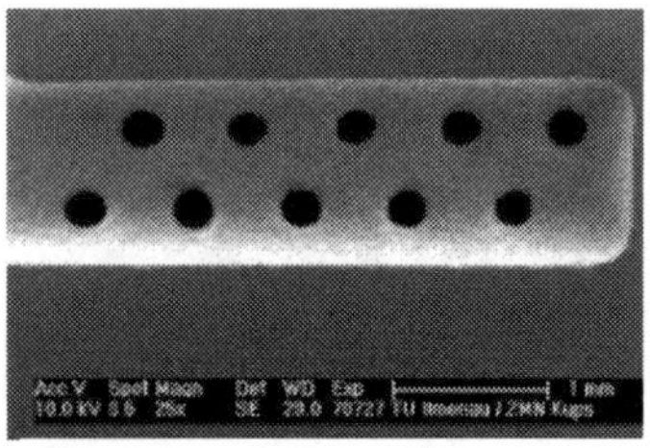

Figure 1: jet structure of micro mixer, which are produced by micro-abrasive compressed air rays.

By the use of micro-reactors these variables can be moved to the direction of higher response rate, particularly well succeed in this highly exothermetic reactions. However, many endothermic reactions at extreme reaction conditions are still to be investigated. Especially for the study of reaction kinetics of micro-reactors were developed, what permit the reaction temeratures up to 1000°C and pressures up to 200 bar. The micro-reactors consist of fusion bonded borosilicate glass(maximum operating temperature 450°C) or quartz glass(operating temperature 1000°C). The structures are produced by micro-abrasive compressed air rays, Figure 1 shows a SEM micrograph of a jet structure, as they are used for a micro mixer. Ports and temperature control prepare the largest problems at the extreme process conditions. When teflon tubing used align themselves with high temperatures and pressures, there are still problems at 1000°C even clamping ring joints made up of stainless steel(eg. WNr. 1.4571) in combination with metal pipes. The task can be resolved by micro-reactors with a cold(up to 200°C) and a hot(450 °C - 1000°C) zone. It is possible to implement hot and cold zones, conditionally due to the low heat conductivity of glass, l_{Glas} » 1,2-1,5 1,2 $W(mK)^{-1}$. Additionally, the reaction can be intensified by microwave for metalfree. The reactor can be also operated in a Ultrasound cleaning bath. This has been particularly proved its worth, which the particle afflicted reactions are as a measure to antifouling.

Due to continuous record of mechanical energy the particles have no time to agglomerating.

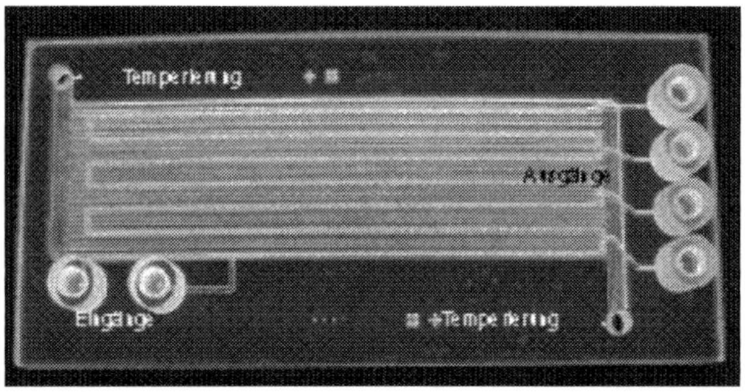

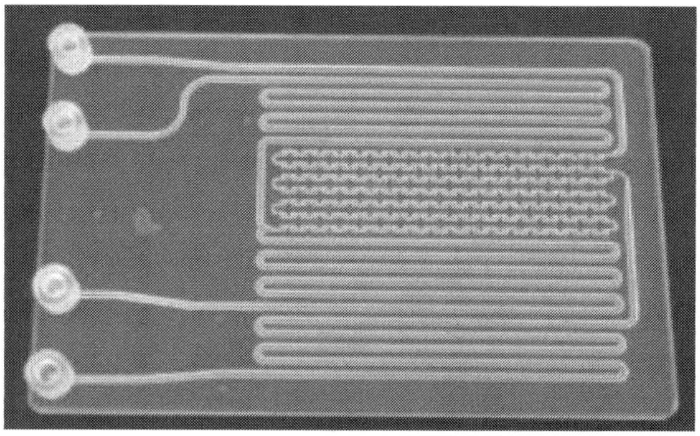

Figure 2: two micro-reactors, above:

micro-reactors with two inputs of dwell route and four output and intergrated temering.

For both is the entire dwell route with mixed element edit in order to counteract decomposition. The outputs allow for different dwell route as the sampls.

Right: micro reactors with three inputs, for a two-stage reaction and for direct temperature on hot plate or immersion in a thermostat. All three educts are prewarmed in the reactor.

The figure 2 shows two micro-reactors made up of borosilicate glass.

The figure 2 at top display us a reactor with integrated tempering and the thermostat is directly connected. Through the four outputs the samples can to be taken according to various dwell so as the turnover to assess. In particularly low or high temperatures the direct temperature will be affected. The reactor will be tempered direct in a bath or on a hot plate. Only the reaction field is heated, the ports remain largely cold due to the low heat conductivity in the glass and the optional cooling. The distance between the ports and the warm reaction field is chosen in such a way that feed line and sealing material thermally not overloaded.

Web Page: WWW.ltf-gmbh.de

45g Styrene Production Via a Highly Selective Partial Oxidation Route Enabled by Novel Catalyst Substrates in Microchannel Reactors

Anna Lee Tonkovich, Kai T. Jarosch, Timothy J. LaPlante, Richard Q. Long, and Thomas Hickey.
Velocys, Inc., 7950 Corporate Boulevard, Plain City, OH 43064

Commercial styrene production requires large capital investment and results in high operating costs to supply heat for the endothermic dehydrogenation reaction with superheated steam. An alternative chemistry has long been suggested in the literature for an exothermic partial oxidation route from ethylbenzene to styrene, but high selectivity at commercially-attractive conversion rates has not been realized. The challenge for this reaction is obtaining the desired selectivity, resulting from more than 30x difference in heat of reaction for styrene production versus deep oxidation.

Recently, the performance barrier for oxidative styrene production has been overcome by merging a new class of catalysts with a novel thermally-effective and low-tortuosity catalyst substrate. In a single stage microchannel reactor, selectivity to styrene over 95% has been demonstrated with an ethylbenzene conversion exceeding 70%. Comparative experiments were conducted for the same catalyst without the novel substrate; the catalyst as packed in a narrow channel reactor demonstrated thermal runaway above 40% conversion. An evaluation of commercial economics based on this improved oxidative synthesis route illustrates the substantial improvement in economic return for this process.

Three-phase mass transfer in pillared micro channels

S.R.A. de Loos[1], J. van der Schaaf [1], M.H.J.M. de Croon[1], T.A. Nijhuis[1] ,
R.M. Tiggelaar[2], J.G.E. Gardeniers[2], J.C. Schouten[1]
[1]*Dept. of Chemical Engineering and Chemistry, Eindhoven University of Technology,*
P.O. Box 513, 5600 MB Eindhoven, the Netherlands
[2]*MESA+, University of Twente*
P.O. Box 217, 7500 AE Enschede, the Netherlands

E-mail address corresponding author: S.R.A.d.Loos@tue.nl

Keywords: Microreactors, Microchannel, Taylor flow, Oxidation reaction

Continuous processing of fine chemicals in micro fabricated reactors offers numerous advantages compared with batchwise production in larger sized reactors. For example, gas-to-liquid and liquid-to-solid mass transfer rates are considerably increased in micro channels because of the high surface area to volume ratios. Furthermore, the plug flow of reactants results in much better control of the residence time distribution which leads to higher selectivities and yields. Finally, the excellent heat management prevents hot spot formation and provides novel windows of operation. This means that reactions can be performed safer and more sustainable.

Much research has been done on two-phase flow patterns in micro structured reactors [1, 2]. In these narrow channels, the channel orientation is not relevant as the gravitational force is almost negligible and the flow is mainly determined by capillary and pressure forces. This results in the disappearance of stratified flow and different flow regimes are observed [2].

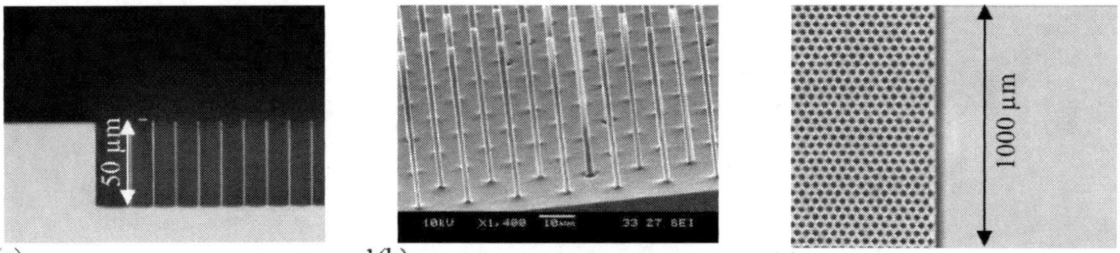

1(a) 1(b) 1(c)

Figure 1: SEM pictures of micro pillars (3 μm in diameter) etched in the microchannel (1 mm x 50 μm x 6.6 cm) to enlarge the surface area for catalyst deposition; sideview (a) and (b) topview (c).

A recent trend in the design of three-phase microreactors for catalyzed reactions focuses on the use of structured catalyst supports, i.e. a structured geometry with a high surface area with a metal catalyst, like Pd, deposited on that structure and/or on the micro channel walls. The catalytic surface area of these structured microreactors is sufficiently large to accommodate fast reactions. However, for slow reactions at low temperatures, still larger surface areas are required. Therefore, surface enhancing structures have to be etched on the microstructured support to further enlarge the surface area to volume ratio (Figure 1).

A proper design of a micro structured multiphase reactor requires a comprehensive understanding of the three-phase flow hydrodynamics. The pressure drop, the liquid hold-up and the flow regimes and their transitions need to be known as a function of the micro channel and structure geometries. Moreover, the relation of the hydrodynamics to the gas-liquid and liquid-solid mass transfer needs to be established. In Figure 2 an overall picture is given of the relevant mass transfer steps for the structured microreactor in Figure 1.

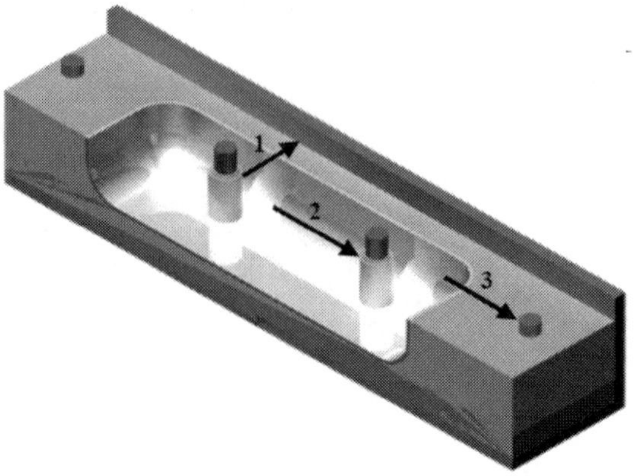

Figure 2: Schematical representation of the mass transfer steps during Taylor flow using a pillared micro channel: gas-film-wall transfer (1) gas-film-pillar transfer (2) and gas-slug-film-pillar transfer (3).

In this work, microchannels having a length of 6.6 cm, a width of 1000 μm and a depth of 50 μm etched in silicon wafers were used. To increase the catalytic support area from 4.2 x 10^4 m_{sup}^2 m_{reac}^{-3} to 9.5 x 10^4 m_{sup}^2 m_{reac}^{-3}, micro pillars of 3 μm in diameter and 50 μm in length were fabricated in the second axial length part of these channels (Figure 1).

The flow channels and pillars were etched in silicon in one step using the Bosch process and powder blasting of the wafer was done to create inlet and outlet holes to the flow channels. Finally, anodic bonding with a Pyrex wafer was used to complete the chips. Making pillars in microchannels requires therefore no extra step in the synthesis.

To introduce the gas and liquid to the micro channel, a cross shaped mixer was used. Nitrogen/water systems were investigated for superficial gas velocities of 0.3 to 15 m/s (Re_G = 7.5 - 375) and superficial liquid velocities of 0.01 to 1.67 m/s (Re_L = 0.95 - 158). Video analysis of the two-phase flow in these water/nitrogen systems was performed using a Zeiss Axiovert Microscope coupled to a high-speed camera (maximum 10,000 frames s^{-1}). Figure 3 shows a number of observed flow patterns at different gas and liquid velocities. The pillar structure hardly influences the two-phase flow and the observed flow patterns and regimes are similar to their counterparts in non-pillared microchannels [2-4].

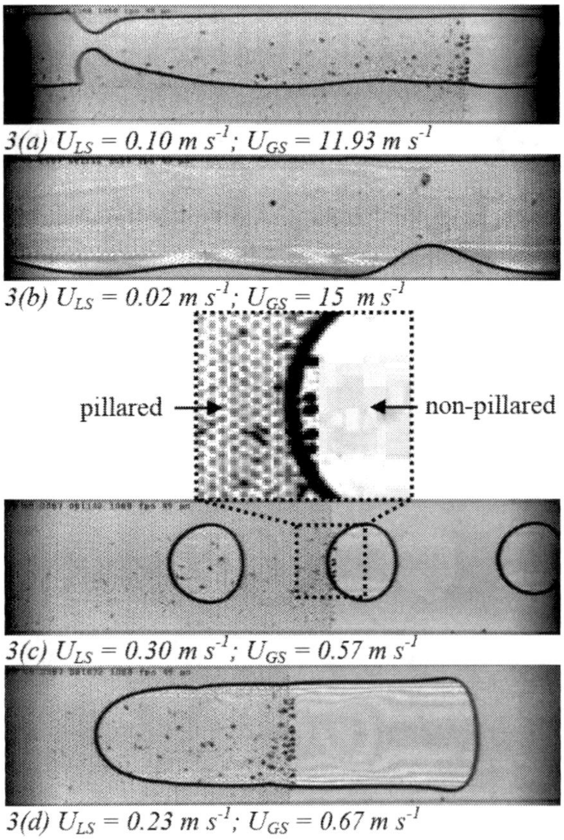

Figure 3: Pictures of two-phase H_2O/N_2 flow patterns in a 1000 x 50 μm rectangular micro channel with 3 μm diameter pillars (a) slug-annular flow; (b) annular flow; (c) bubbly flow; (d) Taylor flow.

Furthermore, specifically for the bubbly and Taylor flow regime (Figure 3(c,d)), the rates of overall mass transfer were estimated using the mass transfer model configuration of Figure 2. A very thin liquid film is present around the pillars. The capillary force rapidly squeezes the liquid in the film to the top and the bottom of the channel. The film is refreshed by liquid when the bubble has passed. This will result in very high gas-liquid and liquid-solid mass transfer rates due to the continuously alternating gas-liquid environment exposed to this film and the short diffusion lengths.

The oxidation of sodium formate over a Pd catalyst was chosen as a model reaction. This reaction is kinetically (pseudo) half order in oxygen. The intrinsic kinetics is very fast and the overall reaction rate is only limited by mass transfer resistances. This overall reaction rate can therefore be further increased by increasing the gas-liquid and liquid-solid mass transfer interfacial areas.

Palladium was deposited on the silicon pillars by means of an ion-exchange procedure with silanol groups present on the surface. In normal slurry bubble reactors, overall mass transfer coefficients of 0.2 s^{-1} were obtained using 1% Pd on carbon particles and a liquid volume of $5.0 \cdot 10^{-4}$ m^3. For the pillared micro reactor in this study, overall mass transfer coefficients of the order of at least 1.0 s^{-1} are expected [5].

References:

[1] Bretherton, F.P., J. of Fluid Mech. 10 (1960), 166-188

[2] Triplett, K.A., Ghiaasiaan, S.M., Abdel-Khalik, S.I., Sadowski, D.L., Int. J. of Multiphase Flow 25 (1999), 377-394

[3] Warnier, M.J.F., Rebrov, E.V., de Croon, M.H.J.M., Hessel, V., Schouten, J.C., Chem. Eng. Journal (2007), in Press

[4] Hetsroni, G., Mosyak, A., Segal, Z., Pogrebnyak, E., Int. J. of Multiphase Flow 29 (2003), 341-360

[5] Kreuzer, M.T., Peng Du, Heiszwolf, J.J., Kapteijn, F., Moulijn, J.A., Chem. Eng. Sc. 56 (2001) 6015-6023

83ae A Thick Film Printed Electrochemical Biosensor and Its Application for Testing the Ketone 3-β-Hydroxybutyrate for Potential Diabetic Patient Management

Lei Fang[1], Shih-Han Wang[2], and Chung-Chiun Liu[1]. (1) Chemical Engineering, Case Western Reserve University, A.W. Smith Building, 10900 Euclid Avenue,, Cleveland, OH 44106, (2) Chemical Engineering, I-Shou University, County 840, Kaohsiung, Taiwan

The American Diabetes Association advises that blood ketone testing methods that quantify 3-β-hydroxybutyrate (3HB) will be desirable for the diagnosing and monitoring ketoacidosis for diabetic patient management. Blood ketone refers to 3-β-hydroxybutyrate (3HB), acetoacetate (AcAc) and acetone. The detection of 3HB is suggested for the management of DKA. Amperometric biosensors based on enzymes have raised increasing attention recently due to the high selectivity, sensitivity and convenience of these biosensors. Thick film printing technique has been applied to the fabrication of amperometric biosensors, improving the quality and reproducibility of the biosensors, and reducing the cost. In this study we can show that the detection method using this biosensor involved only a single reaction step without any mediator. The concentration of 3HB could be quantified with enzyme 3HBDH and NAD+ immobilized on the working electrode of this biosensor prototype (Figure 1). This electrochemical biosensor could be used effectively for 3HB detection for potential diabetic patient management. Fig.1. Amperometry of 3HB tested by biosensor under 37.5 °C in the presence and absence of enzyme 3HBDH

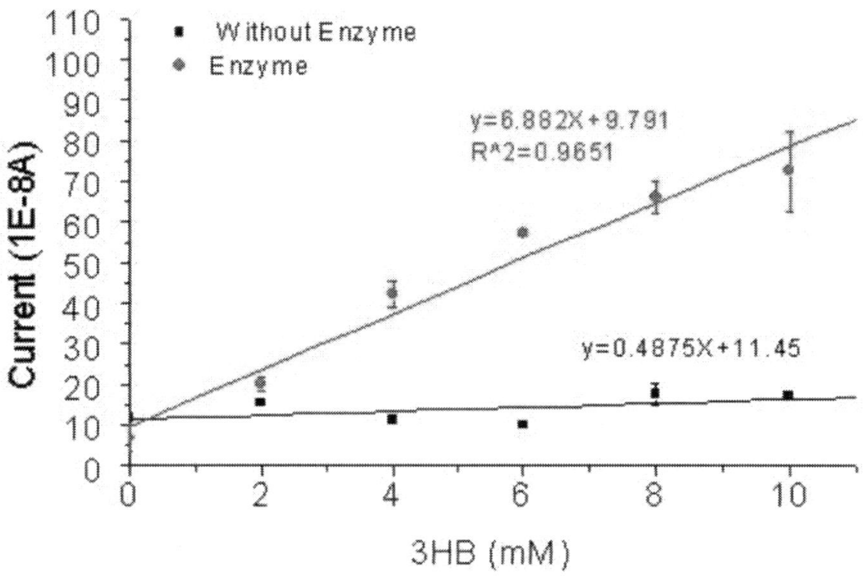

130d Synthesis of Ultrahigh Molecular Weight Polymer in a Catalytic Nanotube Polymerization Reactor

Kyu Y. Choi[1], Joong Jin Han[2], Bo He[3], and Sang Bok Lee[3]. (1) Chemical and Biomolecular Engineering, University of Maryland, Stadium Drive, Bldg 090, College Park, MD 20742, (2) Department of Chemical & Biomolecular Engineering, University of Maryland, Stadium Drive, Bldg 090, College Park, MD 20742, (3) Department of Chemistry, University of Maryland, Stadium Drive, Bldg 090, College Park, MD 20742

Nanofibers of syndiotactic polystyrene have been synthesized in a metallocene catalyst-anchored silica nanotube reactor (SNTR) in an porous alumina template. Very thin sPS nanofibrils (< 10 nm) grown at the catalytic sites on the pore walls aggregate to form intertwined, rope-like nanofibrils with 30-50 nm diameters, which further intertwine into even larger 200-nm diameter polymer nanofibrils. The extrusion of nanofibrils synthesized inside the SNTR was directly observed by scanning electron microscopy, and the individual SNTR containing a single polymer nanofibril was separated and observed by transmission electron microscopy. The syndiotactic polystyrene produced in the SNTR is a delta-form crystal polymer and it has very high molecular weight including a large fraction of 2,000,000 - 5,000,000 g/mol molecular weight and higher crystalline melting point than the sPS synthesized over homogeneous or silica-supported metallocene catalysts.

Controlled Hydrogen Peroxide Decomposition for a Solid Oxide Fuel Cell (SOFC) Oxidant Source with Microreactors

E. Lennon[1], A. A. Burke[2], and R. S. Besser[1]
1. Department of Chemical Engineering and Materials Science
Stevens Institute of Technology, Hoboken, NJ 07030 (elennon@stevens.edu, rbesser@stevens.edu)
2. Chemistry Division, Naval Undersea Warfare Center, Newport, RI 02841

Abstract

It is well established that the catalytic decomposition of hydrogen peroxide (H_2O_2) into water (H_2O) and oxygen (O_2) [$H_2O_2 \rightarrow H_2O + \frac{1}{2} O_2$] is highly exothermic (Heat of Reaction: -98 kJ/mol) and notoriously susceptible to thermal runaway. However, liquid H_2O_2 has high oxygen density making it a particularly attractive oxidant for undersea naval power applications where air-independence and space limitations elicit additional power supply design challenges. Microchemical technologies have inherently higher surface-to-volume ratios which facilitate heat and mass transfer. Small-scale geometry thus offers a mechanism to harness the benefits of H_2O_2 decomposition as an oxidant source while reducing the risk of uncontrollable heat rise. This study investigated two small-scale reactors to illustrate thermal management and oxygen production feasibility during the multiphase decomposition of H_2O_2 as an oxidant for an unmanned undersea vehicle (UUV) solid oxide fuel cell (SOFC). A microscale packed bed (MPB) reactor model with channel radius 0.5 mm and miniature tubular, catalyst-coated (TCC) reactor model with channel radius of 1.5 mm were developed in the finite element modeling program COMSOL. The properties of the multiphase product stream were described by weighted averages of the components in the stream. This approach assumed that the stream maintained a well-mixed, dispersed phase. The heat generated during simulated decomposition in the convectively cooled reactor channels caused temperature rises greater than 400 K (neglecting the energy required to evaporate water). With small-scale channel geometry alone unable to prevent significant heat rise, further cooling was needed to ensure thermal management of the reactor zone. The addition of extended surface areas, improved passive cooling and simulated temperature rises decreased to less than 20 K under the previous reaction conditions. Experimental data collected with a physical TCC reactor demonstrated increased variability in both the temperature and oxygen production data at higher H_2O_2 concentrations. The increased production of oxygen gas from higher H_2O_2 concentrations resulted in long pockets of oxygen gas within the physical reactor. Since the models averaged the instantaneous properties of the products into a single effluent fluid, the physical effect on decomposition due to separated oxygen gas was not evident during TCC simulation. Despite the occurrence of slug flow instead of dispersed flow, results showed improved oxygen generation and thermal control during H_2O_2 decomposition relative to conventionally sized reactors.

Introduction

A major challenge facing the design of air-independent power systems is continuous energy generation. High energy density and potentially low operating costs make fuel cell power systems an attractive option for high endurance, air-independent, undersea vehicle applications [1]. However air-independent fuel cell power systems require an oxidant supply. Hydrogen peroxide (H_2O_2) decomposes into water (H_2O) and oxygen (O_2) [$H_2O_2 \rightarrow H_2O + \frac{1}{2} O_2$] providing a dense source of oxygen per unit volume that makes it a valuable commodity as an air-independent fuel cell oxidant.

The high oxygen density, ease of handling, and commercial infrastructure of H_2O_2 add to its appeal as an oxidant source [2]. Despite these advantages, it is well established that catalytic H_2O_2 decomposition is susceptible to thermal runaway, which has historically limited its application as a power system oxidant [3]. Microchemical systems by virtue of microscale geometry (relevant fluid dimension in subunits < 1 mm), possess high surface-to-volume ratios resulting in heat and mass transfer coefficients capable of inhibiting thermal runaway [4]. Thus, microchemical systems present a theoretical mechanism to harness the aforementioned benefits of a H_2O_2 oxidant source, and at the same time prevent thermal runaway during the exothermic reaction. The aim of this research was to demonstrate oxygen production and thermal management feasibility during the catalytic multiphase decomposition of hydrogen peroxide in a single microreactor subunit. The catalytic multiphase decomposition of H_2O_2 was simulated in two small scale reactors, a microscale packed bed reactor (MPB) and tubular microchannel with catalyst coating (TCC). Additional experimental data was collected from a physical TCC reactor.

Methods

Simulated H_2O_2 Decomposition

I. Model Geometries

The cross sectional geometry of the MPB channel resembled a half moon and had a sub-millimeter radius (0.5 mm). The channel proceeded straight down the length of the reactor (5 cm). Surrounding the channel was a rectangular stainless steel 316 block (7x2x0.4 cm). A Plexiglas (PMMA) cover (7x2x0.1 cm) served as the top seal. The reaction characteristics due to catalyst in the MPB reactor channel were modeled with manganese dioxide (MnO_2) catalyzed reaction kinetics from preliminary experimental kinetic data. The TCC geometry was based on thick-walled ¼ inch stainless steel type 316 tubing. The length of the reactor was 9.7 cm. The inner diameter of the straight reactor channel was 0.3 cm, while the outer diameter was 0.6 cm. The reaction characteristics in the TCC reactor were modeled with kinetics from a silver epoxy layer on the walls of the inner TCC reactor channel.

II. Balance Equations and Assumptions

Mass Balance. Assumptions pertaining to the phases of the reactant and product species of the mass balance were made. The amount of H_2O_2 in the vapor phase was considered negligible. This assumption was supported by the fact that equilibrium vapor concentration of 50% w/w H_2O_2 reactant at 333 K is less than 10% [5]. Additionally, the amount of dissolved O_2 in the liquid was considered negligible compared to the H_2O_2 and water liquid components, and the amount of water vapor in the gas phase was considered negligible compared to the O_2 gas provided the temperature rise remained below the boiling temperature of water (373 K). The mass balance included the diffusion of H_2O_2 into liquid water and H_2O_2., the bulk flow of the evolving fluid mixture (consisting primarily of liquid H_2O_2 solution in water, and O_2 gas) down the reactor channel, and consumption of H_2O_2 resulting from the decomposition reaction. To model the mass balance, the built-in convection and diffusion application mode in the program COMSOL was used. Equation 1 governed the convection and diffusion mass balance of H_2O_2.

$$\nabla\left(-D_{H_2O_2 eff}\nabla C_{H_2O_2}\right)+\mathbf{u}_s\cdot\nabla C_{H_2O_2}=r_{H_2O_2} \quad (1)$$

In equation 1, $D_{H2O2eff}$, was the effective diffusivity of H_2O_2 into the liquid water, H_2O_2 solution. The diffusivity term neglected diffusion interactions with the catalyst. The mass balance solved for the concentration, C_{H2O2}. The variable $\mathbf{u}_s$ represented the superficial velocity of the fluid. The rate of reaction based on the consumption of hydrogen peroxide, $\mathbf{r}_{H2O2}$, was established assuming first order kinetics weighted via an inputted catalyst mass. The water and oxygen production rates were related to $\mathbf{r}_{H2O2}$. A conventional Arrhenius relationship defined the reaction rate constant coupling it to the energy balance (equation 2). The activation energy and frequency factor were set according to the model descriptions.

Energy Balance. The energy balance of the microreactor system included conduction, advective input and output due to the feed and exit streams, exothermic heat generation due to reaction, sensible heat, and the heat of water vaporization. To model the energy balance, the built-in convection and conduction application mode in COMSOL was used. Equation 2 governed the convection and conduction energy balance.

$$\nabla(-k_{eff}\nabla T + u_s c_p \rho \nabla T = (\Delta H_{rxn} r_{H_2O_2}) - SH + (\Delta H_{vap} r_{H_2O}) \quad \textbf{(2)}$$

In equation 2, k_{eff} gave the effective thermal conductivity of the reactor channel. The energy balance solved for the variable temperature, $\boldsymbol{T}$, coupling it to the mass balance equation. The term, $\mathbf{u}_s$, still represented the superficial velocity of the fluid. The term $\mathbf{c}_p$ represented heat capacity and the term, $\boldsymbol{\rho}$, the average density of the fluid in the microchannel. The heat of reaction, ΔH_{rxn}, defined the heat released during the exothermic decomposition reaction. The sensible heat term, **SH,** described the contribution of the heat capacities per product species required to increase the temperature to the boiling point of water. The remaining term of the energy balance, ΔH_{vap}, expressed the energy required to vaporize water based on the rate of water production during the reaction.

*Momentum Balance***.** To model the momentum balance, the built-in non-isothermal flow application mode in COMSOL was used. The non-isothermal flow application solves the Navier Stokes equations for weakly compressible flows (flows with Mach numbers < 0.3). Neglecting the effects of the catalyst in the flow channel and using low initial flow rates maintained the conditions defining weakly compressible fluid flow. The density, $\boldsymbol{\rho}$, was modeled as an average of the liquid and gas comprising the fluid weighted by the mole fractions of H_2O_2, H_2O, and O_2 species respectively and changed as the reaction progressed (equation 3).

$$\rho = y_{H_2O_2}\rho_{H_2O_{2s}} + y_{H_2O}\rho_{H_2O} + y_{O_2}\rho_{O_2} \quad \textbf{(3)}$$

This average density was applied under the simplifying assumption that the generated O_2 gas was homogenously dispersed throughout the effluent liquid. The momentum balance solved for the velocity field giving the superficial velocity of the fluid and the pressure.

Experimental H_2O_2 Decomposition

Effluent oxygen gas rates and temperature data from the physical experimental TCC microreactor were collected. A World Precision Instruments SP2001 syringe pump dispensed the H_2O_2 reactant at volumetric flow rates (VFR) from 0.05 ml/min up to 0.25 ml/min and concentrations of twenty five percent H_2O_2 weight in water (25% w/w), 30% w/w, and 50% w/w. The experimental TCC microreactor was connected to a condenser system. Two type J thermocouples were also placed near the inlet and outlet of the experimental TCC microreactor. A National Instruments USB 9211 digital to analog converter connected to 9162 USB data acquisition device designed specifically to interface with thermocouples transmitted the temperature data to the laptop for temperature data collection and storage.

The effluent multiphase product stream consisting primarily of liquid H_2O and gaseous O_2 flowed to the condenser system. The condenser system consisted of copper coil carrying water cooled to 10 °C surrounding a stainless steel tube through which the fluid being condensed flowed. The O_2 gas proceeded to the oxygen flow meter and the liquid product flowed to the glass vial beneath the condenser. Residual gas from the liquid product collector was directed to the oxygen flow meter as well. The oxygen streams were joined using a T junction and the full oxygen stream then entered the Aalborg oxygen gas flow meter. A NI USB 6008 digital to analog converter transmitted the flow data to the laptop for collection and storage. The program Labview Signal Express controlled both the temperature and flow data recording processes.

Results

Temperature rises larger than 400 K (neglecting the heat of vaporization) were simulated during H_2O_2 decomposition in the convectively cooled reactor channels caused (Figure 1). Small-scale geometry alone was unable to prevent considerable heat rise. To ensure thermal management of the reactor zone extended surface areas were added to the small scale reactor channels to improve passive cooling capability.

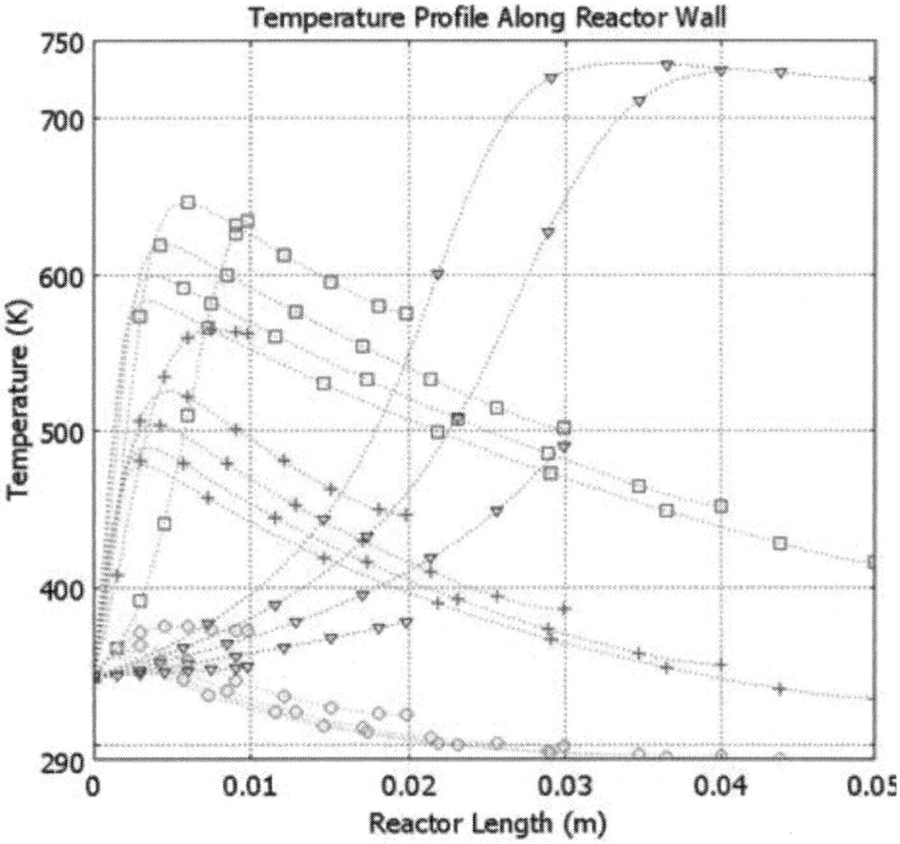

Figure 1: Simulated temperature of reactor channels without extended surface areas and neglecting the heat of vaporization rises at initial volumetric flow rates (m^3/s): 2e-9 (triangles), 2e-10 (squares), 1e-10 (pluses), 2e-11 (circles)

After the incorporation of the extended surface areas on the MPB and TCC reactor models, simulated temperature rises decreased to less than 20 K with equivalent reaction conditions (Figure 2).

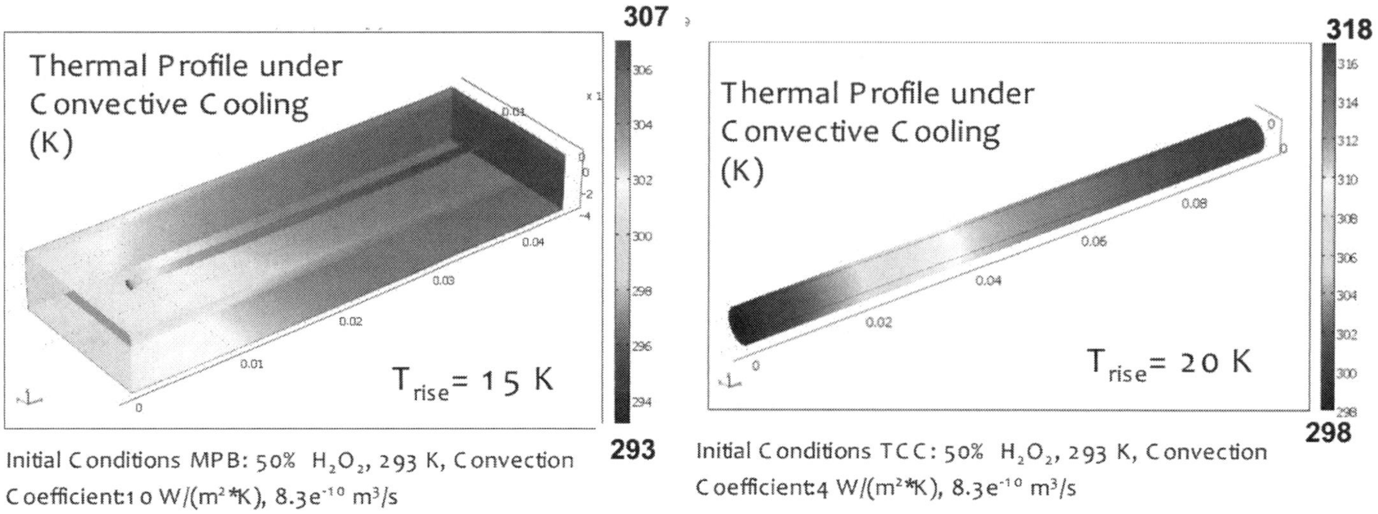

Figure 2: Simulated temperature profiles of MPB and TCC reactors with extended surface areas under given conditions

Experimental data collected with a physical TCC reactor demonstrated increased variability in both the temperature and oxygen production data at higher H_2O_2 concentrations and volumetric flow rates. Using the largest initial volumetric flow rate of 0.15 ml/min, an experimental TCC temperature rise of 56 K was observed, even though simulations projected a temperature rise of only 20 K under equivalent conditions. In the physical TCC reactor the increased production of oxygen gas from higher H_2O_2 concentrations resulted in long pockets of oxygen gas. The steady-state conversion for the experimental TCC under those conditions was approximately 0.4 compared to the simulated conversion of 0.16. Since the models averaged the instantaneous properties of the products into a single effluent fluid, the physical effect on decomposition due to oxygen gas slugs was not present during TCC simulation. Despite the incidence of slug flow instead of dispersed flow, both sets of results showed enhanced oxygen generation and thermal control during H_2O_2 decomposition indicating improvement towards. Despite the incidence of slug flow instead of dispersed flow, both sets of results showed enhanced oxygen generation and thermal control during H_2O_2 decomposition indicating improvement towards satisfying the volume, oxygen production, and temperature control required for the UUV SOFC application. These results are summarized in Table 1.

Table 1. Microreactors' Steady-State Output Summary per Stated Volumetric Flow Rates (VFR) using 50% w/w H_2O_2 Reactant Concentration

Reactor	VFR (ml/min, m^3/s)		Maximum Steady-State Temperature Rise per Subunit (K)	Steady-State Conversion	Rate of Oxygen Generation per Subunit (moles O_2 / min)	Subunits Required for 0.6 moles O_2/min	Total Theoretical Volume (L) (Goal < 2)
Simulated MPB	0.05	$8.3e^{-10}$	5	0.63	$2.75e^{-4}$	2183	15.3
	0.10	$1.7e^{-9}$	6	0.47	$4.20e^{-4}$	1429	10.0
	0.15	$2.5e^{-9}$	7	0.38	$4.99e^{-4}$	1202	8.4
Simulated TCC	0.05	$8.3e^{-10}$	19	0.42	$1.83e^{-4}$	3275	12.3
	0.10	$1.7e^{-9}$	24	0.26	$2.32e^{-4}$	2583	9.7
	0.15	$2.5e^{-9}$	20	0.16	$2.10e^{-4}$	2854	10.7
Experimental TCC	0.05	$8.3e^{-10}$	16	0.37	$2.23e^{-4}$	2687	7.4
	0.10	$1.7e^{-9}$	37	0.30	$5.36e^{-4}$	1120	3.1
	0.15	$2.5e^{-9}$	56	0.39	$1.12e^{-3}$	537	1.5

Additional simulations of both the TCC and MPB reactors projected steady-state conversion levels greater than 95% and volumetric flow rates greater than 0.10 ml/min are necessary to meet the oxygen production levels and subsequent volume goal (Table 2). At these conversion levels more heat is released necessitating further cooling potentially achieved via active water cooling channels. Although the MPB and TCC simulated results and TCC experimental results showed sufficient surface area-to-volume ratios successfully mitigated thermal runaway via passive convective cooling, additional investigation is still required to optimize the microreactor subunits for sufficient conversion levels and the corresponding active cooling measures.

Table 2. Microreactors' Steady-State Projected Output Summary Provided 0.95 Conversion using 50% w/w H_2O_2 Reactant Concentration

Reactor	VFR (ml/min, m^3/s)		Steady-State Conversion	Rate of Oxygen Generation per Subunit (moles O_2 / min)	Subunits Required for 0.6 moles O_2/min	Total Theoretical Volume (L) (Goal < 2)
Simulated MPB	0.05	$8.3e^{-10}$	0.95	$4.14e^{-4}$	1448	4.3
	0.10	$1.7e^{-9}$	0.95	$8.49e^{-4}$	707	2.1
	0.15	$2.5e^{-9}$	0.95	$1.25e^{-3}$	481	1.4
Simulated TCC	0.05	$8.3e^{-10}$	0.95	$4.14e^{-4}$	1448	5.4
	0.10	$1.7e^{-9}$	0.95	$8.49e^{-4}$	707	2.7
	0.15	$2.5e^{-9}$	0.95	$1.25e^{-3}$	481	1.8
Experimental TCC	0.05	$8.3e^{-10}$	0.95	$4.02e^{-4}$	1493	4.1
	0.10	$1.7e^{-9}$	0.95	$8.04e^{-4}$	746	2.0
	0.15	$2.5e^{-9}$	0.95	$1.21e^{-3}$	498	1.4

Conclusions

Significant temperature rise (up to 400 K) occurred across a simulated microchannel reactor when the extended surface area was omitted. Increasing the surface-to-volume ratio to enhance passive convective cooling of the simulated small scale reactors facilitated thermal management during catalyzed H_2O_2 decomposition. Thermal management was thus successfully accomplished with the addition of the high area heat exchanger in the simulated MPB reactor and thick reactor walls in the simulated TCC microreactor, subsequently mitigating the risk of thermal runaway. The momentum balances within the simulated reactors were included under the assumption of homogenously dispersed oxygen gas in a weakly compressible two-phase fluid to characterize the multiphase effluent product stream. Under this assumption, the models demonstrated the promise of thermally controlled H_2O_2 decomposition. However when the simulated and experimental TCC microreactor data were compared, differences between the simulated and experimental TCC data grew as the VFR and concentrations increased. The greatest disagreement between the simulated and experimental TCC data occurred at the highest VFR and concentration, indicating the assumption of homogenously dispersed flow broke down under these conditions. This was further explained by the observation of distinct gaseous oxygen bubble slugs interspersed with liquid slugs exiting the reactor at the higher reactant concentration and VFRs.

The divergence between the simulated and experimental TCC microreactor data demonstrated that the models failed to predict the higher steady-state temperature rises at the highest VFR and concentration. Correspondingly the models predicted lower oxygen molar flow rates than those experimentally observed with TCC microreactor. Despite these discrepancies, the feasibility of oxygen generation from thermally controlled H_2O_2 decomposition in the simulated and experimental small scale reactors for the intended UUV SOFC energy application was assessed. The maximum steady-state temperature rises were all less than 75 K preventing water vaporization. While focusing on thermal control over the microreactor subunits, the coupled generation of oxygen was lowered per subunit. With the oxygen molar flow rates listed in Table 1, the majority of the resulting full subsystem volumes were larger than two liters. Generating 0.6 moles of oxygen per minute with a full MCS subsystem less than two liters in volume is theoretically feasible as shown in Table 2, when conversion exceeds 0.95. These findings present a firm foundation for optimization of the H_2O_2 decomposition small scale reactors in which increases in oxygen production via higher conversion should be investigated further with increased convective cooling by liquid water and integrated active cooling channels.

References

1. United States Department of the Navy. The Navy Unmanned Undersea Vehicle (UUV) Master Plan, 2004. http://www.navy.mil/navydata/technology/uuvmp.pdf

2. Davis, N.S., and Keefe, J.H., "Concentrated hydrogen as a propellant," Industrial Engineering and Chemistry, 48, No. 4, 745–748 (1956).

3. Papadaki, M.; Marques-Domingo, E.; Gao, J.; and Mahmud, T.P., Journal of Loss Prevention in Process Industrie, 06 024 (2005)

4. Ehrfeld, W.; Volker H.; Holger, L., *Microreactors: New Technology for Modern Chemistry*, 1st ed.; pp. 5-8. WILEY-VCH, Weinheim, (2000)

5. Van Laar, J.J., Z. Physik. Chem. 72, 723 (1910). Data table reprinted at <http://www.h2o2.com/intro/properties/physical.html#14>.

101a A Plug-and-Play Microreactor Platform for Small Volume Reaction Optimization

F.P.J.T. Rutjes[1], P.J. Nieuwland[1], Kaspar Koch[1], and J.C.M. Van Hest[2]. (1) Synthetic Organic Chemistry, Radboud University Nijmegen, Toernooiveld 1, Nijmegen, Netherlands, (2) Bio-Organic Chemistry, Radboud University Nijmegen, Toernooiveld 1, Nijmegen, Netherlands

Applications of microsystem technology in organic synthesis are rapidly growing. There are several benefits of scaling down synthetic reactions to the microliter scale, such as extremely efficient mass and heat transport. Furthermore, temperatures can be controlled in a fast, accurate and reproducible way, enabling a higher selectivity in reactions. In addition, working with lower volumes is safer and more cost efficient.

To explore the possibilities of these new technologies, we started a collaboration with the Fraunhofer Institute for Microelectronic Systems in Duisburg (Germany) to jointly design and develop our own microreactor systems aimed at application in organic synthesis. This has resulted in a new plug-and-play automated microreactor system for chemical synthesis, which is currently routinely used by our group to perform approximately one hundred reactions per day in a single chip.

We have used this automated setup to study various catalytic reaction types such as selective protection of alcohol groups, biphasic Pd-catalyzed cross-coupling reactions (e.g. Sonogashira), oxidation reactions (e.g. Swern-Moffatt oxidations at room temperature), organocatalytic reactions (e.g. asymmetric Mannich reactions), but also biocatalytic reactions using crude cell lysates (e.g. enzymatic formation of cyanohydrins involving hydroxynitrile lyases). In all of these cases, minute amounts of catalyst and starting material were sufficient to generate a large number of data points allowing fast and efficient (i) reaction optimization, (ii) determination of scope and limitations of specific reactions, and (iii) determination of pH and temperature profiles of enzymatic reactions. In most cases, there appeared to be no significant differences between the continuous flow microscale setup and batchwise lab-scale experiments, implying that the automated microscale setup can be used to screen reaction conditions for subsequent synthetic application at large scale.

An Experimental Study on Nitration of Naphthalene Using Dynamically Rotating Axis Microreactor

Toshiyuki Ogura, Toshihiko Ohta, Yutaka Takahashi, and Kazihiro Mae
Kyoto research laboratory, The research association of micro chemical process technology, Kyotodaigaku-Katsura, Nishikyo-ku, Kyoto, 615-8530, Japan
Aichi works, Taketoyo plant, NOF corporation, 61-1 Kitakomatsudani, Taketoyo, Chita, Aichi, 470-2379, Japan
Division of Mechanical Engineering, Gradiate School of Engineering, Mie university, 1677 Kurima Machiya-cho, Tsu, Mie, 514-8507, Japan
Department of Chemical Engineering, Kyoto University, Kyotodaigaku-Katsura, Nishikyo-ku, Kyoto, 615-8510, Japan

1. Introduction

Microreactor for chemical synthesis is a promising tool which enables high reactivity, high selectivity, process safety, and so on. On the other hand improvement of productivity is essential for industrial production of chemical materials. Such improvement includes acceleration of reaction besides numbering-up or scaling-up of reactors. Area of interface between substrate and reagent is dominant factor to accelerate liquid-liquid multi-phase reaction such as nitration of aromatic compounds. In case of T-shape mixer strong mixing accelerate reaction at impinging point and retention tube. Even emulsions can be generated at higher flow rate, however, large pressure drop is caused by increasing flow rate and long; 1 to 10 m typically; retention tube. In addition, aggregation of emulsions decreases the effect of elongation of reaction tube.

In this study dynamically mixing microreactor was developed and utilized for nitration of naphthalene. Cylindrical rotor generates shared flow in microchannel and makes micro emulsion, which shorten retention time needed for high conversion rate. This fact can also decrease pressure drop of microreactor system.

2. Microreactor

The photograph of dynamically rotating axis microreactor (DRAM) is shown in Fig. 1. DRAM is composed by three divided components; rotation drive unit, mixing unit, and rotation sensing unit. A motor packed in rotation drive unit rotates a magnet settled on driving shaft. At the mixing unit made of PTFE and ceramic materials, liquid substrate impinges reagent and flow through the gap between inner and outer cylindrical surfaces. The inner cylinder which has a magnet within itself is magnetically coupled and rotates with driving shaft. Shared flow generated at the gap between cylinders brings about strong mixing of the liquids. Minimum gap width is 0.1 mm and total inner volume of the mixing unit is about 3 ml. Rotation speed of the cylinder can be monitored by magnetic sensing device packed in the rotation sensing unit.

3. Experimental

Nitration acid and naphthalene solved by dichlorobenzene were pumped to DRAM and mixed under room temperature. Mixed liquid was conducted to short retention time unit; 0.5 mm in I.D and 100 mm long, and recovered into plenty of distilled water in order to dilute acid concentration and quench reaction. The aromatic phase of the recovered liquid was separated and analyzed by gas chromatography. Nitration acid was the mixture of nitric acid and sulfuric acid. Molar ratio of nitric acid and naphthalene was 10. The degree of nitration reaction was investigated by changing rotation speed of the rotor and total flow rate between 0 and 4000 rpm, and 1and 20 ml/min, respectively.

4. Results and Discussions

The gas-chromatographs of recovered samples are shown in Fig. 2 obtained at the flow rate of 2 ml/min. Nitration of naphthalene proceeds even at the rotation speed of 0 rpm; a large amount of mononitronaphthalene (about 90% of initial material, naphthalene), dinitronaphthalene (1,5- and 1,8-derivatives, about 10%), and nitrated dichlorobenzene were detected. At the higher rotation speed micro emulsions were observed in the recovered liquid and the amount of dinitronaphthalene increased with the rotation speed. The molar ratio of 1,5-dinitronaphthalne to 1,8-dinitronaphthalene was about 0.5, and other dinitro derivatives were quite small (below 0.5%)

and trinitronaphthalene was not detected.

The gas-chromatograph yield of dinitronaphthalene is shown in Fig. 3 with the relationship between total flow rate and rotation speed. The value of yielding is no less than 80% at the small flow rate below 5 ml/min and moderate rotation speed around 1000 rpm. The maximum yield is greater than 90% at 2 ml/min and higher than 1000 rpm. On the other hand, yield gradually decreases below 20% with the increase of flow rate up to 20 ml/min because of the shortening of retention time at the cylinder. Scale-up of the cylindrical parts and increment of rotation speed will work effectively in higher flow rate.

For the case of T-shape mixer 10 m-long tubing was needed for 30% yield of dinitronaphthalene at 10 ml/min. DRAM connected with 100 mm-long tubing accomplishes 30% yield at 1000 rpm. The needless of long retention time unit realizes considerable decrement of pressure drop; about 10 MPa for T-shape with 10 m tubing, 0.06 MPa for DRAM with 0.1m tubing. This brings important advantage in industrial nitration process because of needless to introduce high pressure pump and rigid metal tubing which are not tolerable under corrosive condition.

Acknowledgement

This research was financially supported by the Project of "Development of Microspace and Nanospace Reaction Environment Technology for Functional Materials" of New Energy and Industrial Technology Development Organization (NEDO), Japan.

Fig.1 The photograph of Dynamically Rotating Axis Microreactor (DRAM).

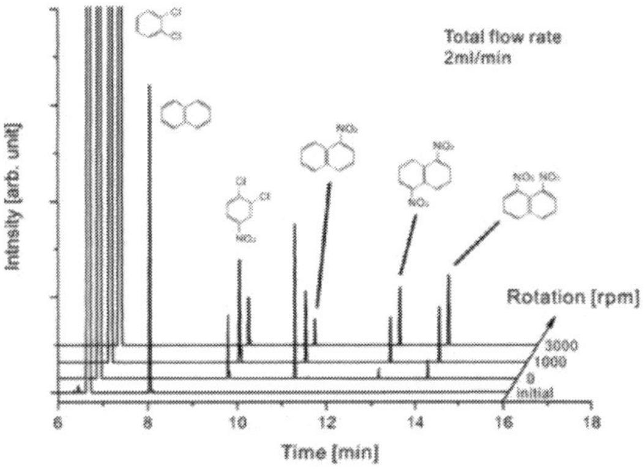

Fig. 2 Gas-chromatographs of nitro derivatives of naphthalene obtained using DRAM for the rotation speed between 0 and 3000 rpm.

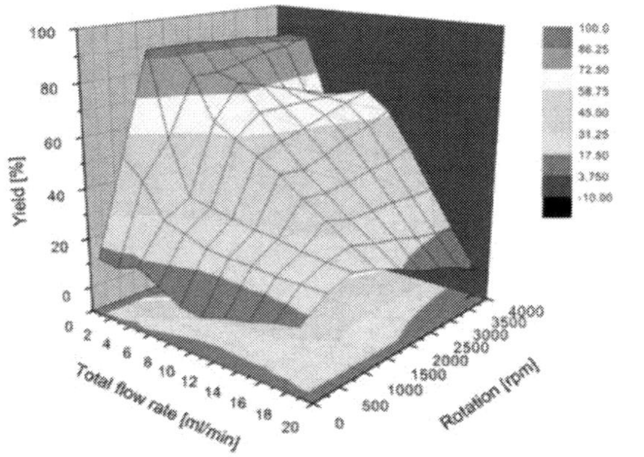

Fig. 3 The nitration property of DRAM; the relationship between total flow rate, rotation speed, and GC-yield of dinitronaphthalene.

103c Gucha-Gucha Extraction Microchip

Akihide Hibara[1], Ko Kasai[2], and Takehiko Kitamori[2]. (1) Institute of Industrial Science, The University of Tokyo, 4-6-1 Komaba, Meguro, Tokyo, 153-8505, Japan, (2) Department of Applied Chemistry, The University of Tokyo, 7-3-1 Hongo, Bunkyo, Tokyo, 113-8656, Japan

Introductions:

Chemical processes utilizing microfluidic systems have been investigated intensively. We have investigated fast solvent extraction process utilizing parallel two-phase flows, which have high specific interface area. By utilizing the two-phase flow, we have demonstrated cobalt ion analysis [1], phase transfer catalysis reaction [2], liquid membrane transport [3], interfacial polymerization for construction of polymer membrane [4], counter-current extraction [5] and so on.

In order to extend applicability of this technique, we have developed a novel extraction microchip having droplet formation, coagulation, and phase-separation parts. In this microchip, organic phase forms droplets in aqueous phase and, therefore, higher specific interface area is expected. Since the two-phase forms disordered state at the first part, which is called as "gucha-gucha" in Japanese, we named the microchip as "gucha-gucha extraction microchip".

Experimental:

Microchannels were fabricated on glass substrates by ordinary lithographic wet-etching technique. In order to make two different depths, two-step etching method was applied [6]. The microchannels were fabricated on a bottom substrate and liquid inlets were fabricated on a top substrate. There two substrates were thermally bonded. Organic and aqueous solutions were driven by syringe pumps and fed into the microchannel via polymer capillary and some connection parts.

Results and discussion:

The gucha-gucha extraction microchip consists of three parts; (1) droplet formation, (2) coagulation of the droplets, and (3) two-phase separation. For the droplet formation part, we designed Y-shape configuration. Two inlet microchannels having a width of 250 μm and depth of 100 μm were connected by shallow Y-shaped microchannel having a width of 50 μm and depth of 10 μm. The output of the shallow microchannel was connected to a microchannel having a width of 250 μm and depth of 100 μm. In order to form oil-in-water (O/W) droplets, the shallow part and output microchannel were washed by sodium hydroxide solution to obtain clean hydrophilic surface. At the output of the shallow channel, organic phase was divided to form droplets. When hexane and water was introduced with 1 μl/min flow rate, droplets having average diameter of 43 μm was formed.

In the second part, the droplets were coagulated to form plug or parallel two-phase flow. From the droplet formation part to the coagulation part, the droplets were flowed in the hydrophilic microchannel. In order to coagulate the organic droplets, the microchannel surface was modified with hydrophobic octadecylsilane group. It the hydrophobic microchannel, the droplet adsorbed to the microchannel wall and the following droplets attached to the adsorbed organic phase. By repeating these processes, the droplet flow was converted to plug or parallel flow.

In the third part, the organic and aqueous phases were separated by utilizing hydrophilic-hydrophobic patterning microchannel. In our previous paper, gas-liquid separation by utilizing the hydrophilic-hydrophobic patterning, where a microchannel having deep and shallow parts was used and only the shallow part was modified with hydrophobic group utilizing capillarity [6]. By utilizing the patterned structure, gas bubbles were purged to the shallow part of the microchannel completely. In the present study, we tried to use the similar structure, but the structure did not work very well. In that structure, viscosity of the organic phase in the shallow part affected the phase separation, which could be neglected in gas bubble purge. Therefore, we connected a deep microchannel to the shallow part. In brief, hydrophilic deep channel (sample flow) was connected to hydrophilic shallow channel (separation part), which connected to hydrophobic deep channel (separated organic flow). This structure worked very well. The organic and aqueous phases in a microchannel having a width of 250 µm and depth of 100 µm could be separated with very high flow rate up to 70 µl/min.

After investigating these three unit operations, these operations were integrated into a single microchip and we have confirmed the flow control as we designed. In order to demonstrate effectiveness of our gucha-gucha extraction microchip, cobalt complex was extracted from aqueous phase to organic phase.

[1] "Continuous-flow chemical processing on a microchip by combining microunit operations and a multiphase flow network", M. Tokeshi, T. Minagawa, K. Uchiyama, A. Hibara, K. Sato, H. Hisamoto and T. Kitamori, Analytical Chemistry 74(7), 1565-1571 (2002).

[2] "Fast and high conversion phase-transfer synthesis exploiting the liquid-liquid interface formed in a microchannel chip", H. Hisamoto, T. Saito, M. Tokeshi, A. Hibara and T. Kitamori, Chemical Communications (24), 2662-2663 (2001).

[3] "Three-layer flow membrane system on a microchip for investigation of molecular transport", M. Surmeian, M. N. Slyadnev, H. Hisamoto, A. Hibara, K. Uchiyama and T. Kitamori, Analytical Chemistry 74(9), 2014-2020 (2002).

[4] "Chemicofunctional membrane for integrated chemical processes on a microchip", H. Hisamoto, Y. Shimizu, K. Uchiyama, M. Tokeshi, Y. Kikutani, A. Hibara and T. Kitamori, Analytical Chemistry 75(2), 350-354 (2003).

[5] "Countercurrent laminar microflow for highly efficient solvent extraction", A. Aota, M. Nonaka, A. Hibara, and T. Kitamori, Angewandte Chemie International Edition, 46(6), 878-880 (2007).

[6] "Surface modification method of microchannels for gas-liquid two-phase flow in microchips", A. Hibara, S. Iwayama, S. Matsuoka, M. Ueno, Y. Kikutani, M. Tokeshi and T. Kitamori, Analytical Chemistry 77(3), 943-947 (2005).

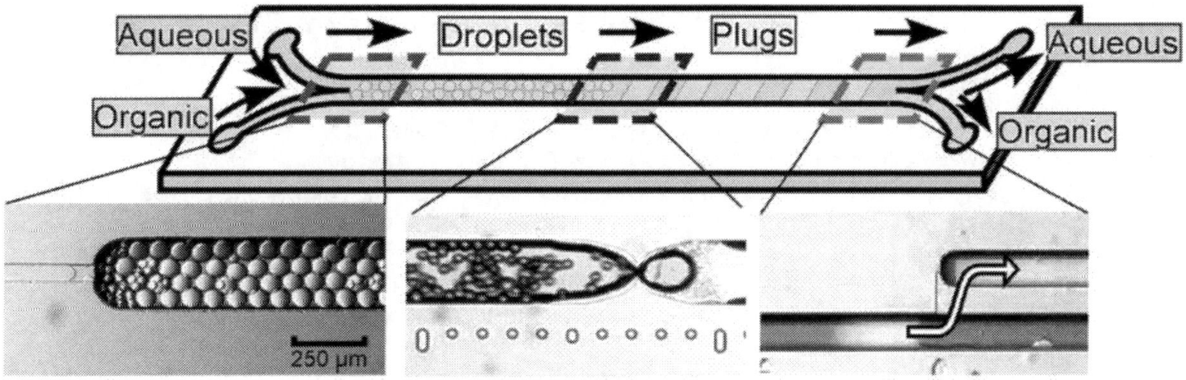
Aqueous
Droplets
Plugs
Aqueous
Organic
Organic
250 µm

46f Multiple Purpose Micro Reaction Plants for Development and Production

T. Dietrich, Andreas Freitag, Ralf Scholz, and Stefan Link. mikroglas chemtech GmbH, Galileo-Galilei-Str. 28, Mainz, D-55129, Germany

Since more than 15 years micro reactors have been developed. It turned out that Micro Reaction Technology (MRT) is a perfect tool for Process Intensification (PI), to characterize chemical reactions (kinetics, thermodynamics), to develop new chemical reactions and optimize conventional reactions, and to produce in a sustainable way: safer and with less reagents, less waste, less energy consumption. Even though these advantages are accepted by scientists, the technology is still not used broadly in the chemical industry. The following talk will explain a new concept of a modular micro chemical plant to use these advantages. This new concept facilitates the multi-purpose use of a micro reaction plant and shows the economical benefits. The described plants have been designed and built to carry out different chemical reactions (gas-liquid or liquid-liquid, at low or higher temperatures, at different pressures, at different flows, one or more step reactions, etc.). For this, the plant consists out of a "base plate" and different "reaction plates". The "base plate" includes all necessary "peripheric" equipment, like pumps, mass flow controllers, sensors, filters, temperature control circuits, electronic control unit, etc. The electronic hardware for controlling is directly combined with these components to facilitate any changes. With these functions the "reaction plates" can be used easily. They include the reactors, sensors and components, which are specific for a given reaction. By connecting the "reaction plate" to the "base plate", the electronic control unit automatically realizes which pumps, sensors and feeding lines are required and starts the required software. With this set-up it is very simple to change from one reaction to another within minutes, run a cleaning cycle for several minutes and start the new reaction. (Fig 1 and 2) Besides changing modules, it is also important to vary parameters within one set-up. For this purpose different tools are implemented: - The electronic control unit allows to program a combination of different experiments with different flow, temperatures, and pressures. - To vary independently flow and residence time, an adjustable residence time module has been developed which allows to use residence times between 1s and several hours. - An autosampler can be integrated to collect product fractions, produced under different conditions. - IR-measurements or pH- and viscosity sensors can be integrated to have an online-analysis of the chemical reactions. After development of a certain reaction, it is possible to use the same plant for production up to several kg per day. With small changes (larger pump head, larger reaction device) or a parallelization of up to 10 reactors, it is possible to achieve up to 1 ton per day (depending on the chemistry). When increasing the production volume, safety might become an issue. Even though micro reactors are very "safe" (low internal volume !) different safety feature have been included in the multi-purpose plants: pressure relief valve, check valves, leakage sensors, software controlled limits for temperature and pressure, etc. Examples will be shown. The micro fluidic devices used in this work are made out of glass, because of its ideal properties, like chemical stability or optical transparency. For the same reason, all modules to run a glass micro reactor, e.g. pumps, valves, tubings, temperature-, pressure-sensors, analytical modules, and safety installations use materials with the same chemical stability. In a conventional chemical production the apparatus is fixed and built out of standard equipment. The chemistry is forced to work in this reactor. In contrast, micro reactors give the opportunity to build a fluidic system which is optimized for the chemical reaction ! Therefore, a flexible material and structurisation process is needed. At mikroglas mainly the photo etchable glass FOTURAN is used. With this material micro reactors, mixers, heat exchangers, dwell devices, and analytical modules have been built. Even very complicated three-dimensional structures can be manufactured by etching different geometries into different layers and bond up to 20 of these layers together to form a micro reactor. Other materials (like stainless steel or titanium) can also be used if required, e.g. for higher pressure. Different examples will be shown during the presentation which show the advantages of the applied technology, e.g. two-phase systems, catalytic reactions, polymerizations, photo reactions, ozonizations, hydrobrominations, etc. The authors like to thank the German Ministry for Education and Research for funding parts of this work and the partners within the project ZOHIR (Bayer Schering Pharma AG, Leibniz-Institute for Catalysis, infrared fiber sensors) for their extensive help. Fig.1 Micro reaction equipment for multipurpose use Fig.2 Flow diagram for a fluorination reaction (screen shot of the SIMATIC control unit)

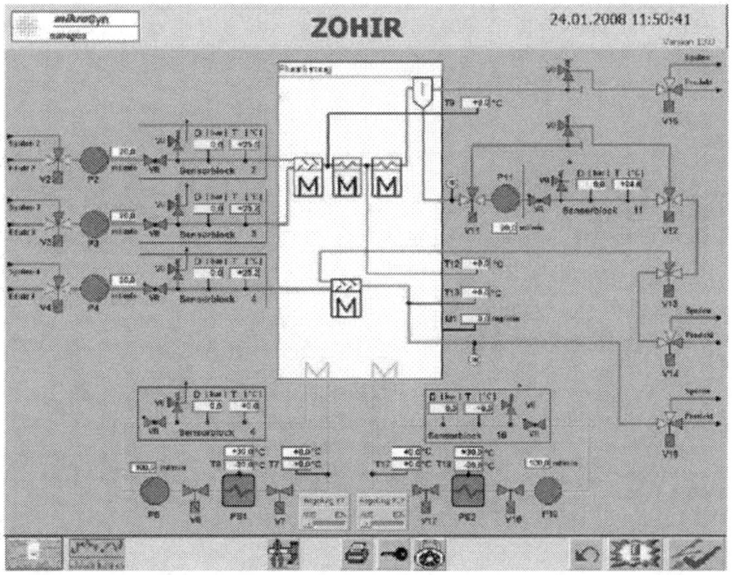

64a Development and Evaluation of Microreaction Experiments and Implementation into University Teaching

Johannes Zimmermann, Bernd Ondruschka, and Annegret Stark. Institute for Technical Chemistry and Environmental Chemistry, Friedrich-Schiller University Jena, Lessingstr. 12, Jena, 07743, Germany

Microreaction technology is the focus of both, small- and middle-sized businesses and research institutes in Germany. In particular the Freestate of Thuringia has developed a leading position. At the Friedrich-Schiller-University of Jena, relevant contents have already been integrated into educational curricula of chemistry and environmental chemistry degree programs. However, due to the very fast development of technological aspects, there is a lack of university experiments and lectures aiming to familiarize students with the advantages and potentials of microreaction technology. Hence, the demands of today's industry with regard to manpower are not answered by the availability of graduates with solid experiences in the microreaction sector. The increasing industrial interests in microreaction technology for continuous production processes (e.g. fine chemicals, pharmaceuticals) has to be reflected in academia by gradually implementing this area of science into university curricula.

Therefore, the “Middle-German Cluster for Microreaction Technology”, a cooperation between the universities of Jena, Ilmenau and Chemnitz, was founded with the financial assistance of the “Federal Ministry of Education and Research” (BMBF) in the context of the larger funding program “Microsystems (2004-2009)”. The main aim of the cluster is to design six microreaction experiments, which can be handled with one consistent periphery of sensor and actuating elements (UV/Vis-spectrometers, syringe pumps). In the center of this device, the six different microreaction experiments can be exchangeably integrated. Hence, every institution will possess a set of six different transportable and well-established microreaction experiments for student education. On the one hand this will generate a very high benefit for the standardization of microreaction education and curricula. On the other hand, it will enhance the number of graduates in microreaction education and their potential to find employment in industrial firms and, therefore, close the gap between actual and desirable market conditions.

For this reason, two microreaction liquid-phase experiments were developed within the above-mentioned cluster with the aim of teaching fluidic phenomena in dependence of residence time, pressure loss (energy dissipation) and mixing quality in dimensionally different microstructured T-mixers. The first experiment deals with the formation of red iron(III)-thiocyanate, which is quantified using a stereo-microscope on the basis of digitalized color detection. The second experiment deals with the kinetics of ionic liquid formation (direct synthesis of symmetric and asymmetric 1,3-disubstituted imidazolium salts) using online UV-Vis analysis. Besides the mixing quality at different flow-rates the process suitability for numbering-up is evaluated.

Since there is no valid and reliable data on the benefit and the learning effects of microreaction experiments on students we have devised a pre-post-follow-up-design including different groups of students into the sample to evaluate the implementation of the above experiments from a didactic point of view. The students will be tested at three points in time applying quantitative as well as qualitative methods (semi-structured interviews, questionnaires, participating observations as well as video documentation).

Overall, the project has the following goals:

• Development of competences of students in the field of microreaction technology.

• Know-how in subject- and structure-optimized experimental content by the instructional personnel.

• Improvement of job opportunities of graduates.

• Provision of well-qualified personnel to small- and medium-sized businesses in the field of microreaction technology.

198g Novel Routes to Fluidized Bed Process Intensification Using a Centrifugal Field

Juray De Wilde, Materials and Process Engineering Dept. (IMAP), Universite catholique de Louvain (UCL), Reaumur, Place Sainte Barbe, 2, Louvain-la-Neuve, B - 1348, Belgium

The operating conditions in conventional, that is, gravitational fluidized bed reactors are determined and limited by the use of earth gravity. Internal mass and heat transfer limitations can be encountered by the impossibility to fluidize smaller size particles, the van der Waals forces becoming too important compared to the weight of the particle bed in the earth gravity field and the counteracting drag force. External mass and heat transfer limitations can be encountered by the impossibility to work at higher gas-solid slip velocities. With heterogeneous (catalytic) reactions, as well internal as external mass or heat transfer limitations may impose limitations on reaction rates and, for example, the activity of the catalyst.

Fluidization in a centrifugal field may allow overcoming the above mentioned limitations and may, as such, allow fluidized bed process intensification. Two novel routes to fluidization in a centrifugal field are presented and their advantages discussed and illustrated.

In rotating fluidized beds in a static geometry, the centrifugal field is generated by injecting the fluidization gas tangentially in the fluidization chamber via multiple gas inlet slots in its outer cylindrical wall. A combined tangential-radial fluidization of the particle bed is obtained by forcing the fluidization gas to leave the fluidization chamber via a centrally positioned chimney. The fluidization gas flow rate influencing both the centrifugal force and the counteracting radial gas-solid drag force in a similar way renders rotating fluidized beds in a static geometry extremely flexible with respect to the fluidization gas flow rate and the gas-solid contact time. In particular, dense operation at high fluidization gas velocities is possible, allowing fast and highly endothermic or exothermic reactions to be carried out in a rotating fluidized bed in a static geometry of much smaller volume than when using conventional fluidized bed technology.

In rotating fluidized beds around a rotating chimney, the centrifugal field is generated by a rotating chimney consisting of multiple blades and positioned centrally in the fluidization chamber. The fluidization gas being forced to leave the fluidization chamber via the chimney, a radial gas-solid drag force counteracting the centrifugal force is generated. Furthermore, both the fluidization gas and the particles obtain a tangential velocity by the action of the rotating chimney. This results in a combined tangential-radial fluidization of the particle bed.

The tangential injection of the fluidization gas in the fluidization chamber via multiple gas inlet slots in its outer cylindrical wall and the centrally positioned rotating chimney can eventually be combined to obtain extremely flexible fluidization technology. In such case, the rotating chimney allows to increase locally, that is, in the vicinity of the chimney, the centrifugal force. Solids losses via the chimney can, as such, be drastically reduced. This is shown to be particularly advantageous when fluidizing small, micro- or nano-scale particles.

Experimental and numerical data on the fluidization behavior with different types of particles, the pressure drop over the fluidization chamber and the rotating chimney, the gas-solid mass and heat transfer characteristics in the fluidization chamber and the solids losses via the chimney are presented and discussed. The advantages of fluidization in a centrifugal field using the novel technologies presented are illustrated with potential applications.

102f Mass Transfer Characterization Inside Reactive Droplets Flowing Throw Microchannels

Flavie Sarrazin, LOF Laboratoire du Futur - Pessac - France, 178 avenue du docteur Schweitzer, Pessac, 33608, France

In the pharmaceutical and fine chemicals industries, process fast design is a determining step for economical viability. To reduce this "time to market" constraint, the product development phase can be reduced by improving the initial step concerned with physical and chemical data acquisition. We are particularly interested in determining chemical kinetics and mass transfer parameters. To do so, a very promising application of microtechnologies is the use of two-phase liquid-liquid flow in microchannels. In this kind of system, microdroplets of reagents are transported at constant velocity in an inert continuous phase so that they can be considered as individual nanovolume batch reactors. One of the determining parameters of this device for kinetics acquisition purpose is the mixing time inside the droplets. That is why we first focused on hydrodynamic characterisation of the droplet inner stream loops by using numerical simulations and micro PIV experiments. Then, we showed that a chemical reaction can be performed inside the droplets, and monitored with high sensibility thanks to Raman micro spectrometry. Fist, we focused on recirculation time inside the droplets. We performed numerical simulations and experimental validation of the hydrodynamics inside droplets in rectangular microchannels. We use a finite-volume/front-capturing method that allows us to perform two- and three-dimensional simulations with a reasonable cost. The numerical method is an interface-capturing technique without any interface reconstruction. Therefore no complex or expensive interface tracking is needed. Furthermore, micro-PIV measurements are used to obtain experimental velocity fields inside droplets which are compared to simulations. As it is shown on figure 1, the two approaches are in good agreement. Thanks to numerical simulations, droplet interface deformation and velocity fields inside both droplets and continuous phase can then be followed. This study leads to important results in droplet deformation and velocity fields in the droplets and in the continuous phase for mass and heat transfer studies. More particularly, the capillary number seems to have a great influence on the liquid/liquid flow hydrodynamics and no to depend on the channel scale. Secondly, we investigate the use of confocal Raman spectroscopy to obtain spatially resolved concentration maps of chemically reactive fluids flowing in two-phase channel flow. As a model chemical reaction, we use the water deuterium exchange reaction between H2O and D2O which is diffusion-controlled and whose reagents and products can be quantified depending on different the Raman signature. We are able to extract Raman images of H2O, D2O and HOD concentrations in the droplets, and we evidence the influence of wiggles on mixing efficiency inside the droplets.

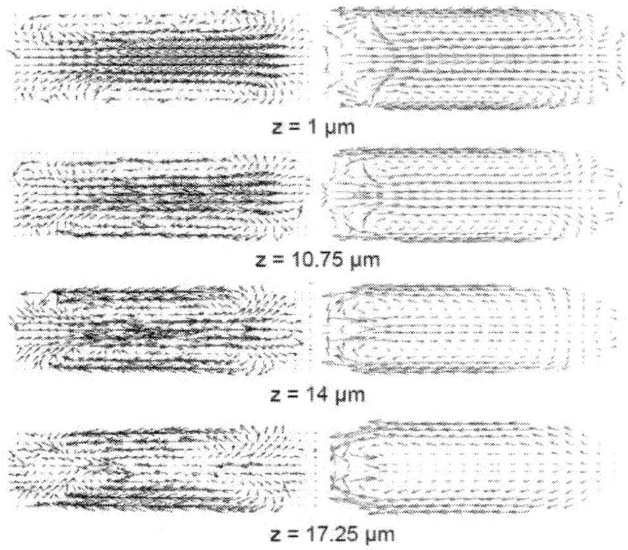

Figure 1: Inner droplet hydrodynamics: comparison between numerical simulation and microPIV acquisition in microchannel of $50 \times 60\ \mu m^2$ section at different distances z from the bottom interface of the droplet.

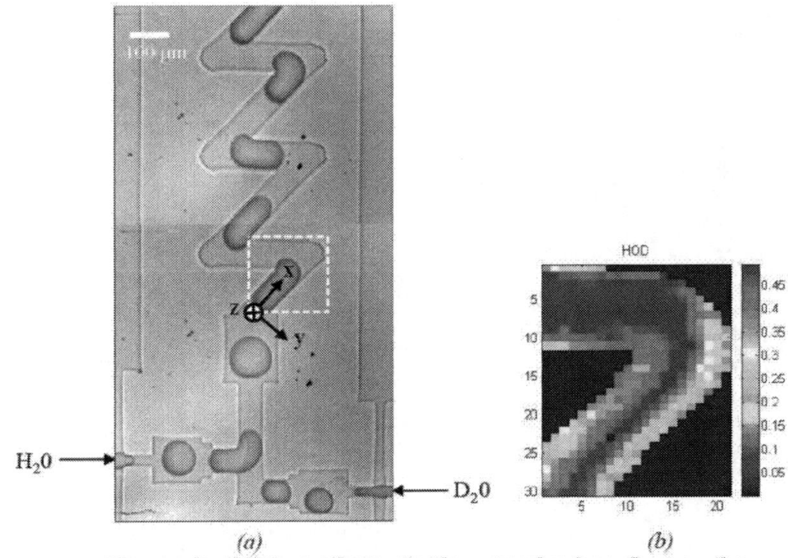

Figure 2: HOD synthesis inside microdroplets flowing throw serpentine microchannels. (a) Reactants meet thanks to coalescence of the droplets and homogenenization in enhanced thanks to bends placed along the flow. (b) Micro Raman spectrometry map of HOD concentration inside the first crossed bend (white square on picture a).

241d Porous Microreactors for Gas-Liquid Contacting

Rob G. H. Lammertink, Jorrit de Jong, Can Aran, David Salamon, Leon Lefferts, and Matthias Wessling. Science and Technology, Mesa+ Research Institute, University of Twente, Enschede, NL-7500AE, Netherlands

The concept for exploiting porous layers (membranes) within microstructured reactors is investigated. A porous structure can be utilized for several functions, including gas-liquid contacting. For this, the porous layer needs to be non-wetting for the liquid phase in order to reduce mass transport limitations. During contacting, the gas phase can diffuse easily through the porous matrix, while the liquid is contained inside the microfluidic channels. The approach for gas-liquid contacting in microfluidic channels is demonstrated by detecting simple carbon dioxide gas dissolution. By measuring the conductivity of the solution dynamically, mass transport kinetics can be obtained. The experimental results were compared to numerical simulations. This concept is demonstrated for both polymeric as well as ceramic microreactors. Ceramic microreactors can be fabricated by a micromolding method followed by a polymer binder burnout and sinter treatment. The resulting alumina structures are open porous with sub micrometer pore sizes. They can be easily modified by using silane chemistry to obtain the desired surface properties.

209c Integration of Microfluidic and Planar Waveguide for Spectroscopic Monitoring of Chemical Reactions

L. Jiang[1], K. P. Gerharadt[2], B. Myer[1], Y. Zohar[2], and **S. Pau**[1]. (1) College of Optical Sciences, University of Arizona, 1630 East University Blvd, Tucson, AZ 85721, (2) Aerospace and Mechanical Engineering, University of Arizona, P. O. Box 210119, Tucson, AZ 85721

Spectroscopy is a powerful analytical tool in studying and detecting chemical bindings on surface. We perform evanescent wave spectroscopy using a waveguide with a solid core and a liquid cladding layer to probe chemical reactions on the waveguide core-cladding interface. Unlike the conventional luminescence detection principles based on confocal microscopy, where light is focused onto the sample, evanescent field excitation utilizes exponentially decaying fields, within a distance of wavelength of the guided modes on the surface of a waveguide. The signal-to-noise ratio is improved by probing only the labeled analyte molecules bound to recognition elements, such as antibodies, oligonucleotides and bioreceptors on the waveguide surface, without background signals from impurity and unbound fluorophores outside the evanescent field. The geometry of the evanescent field waveguide provides good discrimination between bound and unbounded analyte in a molecular level and allows the analysis of both scattering and absorbing samples such as blood, serum or cell extracts, which would otherwise strongly attenuate the excitation light beam and build up spurious signals. In our experiments, both absorption and photoluminescence spectrum are measured, allowing spectroscopic monitoring of antibodies-antigen binding activities. Our device integrates multimode waveguide and microfluidic channel on a silicon substrate. Results from interaction of Cy3-conjugated antigens and anti-mouse IgG interactions will be presented.

Flow reactors for chemical synthesis: An opportunity for process control

Paul Watts

Department of Chemistry, University of Hull, Cottingham Road, Hull, HU6 7RX.

Tel: +44 (0)1482 465471; Fax: +44 (0)1482 466410; Email: P.Watts@hull.ac.uk

Until recently, the greatest research effort in the field of miniaturisation has been in analytical science, with the main aim of this research being to develop so called miniaturised Total Analytical System (μ-TAS).[1] Alongside the continuing development of μ-TAS, a concerted effort has also been underway to establish the benefits that micro reactors can bring to the field of synthetic organic chemistry. In their simplest form, micro reactors consist of a network of micron-sized channels (typically 10-300 μm in diameter) etched into a solid substrate. A number of materials such as silicon, quartz, glass, metals and polymers have been used to construct micro reactors,[2] however important considerations in material choice include chemical compatibility, ease of fabrication and compatibility with detection methods. Glass[3] is a popular choice for synthetic applications since it is chemically inert, enables the use of a range of visible light detection methods and fabrication techniques are well established.

For solution-based chemistry, the channel networks are connected to a series of reservoirs containing chemical reagents to form the complete device. In the micro reactor, reagents can be brought together in a specific sequence, mixed and allowed to react for a specified time in a controlled region of the channel network using electrokinetic or hydrodynamic pumping.[4] Many chemical reactions have now been demonstrated to show improved reactivity, product yield and selectivity when performed in micro reactors compared to those generated using conventional laboratory practices.[4-6] The evidence is that the unique *modus operandi* of micro reactors, namely the low-volume spatial and temporal control of reactants and products in a diffusive mixing environment, in which distinctive thermal and concentration gradients exist, offers a novel method for chemical manipulation and generation of products. Reactions performed in a micro reactor invariably generate comparatively pure products in high yield, when compared to the equivalent bulk reactions, in much shorter times. One of the immediate and obvious applications is therefore in drug and process discovery, where the generation of compounds either with different reagents or under variable conditions is an essential factor.

Performing chemical reactions within a microfluidic system also provides the opportunity to perform real time separations. Hence, integration of a micro reactor device *via* a purification/separation step, with one of the many highly sensitive microchannel-based biological assay systems would provide a drug discovery tool. This level of integrated functionality within one device clearly addresses some of industries' requirements for rapid compound production and screening. Apart from the greatly reduced reaction times demonstrated for the micro reactors, handling times to assay and chemical reagent costs are virtually eliminated with this proposed technology.

The past decade has provided overwhelming evidence that continuous flow methodology is a powerful tool for reaction optimisation; unfortunately however, it is analysis of the samples generated by such systems that can cause a bottleneck, leading to reduced efficiency of the overall system. As a result, sampling rates are frequently reduced and data from these information rich systems is lost. With this in mind, Trapp[7] recently communicated a method that enables the on-line analysis of up to 453 samples hr^{-1} using a single chromatographic column. The technique is based on the fact that the majority of analysis time is spent recording detector noise therefore by increasing the duty cycle of a system, the amount of information can be maximised, whilst minimising analysis time. To enable the rapid, reproducible injection of multiple samples, the authors constructed a six-channel continuous flow split/splitless multiplexing injector, mounted onto a standard split injector of a GC. The samples are then continuously evaporated in the sample ports and injected by short pressure pulses, which compared to liquid injections leads to short sample plugs and again increased separation efficiency. Such a technique clearly has advantages for the monitoring of continuous flow processes, having the potential to remove the lag time currently experienced between sample generation and off-line analysis.

Alternatively, on-line methods can be employed to monitor reactions systems, to date these have included techniques such as electrochemical,[8] fluorescence[9] and the incorporation of evanescent wave-guides.[10] At present however the majority of monitoring takes place off-line using spectroscopic techniques such as HPLC, GC, MS and NMR, developments in the field of flow reactor to MS hyphenation however has the potential to revolutionise the optimisation and monitoring of flow reactions.

In this presentation a number of chemical reactions of industrial interest will be used to illustrate the advantages that flow reactors offer for the rapid optimisation of reactions, in which the product is typically produced in both higher yield and purity. However with regard to the optimisation process the slowest step remains the analysis of the reaction mixtures using traditional laboratory techniques such as gas chromatography (GC) and high performance liquid chromatography (HPLC). This presentation will explore the need for on-line, real time measurement techniques to enable reactions to be optimised more rapidly. Furthermore, the integration of the system with real-time analytical detection will enable *in situ* process control to be achieved.

References

1. Oosterbroek RE, van den Berg A (2003) (eds) Lab-on-a-Chip: Miniaturised Systems for (Bio)Chemical Analysis and Synthesis. Elsevier, Amsterdam

2 Ehrfeld W, Hessel V, Löwe H (2000) Microreactors: New Technology for Modern Chemistry. Wiley-VCH

3 McCreedy T (2000) TrAC 19; 396

4 Fletcher PDI, Haswell SJ, Pombo-Villar E, Warrington BH, Watts P, Wong SYF, Zhang X (2002) Tetrahedron 58: 4735

5. Jahnisch K, Hessel V, Lowe H, Baerns M (2004) Angew. Chem. Int. Ed 43: 406

6 Schwalbe T, Autze V, Wille G (2002) Chimia 56: 636

7 Trapp O (2007) Angew. Chem. Int. Ed 43: 406:

8 Zhan W, Alvarez J, Crooks RM (2003) Anal. Chem 75: 313

9 Gosch M, Blom H, Holm J, Heino T, Rigler R (2000) Anal. Chem 72: 3260

10 Pandraud G, Koster TM, Gui C, Dijkstra M, Van den Berg A, Lambeck PV (2000) Sens. Actuators A 85: 58

Studies for a Catalytically Heated Hexadecane Steam Reformer

J. Thormann[1], P. Pfeifer[1], U. Kunz[2] and K. Schubert[1]
[1]Institute for Micro Process Engineering, Forschungszentrum Karlsruhe GmbH, Germany
[2]Institute of Chemical Process Engineering, Clausthal University of Technology, Germany

Introduction

The reforming of diesel fuels for hydrogen generation for fuel cell systems is advantageous due to the common availability of diesel fuel, the possible application in auxiliary power unit generators and the still growing market segment of diesel cars in Europe. Microchannel reactors offer high heat and mass transfer coefficients and are thus able to cope with sudden hydrogen demand, i.e. fast start-up and load changes in systems with a fluctuating power burden.

For this study the surrogate diesel fuel hexadecane was chosen. We investigated the kinetics of the hexadecane reforming and of the consecutive water gas shift reaction by applying an electrically heated microstructured test reactor. The microchannels were coated with a ceria support layer using a sol-gel method and were impregnated with a rhodium catalyst providing good catalyst lifetime and superior coking tolerance compared to our previous results [1, 2]. Based on the results obtained in the electrically heated system, a prototype of a microstructured reactor heated by catalytic combustion (e.g. of anode off-gas of a fuel cell) was built. The system (shown in Figure 1 below) was designed to operate in a self-sustaining mode, i.e. to deliver heat for the endothermic steam reforming (at 800°C) and to balance the heat loss. Its specific operation behaviour was investigated within the present study.

Experimental Methods

The reformer part of the system reactor contains 120 foils with straight microchannels (Length 85mm, width 320µm, depth 110µm), which are coated with a Rh/CeO_2 catalyst system. Adjacent to these foils are located the foils where the combustion air enters, and the catalytic combustion of anode off gases, represented here by hydrogen, takes place. The anode of gas (or hydrogen) is fed along the reaction zone to the microchannels of the combustion foils (30 foils with s-shaped channels, straight length

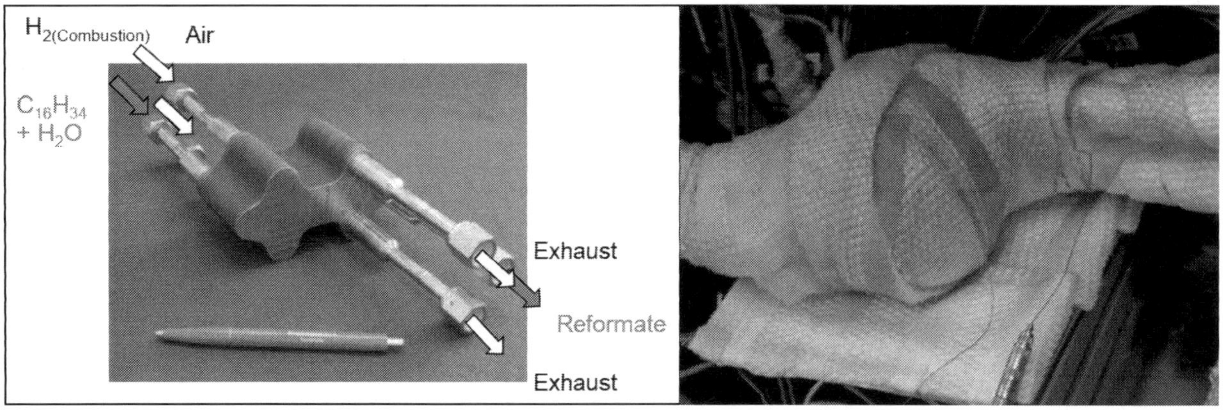

Fig. 1: Hexadecane steam reformer design and reactor under operation (catalytically heated up to 800°C operation temperature)

30mm, width 320µm, depth 110µm). As catalyst for the combustion, a mixture of Rh/Pd was applied on the micro channels by incipient wetness impregnation. The temperature of the system reactor was measured by two thermocouples placed in the middle of the foil stack.

For the complete gas analysis, an Agilent GC 6890 gas chromatograph with a flame ionization detector and a thermal conductivity detector was used to quantify the concentrations of hydrogen, oxygen, carbon monoxide, carbon dioxide, nitrogen, water and hydrocarbons up to C_{16} molecules. In order to investigate changes in the gas composition during dynamic operation conditions, a chemical ionization mass spectrometer (CI-MS, Manufacturer V+F) was used. It consists of an inertizer unit for delivering a constant sample flow, an Airsense 2000 unit for the ionization of components larger than 4 amu, and an H-sense 2000 to analyze the hydrogen.

All the experiments presented in this study were performed with the diesel surrogate hexadecane in order to perform steam reforming with a pure fuel because the operation conditions and the kinetics have been optimized in an electrically heated test reactor for this fuel earlier.

Experimental Results

The steady state operation of the system reactor was investigated before switching to dynamic conditions. The results from a steady state operation are presented in Figure 2. It can be seen that the catalytic combustion of $3.5 l_N$/min hydrogen with an air flow of $10 l_N$/min (λ = 1.2) maintains an operating temperature of 800°C when hexadecane is reformed with $1.5 l_N$/min steam according to a molar steam to carbon ratio of 4. The composition of the product gas was determined on the basis of the water and nitrogen free product flow. The reformate consisted of 75 vol% hydrogen, 18 vol% carbon dioxide, 6.5 vol% carbon monoxide and only traces of methane. The measurements done by the gas chromatograph confirmed a hexadecane conversion of 100%. The reforming was very stable and neither catalyst deactivation on the reforming nor on the combustion side of the reactor could be observed during five hours of operation in total.

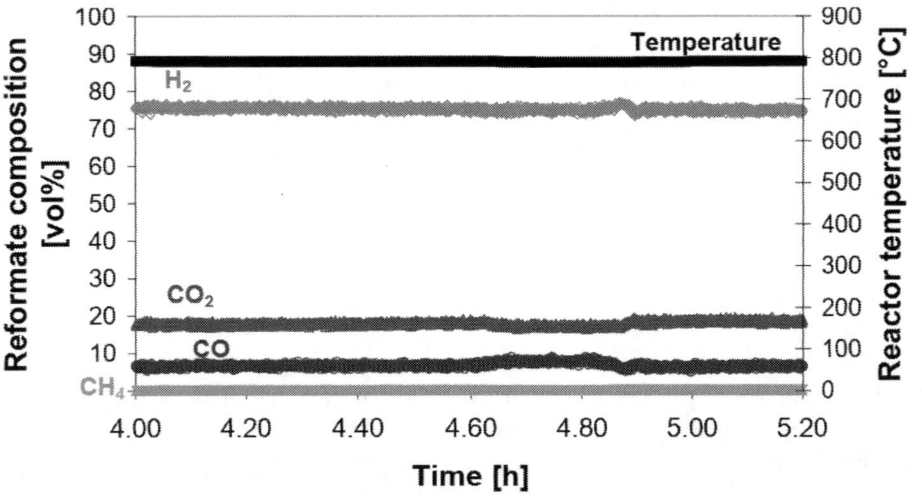

Fig. 2: Product gas composition of the system reactor during stationary operation (heated by catalytic combustion of $3.5 l_N$/min hydrogen with an air flow of $10 l_N$/min)

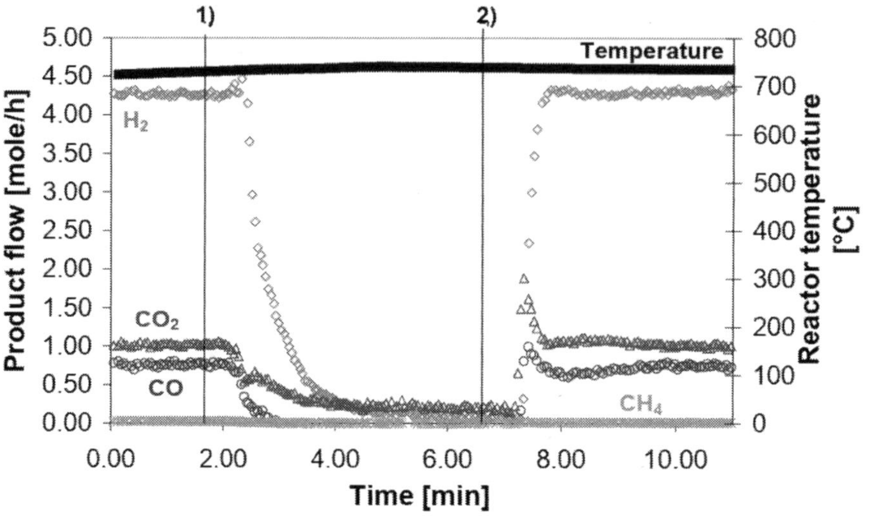

Fig. 3: Reforming behaviour while 1) switching from full load to zero load operation and 2) vice versa; Hexadecane reforming with an S/C ratio of 4 and catalytic combustion of $2.6 l_N$/min hydrogen with an air flow of $10 l_N$/min

To examine load changes and the start of the reforming operation, the reactant flow was switched from the reactor to the bypass line and back (See Figure 3, point of time 1 and 2). The reformate concentrations of hydrogen, carbon monoxide and carbon dioxide decreased considerably fast during switching to bypass (point of time 1). However, due to approx. 3 meters of tube between reactor and analyzer, there is an initial time delay for each component, before a real change can be detected (36s for hydrogen and 27s for CO_2 as well as CO). After this time delay the hydrogen decreased to 10% of the original level within 64s, CO_2 within 73s and CO within 29s. A major evidence which can also be observed from Figure 3 is, that the reactor temperature cannot follow the fast load change because it does not vary significantly within the time scale of four minutes. A long term experiment (not shown) revealed that, when the reforming stream is switched off and the catalytic combustion is maintained without any changes, the reactor temperature increases within 2 hours from 733°C to 815°C (no demand for heat for the endothermic reforming reaction and for the heating of the reactant gases). At the point of time 2 in Figure 3, where the reactant flow was switched back through the reforming channels, it can be calculated by subtracting the initial time delay of analytics that the reactor needs 32 to 36s to reach 90% of the final concentrations and to feed a fuel cell with hydrogen. The concentration of methane stayed at a low level, not largely influenced by the load switching.

Conclusion

The reforming of hexadecane fuel can be performed in a microstructured reactor which is heated by the catalytic combustion of anode off-gases up to 800°C. The catalyst system Rh/CeO_2 realizes high reforming conversion and a reformate with high hydrogen concentration (around 75 vol%) without any sign of deactivation. The dynamic experiments revealed that the reforming reaction establishes this high and constant hydrogen flow already within a short period of half a minute after the start of the reforming. Under the given circumstances of the experiments a high flow of hydrogen (representing the anode off gases of a fuel cell) was needed for the catalytic combustion.

An analysis of the heat demands for the heating up of the reforming gases and the endothermic reaction confirms a large heat loss to the surroundings (five times higher than the necessary heat for the reforming). Thus, for the concept of a reforming system the thermal management must be considered especially concerning possibilities of thermal integration and insulation.

References

[1] Thormann, J.; Pfeifer, P.; Kunz, U.; Schubert, K.: Reforming of diesel fuel in a micro reactor for APU systems, Chem. Eng. J. 135S1 (2007) suppl. 1, pp S74-S81

[2] Thormann, J.; Pfeifer, P.; Kunz, U.; Schubert, K.: Reforming of Diesel Fuel in a Micro Reactor, International Journal of Chemical Reactor Engineering 6 (2008) P1

A microfluidic calorimeter for the measurement of the enthalpy and the kinetic of chemical reactions from 5 to 90 °C IMRET-10: 10th International Conference on Microreaction Technology Sensing and Process Analytics (T5014)

Cindy Hany, LOF, Université Bordeaux 1, UMR CNRS-Rhodia-UB1 5258, Pessac, France
Christophe Pradere, TREFLE, Université Bordeaux 1, UMR 8508 CNRS-ENSAM-UB1, Talence, France
Jean Toutain, TREFLE, Université Bordeaux 1, UMR 8508 CNRS-ENSAM-UB1, Talence, France
Mathieu Joanicot, LOF, Université Bordeaux 1, UMR CNRS-Rhodia-UB1 5258, Pessac, France
Jean-Christophe Batsale, TREFLE, Université Bordeaux 1, UMR 8508 CNRS-ENSAM-UB1, Talence, France

Abstract

The aim of this work is to present a new device for the measurement of the kinetics and the enthalpy of very dangerous (or exothermic) chemical reactions. The main idea is to use a microfluidic chip where the microchannel acts as a chemical reactor. A thermopile (Peltier element) measures the global heat flux dissipated by the chemical reaction. A high thermal conductive plate (heat spreader) is used between the microreactor and the thermopile. This heat spreader avoids severe temperature gradient in the reactor. Due to the size of the microchannel a small volume of reactants (μl) is used. As the chemical reagents are injected by pump syringe, continuous experiments are performed with very good control of the reagents mixing. The costs of the microfluidic chip and the thermopile is pretty low. This device can be combined with an infrared camera for local characterization.

Keywords: Microfluidic, thermal and chemical analysis, flow calorimeter, enthalpy.

Introduction

The knowledge of kinetics and thermodynamics of chemical reactions is important for chemical process development and safety. We know that both phenomena, heat transfer and chemical kinetics, are present in chemical reactions. One way to control and measure those phenomena is calorimetry. The characterisation of exothermic chemical reaction is well developed and different kinds of calorimeter already exist.[1] However, those are not simultaneously adapted to the use of microvolumes, high thermal sensitivity, good control of the reagent mixing and isothermal conditions.

To this end, the recent development of MEMS (Micro-Electro-Mechanical Systems) allows the investigation of new approaches to measure the enthalpy of chemical reactions. The main advantage is the small size of the microchannels which permits the mixing control of the reactants. Due to the microchannel size, the flow is laminar with low Reynolds number (<<100). It is also convenient to use a small amount of products and to be able to easily and quickly test a large number of reaction configurations.[2] Nevertheless, the measurement of temperature at a small scale represents the main difficulty. In fact, the temperature information in such devices is among the most important parameters. However to date, only a few studies were developed on flow microcalorimeters. Such investigations used localised temperature microsensors.[3, 4, 5] The method developed here, measures the global heat flux dissipated by the microchannel

The main purpose of this paper is to present such a new microcalorimeter. Firstly, the experimental device principle is described. Then, an analytical model is presented in order to understand the thermal behaviour. The importance of the thermal properties of the wafer of the microfluidic chip is analysed. The relation between the heat flux generated by the chemical reaction and the thermal response is investigated. Secondly, a thermal calibration (by Joule effect) is presented in order to verify the features of the microcalorimeter. Then, the enthalpy of a chemical strong acid base reaction is determined in two flow configurations (co-flow and droplets). Furthermore, the conversion rate of this chemical reaction can be estimated for different residence times and temperatures. Finally, we demonstrate that calorimetric titration experiments are also possible with this device.

Experimental device

Calorimeter description

We have developed a differential microcalorimeter (figure 1), where one microfluidic chip is used for chemical reaction process (1) and the other one (1r) as a thermal reference.[6]

The microfluidic chip is built with a PDMS (PolyDiMethylSiloxane) matrix, containing a negative channel network. This PDMS matrix is glued onto silicon wafer after UV-ozone activation. As silicon is a good thermal conductor, the chemical reaction occurs under isothermal conditions in the microchannel.

The thermal heat flux measured is the differential voltage (Seebeck effect) between the thermopiles (3) placed under the chemical reaction (1) and the reference one (1r). The calorimetric sensitive area is a 50 mm X 50 mm square (four thermopiles series-connected). The voltage is measured with a numerical voltmeter (Agilent 34970 A). The fluids are injected by a syringe pump (Cetoni Nemesys). The measurement temperature (*Ts*) is set by a thermostatic flow circulation (5). A PC using a Labview program controls all these steps via an RS232 port.

In order to minimize perturbations of the baseline caused by the environmental thermal variations, the thermopiles are placed on a brass block (4) of 20 cm X 20 cm X 2.5 cm used as an isothermal reservoir. Then a stainless plate (5) is put below the brass block. This plate is thermally regulated via a thermostatic flow circulation from *Ts*=5 to 90 °C. This allows to performed experiments at various temperature depending on the fluid properties. All parts of the calorimeter are heat shielded by insulating foam (6). Furthermore, the inlet fluid is preheated by passing through the brass block via a tubing groove. As a result, the chemical reaction begins at the microcalorimeter temperature (*Ts*).

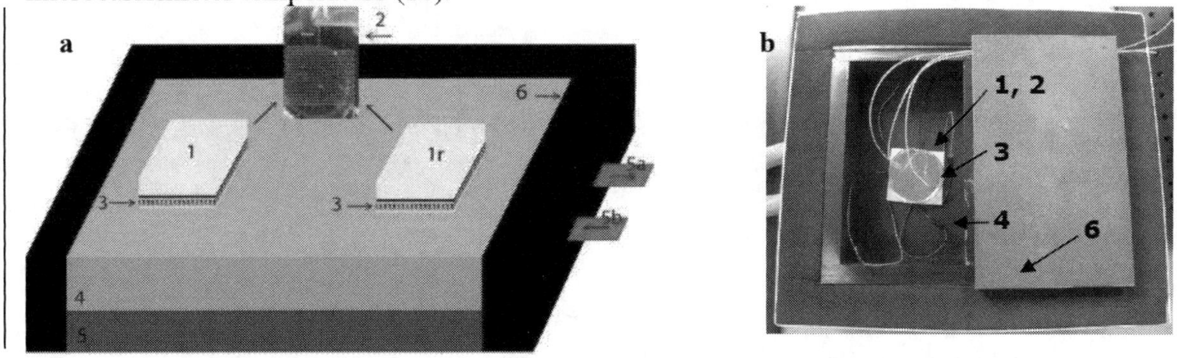

Figure 1: (a) scheme of the microcalorimeter; (b) picture of the microcalorimeter; 1: microfluific chip composed of PDMS resin and silicon wafer, 2: microchannel imprinted in the PDMS, 3: thermopile with ceramic plates and thermoelectric element, 4: isothermal brass block, 5: stainless plate, 5a and 5b: thermostatic flow circulation, 6: insulating foam

Thermal modelisation of the microfluidic chip

In order to modelize the thermal behaviour of the microfluidic chip, the 2D transient thermal problem (figure 2) is considered. The system is made of three homogeneous media of different thermal properties: medium 1 is the thermopile used as fluxmeter, medium 2 is the wafer of the microfluidic chip and medium 3 is the elastomeric PDMS where the microchannels for chemical reaction are imprinted.

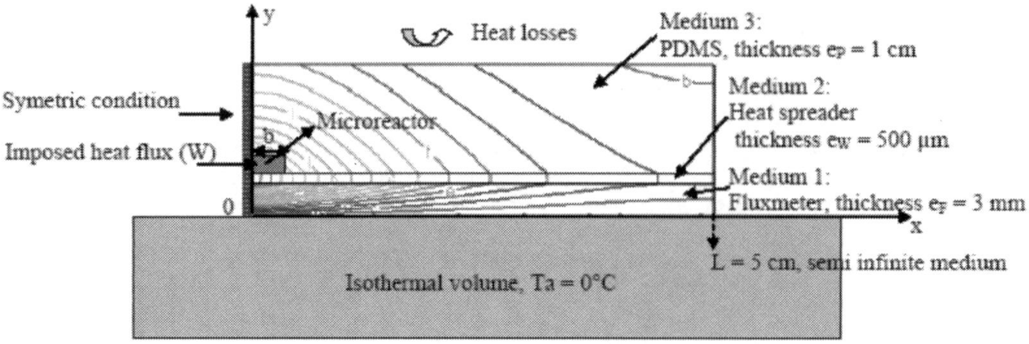

Figure 2: the semi infinite thermal problem

The 2D transient model (Eq. 1, 2, 3) allows to calculate the temperature field in the microfluidic chip when a chemical heat flux is generated in the microreactor (figure 2).

$$\frac{\partial^2 T_i}{\partial x^2} + \frac{\partial^2 T_i}{\partial y^2} = \frac{1}{a}\frac{\partial T_i}{\partial t}, i = 1, 2 \text{ and } 3 \quad (1)$$

Initial and spatial conditions in the x-direction remain finite: $t=0; T_i(t)=0; x=0; \frac{\partial T_i}{\partial x}=0; x=L; T_i$ whereas the y-direction spatial conditions are as follow for each medium:

Medium 1: Thermopile

$$\begin{cases} y=0; T_1=0 \\ y=e_1; -\lambda_1 \frac{\partial T_1}{\partial y} = -\lambda_2 \frac{\partial T_2}{\partial y}; T_1 = T_2 \end{cases}$$

Medium 2: Wafer

$$\begin{cases} y=e_1; -\lambda_1 \frac{\partial T_1}{\partial y} = -\lambda_2 \frac{\partial T_2}{\partial y}; T_1 = T_2 \\ y=e_1+e_2; -\lambda_2 \frac{\partial T_2}{\partial y} + \Phi(x,t) = -\lambda_3 \frac{\partial T_3}{\partial y}; T_2 = T_3 \end{cases}$$

Medium 3: PDMS layer

$$\begin{cases} y=e_1+e_2; -\lambda_3 \frac{\partial T_3}{\partial y} = -\lambda_2 \frac{\partial T_2}{\partial y} + \Phi(x,t); T_3 = T_2 \\ y=e_1+e_2+e_3; -\lambda_3 \frac{\partial T_3}{\partial y} = hT_3 \end{cases} \quad (2)$$

where $_i$ is the subscript of each media, T_i is the temperature (K), λ_i is the thermal conductivity ($W.m^{-1}.K^{-1}$), a_i is the thermal diffusivity ($m^2.s^{-1}$), h is the heat losses ($W.m^{-2}.K^{-1}$), e_i is the thickness (m) and Φ is the total heat flux (W).

First, we assume that the chemical heat flux can be represented as a spatial and temporal square, $\Phi(x, t)$, along the plate (Eq. 3).

$$\Phi(x,t) = \begin{cases} QH(t) & 0 < x \le b \\ 0 & b < x < L \end{cases} \quad (3)$$

where Q is the heat flux (W), H is the Heaviside function and b is the width of the thermal flux. Then, we suppose that the media are semi infinite in the x direction. We also assume that heat losses by convection are located at the top of the elastomeric PDMS. Finally, the thermal reservoir positioned at the base of the fluxmeter permits to impose a well known temperature (Ts) on the calorimeter.

These equations are solved by the quadrupole method.[7, 8] With this simple analytic model, it is possible to accurately analyze the thermal heat transfer in the microfluidic chip. The thermal properties of the chip wafer and its size are very important to design the chip properly. Usually, glass or silicon wafers are used. Figure 3a shows the effect of the thermal properties of the wafer on the microreactor temperature level. For example, a heat flux of 1 W is dissipated through the channel to a wafer thickness (500 µm).The maximum temperature variation depends on the wafer. The highest temperature for a glass wafer is about 1.5 K, whereas it is only 250 mK for silicon wafer (6 times smaller). Thus, with the aim of keeping the microreactor as isothermal as possible, it is better to work with the silicon wafer.

Another interest of this analytical model is to study the heat flux distribution in transient conditions. First, the response time of the system is analysed. So the heat flux superimposed at the top of the silicon plate is still a spatial square (b = 1 mm long) and constant in time. The heat flux received by both the fluxmeter and the PDMS layers can be compared (figure 3b). The heat flux on the thermopile is very important in relation to the flux through the PDMS. The calculation of the percent ratio (figure 3c) shows that 99 % of the heat flux generated by the chemical reaction goes through the silicon.

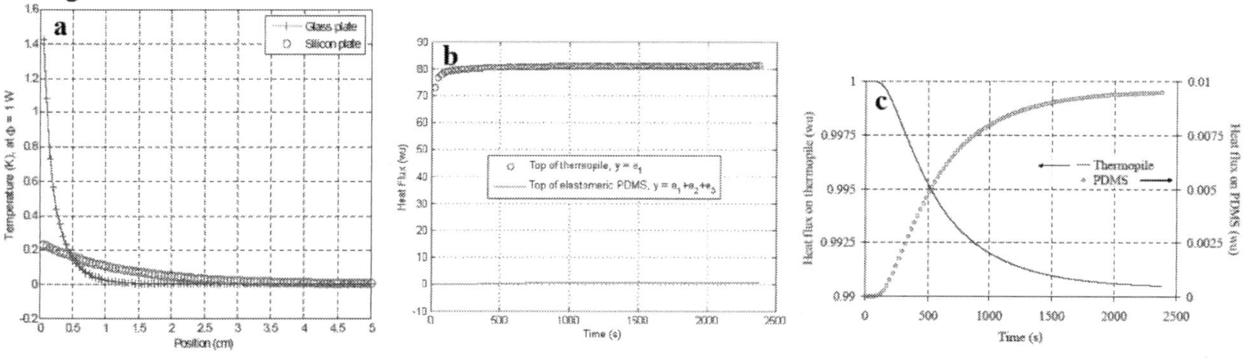

Figure 3: (a) steady state temperature on fluxmeter ($y=e_1$) for different heat spreader of 500 µm thick; (b) transient heat flux on both thermopile and PDMS layer; (c) transient ratio of flux received by the thermopile and the PDMS layers

Calibration and performance

Electrical calibration

This device is calibrated with an electrical heat source. For that purpose, a calibration chip is manufactured with an electrical resistance. This resistance is realized by gold deposition and located between the silicon wafer and the PDMS. This calibration chip is placed in the reaction part and the microfluidic chip without flow in the reference part. Various electrical power from 100 µW to 1 W are applied to the system thanks to a power supply (Agilent E3631A). Then, the dissipated heat flux is measured by the thermopile as function of time. For each of them, electrical power was set on until steady state was reached. Then, the power supply was stopped (figure 4a).

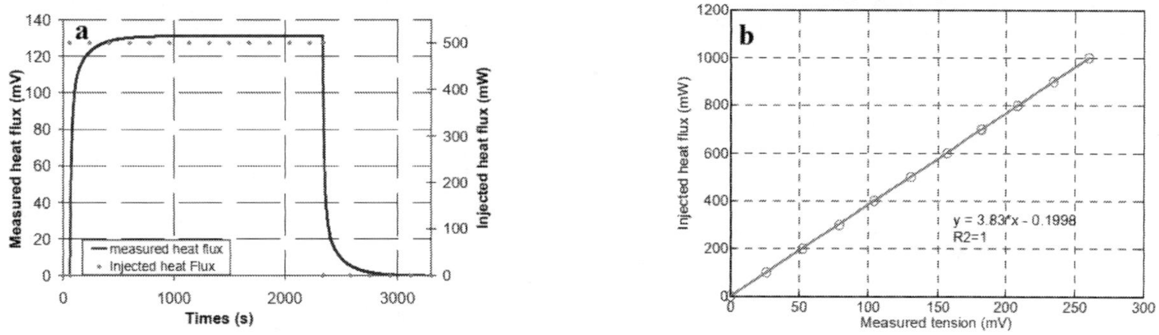

Figure 4: (a) heat flux measured versus time; (b) injected heat flux versus measured tension at *Ts*=23°C

This procedure allows the measurement of two values of the heat flux, one during the rising part of the response, the other one during the decreasing part. The measured flux (mV) is then compared with the injected flux (mW). This determines the conversion coefficient: $\alpha = 3.83 \pm 0.38$ W/V (figure 4b). When a chemical reaction occurs in the microcalorimeter, the heat flux is measured in volt and this coefficient (α) is used to deduce the chemical heat flux (W). The fluxmeter properties changes with temperature so this coefficient should be estimated at each measurement temperature. We have also determined α at *Ts*=5°C and *Ts*=50°C.

The incertainty of this device depends on the sensibility of the thermopile (3 μV), the pump syringe (1μL/h), the numerical voltmeter (1 μV), the power supplier (1 μW), the resistance of the calibration chip (0.1 Ω) and the solution concentration (0.1mmol/L). With the knowledge of those sensitivities we can estimate the relative error of the measurements. The heat flux is measured with 4% precision and the enthalpy is determined with 4.5% precision.

Chemical application

Chemical enthalpy measurements

A promising application of microtechnologies is the use of one phase liquid flow (co-flow) or two-phase liquid-liquid flow (droplets flow). In the latter, microdroplets of reagents are transported at constant velocity in an inert continuous phase. So, they can be considered as individual nanovolume batch reactors. A very exothermic reaction between a strong acid (HCl) and base (NaOH) is performed in these two flow configurations. The concentration of the two products is equal to 0.2 M. The microfluidic chip used has the following dimensions: length 32 cm, width 250 μm and thickness 250 μm. The selected substrate (thickness 500 μm) is made of silicon with a 5 cm diameter. The experimental procedure followed during measurements is similar to the one described for the calibration (figure 4a). First, the transient thermal response measured by the fluxmeter is recorded until steady state is reached. Then, the flow is stopped. At last, the thermal response, corresponding to the decrease of the temperature, is recorded. For all experiments, the stoechiometric proportions are applied. The measured heat flux can be described by Eq. 4.

$$\phi = \Delta H_r FCX \qquad (4)$$

where ϕ is the heat flux (W), ΔH_r is the enthalpy (J/mol), F is the flow rate (L/s), C is the concentration (mol/L) and X is the conversion rate. As the reaction is total, the heat flux released by the reaction is directly proportional to the molar flow rate. Thus, the reaction must be complete (X=1) in the microreactor. For this reason, the residence time has to be long in comparison with the reaction time.

Liquid phase flow

We studied the acid-base reaction with one liquid phase flow: co-flow. This experiment is realised at Ts=23 °C for several flow rates ranging from 250 to 1250 μL/h. The mixing is realized by species diffusion. Measurements performed by an Infrared camera allow us to show that the reaction occurs at the interface between the two reagents (figure 5a). The heat flux evolution is recorded as a function of time at various flow rates (figure 5b). Then the heat flux released by the reaction is compared to the molar flow rates. The evolution is linear and can be minimized. The slope we obtained provides the reaction enthalpy (figure 5c).

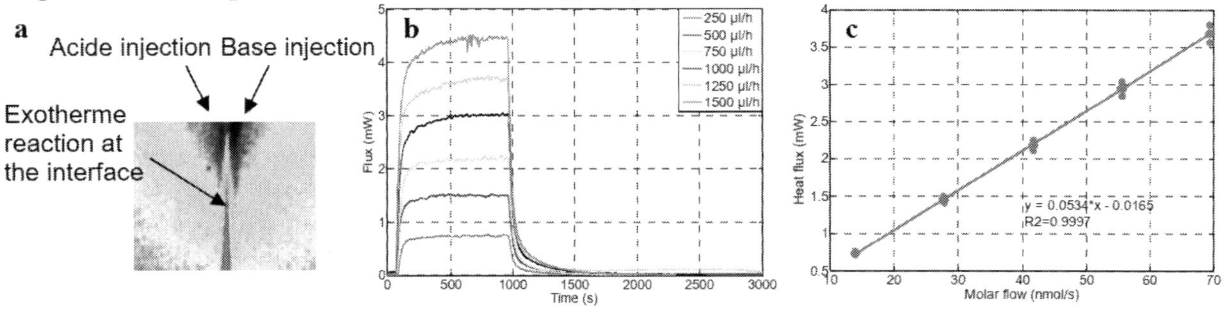

Figure 5: (a) Infrared image of acid base co-flow; (b) measured heat flux versus time at reagents different flow rates in co-flow at *Ts*=23°C; (c) measured heat flux versus molar flow rates in co-flow at *Ts*=23°C

The value of the enthalpy obtained is 53.4 ± 2.4 kJ/mol, which is closed to the value of 56 kJ/mol provided by the literature.[9] The variation between the theoretical and experimental value is 5 %.

Liquid-liquid phase flow (droplets)

Figure 6a shows the temperature field of the liquid-liquid phase flow for an acid-base chemical reaction. The chemical reaction occurs in the droplets (microreactors). The heat diffusion along the channel is due to the diffusion in the wafer. The inert continuous flow is silicone oil with a viscosity 20 mPa.s. Two different sets of experiments are carried out. First, the droplets volume is changed by variation of oil flow rate with a constant reagent flow rate. As the reagent flow rate is constant, the reagent quantity introduced in the microcalorimeter is constant. Figure 6b represents the released heat by the chemical reaction which is the same for all oil flow rates. Secondly, the ratio between the reagents flow rate and the oil flow rate is constant and equal to 0.5. The released heat increases with the reagent flow rate. The reaction enthalpy can be determined by plotting the heat flux versus the reagent molar flow rate as it is shown in figure 6c. The reaction enthalpy measured is 58.3 ± 2.62 kJ/mol. This value is in agreement with the value of 56 kJ/mol found in the literature[9]. This result is closed to the value determined by this microcalorimeter in co-flow.

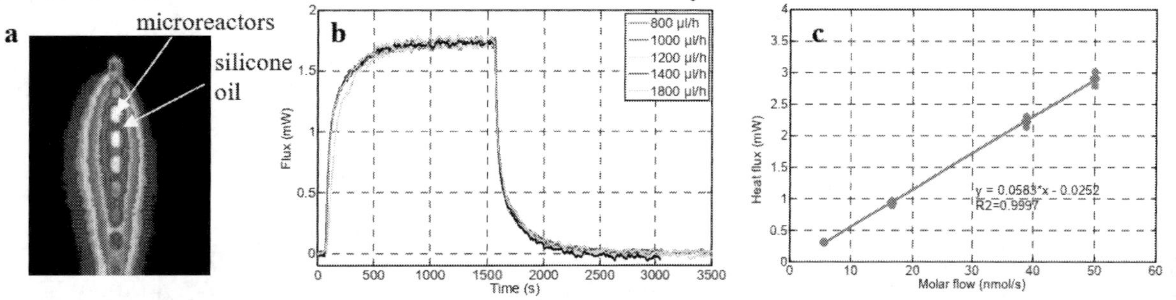

Figure 6: (a) Infrared picture of a acid-base droplets flow, the reagents contact first occurs in co-flow, then oil injection allows to get droplets (microreactors); (b) measured heat flux versus time at constant reagents flow rate and various oil flow rates in droplets flow at *Ts*=23°C; (c) measured heat flux with constant ratio (reagent flow rate versus oil flow rate) versus the reagent molar flow rates in droplets flow at *Ts*=23°C

Reaction kinetics

In this study, the reagents are introduced in co-flow. In such microfluidic chip the Reynolds number is low (<<100), so we assume the flow is laminar. The reagents mixing is performed by species diffusion. The strong acid-base reaction is a very fast reaction, thus chemical reaction is limited by the species diffusion. The interdiffusion area was observed and studied by several investigations.[10, 11, 12] The diffusion length, which corresponds to the mixing length, can be described by Eq. 5.

$$\frac{\partial C}{\partial t}+\bar{U}\nabla C = D\nabla^2 C \qquad \sigma = 2\sqrt{Dt} = 2\sqrt{D\frac{X_m}{U}} \rightarrow X_m = \frac{1}{4}\frac{w^2U}{D} \text{ or } Pe = \frac{wU}{D} \rightarrow X_m = \frac{1}{4}Pew \tag{5}$$

where σ is the diffusion width (m), D is the diffusion coefficient ($m^2.s^{-1}$), C is the concentration ($mol.m^{-3}$), U is the flow velocity ($m.s^{-1}$), X_m is the diffusion length (m), t is the time (s), w is the channel width (m) and Pe is the Peclet number.

In order to study the reaction conversion rate, the reaction should be uncompleted in the microreactor. The residence time has to be lower than the reaction time. Therefore a microfluidic chip is designed with 100 µm width, 80 µm depth and2 mm length.

The acid-base reaction is studied in the microcalorimeter at different flow rates from 50 to 1000 µL/h (figure 7a), and in stoechiometric proportions at *Ts*=5, 23 and 50°C. Then, the conversion rates are determined by Eq. 4 and plotted versus the residence times for different temperatures (figure 7b). The acid-base reaction is a second order reaction involving two species (HCl and NaOH). The concentration evolution of one species can be described by Eq. 6.

$$\frac{dC_{HCL}}{dt} = kC_{HCl}C_{NaOH} \text{ or } C_{HCl} = C_{0HCl}(1-X) \text{ and } C_{NaOH} = C_{0HCl} - C_{0NaOH}X \rightarrow X = \frac{1-Me^{(1-M)kC_{0HCl}t}}{1-e^{(1-M)kC_{0HCl}t}} \tag{6}$$

where C is the concentration (mol/L), C_0 is the initial concentration (mol/L), M is the ratio C_{oNaOH}/C_{oHCl}, X is the conversion rate, t is the time (s) and k is the kinetics constant. The experimental data are fitted by Levenberg Marquardt algorithm as it is shown in figure 7a. The theoretical data are in good agreement with the experimental ones. Then, the kinetics can be estimated by using the Arrhenius law at different temperatures.

On figure 7b, the mixing time is estimated for X=0.95. We can verify that the diffusion mixing is faster at high temperature. Concerning experiments at room temperature, the diffusion coefficient of H^+ can be determined by using Eq. 5. The experimental mixing time is about 0.145 s (figure 7b), thus D_{H+} is equal to $8.6.10^{-9} m^2.s^{-1}$. This value is in agreement with the literature value[13]: $D_{H+}=9.10^{-9} m^2.s^{-1}$.

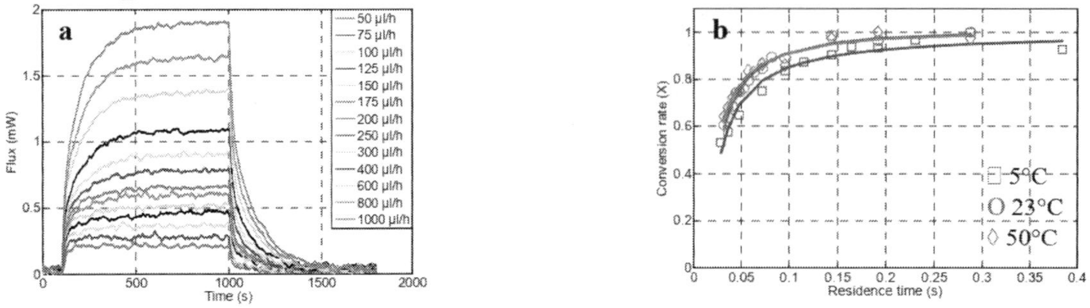

Figure 7: (a) measured heat flux versus time at different flow rates in co-flow at *Ts*=5°C; (b) conversion rate versus residence time at different setting temperatures (*Ts*): points are experimental data and lines are the fits

Calorimetric titration

We use the microfluidic chip designed for the enthalpy measurement. An HCl solution is considered as the unknown concentration solution and has a concentration equal to 0.1 mol/L. This HCl solution is titrated by a NaOH solution. The concentration of this NaOH solution is well known and equal to 0.2 mol/L. The experiments come out the following way: the NaOH flow rate is constant and equal to 400 µL/h, then the HCl flow rate increases steadily until the measured heat flux reaches a constant, as it is shown in figure 8. The reaction occurs in stoechiometric proportions when the heat flux steady state is reached. Thus, the equivalent flow rate is determined at the beginning of this steady state.

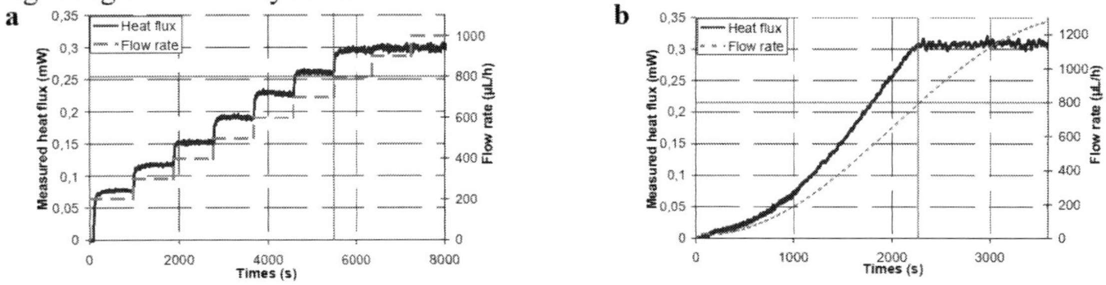

Figure 8: measured heat flux versus time at different flow rates at *Ts*=23°C; (a) flow rate changes each 15 min; (b) flow rate changes each 60 s.

According to the two experiments, the maximum heat flux is measured when the HCl acid flow rate is equal to 800 µL/h. So the stoechiometry of the chemical reaction is obtained at the expected value of the HCl concentration (i.e. 0.1 mol/L).

Conclusion

A microfluidic calorimeter under continuous flow has been developed and validated for the measurements of both kinetic and enthalpy of chemical reaction. In this paper, a validation on strong acid-base chemical was performed. The experimental results obtained were very closed to the values of the literature (< 5%). As we demonstrated, this isothermal microcalorimeter allows chemical characterizations (enthalpy, conversion rate, and titration) in droplets and co-flow. The droplets flow is very useful in case of very high exothermic reaction. In fact, the small volume and the important exchange surface of the droplets give very low temperature increases. These advantages are very interesting for safety studies.

The main advantages of the device are summarized:

- the manufacturing of microfluidic chips makes it possible to obtain isothermal conditions with various geometries of reactor (sizes and volumes).
- the fluxmeters offer a good sensitivity and a direct measurement of heat flux. The thermopiles are independent of the chip and not very expensive. Working in steady state removes all problems involved in its inertia.
- thanks to the injection system, it is possible to sweep a large range of residence times and concentrations.
- the required time to estimate the reaction enthalpy is around 2 hours which is very short.

Finally, this device could be coupled with an InfraRed camera in order to perform local chemical analysis.[14] The use of both methods will allow the study of various chemical reactions in order to obtain thermochemical data.

References

1. Zoog A., Stoessol F., Fisher U. and Hungerbühler K. (2004), "Isothermal reaction calorimetry as a tool for kinetic analysis," *Thermochimica Acta*, 419, 1-17.
2. Tabeling P. (2003), "Introduction à la microfluidique," *Collection Echelles*, Belin.
3. Köhler J.M. and Zieren M. (1997), "Chip reactor for microfluidic calorimetry," *Thermochimica Acta*, 310, 25-35.
4. Ernst H., Jachimowicz A. and Urban G.A., "High resolution flow characterization in Bio-MEMS," *Sensor and actuators A*, 100, 54-62.
5. Schneider M.A., Maeder T., Ryser P. and Stoessel F. (2004), "A microreactor-based system for the study of fast exothermic reactions in liquid phase: characterization of the system," *Chemical Engineering Journal,* 101, 241-250.
6. Bastsale J.C., Pradere C., Joanicot M., Toutain J. and Gourdon C. (2006), "Procédés de détermination d'au moins un paramètre d'une transformation physique et/ou chimique, dispositif et installation correspondants", patent, 2897156
7. Fudym O., Pradere C., Batsale J.C. (2007), "An analytical two-temperature model for convection-diffusion in multilayered systems: Application to the thermal characterisation of microchannel reactors," *Chemical Engineering Science,* volume 62, issue 15, p. 4054-4064.
8. Maillet D., Andre S., Bastsale J.C., Degiovanni A. and Moyne C. (2000), "Thermal quadrupoles: solving the heat equation through integral transforms", *Wiley and sons.*
9. Lide D.R. (1991-1992), "Handbook of Chemistry and Physics", *CRC Press*, 72ND Edition, p. 5-1.
10. Kamholz A.E., Weigl B.H., Finlayson B.A. and Yager P. (1999), "Quantitative analysis of molecular interaction in a microfluidic channel: the T-sensor", *Analytical Chemistry*, 71, 5340-5347.

11 Ismagilov R.F., Rosmarin D., Kenis P.J.A., Chiu D.T., Zhang W., Stone H.A. and Whitesides G.M. (2001), "Pressure-driven laminar flow in tangential microchannels: an elastomeric microfluidic switch", *Analytical Chemistry,* 73, 4682-4687.

12 Salmon J.B., Dubrocq C., Tabeling P., Charier S., Alcor D., Jullien L., Ferrage F. (2005), "An approach to extract rate constants from reaction – diffusion dynamics in a microchannel", *Analytical Chemistry,* 77, 3417-3224.

13. Lide D.R. (2004-2005), "Handbook of Chemistry and Physics", *CRC Press*, 85ND Edition, p. 5-95.
14. Pradere C., Joanicot M., Batsale J.C., Toutain J., Gourdon C. (2006), "Processing of temperature field in chemical microreactors with infrared thermography," *Quantitative InfraRed Thermography (QIRT),* Vol.3, n°1, p. 117-135.

Deposition of zeolitic coatings onto FeCrAlloy microchannels: washcoating vs. in situ growing.

Asier Eleta[1], Luciano Costa[1], Mario Montes[1], José Antonio Odriozola[2].

1) Departamento de Química Aplicada , Facultad de Ciencias Químicas, University of the Basque Country, Apdo. 1072, 20080 San Sebastián, Spain. e-mail: mario.montes@ehu.es

2) Instituto de Ciencia de Materiales de Sevilla, Centro Mixto CSIC-Universidad de Sevilla, Avda.Americo Vespucio s/n 41092 Sevilla. Spain. e-mail: odrio@icmse.csic.es

1. Introduction

Due to their versatility in pore structure, chemical composition and acidity, zeolites have been widely used in several catalytic processes such as oil refining, fine chemistry and environmental applications [1, 2]. Usually those materials are used by packing them in particles, forming a fixed bed and passing the flow through them. A large effort has been employed testing alternatives to fixed bed due to the higher pressure drop that this kind of configuration produces, especially when high flows are used. In that way, monoliths are the alternative to solve this problem [3]. Among all type of structured reactors, systems with micron size channels (microreactors) have received attention in the recent years. The narrow channels and thin walls make them especially efficient for processes that are mass and heat transfer limited [4].

Metallic structures have interesting properties comparing to those made of ceramic materials such as higher mechanical resistance, thermal conductivity and lower pressure drop. Ferritic alloys containing aluminum can generate alumina scale protecting coatings with excellent properties for anchoring the catalytic coating [3].

The aim of this work is to study the deposition of hydrophobic MFI type zeolite onto previously calcined FeCrAlloy microchannels. Two coating methods were compared [5]: Washcoating from a stabilized slurry and in situ growing of the zeolite on the metallic surface introduced in the synthesis medium.

2. Materials and experimental techniques

2.1. Materials

Metallic supports: Fecralloy plates and monoliths (Figure 1a and 1b) were manufactured and calcined at 900°C for 22h for generating a rough and non-porous alumina layer on the surface (Figure 1c). Plates were used to carry out preliminary tests for both deposition methods. Regarding to the monoliths, the dimensions of the channels are in the range of microns (channel width of 550 microns) and they could be considered as a good approach to microchannel reactors (Figure 1b).

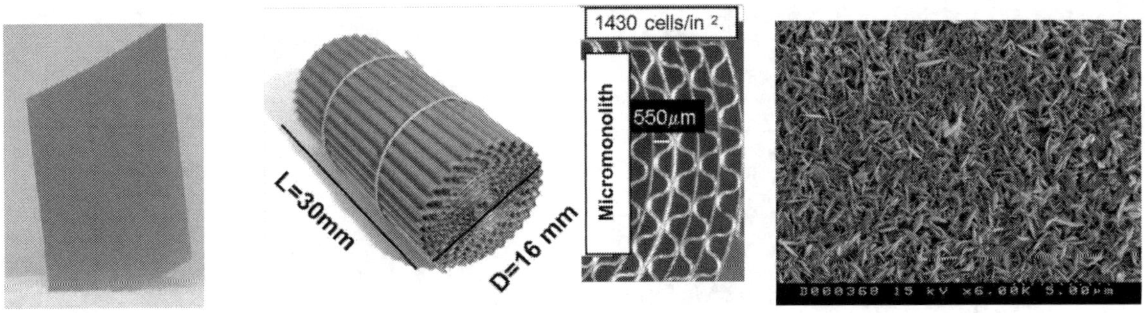

a b c

Figure 1. Fecralloy plates (a) and monoliths (b). Micrograph of Fecralloy surface full of alumina whiskers after calcination at 900ºC and 22h.

Deposition by Washcoating: Commercial zeolite CBV 28014 (provided by Zeolyst) was used as received (Si/Al: 280, S_{BET} : 370 m^2/g). Polyvinyl alcohol (PVOH 4/88, Fluka) and colloidal silica (Ludox AS30, Dupont) were used as additives for obtaining stable zeolite slurries.

Deposition by in situ synthesis: Tetraethyl ortosilicate (TEOS, Fluka) was used as silicon source, sodium hydroxide (NaOH, Fluka) as mineralizing agent and tetrapropylammonium bromide (TPABr, Fluka) as template.

2.2. Deposition by washcoating

Metallic supports (plates and monoliths) were washcoated by dipping and withdrawing them in stabilised zeolite slurry at constant speed (3 cm/min). Excess slurry in monoliths was removed by centrifugation. After that, the supports were dried at 80ºC for 1h. Plates used in the preliminary tests were coated in this way just one time while for the monoliths this process was 7 times repeated. Finally all the supports were calcined at 500ºC for 2 h with heating rate of 1ºC/min.

2.3. In situ synthesis

To carry out the synthesis of zeolitic materials the following equipment was used. A Selecta force-air oven (Digitronic model) was modified adding a stainless steel axis across one of the oven's wall. Externally, the axis was connected to an electronically controlled drive engine, while the inside part bears a hexagonal stainless steel piece with moorings, where six teflon-steel autoclaves can be fixed (Figure 2a). The 45° angle between the autoclaves and the rotary axis assured the rocking movement of the liquid inside the autoclave. The monoliths are fixed in teflon supports inside the vessel (Figure 2b) to avoid their movement while rotation and in this way to force the reactive liquid to pass through the monolith channels.

Metallic supports were introduced in autoclaves together with a 24h hour aged synthesis gel of molar composition $20SiO_2$:5TPABr:5NaOH:$10000H_2O$ and placed into the oven [6]. Around 45 g of this gel were placed in each autoclave for each reaction. In situ growing was carried out under hydrothermal synthesis at 150ºC for 24h with orbital agitation of 6 rpm.

a b

Figure 2. Pictures of the oven (a) and autoclaves used for deposition by synthesis.

Two different in situ syntheses of hydrophobic zeolite were carried out. Direct synthesis using bare oxidized metallic supports, and seeded synthesis in which an additional step was included before the first reaction, where the support surface was washcoated with colloidal silica, template (TPABr) or submicron zeolite seed.

Zeolite seeds were obtained by hydrothermal synthesis during 8h and 125ºC of a gel with a molar composition of $4SiO_2$:1TPABr:1NaOH:1000H_2O. Previously this gel was aged for 24h under vigorous agitation [6]. After the synthesis, the seeds were washed repeating for 3 times separation by centrifugation and redispersion in distilled water. Finally, the solid content of this seed suspension was adjusted to 1%. Supports were washcoated with this seed, dried at 80ºC for 1 h and calcined at 500ºC for 4h.

Pre-coating of supports with SiO_2 was carried out washcoating them with diluted Ludox AS30 colloidal silica (4 water:1 Ludox), drying the supports at 80ºC for one hour and finally calcining them at 500 ºC for 4h. Pre-coating with TPABr was carried out washcoating the supports with a 1M TPABr disolution and drying the supports at 60 ºC for 1 hour [7].

After the synthesis, the supports were washed with distilled water, dried at 80ºC for 1h and calcined at 500ºC for 4 h with a heating rate of 1ºC/min to remove the organic template. This synthesis process was 5 times repeated for monoliths.

2.4. Characterization

Specific surface measurements were carried out by N_2 adsorption in a Micromeritics ASAP 2020. N_2 and He gases were used (Air Liquide, quality 99,999%). A homemade cell was used allowing the measurement of the whole monolith.

Zeolite crystallinity was analyzed by XRD. Philips PW 1729 bean generator was used, connected to a PW 1820. The signals were taken by a Philips 3710 diffractometer and processed by a PC-APD diffraction software. Kα of copper (λ=1,5406 Å) filtered with niquel was used as radiation.

The morphology of the supports was studied with a HITACHI S2700 scanning electron microscope. Optical images were taken with a Leica microscope (S8APO+DFC 280) and processed with IM50 software.

Viscosity of zeolite slurries was measured with Haake VT500 device equipped with NV sensor. The viscosity value used was the final constant value of the asymptotic curves obtained at high shear rates. Adherence test of the supports was carried out immersing them in petroleum ether and under intense sonication for ½ h [8]. Results of adherence test are reported as percentage of coating remaining after sonication, drying and calcination.

3. Results and discussion

3.1. Deposition by washcoating

3.1.1 Preliminary tests with plates

4 Slurries were prepared mixing together during 24 h water, commercial zeolite, polyvinyl alcohol and colloidal silica. Table 1 summarizes the formulations and viscosities of those slurries, the mass adhered in plates after one coating and the result of adherence test.

Table 1. Formulations, properties of slurries and final characteristics of the catalyst layer obtained in washcoated fecralloy plates.

Slurry	Water (%)	Zeolite (%)	SiO_2 (%)	(%)PVOH	Visc.(cp)	Adh.mass (mg)	Adh. Test (%)
A	70	30	-	-	1	1.9	15
B	64	30	6	-	4	3.7	68
C	64	30	-	6	1	9.3	19
D	58	30	6	6	11	21	17

A minimum slurry viscosity of 10-20 cp is recommended to produce a proper and homogeneous coating [9]. Slurry A was not stable and settled very fast due to the low viscosity. This low viscosity was also the reason of obtaining a very thin and not homogeneous layer that detached easily when plates were manipulated.

PVOH was added to increase viscosity and producing a more stable slurry (B). The viscosity of this slurry was lower than the range recommended in literature, but a homogeneous layer was formed when plates were dip-coated. It was observed that PVOH acted also as a binder increasing the adherence of the coating before calcination. Nevertheless, after calcination, PVOH is removed leading to an easy detachment of the coating layer.

When colloidal silica was used, the viscosity of the slurry (C) was also low, leading again to a fast sedimentation. When plates were coated with this slurry, more adhered coating was obtained because of the sticking effect of the colloidal silica, but due to its low viscosity, the coating was not homogeneous.

The combination of both additives (slurry D) was necessary to obtain a homogeneous and well adhered zeolite layer after calcination. The viscosity in that case was in the recommended range and the slurry was stable to carry out washcoating processes.

Results in adherence test showed high mass lost in all cases. But it was clear that as additives increased the adhered mass, the final load (after adherence test) was higher even with this low adherence.

3.1.2 Washcoated monoliths

Seven repeated coatings were carried out for each monolith presented in table 2 (M7 to M13). The influence of the SiO_2 amount (3, 6 and 12%) in the final adherence and textural properties of the catalyst layer was studied. Zeolite (30%) and PVOH (6%) content were kept constant in the formulations tried. Table 2 and Figure 3 summarize the coating results.

Table 2. Formulations and final characteristics of the catalyst layer obtained by washcoating over monoliths

Monolith	$\%SiO_2$ slurry	Viscosity of slurry (cp)	Adhered mass after calcinaltion (mg)	Adhered mass/area (mg/cm^2)	BET area (m^2/g)	Adh.Test (%)
M7	3	10	775	1.64	326	92.9
M8	3	10	690	1.46	325	
M9	6	11	770	1.65	303	94.8
M10	6	11	795	1.69	310	
M11	6	11	724	1.54	307	
M12	12	32	1143	2.43	291	97.9
M13	12	32	1089	2.31	289	

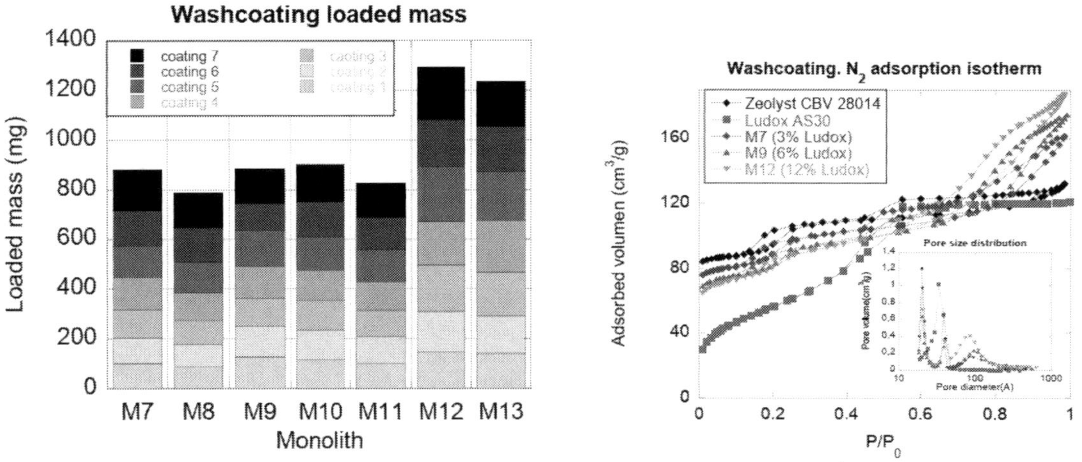

a b

Figure 3. a) Evolution of the loaded amount during 7 washcoating processes for monoliths. b) Textural properties for powder zeolite (black), dried and calcined colloidal silica (purple) and M7 (blue), M9 (green) and M12 (orange) monoliths.

Increasing the amount of SiO_2 from 3 to 6%, viscosity varied slightly and so, the final mass loaded was quite similar (Monoliths M7 to M11). But increasing this amount up to 12%, the viscosity also increased and then, the loaded amount (Monoliths M12 and M13).

The higher the amount of SiO_2, the better the adherence, but the textural properties of the coating differ from that of the pure zeolite. As it is observed in Figure 3b, increasing the amount of colloidal silica, the final coating was less microporous and there was a decrease in BET area due to the lower area of colloidal silica. But some mesoporosity was generated with pore diameter of around 100 Armstrongs. This mesoporostity belongs neither to colloidal silica nor to zeolite crystals and it is probably created during drying and calcination process, when some hollows are formed in between the zeolite crystals, cemented by the colloidal silica. It was observed that increasing the amount of colloidal silica, this mesoporosity also increased.

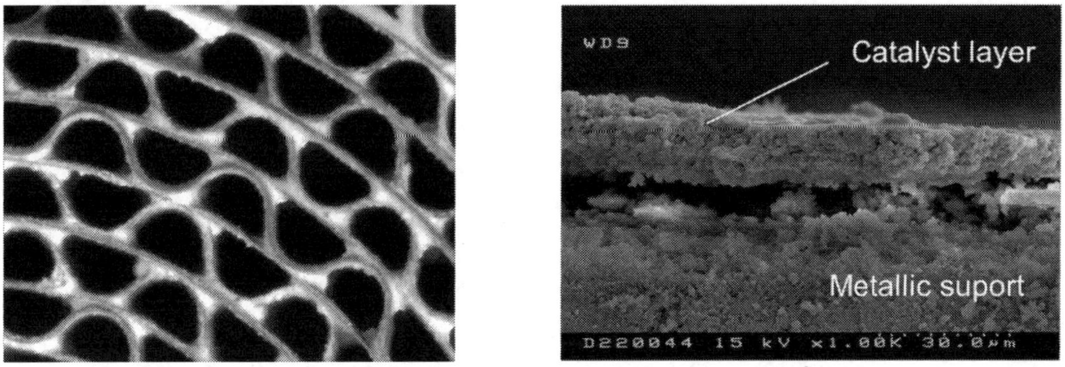

a b

Figure 4. Optical image (a) and SEM micrograph of the coating of M10 monolith.

Figure 4a shows material accumulations in the corners of the channels that are typical of washcoating method. The adherence in all monoliths was good with mass loss lower than 8 % and layers of about 1.6 - 2.4 mg zeolite /cm^2.

It should be noted the difference in adherence between plates and monoliths. The reason of the better adherence in the case of the monoliths could be related with geometrical constrains of the narrow microchannel not present on the flat surface of the plates.

3.2. Monoliths coated by in situ synthesis

3.2.1 Preliminary tests with plates

In situ synthesis over bare oxidized metal plates showed non-homogeneous coating with some zones completely covered while some others were partially or not covered with crystals (Figure 5a). The crystal size was quite homogeneous and regular with crystal length of around 20 microns.

a b c

Figure 5. Surface of a bare oxidize fecralloy plate after synthesis (a). Upper (b) and side view (c) of a fecralloy plate, previously coated with Colloidal silica, after the synthesis.

When synthesis was carried out with plates pre-coated with SiO_2 or TPABr, it was possible to cover the surface with a homogeneous and continuous layer. Crystal sizes in this case were more heterogenous ranging from 1 to 15 microns (Figure 5b and c). Final load varied from 0.6 to 1.6 g/cm^2 and XDR difractograms confirmed the presence of pure MFI zeolite over both the plates and in the reactive medium.

When fecralloy plates were pre-coated with silicalite submicron seeds before the reaction, the final result was not reproducible. It was possible to cover properly the surface in some experiments, but in others, similar heterogeneities like those obtained with bare oxidized plates were achieved. This could be attributed to an heterogeneous seeding of the plates observed by SEM, suggesting that the hollows between alumina whiskers were too small to anchor physically 200 nm size zeolite seeds. Taking into account the hard work to prepare the submicron seed compared to the easy use of commercial colloidal silica or TPABr solutions, the strategy of pre-coating with zeolite seeds was abandoned.

3.2.2 Monoliths coated by in situ synthesis

Bare oxidized monoliths and pre-coated with SiO_2 or TPABr were used in this series that is summarized in table 3.

Table 3. Synthesis methods and final characteristics of the zeolite layer obtained by in situ synthesis over calcined FeCrAlloy monoliths.

Monolith/Autoclave (*)	Pre-coating (1st reaction)	Adhered final mass (mg)	Adhered mass/area (mg/cm^2)	BET area (m^2/g)
M1/1	No	664	1.41	381
M2/2	No	698	1.48	378
M3/3	Pre-coated with SiO2	627	1.33	384
M4/4	Pre-coated with SiO2	724	1.54	376
M5/5	Pre-coated with TPABr	633	1.34	375
M6/6	Pre-coated with TPABr	697	1.48	379

(*) M1 monolith was placed during the 5 reactions in autoclave number 1 and so on.

As it was observed in test with plates, the effect of a pre-coating with SiO_2 or TPABr was to direct the zeolite growing preferentially to the metallic surface. After the first reaction, the amount of zeolite loaded in pre-coated monoliths (M3 to M6) was higher than in bare oxidized monoliths (M1 and M2) (Figure 6a).

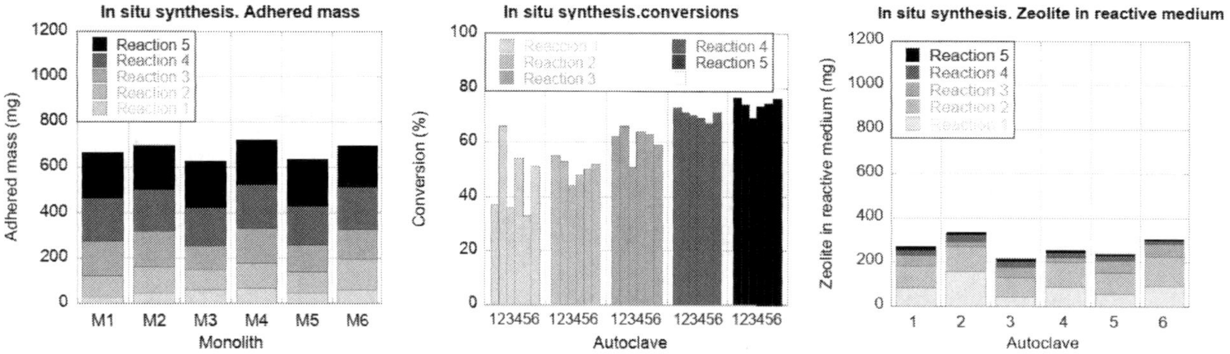

a b c

Figure 6. Evolution of the loaded mass, total conversion and the zeolite amount generated in reactive medium during 5 syntheses for Monoliths M1 to M6/Autoclaves 1 to 6.

But this first reaction was not enough to cover the surface of the monoliths, due to the higher surface of the monoliths (480 cm^2) compared to the surface of the plates (6 cm^2). It was necessary to carry out additional reactions to cover the surface with a zeolite film. In this sense, another four reactions were carried out with each monolith. The zeolite amount deposited during each of those syntheses was more or less similar for all monoliths, showing that the effect of the pre-coating with SiO_2 or TPABr on the first reaction disappeared. This happened because the crystals deposited in the first reaction acted as seeds nucleating the growing of the second reaction and so on. Increasing the number of syntheses, both, the loaded amount in every monolith and the global conversion (taking into account the zeolite deposited plus the zeolite generated in reactive medium) increased (Figure 6a and b) while the zeolite generated in reactive medium decreased (Figure 6c). This means that repeated synthesis increase the conversion and the selectivity to the heterogeneous nucleation on the monolith surfaces versus the heterogeneous nucleation in the liquid bulk producing zeolite powder.

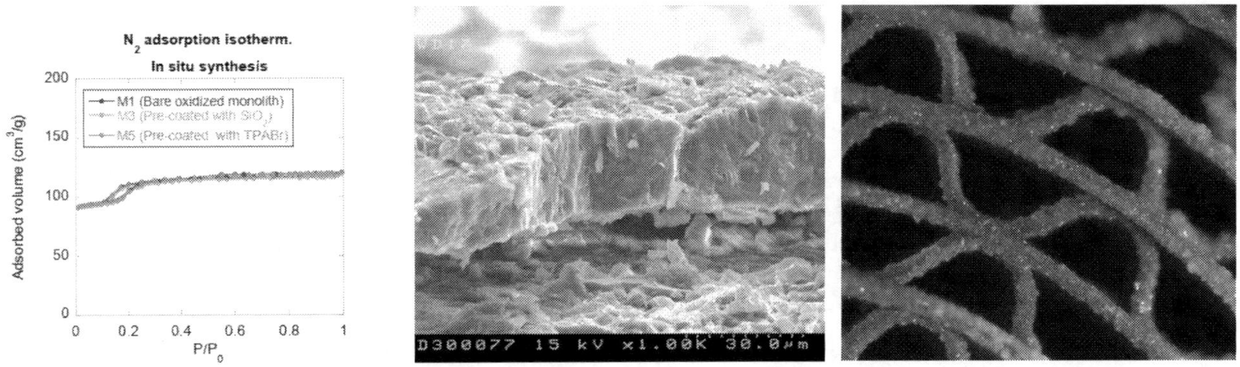

a b c

Figure 7. a) Textural properties of zeolitic layer obtained by in situ synthesis over metallic monoliths. b) SEM micrography of the profile of zeolite layer obtained by in situ synthesis (Monolith M5). c) Upper view of the channels of M5 monolith.

As it can be observed in Figure 7c, after 5 reactions the surface of the monoliths was completely covered with a transparent zeolite layer. The N_2 adsorption measurements showed a BET area of about 380 m^2/g in all cases and a isotherm shape corresponding to a pure microporous material (Figure 7a). The XRD confirmed that MFI type zeolite was obtained. The zeolite film is formed by completely interpenetrated crystals (Figure 7b), forming a continuous layer. Zeolite layers of around 1.5 mg/cm^2 were obtained after 5 reactions in all monoliths.

The result of adherence test was excellent, as no detachment at all was observed after it. This result showed that this layer is strongly adhered to the metallic surface and the consistence of the layer is also really good.

It could be concluded that no previous surface treatment is necessary to direct the crystal growing to the metallic surface. Just by means of consecutive reactions it was possible to achieve a proper catalyst layer, as the crystals deposited in every reaction acted as growing points in the next reaction, improving every reaction both the conversion and the selectivity to grow on the metallic surface.

4. Main conclusions

Completely covered Fecralloy micromonoliths were obtained with both coating methods. Surface coating by washcoating seems to be more interesting because is simpler and faster than in situ synthesis. Besides, the use of silica as binder affects the textural properties of the zeolitic coating, generating some inter crystalline mesoporosity that could be interesting for improving material transport properties. Randomly oriented zeolite crystal layer was obtained by washcoating while in situ growing produced more oriented layer. The adherence in all cases was good, being excellent for monoliths coated by in situ synthesis.

5. Acknowledgements

Financial support of Spanish Education Ministry (MAT2003-06540-CO2) and A. Eleta fellowship (FPI program) are gratefully acknowledged.

6. References

1. Garrot, B.; Simonot-Grange, M. H.; Clausse, B., Use of dealuminated faujasite Y and ZSM-5 zeolites for VOC's adsorption. Application of air cleaning. *Porous Materials in Environmentally Friendly Processes* **1999,** 125, 683-690.
2. Meininghaus, C. K. W.; Prins, R., Sorption of volatile organic compounds on hydrophobic zeolites. *Microporous and Mesoporous Materials* **2000,** 35-6, 349-365.
3. Avila, P.; Montes, M.; Miro, E. E., Monolithic reactors for environmental applications - A review on preparation technologies. *Chemical Engineering Journal* **2005,** 109, (1-3), 11-36.
4. Wan, Y. S. S.; Chau, J. L. H.; Gavriilidis, A.; Yeung, K. L., Design and fabrication of zeolite-based microreactors and membrane microseparators. *Microporous and Mesoporous Materials* **2001,** 42, (2-3), 157-175.
5. Meille, V., Review on methods to deposit catalysts on structured surfaces. *Applied Catalysis A-General* **2006,** 315, 1-17.
6. Zhang, X. F.; Liu, H. O.; Yeung, K. L., Influence of seed size on the formation and microstructure of zeolite silicalite-1 membranes by seeded growth. *Materials Chemistry and Physics* **2006,** 96, (1), 42-50.
7. Rebrov, E. V.; Seijger, G. B. F.; Calis, H. P. A.; de Croon, M. H. J. M.; van den Bleek, C. M.; Schouten, J. C., The preparation of highly ordered single layer ZSM-5 coating on prefabricated stainless steel microchannels. *Applied Catalysis a-General* **2001,** 206, (1), 125-143.
8. Yasaki, S.; Yoshino, Y.; Ihara, K.; Ohkubo, K. Method of manufacturing an exhaust gas purifying catalyst. US 5208206, **1991**.
9. Tao, T;Torns, J.M.;Usiak, B.P. In situ theta alumina coated monolithic catalyst support. US 6903051 B2, **2005**.

83t Nitration of Toluene in a Microreactor Using Mixed Acid

Raghunath Halder and Adeniyi Lawal. New Jersey Center for Microchemical Systems, Dept. of Chemical, Biomedical and Materials Engineering, Stevens Institute of Technology, Castle Point, Hoboken, NJ 07030

Highly exothermic reactions such as aromatic nitration can be carried out in a highly efficient manner in microreactors due to the high heat and mass transfer rates. In this study, nitration of toluene was conducted in a microreactor immersed in a constant temperature bath using mixed acid as nitrating agent. The effects of residence time, reaction temperature, and mixed acid composition on the reactor performance were studied and optimum operating conditions were determined. Studies were also made in a 25ml PARR batch reactor to compare the performance in the microreactor with that of the conventional batch reactor. The results indicate that reaction rate for mixed acid nitration in microreactor is an order of magnitude higher than that in the batch reactor. Also, the amount of by-products formed in the microreactor was much lower than that of the batch reactor. From the experimental data obtained in the microreactor, the kinetic rate expression for the reaction was developed.

160a Panelists

Barbara Jeol-Pieters, Yole Développement, 45 rue Sainte-Geneviève, Lyon, F-69006, France, **Richard Gray**, Syrris, Inc., 29 Albion Place, Charlestown, MA 02129, Thomas Bayer, A&d as Pa, Siemens AG, Industriepark Höchst, Bldg. G 811, Frankfurt, Germany, Juergen Brandner, Institute for Micro Process Engineering IMVT, Forschungszentrum Karlsruhe, Hermann-von-Helmholtz-Platz 1, D-76344 Eggenstein-Leopoldshafen, Germany, Martin Johnson, Chemical Process Research & Development, Eli Lilly &Co., Inc., 1400 W. Raymond St., Indianapolis, IN 46221, Jan J. Lerou, Velocys, Inc., 7950 Corporate Boulevard, Plain City, OH 43064, Norbert Kockmann, Process Optimization, Lonza, Lonzastr., Visp, 3930, Switzerland, Alberto Simoncelli, F&HC Engineering Process Breakthrough, P&G Technical Centres Limited, Whitley Road, Longbenton, Newcastle upon Tyne, NE12 9TS, United Kingdom, Michael Pollastri, Centre for chemical methodology and library development, Boston University, 590 Commonwealth Avenue, Boston, MA MA 02215, Flavie Sarrazin, LOF Laboratoire du Futur - Pessac - France, 178 avenue du docteur Schweitzer, Pessac, 33608, France, and Volker Hessel, Institut für Mikrotechnik Mainz GmbH, Carl-Zeiss-Straße 18-20, Mainz, D-55129, Germany.

This session provides short statements of the participants on the panel followed by a panel discussion open for all attendees which is moderated. After a short introduction on the use of Micro Process Engineering, attendees will interact both with industrial users and solutions providers in analysing reached performance (also in terms of costs) and existing offers, their advantages and drawbacks. We will discuss to which extent and where the technology will go into production. We will then analyse what impact MicroProcess Engineering will have on the chemical industry by cross checking needs and expectations from potential users with the advantages and added value of the technology and challenges to overcome.

65f Development of a Cross Flow Micro Heat Exchanger: Design and Analysis

Amol A. Kulkarni[1], Vivek V. Ranade[1], G. K. Sadekar[2], Chaitanya M. Karale[1], and Nishikant Mandal[3]. (1) Cepd, National Chemical Laboratory, Dr. Homi Bhabha Road, Pashan, Pune, 411008, India, (2) L&T Ltd., Mumbai, India, (3) Hed, L&T Ltd., Mumbai, India
Abstract Micro heat exchangers (ƒÝHEx) form an important component of the modular microplants that usually includes micromixers, microreactors, etc. Recently the advantages of miniaturization are seen to realize efficient process development and process intensification. Smaller dimensions help in achieving higher effective transport area, practically isothermal conditions in the reaction system and also suitability towards extreme conditions. Being smaller in size, these devices handle very small volumes of reacting liquids. As a result, they offer a relatively smaller time scale and hence are more suitable for the fast reactions. Being lightweight and compact, they facilitate rapid heat and mass transport, offer extremely precise control of process conditions, and deliver high performance. In comparison to the relatively a large body of literature available on the micromixers, micro reactors, very few papers deal with the micro heat exchangers. The micro heat exchangers (Alm, 2005; Schubert, 2001) are different from the conventional compact heat exchangers (Shah, 2006). In general, the micro heat exchangers are usually used for heating/cooling the reacting media by passing it directly through the heat exchanger, which is similar to the conventional heat exchanger applications in chemical plants. Although it offers extremely high heat transfer area density (> 5,000 m2/m3), due to the smaller dimensions they offer relatively high pressure drop and hence the internal design of heat exchanger and the fluid volume capacity limits the range of operational flow rates. The limited range of flow rates mainly restricts the operation to be laminar and hence structuring of the flow area becomes one of the very few ways of generating local turbulence. The most important design aspects of a ƒÝHEx are the design of flow area, distribution of fluid on the flow area and the accuracy in the connection of alternate plates that minimizes the risk of mixing of two fluids. First two of these issues raise the need of the detailed CFD simulation of the flow on a heat exchanger plate and identify better plate geometries that would help in enhancing the heat transfer coefficient (U). Although a few design configurations for the micro heat exchangers exist, here we have chosen the cross flow type of heat exchanger which is relatively simple to build and has potential to be a microreactor. A brief account of the experiments and the CFD simulations of the micro heat exchanger are given below. The performance of the in-house designed and fabricated heat exchanger was further compared with a commercially available heat exchanger for identical experimental conditions. The designed ƒÝHEx was fabricated in SS316 using precision micro-machining. The design of a single heat exchanger plate included three sections: a raised portion on the top of the plate with a recessed volume for flow of heat transfer fluid, a flat section that mainly acts as a flange for stacking of the plates and finally a recessed portion at the back of the raised portion that makes it sure that the raised portion of the next plate would fit in it. This approach helped in avoiding the mixing of fluids between the alternate plates. Typical heat exchanger plate is shown in Figure 1A. On each plate, the size of the region for fluid flow is 20mm x 20mm x 0.5mm. Thickness of a typical plate with heat transfer fluid is 2mm, while the end plates are 8mm thick to facilitate suitable projections for the standard Swagelok fittings. Experiments were carried out by passing the process fluid (40 - 80C) and the cooling fluid (26C) through the inlet ports for at different flow rates. Four thermocouples RS-32 were used for online temperature monitoring, one each at each of the inlet and outlet ports. The LMTD was estimated using suitable correction factor for cross flow and the heat transfer coefficient was obtained for different experimental conditions. The variation in the estimated heat transfer coefficient with inlet flow rate of the cold fluid for a heat exchanger with 5 plates is shown in Figure 2 and compared with the performance from a commercially available heat exchanger (IMM-CR). The range of heat transfer coefficients observed here

was similar to that reported in the literature and higher than that for conventional heat exchangers. CFD simulations were carried out to study the flow pattern over the heat exchanger plates and also to identify the strategy to simulate the heat transfer in the microscale dimensions. In order to achieve the above objectives, geometry of the system and subsequently the grid was generated using GAMBIT 3.6. In order to facilitate the suitable meshing of the geometry, the entire geometry was divided in different volumes having height of 0.5mm each. The flow volume and the solid volumes were meshed appropriately with suitable grid on the common surfaces. A typical grid for the flow volume is shown in Figure 1B. The simulations were carried out using FLUENT 6.3.9. The flow was solved assuming the laminar viscosity model and the conjugate heat transfer between the liquids (cooling and process fluids on alternate plates) was solved to obtain the temperature profiles. Simulations were carried out for a range of flow rates and inlet temperatures of the two fluid streams. Pressure drop across the fluid streams and also the temperatures of the outlet streams of both the fluids was monitored. Typical simulation results in terms of the velocity distribution and the temperature distribution for a heat transfer plate are shown in Figure 3. The difference in the experimental and simulated heat transfer coefficient for the case of ƒÝHEx with just 2 plates obtained at different inlet flow rates was less than 4.2%. Further simulations for different inlet flow rates, fluid temperatures, and different fluids and with more number of plates will be carried out. References: Alm, B.; Knitter, R. and Haubelt, J. (2005) Chem. Eng. Tech., 28(12), 1554 Schubert, K. et al. (2001) Microscale Thermphysics, 5, 17 Shah, R. (2006) Heat Trans. Eng. 27(5), 3

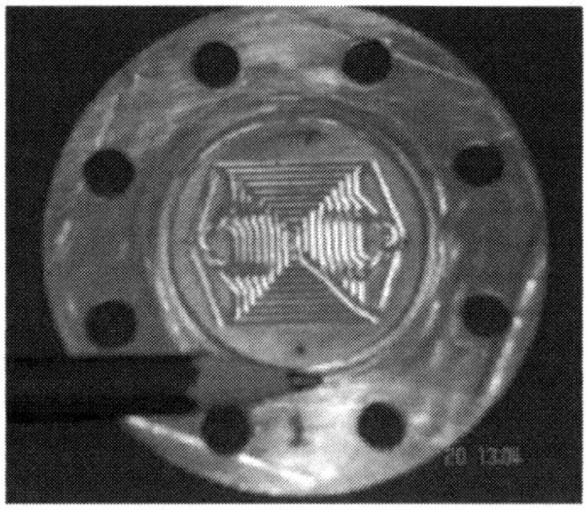

Figure 1: Typical exchanger plate

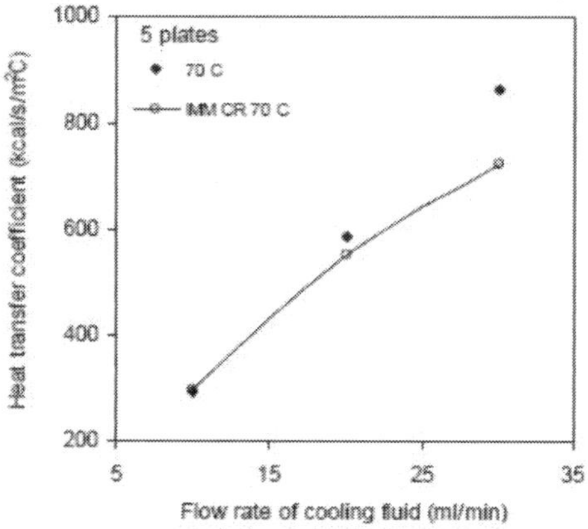

Figure 2: Experimental results

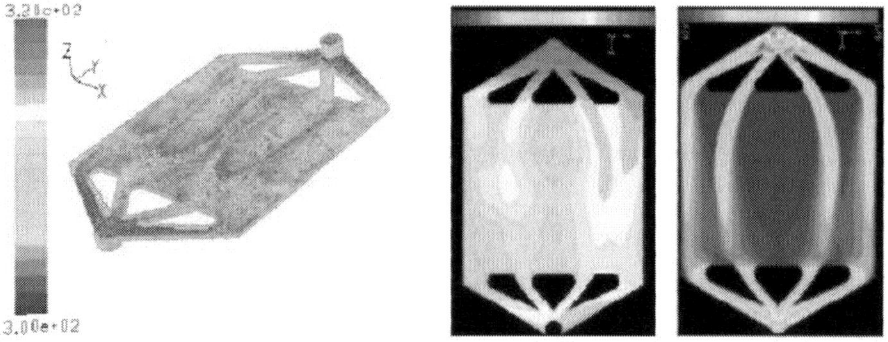

Figure 3: Simulated temperature variation over a grid and the contour plots for the temperature and velocity fields over a plate.

103a Microchip-Based Sensitive and Fast Gas Analysis System by Gas/ liquid Two Phase Flow and Thermal Lens Microscope

Kazuma Mawatari[1], Arata Aota[1], Yoshikuni Kikutani[1], Nobuyuki Hanaoka[2], and Takehiko Kitamori[3]. (1) Microchemistry group, Kanagawa Academy of Science and Technology, 3-2-1 Sakado, Takatsu, Kawasaki, Kanagawa, Japan, (2) Institute of Microchemical Technology, 3-2-1 Sakado, Takatsu, Kawasaki, Kanagawa, Japan, (3) Department of Applied Chemistry, Shool of Engineering, The University of Tokyo, 7-3-1 Hongo, Bunkyo, Tokyo, 113-8656, Japan

1. Introduction

In modern semiconductor industry, ultra-clean environment is increasingly desired. Especially, control of ammonia gas concentration in a clean room is quite important for fine patterning process because ammonia is a basic gas that can form salt with acidic gases and the salts deposited on a photomask turn into haze and deteriorates photo-patterning. In the next decade, control of ammonia concentration in 0.1 ppb level will be required as is shown in a semiconductor road map. In addition, a portable and continuous monitoring system is highly required. However, conventional methods which utilized macroscale chemical processes did not meet these demands due to long chemical processing time for concentration of ammonia gas in liquid and the large system size. Previously, we reported a basic method for sensitive gas (formaldehyde) analysis on a microchip with gas/liquid two phase flow and TLM (thermal lens microscope) detection. In this work, we further developed a sensitive (0.1ppb w/w) and fast analysis method on a microchip for ammonia gas utilizing microchip technology. The basic performance of the proposed method was investigated.

2. Experimental

We designed a glass microchip which included gas/liquid two phase flow formation, gas/liquid adsorption and concentration, gas/liquid separation, colorimetric reaction, and TLM detection processes. All these processes were integrated into a single microchip.

3. Results and discussion

In order to realize concentration (~100 times in weight unit) for sensitive detection, gas/liquid flow condition was investigated. Investigating liquid evaporation, volume flow rates of gas and liquid were set at 100mL/min and 3.7μL/min (~1μL/min after evaporation), respectively. In this condition, gas flow through the center of the microchannel, while liquid flow around the channel wall due to the viscosity difference. Next, gas/liquid adsorption and concentration was investigated with these flow conditions. For efficient gas adsorption, 25mM oxalic acid was added to the water. After 15ms contact time of gas with liquid, 97% ammonia gas was effectively adsorbed into the liquid due to the large surface-to-volume ratio. High concentration (97 times in weight unit) was achieved. Then, gas-liquid separation method was investigated. A reservoir (diameter of 2mm) was formed on the microchannel, and gas permeable PTFE membrane (pore size 5μm) was covered on the reservoir to avoid liquid leakage. The gas was effectively removed from the reservoir, and the liquid was introduced to downstream for colorimetric reaction and TLM detection. Finally, colorimetric reaction (indophenol method) and TLM detection were investigated. Two step colorimetric reaction (10 min) and TLM detection (633 nm excitation) was conducted. From the calibration curve, lower limit of detection (LOD) of approximately 1ppb ammonium solution was obtained. By considering these performances, 0.01ppb sensitivity in less than 15 min was obtained, and the applicability was verified.

ADVANTAGES OF ON-SITE MICROREACTORS FROM SAFETY VIEWPOINT

Fatemeh Ebrahimi, Eero Kolehmainen, Terhi Virkki-Hatakka, Benguang Rong, and Ilkka Turunen. Lappeenranta University of Technology, Lappeenranta, Finland

Introduction

On-site production enables the use of products at the place of consumption, thus avoiding the transportation and storage of dangerous materials. These factors improve the safety and may also improve economics.

Safety can be increased in microreactor by minimizing the amount of reaction medium, the amount of head space above the reaction medium, the storage requirements of the final product, and generating a minimum amount of a product on demand only as needed. Also, the efficient heat transfer, resulting from high surface area to volume ratio, increases safety e.g. by minimizing the risks of run-away reactions.

In this study, safety advantages of on-site production in microreactors were studied and production of peracetic acid was used as an example process. Peracetic acid is an unstable substance and therefore safety aspects in the production have to be emphasized [1]. Safety advantages of microreactors were studied by comparing of conventional and microprocesses with different safety evaluation methods. Conventional production process with the capacity of 170 kg/h and an on-site microprocess producing 10 kg/h peracetic acid for municipal wastewater treatment were studied. Preliminary design of these processes was carried out.

There are many methods for the safety evaluation. Information requirements of the methods are different, also the results produced vary. In this study, four safety methods were applied:

1. Reaction matrix
2. Dow's Fire and Explosion Index
3. Inherent Safety Index
4. Worst case and consequence analysis

These methods are applied in different stages of design. In the preliminary design phase, the available information is limited. At this stage, however, the influence of the decisions is the most important because they determine major features of the process. The reaction matrix can be prepared based on the earliest knowledge of proposed reaction paths, raw materials, intermediates and all the other materials [2]. Dow's Fire and Explosion Index is a suitable safety analysis method for predesign (conceptual design) stage [3]. Inherent safety index [4,5,6] can be applied quite well in the R&D and process predesign stages. Worst case and consequence analysis is a method to recognize the most harmful events and analyze their consequences.

Production of peracetic acid

Peracetic acid is an aliphatic peroxy acid. It is a strong oxidant, which is used e.g. as a disinfection and bleaching agent. Peracetic acid can be prepared from acetic acid and hydrogen peroxide. The equilibrium reaction is catalyzed by sulfuric acid. Water is also present in the system. The reaction scheme is below.

$$CH_3COOH + H_2O_2 \xleftrightarrow{H_2SO_4} CH_3COOOH + H_2O \quad (1)$$

For many oxidations, only low concentrations of peracetic acid are needed and equilibrium mixtures can be used directly [1]. For example, in wastewater disinfection, the concentration of 15 mg/l (peracetic acid in wastewater) leads to efficient elimination of microbes [7,8]. In production and handling of peracetic acid, attention should be paid to conditions and concentrations to minimize the risk of thermal decomposition.

Conventional process

Usually peracetic acid is produced in a stirred tank reactor, which is connected to a distillation unit. Sulfuric acid is first fed into the reactor. After that, acetic acid and hydrogen peroxide are added. The mixture is heated up to 60 °C at which the reaction takes place. After the reaction stage, vacuum of 3...17 kPa is drawn to the reactor. Vaporization of the reaction mixture begins. Acetic acid, hydrogen peroxide and peracetic acid are separated from sulfuric acid in distillation column. Sulfuric acid remains in the reactor.

Microscale process

The microscale process represents a continuous production mode. A simple block diagram of the continuous peracetic acid process is shown in Figure 1. The process includes of two mixing steps. In the first mixing step (mixing I), acetic acid and sulfuric acid are mixed. Next, hydrogen peroxide is added to the mixture (mixing II). The reaction takes place continuously in the reactor (reaction). After the reactor, the equilibrium mixture containing acetic acid, hydrogen peroxide, peracetic acid, sulfuric acid and water is ready for further use or treatment.

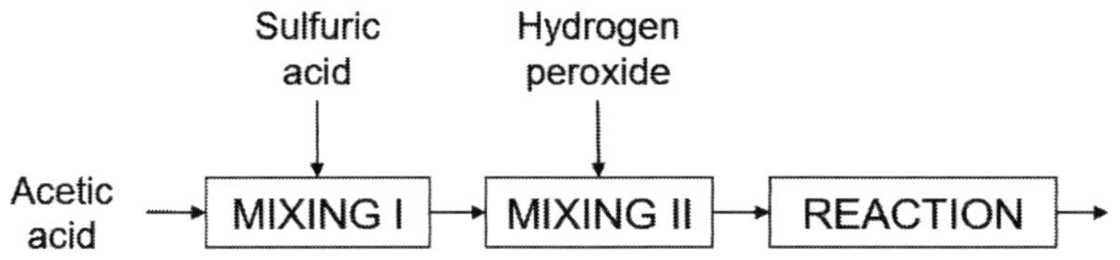

Figure 1. Block diagram for the continuous microscale process.

The microscale reactor concept consists of a multichannel reactor unit, which includes submillimeter sized channels (0.25 mm). In this reactor, more efficient heat transfer can be achieved than in a stirred tank reactor. Therefore, reaction can be performed safely in higher temperatures with higher reaction rate. Microreactors can be built as mechanically strong and they tolerate higher pressure.

For comparison, detailed designs with PI-diagrams for both processes were performed.

Safety criteria

Reaction matrix

The reaction matrix is an interaction matrix, which is a useful tool to predict desired and undesired reactions and interactions between all the materials in the process. It is particularly valuable in the development stage of a new chemical process to identify undesired chemical interactions [2]. In this study, results show similar incompatibilities in both processes. The essential factors such as size of equipment and quantity of chemicals are not taken into account in this method. The reaction matrix does not reveal any differences in the safety of the microreactor and the conventional process.

Dow's Fire and Explosion Index

Dow's Fire and Explosion index is a stepwise evaluation of the realistic fire, explosion and reactivity potential of a process. The accurate plot plan of the plant and the process flow diagram are needed in the evaluation. The evaluation is based on the penalties and factors, which are judged according to reaction and substance characteristics, material handling, operational conditions and equipment characteristic. The obtained index and factors are used in determining the area of exposure and probable property damages [9]. The summary of the process risks and the comparison between microprocess and conventional process is presented in Table 1.

Table 1. Dow's Fire and Explosion Index for the conventional batch process and continuous on-site microprocess.

Process Unit Risk Analysis	**Conventional process**	**On-site microprocess**
Material Factor for peracetic acid	40	40
Fire & Explosion Index(F&EI)	***280.4***	***122.4***
Radius of Exposure [ft]	240	105
Area of Exposure [ft^2]	$2 \cdot 10^5$	$0.5 \cdot 10^5$
Value of Area of Exposure [$]	$5 \cdot 10^6$	$5 \cdot 10^5$
Damage Factor [-]	0.98	0.80
Actual Maximum Probable Property Damage [$]	$5 \cdot 10^6$	$4 \cdot 10^5$
Maximum Probable Days Outage day	75	15

This method indicates that in this case the on-site microprocess is much safer than the conventional process. The index is lower and area and value of exposure are smaller for microprocesses. The different sizes of equipment of these two processes have not been taken fully into account in this method. The method is based on conventional processes and large volume production. Often the quantity of chemicals in microreactors is substantially smaller, therefore some definitions of the penalties are not covering such small sizes of microreactors.

Inherent Safety Index

Heikkilä [4,5,6] has published an inherent safety index method for conceptual chemical process design. A numerical index for a studied process is determined, and the total safety of the process is then evaluated by comparing it with a reference process.

In the original method, the total inherent safety index is divided into chemical inherent safety index I_{CI} and process inherent safety index I_{PI}, which both represent a sum of inherent safety subindexes.

$$I_{TI} = I_{CI} + I_{PI} \tag{2}$$

Since the comparison is made now between two process types producing the same product via similar reaction route, the sum of chemical inherent safety subindexes, i.e. *reaction heats, flammability, explosiveness, toxicity, corrosiveness and chemical interaction indexes,* will be similar for both processes. Therefore, the differences between total inherent safety indexes of the processes are explained only by process inherent safety index:

$$I_{PI} = I_I + I_{T,max} + I_{p,max} + I_{EQ,max} + I_{ST} \quad (3)$$

where I_I inventory (ISBL or OSBL, inside or off-site battery limit area)

$I_{T,max}$ process temperature

$I_{p,max}$ process pressure

$I_{EQ,max}$ equipment safety

I_{ST} process structure

The inherent safety indexes based on index determination tables [5] for both conventional batch process and continuous on-site microprocess are shown in Table 2. Lower index values mean inherently safer process.

Table 2. Process inherent safety subindexes for the conventional batch process and continuous on-site microprocess.

Safety subindex	**Conventional batch process**	**Score**	**Continuous on-site microprocess**	**Score**
Inventory (OSBL), I_I	100 - 500 t	2	0 -10 t	0
Process temperature, $I_{T,max}$	60 °C	0	80 °C	1
Process pressure, $I_{p,max}$	1 bar	0	5 bars	1
Equipment safety (max of ISBL or OSBL), $I_{EQ,max}$	reactor	2	reactor	2
Process structure, I_{ST}	sound engineering practice	1	no data or neutral	2
Total Inherent Safety Index, $I_{TI} = I_{CI}$		5		6

This method indicates that the conventional batch process is safer than on-site microprocess. In determining of equipment safety index in Table 2, the type of reactor has not been specified. The inherent safety index does not take into account the size of the reactor that inevitably makes microprocess inherently safe due to small internal volumes and efficient temperature control. It is clear that the reactor technology has significant effect on inherent safety of the process.

Process structure subindex in Table 2 favors the standard and well known processes and equipment. Novel process solutions such as microreactors suffer in this method, although they might be considered generally safer technology. Also, the method does not take into account the risks in chemical transport that is eliminated mostly in on-site production.

Because of the above mentioned aspects, the results may be misleading. It can be concluded that the inherent safety index is more suitable for evaluating safety of conventional technology than novel processes.

Worst case and consequence analysis

Worst case scenario analysis can be used for the study of the worst cases and the associated consequences. For the peracetic acid process, the following worst cases are recognized. The worst cases relate to materials, operation condition, process equipment, transportation and other factors:

1. Materials
    - Worst case: Release
    - Consequences: Release of sulfuric acid, acetic acid, hydrogen peroxide and peracetic acid

2. Operating conditions: Operating temperature and pressure
    - Worst case: Malfunction of heat exchanger
    - Consequences: reaction heat accumulation, run-away reaction, pressure rise, explosion, release of materials

3. Process equipment
    - Worst case: Corrosion
    - Consequences: Damage of equipment, release of materials and impurities

4. Raw material and product transportation
    - Worst case: Damage of material tank
    - Consequences: Release of material to environment, pollution

5. Other unpredictable external factor
    - Worst cases: Flood, tornado, earthquake, terrorism
    - Consequences: Release of material, explosion, fire

Usually worst cases refer to releases. Released hazardous or toxic materials may cause serious hazards to the area, which the materials reach [10]. For peracetic acid production processes, safety and hazard assessment of worst cases and consequences are considered. The comparison is presented in Table 3 by using the following safety grades:

5 - Very Serious Hazard
4 - Serious Hazard
3 - Medium Hazard
2 - Small Hazard
1 - No Hazard

Table 3. The comparison of the worst cases and consequences for the process alternatives.

Worst cases and consequences	**Conventional batch process**	**Continuous on-site microprocess**
1. Materials release	4	2
2. Operating conditions	2	2
3. Process equipment	4	3
4. Transportations	5	1
5. Unpredictable factors	4	2
In Total	19	10

Due to the intrinsic advantages of the on-site manufacturing of peracetic acid with a microprocess, it can be expected that the safety of the on-site microprocess would be clearly higher compared to the centralized conventional batch process. Therefore, the hazardous consequences of some worst cases are also expected to be significantly milder for the on-site microprocess than for the batch process. For example, Table 4 presents the brief comparison for the worst cases in fire, explosion and materials release.

Table 4. The worst cases and consequences for the process alternatives.

Worst cases and consequences	**Conventional batch process**	**Continuous on-site microprocess**
Worst case in fire	High detonation High damage of property More employees in danger	Small detonation Small damage of property Less employees in danger
Worst case in explosion	Big area of detonation High lost of capital More employees in danger	Small detonation of properties Small lost of capital Less employees in danger
Worst case in all materials release	Big area influence More employees in danger	Small area influence Less employees in danger

Based on the worst cases analysis for the peracetic acid production, the inherent safety improvement is one of the major driving forces to develop the on-site microprocesses to replace the conventional inherent unsafe processes.

Conclusions

The microreactor technology is an effective tool in small scale synthesis for hazardous chemicals and reactions, which are challenging from the viewpoint of scale-up and transportation. On-site production of chemicals may eliminate the transportation and storages of dangerous chemicals that is clearly beneficial from safety viewpoint.

In this study, the safety advantages of microreactors in on-site production are studied. The production of peracetic acid was selected as an example process. Peracetic acid is used as disinfection and bleaching agents. It is an unstable compound and therefore safety aspects in production are of great importance.

The safety study was performed by comparing on-site microprocess to conventional batch process, which operates centralized. The safety aspects were evaluated by means of four methods, which were reaction matrix, Dow's Fire and Explosion Index, inherent safety index and worst case and consequence analysis. The reaction matrix shows systematically the interactions between all the compounds and material involved in a process. The aim is to find out possible incompatibilities between materials. Basically, this method does not reveal any differences between production modes. The method does not take into account the differences in sizes and transportation needs between processes. Dow's Fire and Explosion Index proved the safety advantages of on-site microprocesses. The advantages were based on the smaller liquid volumes of the processes and storages. However, the method does not take completely into account dramatically smaller internal volumes of the on-site microprocess compared to conventional production. Inherent safety index favors clearly the conventional production technology indicating that it is safer than microprocess technology. The result is misleading and therefore the method is not reliable here. Worst case and consequence analysis indicated clearly higher safety for on-site microprocess compared to the conventional process. By means of this method, it was possible to emphasize the specific characters of the processes such as transportation needs and size difference. It is also suitable for the evaluation of totally novel processes.

Most of the conventional methods for safety evaluation are based at least partly on experience. Therefore, they are not at their best in the connection of novel technology, such as microreactors.

Microreactors have plenty of advantages from safety viewpoint, when compared to conventional production technologies. These advantages are mostly based on radically smaller reaction volumes and on effective heat transfer, which allow a fast temperature control. The safety advantages of microreactors can be effectively utilized in on-site production.

References

1. Swern, D. (1970). *Organic Peroxides*. Vol. 1, Wiley Interscience, New York London.

2. Mosley, D.W., Ness, N., Hendershot, D.C. (2000). Tools for understanding reactive chemical hazards early in process development, *34th Annual Loss Prevention symposium*.

3. Hurme, M, Rahman, M. (2005). Implementing inherent safety throughout process lifecycle, *Journal of Loss Prevention in the Process Industries*, 18, pp. 238-244.

4. Heikkilä, A.-M., Hurme, M., Järveläinen, M. (1996). Safety considerations in process synthesis, *Computers and Chemical Engineering*, 20, pp. s115-s120.

5. Heikkilä, A-M. (1999). Inherent safety in process plant design. *PhD Thesis. Technical Research Centre of Finland, VTT publications 385, Espoo.*

6. Rahman, M., Heikkilä, A-M., Hurme, M. (2005). Comparison of inherent safety indices in process concept evaluation, *Journal of Loss Prevention in the Process Industries*, 18, pp. 327-334.

7. Kitis, M. (2004). Disinfection of wastewater with peracetic acid: a review, *Environmental International*, 30, pp. 47-55.

8. Koivunen, J., Heinonen-Tanski, H. (2005). Peracetic acid (PAA) disinfection of primary, secondary and tertiary treated municipal wastewater, *Water Research*, 39, pp. 4445-4453.

9. Dow's Fire & Explosion Index Hazard Classification Guide. (1994). *AIChE technical manual, 7th edition, American Institute of Chemical Engineers*, New York.

10. Kleindorfer, P.R., Belke, J.C., Elliott, M.R., Lee, K., Lowe, R.A., Feldman, H.I. (2003). Accident epidemiology and the U.S. chemical industry: Accident history and worst-case data from RMP info, *Risk Analysis*, 23, No.5, pp. 865-881.

Supported Gold Catalysts for CO Oxidation in Microchannel Reactors

Leidy Marcela Martínez T[1], Francisca Romero-Sarria[1], María Isabel Domínguez[1], Miguel Ángel Centeno[1], José Antonio Odriozola[1] and Mario Montes[2].

(1) Inorganic Chemistry Department and Materials Science Institute, University of Sevilla-CSIC, Av. Américo Vespucio, 49, 41092 Sevilla, Spain.
e-mail: odrio@us.es
(2) Applied Chemistry Department, Chemical Engineering Group, Faculty of Chemistry, University of the Basque Country, Apdo. 1072, 20080 San Sebastián, Spain.
e-mail: mario.montes@ehu.es

1. Introduction.

The use of renewable sources to produce hydrogen that will feed fuel cells requires the oxidation of the produced CO during the reforming process in order to avoid catalyst poisoning. Water-gas-shift and CO oxidation reactions take care of reducing the CO concentration at the reformer-outlet to acceptable levels for the fuel cell (below 10 ppm). Besides this, miniaturization and the requirements for portable devices (cell phones, laptops ...) have pushed the research on microchannel catalytic reactors for these reactions. Gold catalysts are highly active and selective in low temperature CO oxidation and WGS reactions. Gold activity depends on the size and shape of gold particles,[1] the gold oxidation state and nature of the support,[2-5] which have been classified as inert (Al_2O_3, SiO_2, MgO) or active ones (Fe_2O_3, CeO_2, MnO_2) in function of their reducibility.[6] When using a catalytic microreactor both the activity and selectivity of the catalyst, as well as the reactivity of the metallic alloy in the reaction atmosphere, may determine the life cycle of the catalytic device. In this work, we study the oxidation of CO over CeO_2 and MnO_2 (cryptomelane) supported gold deposited on microchannels built on austenitic (AISI 304) and aluminium-alloyed ferritic (FeCrAlloy®) stainless steels. In this way, we test the chemical compatibility between the oxides forming the steel scale and the catalytic deposits, as well as the possible modifications induced in the catalytic performances due to the steel nature. The catalytic results obtained in the microchannel devices are compared with the ones obtained for the un-deposited powders. Two catalysts have been prepared (1%Au/CeO_2 and 1% Au/cryptomelane) and microchannel reactors built from AISI 304 and FeCrAlloy® sheets coated and tested in CO oxidation. The powders catalysts and the fresh and reacted monoliths devices have been characterised. The coated microchannel reactors were characterised by BET, XRD, TEM and adherence tests. The 1%Au/CeO_2 catalyst results in total CO oxidation conversion at lower temperatures than 1%Au/cryptomelane. However, at room temperature, the Au/cryptomelane catalyst shows higher conversion than the Au/CeO_2 solid. Moreover, differences of behaviour were detected for the 1%Au/cryptomelane in successive cycles of reaction.

2. Experimental

Gold/ceria colloid was prepared using a non-conventional synthesis procedure for supported gold catalysts described elsewhere.[7] A commercial colloidal solution of cerium acetate (Nyacol) was used as the ceria precursor. The adequate amount of gold acetate (Alfa Aesar 99.99%) to obtain a gold concentration of about 1% weight in the final solid was added to the colloidal solution and the mixture was stirred to homogenize (pH = 3, viscosity= 3 cP, isoelectric point= 6.3). The use of this metallic precursor is an innovation and prevents the damage that chlorine would cause to the metallic surface, especially to the austenitic steel, if the gold precursor usually employed, hydrogen tetrachloroaurate,[8,9] were used.

Cryptomelane was prepared by the reflux method.[10,11] A dissolution of $KMnO_4$ is slowly added to the mixture formed by the dissolution of $Mn(CH_3COO)_2$ and $K(CH_3COO)_2$ in glacial acetic acid. Later, the dissolution is kept under reflux for 24 hours, filtered and washed until the elimination of the acetic acid. To prepare cryptomelane colloid, the synthesized solid was used without the drying step in order to favour its dispersion in water. Cryptomelane (14 g) was dissolved in 140 ml of deionised water under constant agitation to obtain a concentration in solid of 10 % in weight (pH=3.3, viscosity=5.2 cP, isoelectric point=5.0). As in the case of the gold/ceria colloid, the gold/cryptomelane colloid was prepared by mixing the colloidal suspension of cryptomelane with gold acetate in the necessary quantities to obtain concentrations of gold of 1 % in weight with regard to the support weight, then, the obtained suspensions were left in agitation during 24 h.

For the manufacture of the monoliths, two types of stainless steel were used: austenitic (AISI 304) and ferritic (FeCrAlloy), both supplied by Goodfellow. Cylindrical monoliths were made rolling together flat and corrugated steel sheets to form spiral-type monoliths with the following characteristics: L=30 mm, d=16 mm, V=6 cm^3, cells density=55 cells/cm^2. Before the coating of the catalyst, the metallic monoliths were submitted to a thermal treatment to produce a surface composition and roughness convenient to assure the catalyst adherence. Thus, AISI 304 and Fecralloy monoliths were heated in air at 900 °C for 1 and 22 h, respectively.[12,13] Monolith coating was performed according to a procedure previously described.[14] This procedure was repeated twice in order to increase the quantity of deposited catalyst and to reach similar weights (≅ 100 mg). The powdered catalysts were obtained from the prepared colloids after drying at 80°C and calcining for 2h at 450°C (gold/cryptomelane) or for 4h at 300°C (gold/ceria).

Chemical composition of the obtained catalytic systems was analysed by X-ray fluorescence spectrometry (XRF), employing a Siemens SRS 3000 sequential spectrophotometer with a rhodium tube as source of radiation. XRF measurements were performed onto pressed pellets (sample including 10 wt.% of wax). The textural properties were studied by N_2 adsorption measurements at liquid nitrogen temperature. The experiences were performed in a Micromeritics ASAP 2010 equipment, using a cell where the complete monolith was introduced. Before analysis, the samples were degassed for 2 h at 150°C in vacuum. X-ray diffraction (XRD) analysis was carried out on a X'Pert Pro Philips. Diffraction patterns were recorder with Cu Ka radiation (40 mA, 40 kV) over a range of 20–80°. The adherence of the coating was measured by the ultrasound test.[15] The adhesion is evaluated by calculating the amount of coating lost after dipping samples in a beaker containing acetone placed in an ultrasonic bath (Cole Parmer, 47 KHz, 130W) for 30 min.

SEM micrographs were taken using a ZEISS ULTRA 55. Transmission electron microscopy (TEM) observations were carried out in a Philips CM200 microscope operating at 200 kV. The samples were dispersed in ethanol by sonication and dropped on a copper grid coated with a carbon film.

The CO oxidation catalytic tests were carried out in a conventional continuous flow U-shape glass reactor working at atmospheric pressure. The monolith is placed over glass wool and carborundum (Prolabo). A thermocouple in contact with the monolith assures the right measure of temperature. The reaction products were analyzed by mass spectrometry, using a Balzers Thermostar benchtop mass spectrometer controlled by the software Balzers Quadstar 422 with capabilities for quantitative analysis. The light-off curves of CO oxidation (300°C or 450°C for ceria and cryptomelane respectively, 5°C/min) were obtained with a mixture 3.4% CO and 21% O_2 balanced by He at a total flow rate of 42 ml.min^{-1}. Empty reactor (without catalyst) shows no activity under such conditions. The catalysts were pre-activated *in situ* during 1 h at 300°C (or 450°C) with a mixture of 21% O_2 in He at a flow of 30 ml.min^{-1} and then stabilized at room temperature before the light-off curve started.

3. Results and discussion

For ensuring adherence of the catalytic layer to the metallic surface thermal oxidation of the metallic alloy have been carried out. Upon oxidation the alloy are covered by an oxide layer constituted by α-Al_2O_3 (corundum) in the case of the FeCrAlloy® monoliths and by a mixture of α-Cr_2O_3 and $Mn_{1+x}Cr_{2-x}O_{4-x}$ in the case of the AISI 304 monoliths. In addition to these main components the presence of $YAlO_3$ (perovskite) and martensite (α'-Fe) are observed for the oxidized surfaces of FeCrAlloy® and AISI 304 monoliths, respectively. A detailed description of the formed phase upon oxidation has been reported elsewhere.[16]

The oxidation of the alloys results in an increase in the surface roughness for both alloys being higher the attained roughness in the case of the surface, the formation of the alumina layer implied an increase of the surface roughness from 0.9 to 2.7 µm while the formation of the chromium-rich scale in the AISI 304 alloy implied an increase of the surface roughness from 0.5 to 1.3-1.4 µm. The higher roughness has to be associated with the morphology of the oxide crystals forming the scale. The α-Al_2O_3 crystals show a plate-like morphology pointing outwards the surface (figure 1), while the chromia scale show homogeneous irregular shaped crystals. As stated by the SEM images of the cross section the higher roughness is directly related with the bigger thickness of the oxide scale ca. 2.5 µm for the FeCrAlloy® monoliths and ca. 0,75 µm for the AISI 304 monoliths. However, the high oxidation temperature resulted in surface areas for both monoliths below 1 m^2/monolith.

The colloids used to deposit Au/crytomelane catalysts on the monoliths were analyzed before and after deposition. XRF results showed a gold concentration of 0.91 wt.% (selected nominal value was 1.0 wt.%). The XRD analysis of the powdered Au/cryptomelane catalyst reveals the presence of the cryptomelane phase (KMn_8O_{16}) and metallic gold. From the XRD diagram and using the Scherrer equation, the size of gold crystallites in the powdered catalyst has been estimated in 20 nm. After coating the metallic monoliths with the catalyst the gold crystallite size remains constant.

Table 2 presents the textural properties of the gold/cryptomelane monoliths as well as the ones corresponding to similar monoliths coated with cryptomelane alone. All the catalysts present N_2 type IV adsorption–desorption isotherms with hysteresis type H2 (not shown), characteristic of mesoporous solids with heterogeneous size and shape pore distribution. For both metallic substrates, the introduction of gold causes the decrease of the BET surface area, the pore volume and the pore diameter. As cryptomelane porosity is mainly originated from interparticle volume, this pore volume decrease suggests that gold nanoparticles fill the interparticle space reducing the surface area, pore volume and mean pore diameter. The coating weight loss during the adherence test is smaller than 1% by weight for FeCrAlloy substrate and smaller than 1.5% for AISI 304, showing excellent adhesion of the catalyst to the metallic substrate.

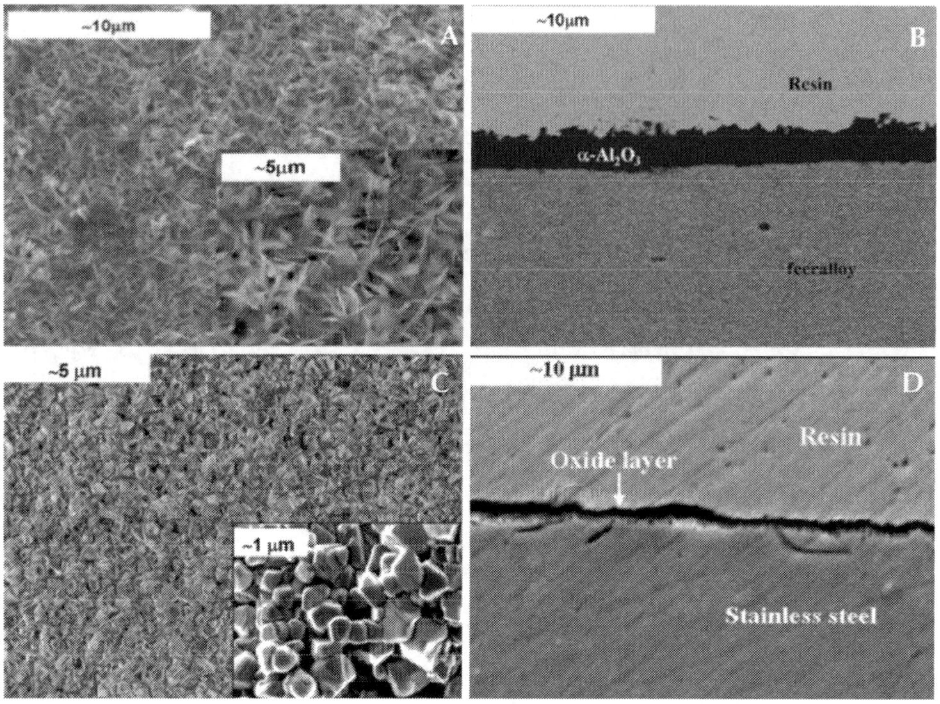

Figure 1. SEM micrographs of air-oxidized metallic alloys. A) Top view of FeCrAlloy® specimen at 900°C for 22h; B) Mirror polished cross-section of FeCrAlloy® specimen at 900°C for 22h; C) Top view of AISI 304 specimen at 900°C for 1h; D) Mirror polished cross-section of AISI 304 specimen at 900°C for 1h.

Table 2: Textural properties of the 1% gold loaded cryptomelane (MnO_x) catalysts

Characteristics	Powder		Fecralloy		AISI 304	
	MnO_x	Au/MnO_x	MnO_x	Au/ MnOx	MnO_x	Au/ MnO_x
Deposited coating (mg)	--	--	98	102	101	102
S_{BET} (m^2/monolith)	--	--	9	7	10	8
S_{BET} (m^2/g catalyst)*	115	72	92	75	99	79
Pore volume (cm^3/monolith)	--	--	0.05	0.02	0.04	0.03
Pore volume (cm^3/catalyst)*	0.47	0.23	0.47	0.16	0.40	0.27
Pore diameter (Å)	180	142	171	144	174	161

* Re-calculated values considering only the weight of the catalyst and not the total weight of the monolith

Transmission electron microscopy has been used for determining the particle size of the manganese oxide support and the gold catalyst. The metallic substrate does not influence the gold particle size. Gold particle are observed over a wide range of particle sizes appearing even particles as bigger as 160 nm. The gold particle size distribution has been obtained using FESEM by combining images obtained from secondary and backscattered electrons (figure 2).

From these images, it is clear the high dispersion of the gold particles in the surface. The size distribution histogram was obtained for an area of 17 mm x 13 mm, 1244 particles were analized using the software Micrograph (DM). The average size calculated from this distribution is 20 nm, what is in agreement with the value obtained applying the Scherrer's equation.

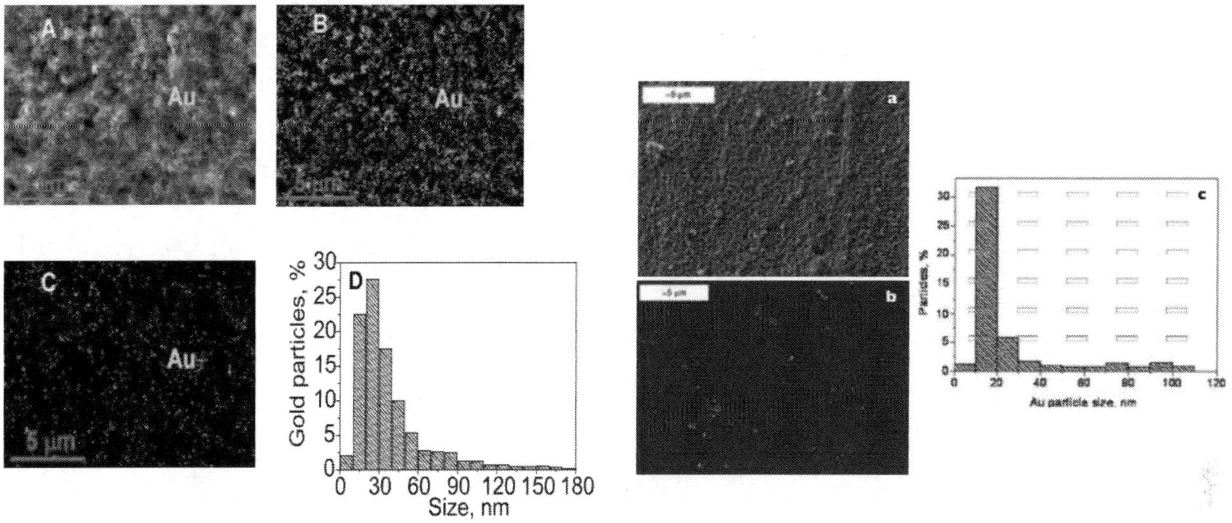

Figure 2. Left panel: FESEM top view of the AISI 304s steel surface coated with 1%Au-cryptomelane. Right panel: FESEM top view of the AISI 304 steel surface coated with 1%Au-CeO_2 (a) secondary electrons detector, (b) backscattered electrons detector, (c) histogram of the distribution of Au size particle

The colloids used to deposit Au/CeO_2 on the monoliths were analyzed before and after deposition. The ICP-OES results showed a gold concentration of 0.84 wt.% (selected nominal value was 1.0 wt.%). Besides this, the absence of supplementary compounds after depositing the catalyst indicates that the oxide layer generated during the thermal treatment is not affected by the catalyst deposition procedure. The textural properties of the catalysts coated with CeO_2 and Au/CeO_2 on the two metallic subtrates are summarized in table 3. The BET surface areas of the monoliths coated with CeO_2 have similar values for both materials. The average pore radius of 4.1 nm and 3.5 nm for the CeO2 layer on AISI 304 and Fecralloy steel respectively indicates that the catalytic layers are mesoporous. The addition of gold increases the BET surface areas and the pore volume. An excellent adherence of the layer created during the thermal treatment has been observed in all the cases.

The Scherrer equation allowed a rough estimation of crystallite sizes of $\approx$ 20 nm using gold diffraction peaks. Figure 2 shows FESEM micrographs using secondary electrons and backscattered electrons detectors respectively for 1%Au/CeO_2 deposited over AISI 304. The Au particles can be clearly distinguished from the background (figure 4b) allowing the estimate of the gold particle size distribution (figure 4c). The total number of Au particles is 281. Most of them have a size between 10-20 nm, but particles with a size below 10 nm have also been found.

In order to determine the reproducibility of the reaction and the regeneration capacity of the gold/cryptomelane catalysts, a sequence of four activity tests was realized. After each test, the heating was stopped and the reactor cooled in the reaction atmosphere until room temperature was reached, then and after the stabilization of the CO signal, a new light-off curve started. Figure 3 shows the ignition curves obtained. As it was expected by the similarity in the structural and textural characteristics of the powder and monolithic catalysts, no influence of the metallic substrate in the activity is observed. The presence of gold improves the catalytic activity of cryptomelane. This result demonstrates once again the strong influence of the gold in this type of reactions. Nonetheless, in any case the complete conversion to CO_2 was reached at temperatures lower than 100°C. It is necessary to emphasize that, at the beginning of the reaction, the complete conversion is reached over the powdered and the AISI 304 catalyst. Nevertheless, the activity falls down drastically within a few minutes of reaction up to approximately 10 % of conversion and then it recovers with the increase in the temperature. On the other hand, as the cycles of reaction increase, the catalysts experience an improvement in the catalytic activity, being lower temperatures necessary every time to oxidize CO, and even good conversion is obtained at room temperature.

Table 3: Textural properties of the 1% Au-CeO_2 catalysts

Characteristics	AISI 304		FeCrAlloy	
	CeO_2	Au-CeO_2	CeO_2	Au-CeO_2
Catalyst deposited (mg)	96.8	103.4	101.0	104.2
Adherence test (% Weight loss)	1.0	1.1	0.9	1.0
S_{BET} (m^2/monolith)	10	13	10	11
S_{BET} (m^2/g catal)	100	121	98	104
Pore volume (cm^3/monolith)	0.014	0.011	0.012	0.008
Pore volume (cm^3/g catal)	0.130	0.113	0.123	0.075

The catalytic activity of pre-treated monoliths (FeCrAlloy and AISI304) was tested in order to properly determine the activity of the coated catalysts. These results are not showed in this paper. The pretreated monolith of AISI 304 started to convert CO at 400°C, and at 500°C the conversion is lower than 5%. This activity is associated to the oxide layer formed during the thermal treatment.[17] The FeCrAlloy monolith after thermal treatment showed a higher activity in CO oxidation. In this case, the conversion of CO starts at 150°C and reach 10% at 300°C. At 500°C, the CO conversion was 47%. This fact evidences that the alumina on this material is more active than the oxides formed on the AISI 304 steel. Another factor influencing the catalytic activity is the higher roughness created on the FeCrAlloy compared with the AISI304 steel, which improves the gas-solid contact during the reaction.

The activity of the monoliths coated with Au/CeO_2 is shown in figure 4, where the effect of the catalyst is clearly observed (comparing with the data about the pretreated monoliths). The catalyst coated on a monolith of AISI 304 steel, the conversion become total at temperatures closed to 110°C. When the metallic subtract is FeCrAlloy, the total conversion of CO is only observed at temperatures higher than 130°C.

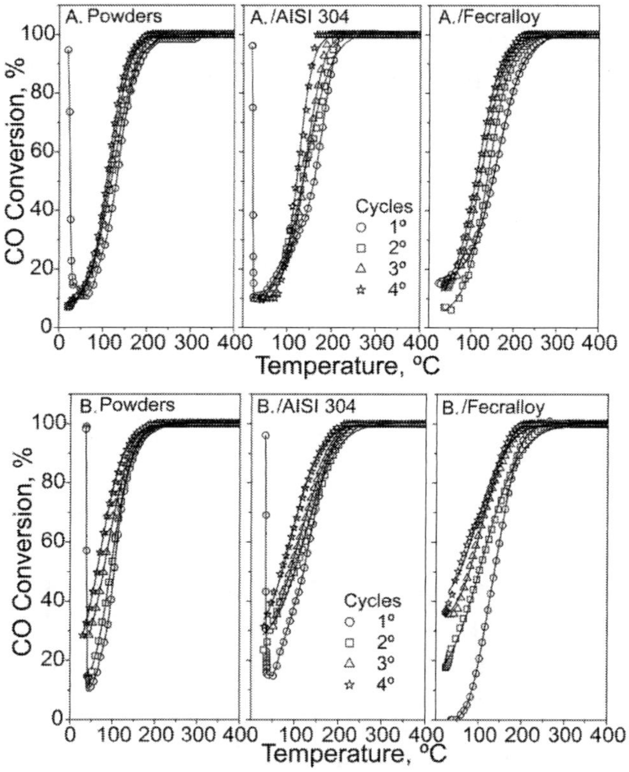

Figure 3. Light-off curves for 1%Au/Cryptomelane in powder and coated on AISI 304 and Fecralloy.

From this, we conclude the AISI 304 steel is a good alternative to the fabrication of metallic monoliths coated by Au-CeO_2. The AISI 304 steel is cheaper than FeCrAlloy and may be used at temperatures as high as 750°C. The use of FeCrAlloy is mandatory when catalytic reaction process in the limit of temperature (700-750°C), but considering the condition of CO oxidation reaction, the AISI 304 steel seems to be the more appropriate in the case of gold-ceria catalysts.

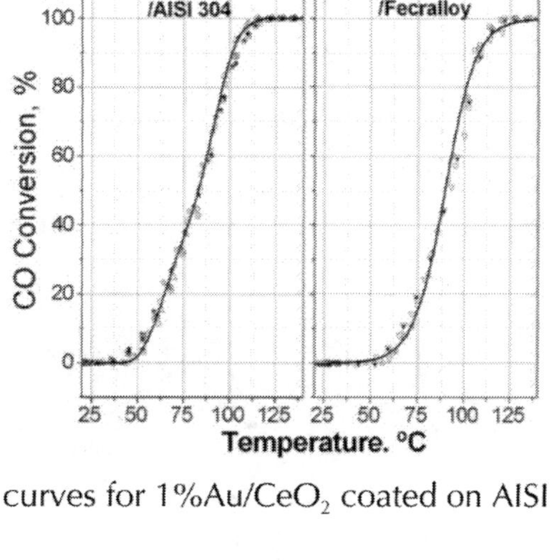

Figure 4. Light off curves for 1%Au/CeO_2 coated on AISI 304 and Fecralloy

The activity of the catalysts has been attributed to different factors and in some aspects the results reported in literature are contradictory. Metallic gold are generally considered the active species for the oxidation reaction,[14] although oxidized species have also been proposed to contribute to this reaction.[17-20] The activity of monoliths with Au-CeO_2 can be related no only to the redox properties of the CeO_2 phase but also to the modifications that it suffers in presence of Au. Gold can modify the CeO_2 properties by breaking the Ce-O bond and increasing the formation of CO_2.[14,15]

The influence of the metallic subtract has been proved in this work, and therefore, we propose that others factors should to be considered to explain this behavior. Among these factors are: the morphology of the metallic substrate and the support in terms of porosity and roughness, the presence of elements after pre-treatment procedures of the metallic monoliths and the interactions metal-support.

Acknowledgments

Financial support by Spanish Ministerio de Ciencia y Tecnología - MCYT (MAT2003-06540-CO_2), CYTED Programme (PI0269), Programme AlBan-European Union Programme of High Level Scholarships for Latin America (L.M. Martínez T Ph.D. UE E04D046878CO Scholarship). F. Romero-Sarria thanks the MEC by her contract in the "Ramón y Cajal" research program. L. M Martínez would like to acknowledge to Juan Almagro (ACERINOX S.A Spain) for his help with the SEM images.

References.

1. M. Haruta, Catal. Today 36 (1997) 153.
3. P. Concepción, S. Carrettin, A. Corma, Appl. Catal. A, 307 (2006) 42.
4. U. R. Pillai, S. Deevi, Appl. Catal. A 299 (2006) 266.
5. F. Arena, P. Famulari, G. Trunfio, G. Bonura, F. Frusteri, L. Spadaro, Appl. Catal. B 66 (2006) 81.
6. S. Minicò, S. Scirè, C. Crisafulli, A. M. Visco, S. Galvagno, Catalysis Letters 47 (1997) 273.
7. M.M. Schubert, S. Hackenberg, A.C. van Veen, M. Muhler, V. Plzak, R. J. Behm, J. Catal. 197 (2001) 113.
8. F. Romero-Sarria, L. M. Martínez, M. A. Centeno, J.A. Odriozola, J. Phys. Chem. C 111 (2007) 14469.
9. M. I. Domínguez, M. Sánchez, M. A. Centeno, M. Montes, J. A. Odriozola; Appl. Catal. A, 302 (2006) 96.
10. M. A. Centeno, K. Hadjiivanov, Tz. Venkov, Hr. Klimev, J. A. Odriozola; J. Mol. Catal. A, 252 (2006) 142.
11. J. Luo, Q. Zhang, A. Huang, S.L. Suib; Micropor. Mesopor. Mater., 35 (2000) 209.
12. D. M. Frías, S. Nousir, I. Barrio, M. Montes, L. M. Martínez T, M. A. Centeno, J. A. Odriozola; Appl. Catal. A, 325 (2007) 205.
13. L. M. Martínez T, D. M. Frías, M.A. Centeno, A. Paúl, M. Montes, J. A. Odriozola; Chem. Eng. Journal 136 (2008) 390.
14. L. Chapman, C. Vigor, J, Watton, US Patent, 4331 6314 (1982).
15. L. M. Martínez, M. A. Centeno, A. Paúl, J. A. Odriozola; "Stainless Steel`05 Science and Market". España, Septiembre (2005), 303.
16. S. Yasaki, Y. Yoshino, K. Ohkubo, US Patent 5,208,206 (1993).
17. H. E. Evans. Mat. Sci. Eng. A. 120 (1989) 139.
18. M.I. Domínguez, M. Sánchez, M.A. Centeno, M. Montes, J.A. Odriozola, Appl. Catal. A. 302 (2006) 96.
19. S. Tsubota, T Nakamura, M. Azuma, M. Haruta, Catal. Lett. 51 (1998) 53
20. L. M. Martínez T, D. M. Frías, M. A. Centeno, A. Paúl, M. Montes, J. A. Odriozola, Chem. Eng. J. 136 (2008) 390.

83s µ2 - Microwave Assisted Microreactor Processing: Synthesis of Ionic Liquids

Holger Löwe, Institut für Organische Chemie, Johannes-Gutenberg Universität Mainz, Duesbergweg 10-14, Mainz, 55128, Germany

Examining the opportunity to prepare ionic liquids in a continuous route the synthesis of 1-methyl-3-hexyl-imidazolium-bromide was investigated under different continuous procedures, using the micro-wave irradiation as well. Experiments were performed in a PARR Multiwave 3000TM microwave oven equipped with a ducts for insertion of tube to ensure a continuous flow. As reactor a micromixer-tube setup made of PEEK and PTFE was used. The tube diameter was 1/8" and different residence times could be adjusted by the tube length as well as by the flow rates of the supply pumps.

As a result it can be turned out that a low microwave energy-supply, partially less than 50 W, and a short residence time of approx. 10 min led to high conversion and selectivity of the targeted product. Also at higher microwave energies (>200W) the residence time could be shortened to less than two minute.

83u High Temperature Reactions in Microstructured Reactors – Synthesis of Biphenyl by Pyrolysis of Benzene

Holger Löwe and Daniel Metzke. Institut für Mikrotechnik Mainz GmbH, Carl-Zeiss-Straße 18-20, Mainz, D-55129, Germany

The aim of the previous investigations was to show the feasibility of microstructured reactor processing for high-temperature pyrolysis of benzene to biphenyl at 800°C. Such kind of reactions are mostly of radical nature and therefore very fast but with low selectivity. The main products for this reaction are low molecular hydrocarbons and coke while the formation of biphenyl by establishing a C-C bond is not preferred. To increase the yield of biphenyl the residence time should be short and therefore fast heating and quenching of the reaction mixture is necessary.

All the experimental results show the possibility to use microstructured reactors for biphenyl synthesis. At 800°C yields up to 30% were achieved during the initial time of the pyrolysis. The coke formation prevents tooks place even at the very short residence times for the gas flow in a microreactor. The yield decreases to 8 to 2% similar to the known conventional processes.

83af Investigation of Different Mixers in a Micro Reactor

Maria Miklautsch[1], Sascha Braune[1], Martina Kotthaus[1], Manfred Kraut[2], Rolf Marr[3], Peter Pöchlauer[1], Martin Vorbach[1], Achim Wenka[2], and Klaus Schubert[2]. (1) DSM Fine Chemicals Austria, St. Peter-Strasse 25, Linz, 4021, Austria, (2) Institute for Micro Process Engineering (IMVT), Forschungszentrum Karlsruhe GmbH, Hermann-von-Helmholtz Platz 1, 76344 Eggenstein-Leopoldshafen, Germany, (3) Department of Chemical Engineering and Environmental Technology, Graz University of Technology, Inffeldgasse 25/C/II, Graz, A-8010, Austria

Typical applications for microreactors are very fast exothermic reactions as for example Ritter reactions. Other areas are reactions with safety issues. A production scale microreactor has already been successfully implemented in the commercial production of an organic intermediate at DSM Linz in 2005. [1]

For further improvement of this Ritter reaction in the microreactor five different mixer types have been studied in the lab. The original cyclone mixer is compared with a double cyclone mixer, a Y-type mixer, a bi-laminating mixer and a five fold cyclone mixer. Additionally the effect of the mixing quality is studied to demonstrate the assumption that higher mixing quality will increase selectivity and thus the yield of the reaction.

The investigation gives a detailed idea of the impact of different flow rates and the effect of the temperature control in the first highly exothermal reaction step.

[1] Poechlauer, Peter; Kotthaus, Martina; Vorbach, Martin; Deak, Martin; Zich, Thomas; Marr, Rolf. Stepwise execution of exothermic reactions with participation of carbon cations. PCT Int. Appl. (2006) WO 2006125502 A1

209b In-Flow Reaction Monitoring by NMR on Nanoliter Samples in a μ-Fluidic Chip

J. Bart[1], A.J. De Vries[2], P. Nieuwland[3], J.W.G. Janssen[2], P.J.M. van Bentum[2], A.P.M. Kentgens[2], and J.G.E. Gardeniers[1]. (1) Mesoscale Chemical Systems, MESA+ Institute for Nanotechnology, University of Twente, P.O. Box 217, Enschede, NL-7500 AE, Netherlands, (2) Solid State NMR, Institute for Molecules and Materials, Radboud University, Toernooiveld 1, Nijmegen, NL-6525ED, Netherlands, (3) Synthetic Organic Chemistry, Institute for Molecules and Materials, Radboud University, Toernooiveld 1, Nijmegen, NL-6525ED, Netherlands
The growing interest in chemistry on small sample volumes performed on chip sets the demand for analytical methods which can handle corresponding volumes. Nuclear Magnetic Resonance (NMR) is a powerful analytical tool not only for determining complex (bio)molecular structures, but also for monitoring molecular dynamics and reaction kinetics. Despite its versatility, NMR analyses currently require relatively large amounts of the relevant molecules because of its intrinsic low sensitivity. We report on the design and measurement results of a NMR microchip that can analyze nanoliter samples. Initial measurements have shown a very high sensitivity combined with an excellent resolution.

NMR is used to study the properties of molecules containing nuclear spins by applying a magnetic field and observing the precessing of excited spins after irradiation with an electromagnetic field. The chemical shift of resonance peaks in NMR spectra is correlated to the electron configuration around a specific group in a molecule. This means that peaks in a spectrum can be directly assigned to a chemical group. Furhtermore, fine structures like J-coupling and dipolar coupling give a measure of the magnetic interaction and distance between two nuclei. NMR is a non invasive technique and provides direct quantitative information about the molecule under study.

Currently, NMR on small sample volumes is performed by means of small solenoids wrapped around a capillary [1], or planar coils on glass chips containing microfluidic channels, [2] [3] [4]. Despite many efforts, these approaches have not yet penetrated mainstream NMR spectroscopy. The most important problem is the fact that the nearby copper windings of the rf-coil tend to induce static field distortions that limits the resolution and (indirectly) the signal to noise performance. We designed a novel route towards microchip integrated NMR analysis, [5]. The basic element in the design is an rf stripline which can be defined in a single layer lithographic process and which is fully scalable to smaller dimensions. This stripline functions as the rf 'coil' which excites the spins. Due to the non-complicated copper geometry, the static field distortion is very low, resulting in a high resolution detection system. A microfluidic channel is integrated, which makes this chip feasible for in-flow measurements, (see figure 1). A probe was designed which contains the fluidical and electrical connections. A microreactor chip was implemented in this probe, which was directly connected to the NMR-chip via a short capillary, (see figures 2-4). All our measurements were carried out in a 600 MHz superconducting wide-bore magnet.

Single shot experiments showed unsurpassed sensitivity and resolution. Initial measurements with a non-integrated system containing only 12 nl sample have been carried out succesfully, [5]. In our fully integrated microfluidic chip containing 600 nl of pure ethanol, a Signal to Noise Ratio (SNR) of 5500 and a linewidth of 0.0017 ppm (1 Hz) full-width-at-half-maximum (FWHM) were achieved. Typical J-multiplets were resolved down to the baseline. This high SNR is especially necessary for in-flow kinetics monitoring, where samples stay for a limited time in the detection area.

NMR has been coupled to separation methods such as liquid chromatography (LC) [6] and capillary electrophoresis (CE) [7], allowing hyphenated analyses. Furthermore, microcoil NMR was used as a method to follow reaction kinetics, [4] [8] [9]. However, due to its intrinsically low sensitivity, these in-flow monitoring experiments remain to be a challenge, especially for low sample volumes. With our chip we were able to perform reaction monitoring on 600 nl sample volumes with minimal averaging (4x). For the monitoring experiments, a Micronit microreactor chip was implemented in the probe. Real-time monitoring of the carbamate formation from toluene diisocyanate (TDI) (0.5M in toluene) and pure ethanol in the microreactor chip by NMR was demonstrated. By changing the relative flow rates, the product formation could be studied as a function of ethanol stoichiometry. Figure 5 shows spectra recorded at different flow-rates. The product peaks clearly rise as a function of the relative compound concentration. From these spectra the conversion as function of the relative concentration can be fastly determined. Other reactions such as Diels-Alder reactions, alcohol protections and oxidation reactions, which are performed in microreactor chips are currently under study using the NMR microfluidic device.

References

[1] D.L. Olson, M.E. Lacey, J.V. Sweedler, Anal. Chem. 70, 645-650, 1998.

[2] J.D. Trumbull, I.K. Glasgow, D.J. Beebe, R.L. Magin, IEEE Trans. Biomed. Eng. 47, 3-7, 2000.

[3] C. Massin, F. Vincent, A. Homsy, K. Ehrmann, G. Boero, P.A. Besse, A. Daridon, E. Verpoorte, N.F. de Rooij, R.S. Popovic, J. Magn. Reson. 164, 242-255, 2003.

[4] H. Wensink, F. Benito-Lopez, D.C. Hermes, W. Verboom, J.G.E. Gardeniers, D.N. Reinhoudt, A. van den Berg, Lab Chip 5, 280-284, 2005.

[5] P.J.M. van Bentum, J.W.G. Janssen, A.P.M. Kentgens, J. Bart, J.G.E. Gardeniers, J. Magn. Reson. 189, 104–113 2007.

[6] D.L. Olson, J.A. Norcross, M. O'Neil-Johnson, P.F. Molitor, D.J. Detlefsen, A.G. Wilson, T.L. Peck, Anal. Chem. 76, 2966-2974, 2004.

[7] D.L. Olson, M.E. Lacey, A.G. Webb, J.V. Sweedler, Anal. Chem. 71, 3070-3076, 1999.

[8] L. Ciobanu, D.A. Jayawickrama, X. Zhang, A.G. Webb, J.V. Sweedler, Angew. Chem. Int. Ed. Engl. 42, 4669-4672, 2003.

[9] Y. Takahashi, M. Nakakoshi, S. Sakurai, Y. Akiyama, H. Suematsu, H. Utsumi, T. Kitamori, Analytical Sciences 23, 395-400, 2007.

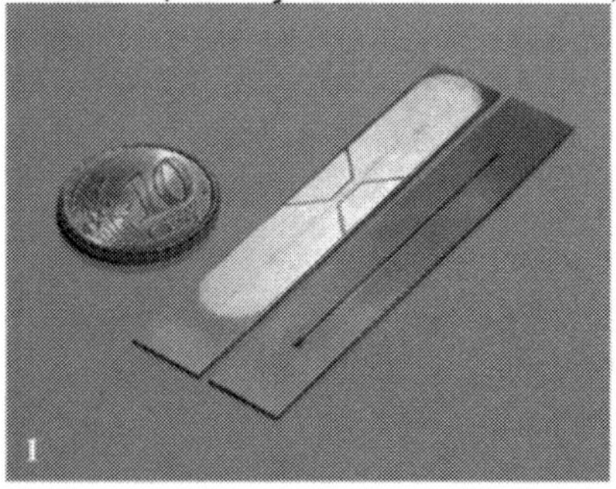

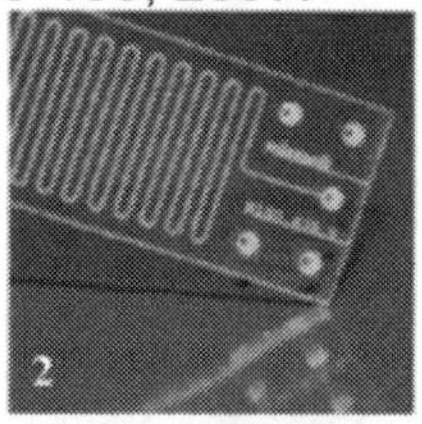

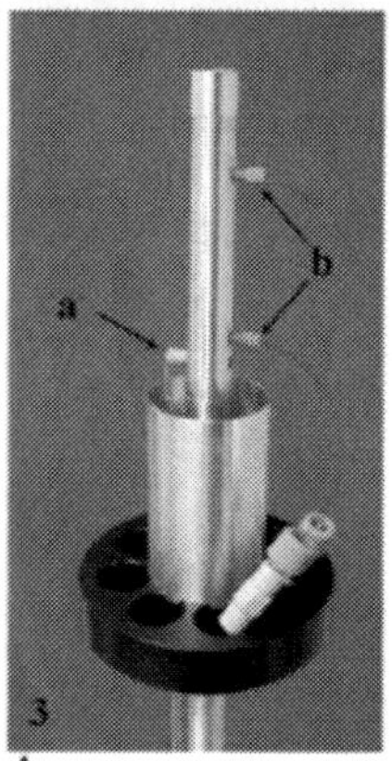

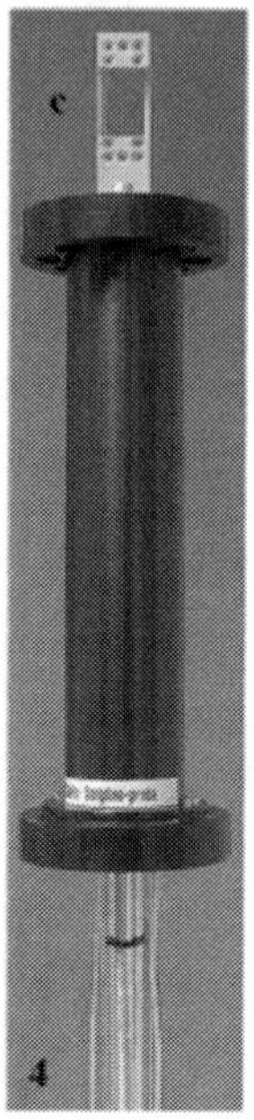

Fig. 1. Stripline chip (unbonded).
Fig. 2. Microreactor chip.
Fig. 3. Top part of the probe. The chip is positioned vertically in the aluminum tube, and connected electrically (a) as well as fluidically (b).
Fig. 4. Probe including shielding. The chip holder for the microreactor is mounted on top (c).

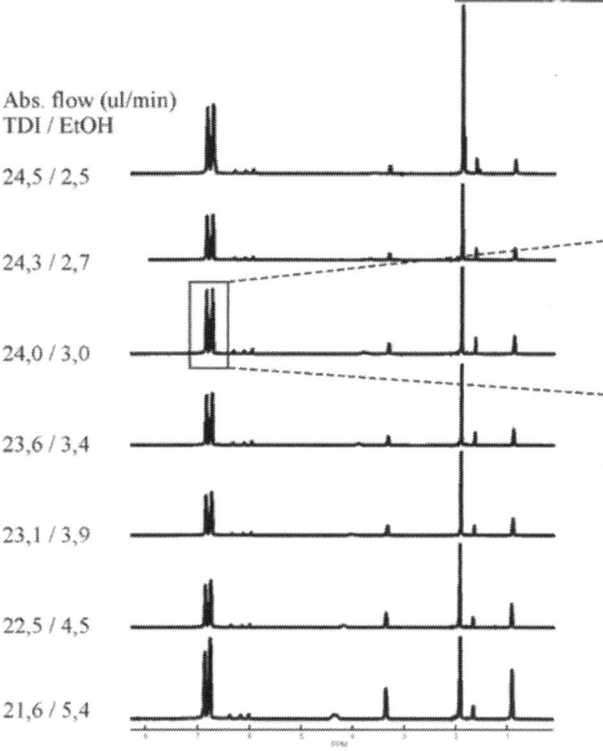

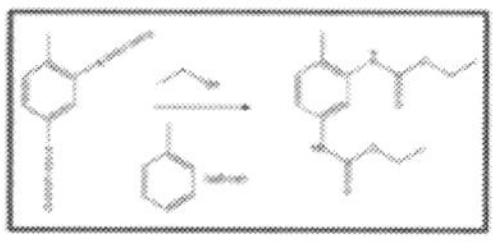

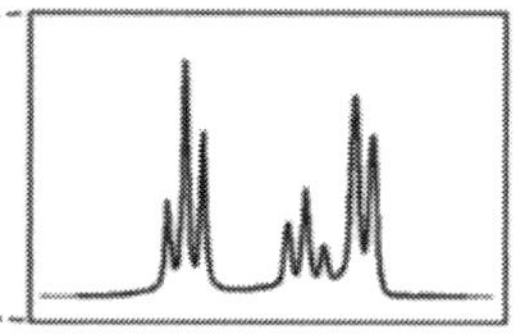

Fig. 5. Monitoring of carbamate formation from TDI (0,5M) and pure ethanol.
By increasing the relative flow rate of ethanol, the product peaks rise.
Inset: magnification of aromatic peaks, showing J-coupling.

Computational Fluid Dynamics Study of Heat Transfer in a Microchannel Reactor for Fischer-Tropsch Synthesis

Gurutze Arzamendi[1], Pedro M. Diéguez[2], Mario Montes[3], José Antonio Odriozola[4], Eduardo Falabella Sousa-Aguiar[5], and Luis M. Gandía[1,*]

[1]*Departamento de Química Aplicada, Edificio de los Acebos. Universidad Pública de Navarra, Campus de Arrosadía s/n, E-31006 Pamplona. Spain*

[2]*Departamento de Ingeniería Mecánica Energética y de Materiales, Edificio de los Pinos. Universidad Pública de Navarra, Campus de Arrosadía s/n, E-31006 Pamplona. Spain*

[3]*Grupo de Ingeniería Química, Departamento de Química Aplicada, Facultad de Ciencias Químicas de San Sebastián, UPV/EHU, Paseo Manuel de Lardizábal 3, 20018 San Sebastián, Spain*

[4]*Instituto de Ciencia de Materiales de Sevilla, Centro Mixto CSIC-Universidad de Sevilla, Avda. Américo Vespucio 49, 41092 Sevilla, Spain*

[5]*CENPES/PDEDS/Petrobrás. Av. Horácio Macedo, 950 – Ilha do Fundão CEP 21941-915 Rio de Janeiro, Brazil*

[*] **Corresponding author.** *E-mail address:* **lgandia@unavarra.es (L.M. Gandía)**

In this work, a Computational Fluid Dynamics (CFD) study with ANSYS CFX software of heat transfer in a microchannel reactor for the low-temperature Fischer-Tropsch synthesis (FTS) is presented. The simulated microreactor is a steel block 17 mm high, 21 mm width, and 21 mm long containing 80 square microchannels (1 mm high, 1 mm width, and 21 mm long) arranged in a cross flow configuration with 40 microchannels in each direction for the transport of syngas and cooling water, respectively. Gas hourly space velocities in the 5,000-30,000 h^{-1} (STP) range have been considered for the syngas stream (H_2/CO molar ratio = 2). An outstanding characteristic of the microreactor is the isothermicity exhibited by the system. It has been found that the synthesis can be conducted at any desired reaction temperature between 210 and 250°C within a wide range of CO conversion levels by setting the operating pressure at the appropriate value between about 5 and 40 atm and adjusting the cooling water flow rate in order to achieve the partial vaporization of water at a temperature slightly below the temperature at which the FTS has to be carried out. The microreactor has been analyzed as a cross-flow heat exchanger, resulting in values of the overall heat transfer coefficient in the 10-50 $W\cdot m^{-2}\cdot K^{-1}$ range, which are typical for flow boiling heat transfer in microchannels at the heat fluxes considered in this study.

Introduction

The Fischer-Tropsch (F-T) synthesis is a key step in the conversion of natural gas and solid feedstocks such as coal or biomass into liquid fuels (synfuels) through the so-called gas-to-liquids (GTL) and coal/biomass-to-liquids (CTL/BTL) processes. Among them, GTL is particularly attractive from the point of view of capital and operating costs, carbon and thermal efficiencies as well as environmental impact [1]. New deposits of natural gas are regularly being discovered; however, very often they are located far from the markets so that the transportation of the gas or the liquefied methane over long distances is too expensive and problematic [2]. On the other hand, for environmental reasons, there are increasing flaring constraints imposed on associated gases from oil fields. Within this context, there is an incipient GTL industry, also favoured by the present high oil prices, with a very large growth potential engaged in the valorization of these resources [1, 2]. Installed GTL capacity is above 40,000 barrels per day (bpd) but in the coming years this could rise to one million bpd [3]. More than half this capacity will be in Qatar, whereas new GTL plants will be also in operation in Nigeria, Egypt and Iran.

The FT synthesis (FTS) consists in the catalytic conversion of syngas into a complex mixture of linear paraffins and α-olefins, and water. The commercial FTS plants have utilized either Fe or Co catalysts although now the trend is towards using low-temperature FTS with Co catalysts in order to increase the middle distillates (gasoline and diesel) yields [4, 5]. The FTS can be viewed as a polymerization reaction where chain growth is the basic feature. Each growth step increases linearly by one the number of carbon atoms of the active hydrocarbon chain. Termination to paraffins by hydrogenation is irreversible whereas termination leading to the formation of olefins is reversible so that these species, especially the heaviest ones, can be readsorbed to continue growing or can be hydrogenated to paraffins. The probability of growth (α) strongly depends on the operating conditions (temperature, pressure, H_2/CO ratio and residence time) as well as on the catalyst properties [6, 7]. The FT reactors used commercially can be classified into two broad categories: high temperature (HTFT, 320-350ºC) and low temperature (LTFT, 220-250ºC) reactors [8]. The HTFT reactors include the fluidized and circulating fluidized bed reactors that are characterized by the absence of a liquid phase inside the reactor. In contrast, in the LTFT reactors which include the multitubular fixed-bed (Shell) and the slurry bubble column (Sasol), the catalyst is wet by a hydrocarbon liquid phase. All of these reactors suffer from several drawbacks and the existence of conflicting requirements makes them the result of a nonoptimal compromise, as analyzed by de Deugd et al. [6]. Microreactors or micro-structured reactors have the potential of solving many of the problems of the conventional FT reactors. It should be noted that present FTS reactors productivity is limited by their heat-removal capacity since the synthesis reaction is largely exothermic (ΔH_R = -165 kJ/mol_{CO}). One of the outstanding characteristics of microreactors is the heat transfer enhancement that results from the increase of the heat transfer coefficients and the heat exchange surface area per unit volume with decreasing the channel hydraulic diameter. Therefore, higher reactants concentrations and catalyst loadings are possible for exothermic reactions thus improving productivity [9]. Moreover, it is possible to achieve the virtually isothermal operation thus preventing the formation of hot spots which means a gain in operational safety as well as in selectivity for middle distillates since the chain growth probability decreases as temperature increases. Selectivity is also favoured by the very short diffusion lengths prevailing in microreactors thus preventing CO depletion which results in shorter chains. Like for monolithic reactors [6], microreactors can be easily stacked which allows for staged feeding to compensate nonstoichiometric H_2/CO ratios. Microchannel reactors are being commercialized for the FTS. The microchannel technology developed by Velocys Inc. and Pacific Northwest National Laboratory (Richland, WA) is based on a reactor coupling the exothermic FT synthesis with the partial boiling of water for heat removal [10].

CFD study

In this work, a Computational Fluid Dynamics (CFD) study with ANSYS CFX software of heat transfer in a microchannel reactor for FT synthesis is presented. Microfluidics is characterized by relevant particularities such as the facts that surface forces dominate over body forces and that due to the typical very low Reynolds numbers mixing is governed by diffusion. From the point of view of both the consistency of the results and the computational speed, this software tool has provided satisfactory results for a variety of microfluidic problems [11]. The ANSYS CFX code is based on the finite-volume method for spatial discretization and solving of the Navier-Stokes equations; both structured and unstructured meshes can be applied to the problem geometry and a fully coupled solver is used for flow speed and pressure [12].

The simulated microreactor is a steel block 17 mm high, 21 mm width, and 21 mm long containing 80 microchannels arranged in a cross flow configuration with 40 microchannels in each direction for the transport of syngas and cooling water, respectively. This block can be viewed as the stacking of 8 stainless steel sheets 2 mm thick, 21 mm width, and 21 mm long; in practice, the two outer sheets are 2.5 mm thick. Every sheet is stacked in an alternate position with regard to the one that is below to provide the cross flow configuration. The sheets have 10 parallel square microchannels 1 mm high, 1 mm width, and 21 mm long. The distance between contiguous channels is 1 mm in both directions whereas from every end of a sheet to the first one of the channels the distance is 0.5 mm. The access to and the exit from the channels is through rectangular chambers 17 mm high, 21 mm width, and 10 mm long that are connected to circular pipes (7.2 mm internal diameter) by means of pyramidal-shaped prolongations 10 mm long. Syngas and water are fed to the microreactor through the pipes. Figures 1 and 2 illustrate some aspects of the geometry adopted for the microchannel reactor under consideration.

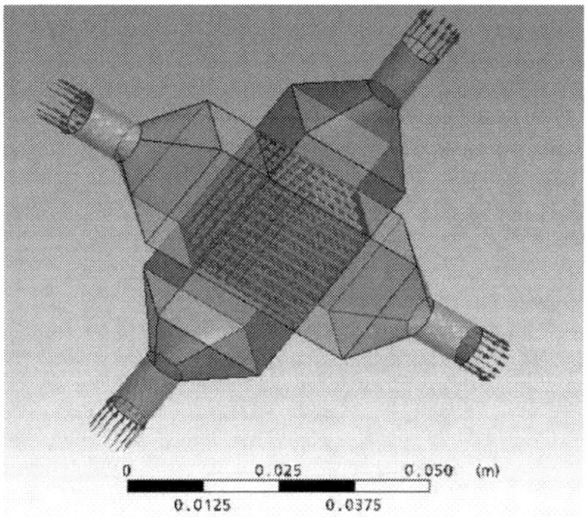

Figure 1. Solid structure of the microreactor

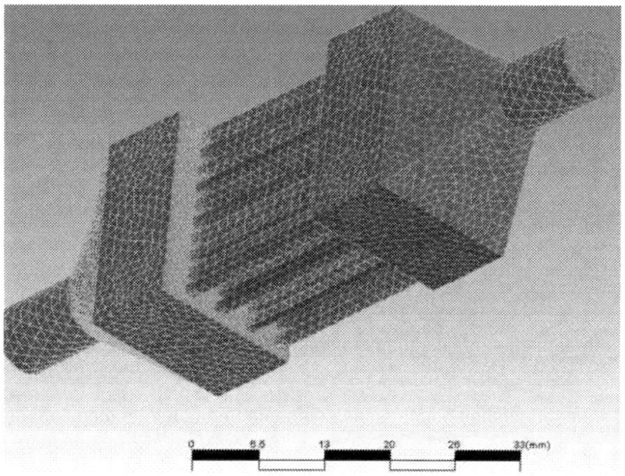

Figure 2. Fluidic region in one of the directions showing the mesh applied

Simulations have been conducted under stationary conditions. It has been assumed that a very thin layer of an active catalyst has been uniformly deposited onto the walls of the microchannels. Gas hourly space velocities (GHSV) in the 5,000-30,000 h^{-1} range at standard temperature and pressure (STP) have been considered for the syngas stream (H_2/CO molar ratio = 2). These GHSV values are well above the ones typical of the FT synthesis conducted in commercial slurry and tubular fixed bed reactors [13]. Taking into account that the available volume for the syngas stream in the microchannels is 0.84 cm^3, the adopted GHSVs imply feeding the microreactor with 70-420 cm^3/min (STP) of syngas. This results in a low contact time of 0.4-0.07 s that obviously require a very active FT catalyst such as the Co-Re/Al_2O_3 prepared by Jarosch et al. [10]. The aim of our work was to study the thermal performance of the microchannel reactor so the emphasis has been put on heat transfer. Therefore, and also due to the very complex kinetics of the FTS [14], this reaction has been modelled as a homogeneous heat source on the inner walls of the syngas microchannels, where the catalyst would be located. Thus, for a given molar flow rate of CO fed to the microreactor ($F_{CO,0}$, mol·s^{-1}) and CO conversion at the reactor outlet (X_{CO}) the heat generated ($\dot{Q}_g$) by the FTS synthesis has been calculated as follows:

$$\dot{Q}_g(kW \cdot m^{-2}) = \frac{F_{CO,0} \cdot X_{CO} \cdot (-\Delta H_R)}{40 \cdot (4 \cdot 1 \cdot 21) \cdot 10^{-6}} = 49107 \cdot F_{CO,0} \cdot X_{CO} \qquad (1)$$

This is of course an approximation because higher reaction and heat generation rates are expected at the microchannels entry. The simulations have evidenced that the velocity of the fluid at the entry is almost identical for the several microchannels, so the residence times should be also very similar. As concerns the cooling microchannels it has been assumed that water enters the microreactor as liquid saturated at the reactor pressure. Water flow rates in the range between 4.16×10^{-6} and 2.08×10^{-3} kg·s^{-1} (0.25-125 g·min^{-1}) have been considered.

Results and Discussion

A noticeable result from this simulation study is the isothermicity exhibited by the microreactor system. An example is shown in Figures 3 and 4 for a simulation conducted at syngas GHSV of 30,000 h^{-1} entering the reactor at 250°C; the system pressure was 20 atm and the CO conversion at the reactor outlet was set at 0.75. Heat removal was carried out with 125 g·min^{-1} of liquid water saturated at 212.9°C. In Figure 3, the temperatures of the solid steel block are depicted. As can be seen, the maximum difference of temperature inside the block is lower than 2°C, and the mean temperature of the solid is about 215°C.

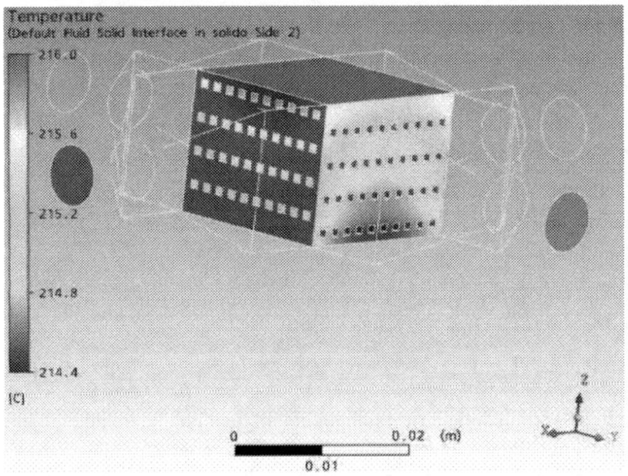

Figure 3. Isometric view of the solid block showing the distribution of temperatures (20 atm, 30,000 h^{-1}, $X_{CO} = 0.75$, 125 g·min^{-1} of liquid water at 212.9°C)

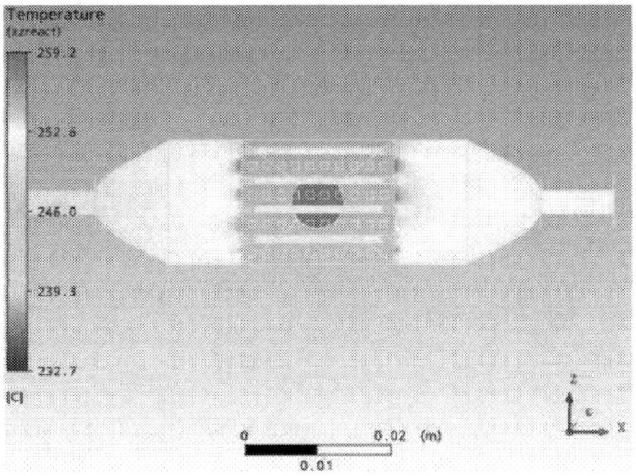

Figure 4. Distribution of temperatures in a plane of the syngas stream (conditions as for Figure 3)

The distribution of temperatures along a plane of the syngas stream is shown in Figure 4. The inlet temperature is 250ºC and the one at the exit section (area averaged) is 252.6ºC. This means that 3.8 $kW{\cdot}m^{-2}$ of heat generated are efficiently removed in order to maintain the temperature at the target value of 250ºC. Due to the relatively high water flow rate used in this case, the steam content of the water at the outlet is only 0.33% by weight. Entry and exit effects give rise to some small points with temperatures slightly higher (259ºC) or lower (240ºC) than 250ºC.

The aim of our study was to analyze the conditions under which the FTS could be conducted at a reasonably high CO conversion under close to isothermal conditions at any desired reaction temperature between 210 and 250ºC. It has been found that isothermicity can be achieved by setting the operating pressure at a value at which the boiling point of water is slightly below the temperature at which the FT synthesis has to be carried out and adjusting the cooling water flow rate. This allows to efficiently removing heat by steam generation.

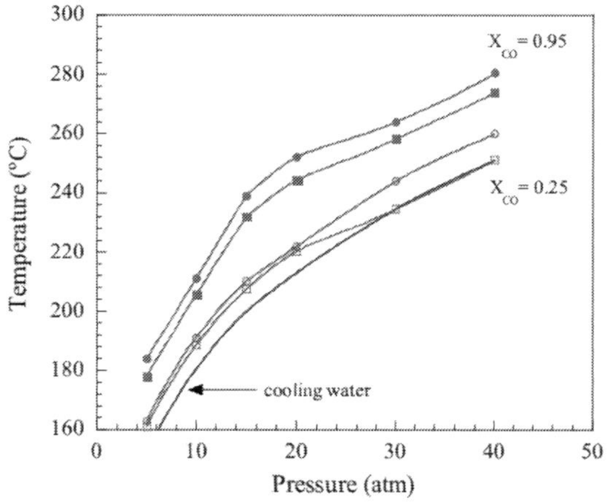

Figure 5. Results of CFD simulations conducted with syngas GHSV of 5,000 h^{-1} and cooling water flow rate of 0,125 $g{\cdot}min^{-1}$ showing the exit temperature of the syngas stream (circles), mean solid temperature (squares), and cooling water temperature (blue line) at two CO conversion levels, 95 (filled symbols) and 25% (open symbols)

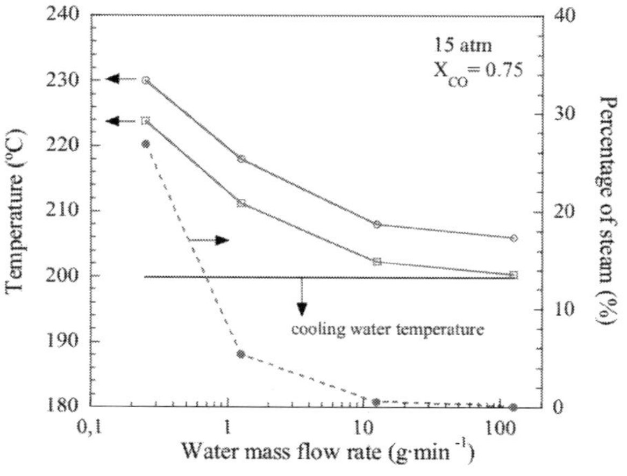

Figure 6. Results of CFD simulations conducted with syngas GHSV of 5,000 h^{-1}, 15 atm and CO conversion of 75% showing the effect of the cooling water flow rate on the exit temperature of the syngas stream (filled circles, solid line), mean solid temperature (squares) and mass fraction of steam generated (open circles, dotted line)

Taking into account that the low-temperature FTS over cobalt catalysts is usually carried out at temperatures in the range between 220 and 250ºC, it can be seen in Figure 5 that the synthesis can be conducted within a wide range of CO conversion levels by setting the operating pressure at the appropriate value between about 5 and 40 atm. The pressure has to be reduced as the temperature decreases and the CO conversion increases. It is well known that in this range the chain growth probability and the selectivity to middle distillates both increase with the operating pressure. As can be seen, at high CO conversions the mean temperature of the solid is closer to the FTS temperature than that of the cooling water.

On the other hand, if the pressure has to be established at some desired value, the FTS temperature can be controlled by adjusting the cooling water flow rate fed to the microreactor. As illustrated in Figure 6 for simulations conducted with syngas GHSV of 5,000 h^{-1}, 15 atm and CO conversion of 75%, it is possible carry out the FTS at temperatures between 210 and 240°C by varying the cooling water flow rate between 0.1 and 100 $g \cdot min^{-1}$. Drop pressure at the relatively high flow rate of 125 $g \cdot min^{-1}$ is as low as 7 Pa; this is due to the laminar regime of flow inside the microchannels. Obviously, the syngas and solid temperatures decrease and approach the refrigerant temperature as the water flow rate increases. Also shown in Figure 6 is the area-averaged weight fraction of steam at the water outlet. In this case, the steam content rapidly decreases from 27% for a water flow rate of 0.25 $g \cdot min^{-1}$ to almost 0 for flow rates above 12 $g \cdot min^{-1}$. The syngas space velocity and conversion level have shown a strong influence in the conditions guarantying isothermicity. As a general rule, the pressure has to be lowered as the GHSV increases in order to control the FTS temperature in the 220-250°C range by adjusting the water flow rate.

The microreactor considered in this study has been analyzed as a cross-flow heat exchanger, and the overall heat transfer coefficient (U) has been calculated as follows:

$$U(W \cdot m^{-2} \cdot K^{-1}) = \frac{\dot{q}}{A \cdot F \cdot \Delta T_{lm}} \quad (2)$$

where A is the heat exchange surface area, which in this case is the same for both the hot and cold sides (A = 0.00336 m^2). The correction factor F is 1 because water vaporization takes place and then, there is no change in the water temperature (T_w). The heat transfer rate ($\dot{q}$) can be easily calculated as latent heat from the data of produced steam. Finally, the temperature difference is calculated as the logarithmic mean for a counterflow heat exchanger. Values of U for the microreactor are in the 10-50 $W \cdot m^{-2} \cdot K^{-1}$ range, which can be considered typical for flow boiling heat transfer in microchannels at the heat fluxes in the range between 0.5 and 4 $kW \cdot m^{-2}$ considered in this study [15].

Acknowledgements

We gratefully acknowledge financial support of this work by Petróleo Brasileiro S.A. – PETROBRAS and Ministry of Education and Science of the Spanish Government (MAT2006-12386-C05-01, 02 and 04).

References

1 Bartholomew, C.H., and Farrauto, R.J., *Fundamentals of Industrial Catalytic Processes*; John Wiley & Sons, Inc. – AIChE, 2006, pp. 461-464.

2 Dry, M.E., and Steynberg, P. Commercial FT Process Applications. *Stud. Surf. Sci. Catal.*, 2004, 152, 406-481.

3 Brumbya, A., Verhelstb, M., Cheret, D. *Cobalt News*, 2007, 2, 5-6.

4 Davis, B.H. *Ind. Eng. Chem. Res.*, 2007, 46, 8938-8945.

5 Riedel, T., Schaub, G. *Topics Catal.*, 2003, 26, 145-156.

6 de Deugd, R.M., Kapteijn, F., Moulijn, J.A. *Topics Catal.*, 2003, 26, 29-39.

7 Schulz, H. *Topics Catal.*, 2003, 26, 73-85.

8 Steynberg, P., Dry, M.E. Davis, B.H., Breman, B.B. Fischer-Tropsch Reactors. *Stud. Surf. Sci. Catal.*, 2004, 152, 64-195.

9 Hessel, V., Hardt, S., Löwe, H. Chemical Micro Process Engineering; Wiley-VCH, Weinheim, 2004, pp. 48-49.

10 Jarosch, K.T., Tonkovich, A.L.Y., Perry, S.T., Kuhlmann, D., Wang, Y. In *Microreactor Technology and Process Intensification* (Y. Wang and J.D. Holladay, Eds.); ACS Symposium Series 914, ACS, Washington, 2005, Chapter 16, pp. 258-272.

11 Glatzel, T., C. Litterst, Cupelli, C., Lindermann, T., Moosmann, C., Niekrawietz, R., Streule, W., Zengerle, R., Koltay, P. *Comput. Fluids*, 2008, 37, 218-235.

12 http://www.designspace.com/products/cfx.asp (accessed February 13, 2008).

13 Cao, C., Wang, Y., Jones, S.B., Hu, J., Li, X.S., Elliott, D.C., Stevens, D.J. In *Microreactor Technology and Process Intensification* (Y. Wang and J.D. Holladay, Eds.); ACS Symposium Series 914, ACS, Washington, 2005, Chapter 17, pp. 273-284.

14 Visconti, C.G., Tronconi, E., Lietti, L., Zennaro, R., Forzatti, P. *Chem. Eng. Sci.*, 2007, 62, 5338-5343.

15 Thome, J.R., Groll, M., Mertz, R. In *Heat Transfer and Fluid Flow in Microhannels* (G.F. Hewitt and G.P. Celata, Eds.); Begell House, Inc., New York, 2004, pp. 60-68.

209d Microanalytical Methods for the Characterization of Microunit Operations

Brian Marquardt[1], Ray Chrisman[1], Rick Gustafson[2], and Renata Bura[3]. (1) Center for Process Analytical Chemistry (CPAC) in Applied Physics, University of Washington, Seattle, WA 98195, (2) Forest Resources, University of Washington, Seattle, WA, (3) College of Forest Resources, University of Washington, Seattle, WA 98195

As microreactor based chemical processes are developed there is the need to characterize other unit operations to insure proper match. Just as microreactors benefit from real-time characterization to speed their optimization, the various other unit operations in the process can benefit from the use of on-line analysis for operation optimization. This presentation will describe the real-time characterization of microscale unit operations for use in bioprocessing. The operation and benefits of using a microscale fluid interface for the Raman spectrographic characterizationof the process stream in real-time will be described.

129d Multiphase Mixing with Disrupted Flow for Enhanced Interfacial Area

Maddalena Fanelli[1], Anna Lee Tonkovich[1], Sean Fitzgerald[1], Christy Burton[1], Mike LaMont[1], Jan Lerou[1], Jennifer Marco[2], Dongming Qiu[1], Laura Silva[1], and Rick Stevenson[3]. (1) Velocys, Inc., 7950 Corporate Boulevard, Plain City, OH 43064, (2) Velocys Inc., 7950 Corporate Boulevard, Plain City, OH 43064, (3) Experimental Operations, Velocys, Inc., 7950 Corporate Boulevard, Plain City, OH 43064

Although micro-scale processing leads to significant heat and mass transfer performance enhancements, not all micro-channels are created equal, and interfacial surface area can be largely controlled by choosing appropriate channel configurations and fluid property combinations. Carbon dioxide mass transfer into water was measured with flow through specially designed channels configured to provide disrupted (or high shear) flow at liquid and vapor phase Reynolds numbers up to 400 and 60, respectively. Mass transfer kLa values as high as 20 s-1 were obtained, corresponding to interfacial specific surface area values higher than 10,000 m2/m3. Hence, experimental work with two-phase flows in microchannels shows that the inclusion of features to provide disrupted flow can substantially increase the vapor-liquid mass transfer over straight microchannels. Since the high shear rate is retained throughout the channel length, bubbly flow can be created and maintained with relative ease. We discuss our experimental values, provide additional experimental results, and compare our findings to published literature results for microchannel vapor-liquid contactors. The links between flow morphology and performance will be used to clarify how we can push the performance envelopes to the limit.

REFORMING OF JP-8 IN MICROPLASMAS FOR COMPACT SOFC POWER

Peter J. Lindner and R.S. Besser, Department of Chemical Engineering and Materials Science, Stevens Institute of Technology, Hoboken, NJ 07030
201-216-5257; rbesser@stevens.edu

Background

One of the major limitations for the United States military today is the need for compact, portable power sources. As advances in weapons, communications, computation, vision, and positioning systems are made, the need for new sources of electrical power in the battlefield increases. Currently this need is met through battery technology, but large quantities of batteries becomes too burdensome for the soldier, and the production demands to supply these batteries cannot always be met. An example of this was illustrated during Operation Iraqi Freedom in 2003 when the standard BA5590 battery (20 W, 154 W-hr) was in such short supply that six manufacturing plants working continuously could not meet demand [1].

Higher energy density is provided by hydrocarbons, converted into a hydrogen rich stream, and fed to fuel cells such as solid oxide fuel cells (SOFCs). Further, if this process can be accomplished without catalysts and their accompanying drawbacks, a long-life, efficient combined fuel processor-fuel cell system would result. This project intends to show that by reforming JP-8, a logistical fuel, through the use of microplasmas rather than catalysts, a compact, portable, reusable power source fed by readily available logistical fuel, could be realized.

Introduction

Microplasmas as a means for reforming hydrocarbons is a process that provides many interesting advantages to other reforming technologies. Microplasmas were used as a reforming process in previous work done at Stevens (in collaboration with FuelCell Energy Inc. of Danbury, CT). During this research non-thermal plasma (using Dielectric Barrier Discharges) was used to reform NATO F-76 logistical fuel and produced some promising results. In this study it was found that the effluent was composed primarily of hydrogen C_1 and C_2 species, and that a significant fraction of the organic sulfur had been converted to hydrogen sulfide [2, 3], all without catalyst in a fast start-up process. These results give promise for reforming JP-8 with microplasmas.

Advantages of Plasmas over Catalysts

Using microplasmas as the process for reforming circumvents the most significant limitations of conventional catalyst-based processes which have been heavily researched to date. Catalyst-based processes have two major limitations; the sensitivity to sulfur compounds and the adverse effects of carbon deposits (coking). Sulfur compounds cannot be tolerated at even the most minuscule concentrations, whereas JP-8 fuel may contain up to 3000 ppm [4]. In order to process JP-8 using a catalyst reformer the fuel would require a separate hydrodesulphurization (HDS) reaction. This would convert the sulfur compounds to hydrogen sulfide (H_2S) that could be removed using a zinc oxide (ZnO) absorber. The addition of the HDS step would not only complicate the system, but this step would require an additional feed stream of hydrogen. The second problem, coking, leads to the deactivation of the catalyst by reducing the available active sites making it cost ineffective to regenerate catalyst. Plasma technology avoids both of these issues. First the sulfur compounds can be broken down into easily aborbed hydrogen sulfide with the plasma, simplifying the overall system design. Second, as no catalysts are required, coking would not limit the reaction rate. For these reasons microplasma technology is used for this project.

Advantages of Microplasmas

Plasma technology offers interesting prospects for hydrocarbon reforming by providing a non-equilibrium environment in which chemically reactive free radical species are generated that accomplish the chemical conversion [6]. By processing with microplasmas, i.e., a plasma confined in at least one dimension to 1 mm or less, it is possible to achieve a much higher power density (exceeding 1 kW/cm^3), higher electron density (exceeding 10^{15} cm^{-3}), and a high surface-to-volume ratio. A high surface-to-volume ratio allows for microplasma systems to have excellent thermal management characteristics and excellent mixing characteristics, producing homogeneous, isothermal reacting volumes [7]. Moreover, microplasmas differ from conventional, large-scale plasmachemical approaches in that electron and free-radical densities are 2 to 3 orders higher in microplasmas and hence they offer greater potential for compact and efficient processing units. Microplasma technology also offers some advantages in the final reactor design, as it will make it easier to size down the entire reactor. Also advantages on scalability from experimental reactions to actual use, and the ability of designing into the system integrated electrodes and heat exchangers. Despite the strong potential for compact fuel processors, little has been reported about this extremely attractive approach.

SOFC Technology's Advantages

Solid oxide fuel cells (SOFC) as the source of power serves to be extremely compatible with reforming JP-8 logistical fuel. A great benefit of SOFCs is their ability to use many various reformates as fuel. Carbon monoxide (CO) a poison to the catalyst in a proton exchange membrane fuel cell (PEMFC), acts as a fuel for a SOFC. Table 1 shows the comparison of the fuel flexibility of SOFCs to PEMFCs, the leading candidate for portable power in fuel cell technology.

Table 1. Ability of SOFCs to process various types of gas-phase feed components. PEMFCs are shown for comparison.

Gas Species	PEMFC	SOFC
H_2	Fuel	Fuel
CO	Poison (>10 ppm)	Fuel
CH_4	Diluent	Diluent/Fuel*
CO_2	Diluent	Diluent
S (as H_2S and thiophenic S)	Poison (>1 ppm)	Poison (>1 ppm)

* = Fuel for SOFC designs with internal reforming capability.

To remove carbon monoxide for a PEMFC, two additional separation units are required, increasing the system complexity, total size, and overall cost. In both cases sulfur is toxic to electrode catalyst and fuel processing catalysts, but will be reformed by the microplasma and removed by a ZnO absorber prior to the fuel cell. Figure 1 shows flow diagram of the overall system design.

Figure 1. Overall System Design

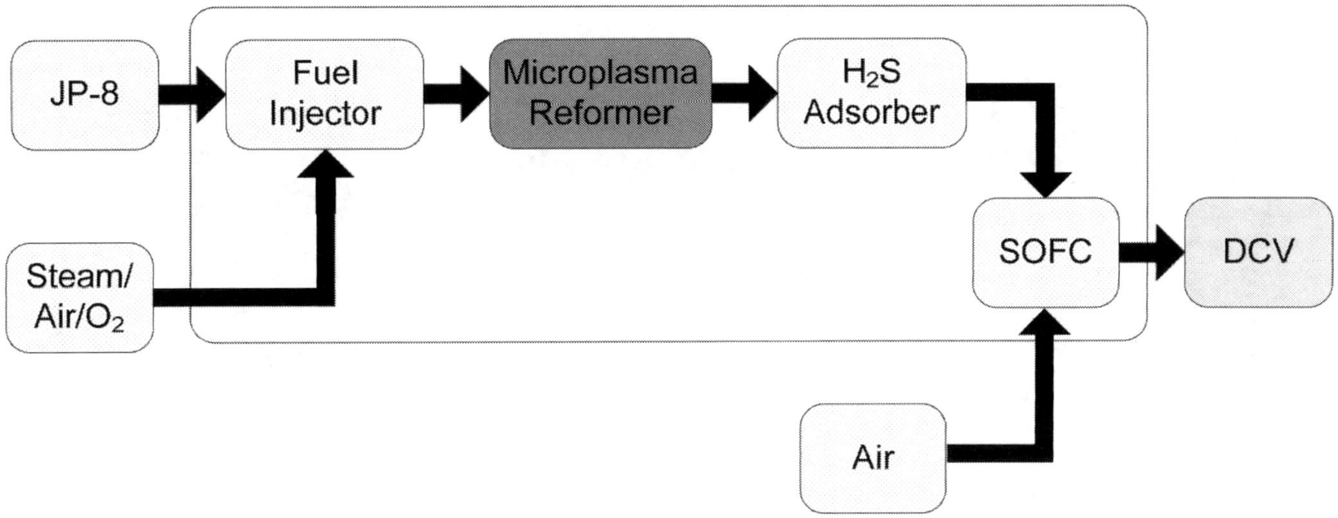

The overall design shows how microplasmas will be used as a means for reforming JP-8 to power a SOFC by providing a simple setup.

Research Accomplishments

To date microplasma reactor chips have been designed and fabricated having different channel geometries using silicon microfabrication technology at the Cornell Nanofabrication Facility of Cornell University in Ithaca, New York. The microplasma reactors will produce a plasma by the method of hollow cathode discharge, and have geometries varying in channel width (25μm, 100 μm, 250 μm), channel depth (25μm, 100 μm, 250 μm), and oxide thickness (2 μm, 4 μm). An example of the microplasma reactors can seen below in Figure 2.

Figure 2. General Microplasma Reactor Chips Dimensions

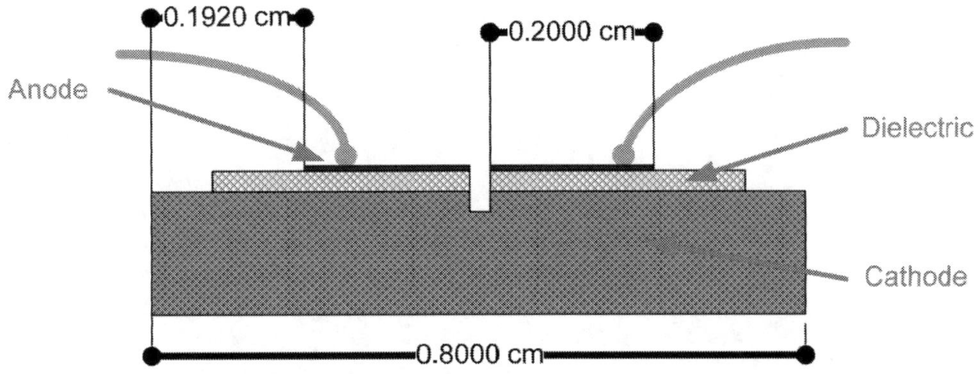

After the fabrication of the microplasma reactors, a vacuum chamber was completed to allow for testing and verification of plasma behavior with various inert gases. Inert plasma experiments have been carried out using gases such as xenon, neon and nitrogen. Images from these experiments can be seen below in Figure 3. Through these experiments, still ongoing, we hope to determine which reactor geometries and plasma operating conditions yield the most efficient plasma ambients for plasma chemistry experiments to follow.

Figure 3. Inert Plasma in Microplasma Reactor Chips

Inert Plasma Experiments

Inert plasma experiments that have been tested are to determine the voltage-current (V-I) trend, as this will establish our power input required for the microplasma reforming process. To date experiments on the microplasma reactors have shown comparability to plasma results in literature when the current is normalized by cathodes volume. Below in Figure 4 shows a plot of this data.

Figure 4. Plasma Data compared with Literature
Neon Plasma at 400 Torr

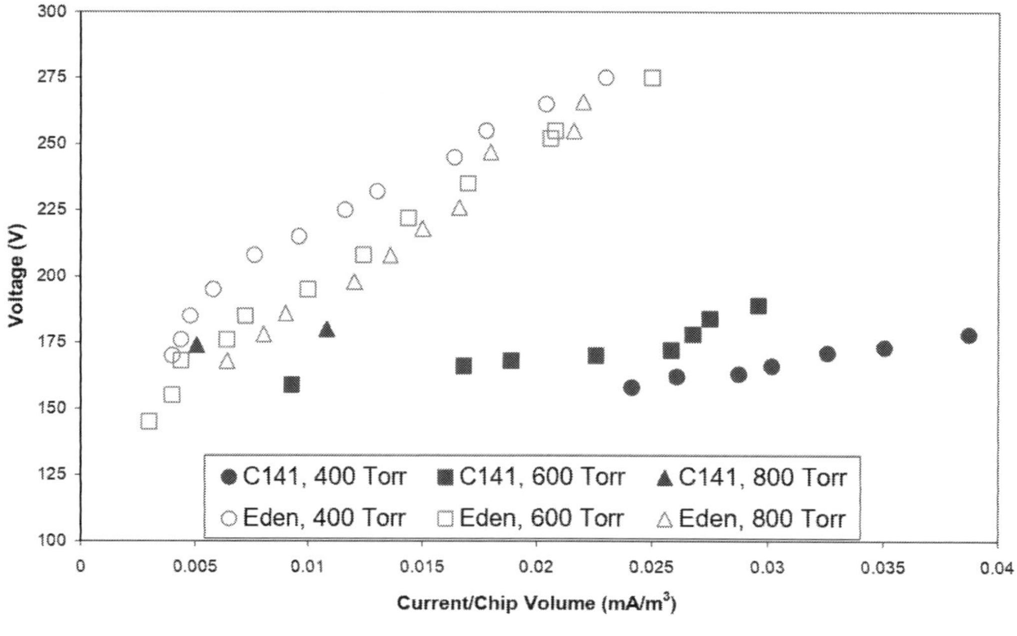

The referenced article [9] uses three types of hollow cathode plasma devices. Reference [9] reports on devices that are micromachined into three different geometries: planer, inverted pyramid and vertical cavity. The vertical cavity is most similar to our design, with dimensions of 50μm wide by 50μm long by 200μm deep giving a total volume of $5{\cdot}10^5 \mu m^3$. The most similar reactor tested had dimensions of 25μm wide by 7850μm long by 250μm deep giving a total volume of $4.9 \cdot 10^7 \mu m^3$. The volume to volume ratio is a factor of ≈100 is evident in the spread between the data sets in the figure. A difference between the data sets is that the microreactors for this experiment tended to show slightly larger current values at the same voltages. This difference could be possible due to differences in plasma efficiency or possibly to the existence of parasitic side plasmas.

Experiments have also been done to show the effects of the reactors channel-width on the V-I trends. For these experiments microplasma reactors with depths of 250μm were tested with varying widths from 25μm to 250μm. Figure 5 shows this data. This experiment shows that as the width of the material increases the voltage required to sustain a plasma increases.

Figure 5. Depth of Microplasma Reactor Channel Comparison
Neon Plasma at 400 Torr in C328 and C141

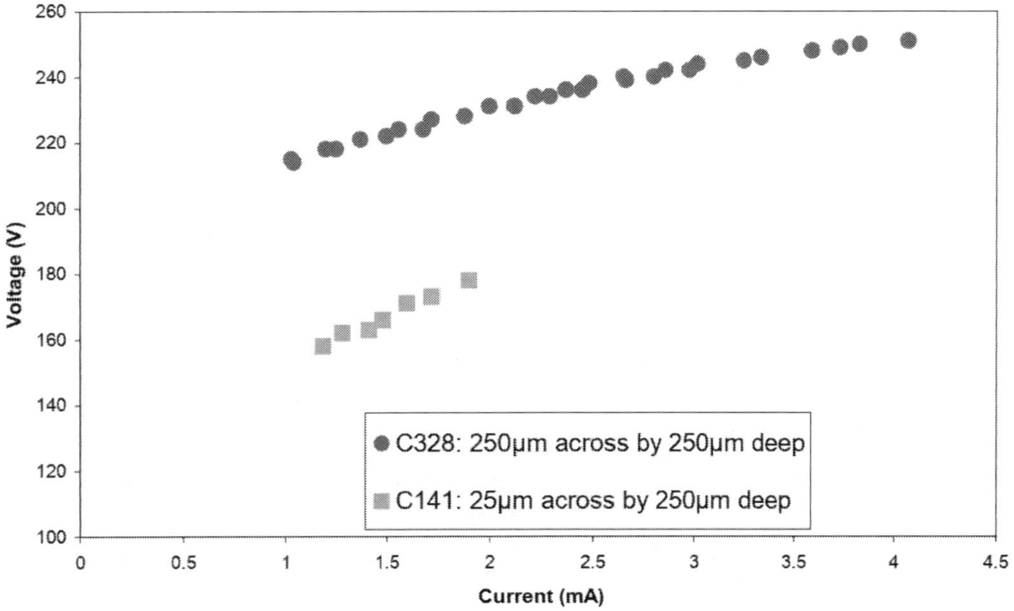

The next experiment included doping neon with small amounts of nitrogen in order to improve plasma stability. During this experiment the nitrogen was shown to dramatically improve the ease with which a plasma is obtained. Moreover, the effects of repeated cycling of the device are observed leading more stable and repeatable operation (Figure 6). At this point it is not determined whether this is simply a conditioning process, which for example could lead to the reduction of parasitic side plasmas, or the onset of a degradation mechanism ultimately leading to failure. More work is planned to explore and understand these observations.

Figure 6. Chip C126 Multiple Experiments
Neon with 0.5% Nitrogen in Chip C126 at 400 Torr

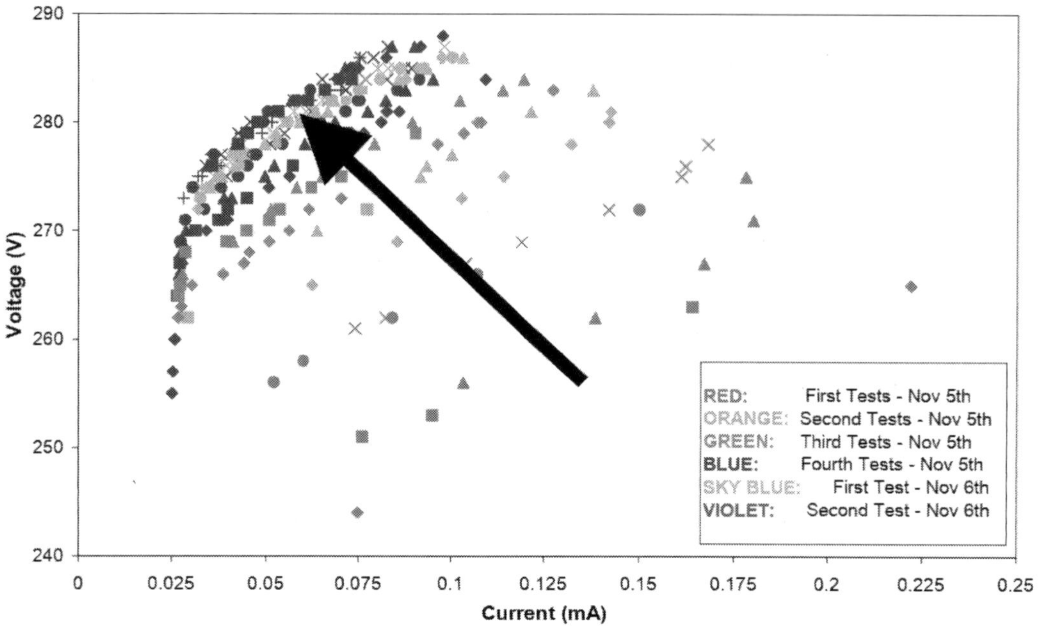

Future Work

Second-generation enclosed microplasma reactors will allow us to extend our experiment to include the interaction of hydrocarbon species with microplasmas. After testing new microreactors to determine plasma efficiency, reaction experiments will be run with key components of JP-8 fuel. We will measure conversions of these components in plasma activated steam reforming, in order to understand what factors are most important for producing hydrogen-rich reformates suitable for fuel cell processing. We plan to present initial results from these chemistry experiments at the conference. Some of these key components are listed below in Table 2.

Table 2. Key Components of JP-8 [8]

Model Compound	Molecular Formula	Boiling Point (°C)	Vapor pressure at 25°C (Torr)
Dodecane	$C_{12}H_{26}$	216.32	0.120
Thiophene	C_4H_4S	84.0	79.507
1-Methylnaphthalene	$C_{11}H_{10}$	244.7	0.068
Methylcyclohexane	C_7H_{14}	100.93	46.354

From experiments with key components it will be possible to determine the most efficient way to reform JP-8 to produces the most amount of fuel (H, CO) while converting the most amount of sulfur compounds into hydrogen sulfide.

Acknowledgements

The authors wish to express their gratitude to the Army Research Office for supporting this project and to various individuals at Stevens Institute of Technology which contributed to the experiments (E. Whittaker, V. Tarnovsky, G. Wohlrab, B. Fraser, Y. Liu, A. Najem, K. Ciampi).

References

1. Lt. Col. John LaTulip, "Lessons Learned from Operation Iraqui Freedom," *Tri-Services Power Expo*, July 15, 2003, Norfolk, VA.
2. K. Becker, Final Report, "Compact Plasma Reactor for Ship Service Fuel Cell System," Office of Naval Research Ship Service Fuel Cell Program, 2004, p. 234.
3. D. Dietz, H. Ghezel-Ayagh, J. Hunt, A. Belkind, K. Becker, and A. Nickens, "Plasma Treatment of a Heated Diesel/Steam Mixture for Use in Ship Services Fuel Cell Systems", Proc. 31st IEEE ICOPS, Balitmore, MD (2004), p.412.
4. B. Cannela "Refining and Distillate Fuel Trends, " 4th *DoD Logistics Fuel Reforming Conference*, October 22-23, 2003 Philadelphia, PA.
5. M. Gallagher, "Fuel Cell Panel," *Tri-Services Power Expo*, July 14-18, 2003, Norfolk, VA.
6. T. Paulmier, L. Fulcheri, "Use of non-thermal plasma for hydrocarbon reforming," *Chem. Eng. J.,* 106, 59-71 (2005), and references therein, especially 1-8.
7. X. Ouyang, L. Bednarova, P. Ho, and R.S. Besser, "Preferential Oxidation of Carbon Monoxide in a Thin-Film Catalytic Microreactor: Advantages and Limitations, *AIChE J.,* 51 (6), 1758-1772 (2005).
8. D. Lide, ed., *CRC Handbook of Chemistry and Physics, 88th Edition (Internet Version 2008),* CRC Press/Taylor and Francis, Boca Raton, FL.
9. J. Chen, S. Park, Z. Fan, J. Eden, C. Liu, "Development and Characterization of Micromachined Hollow Cathode Plasma Display Devices," *J. Microelectromechanical Systems*, (2002).
10. A. Agosta, N.P. Cernansky, D.L. Miller, T. Faravelli, E. Ranzi, "Referente components of jet fuels: kinetic modeling and experimental results," *Exper. Thermal and Fluid Science*, (2004).
11. D. Sadoway, A.M. Mayes, "Portable Power: Advanced Rechargeable Lithium Batteries," *MRS Bulletin*, 27 (8), 590 (2002).

236b Synthesis of Shape-Controlled Polymer Microparticles Using Asymmetric Microfluidic Channels

Takasi Nisisako and Takeshi Hatsuzawa. Precision and Intelligence Laboratory, Tokyo Institute of Technology, R2-6, 4259 Nagatsuta-cho, Midori-ku, Yokohama, 226-8503, Japan

We present a two-dimensional (2D) microfluidic system for the off-chip synthesis of the shape-controlled polymer microparticles. Using asymmetric microfluidic channels, such as T-junction and cross-junction, we could produce phase-separated droplets comprised of two immiscible organic fluids in aqueous phase; these droplets formed Janus or core-shell architecture by the minimization of interfacial free energies among the three liquid phases. Subsequent UV-polymerization outside the fluidic module produced nonspherical polymer particles such as hemispheric particles, and shells with a pore on their surfaces. The size and shape of the droplets and particles were tunable through the variation of flow conditions.

Nonspherical microparticles have myriad applications in various fields, because they possess distinct advantages over their spherical counterparts as a result of their scattering, rheological, coagulation and other properties. A variety of particle shapes have been achieved by the seeded polymerization techniques. However, these methods involve complicated material formulations, limiting their utilities for practical applications. Recently, a few microfluidic systems have been reported for the *in-situ* synthesis of nonspherical polymer particles, such as ellipsoids, rods, disks, and truncated particles.[1] Photolithography in a microstream of polymer precursors has also been proposed for the fabrication of particles with arbitrary 2D shapes.[2] However, all these microfluidic approaches share several drawbacks, such as the risk of clogging in the channels, increased pressure loss, limited controllability in the 3D shape of the particles, and low production throughput.

To overcome the above-mentioned limitations, the author previously proposed the off-chip synthesis of shape-controlled polymer particles through the formation of biphasic droplets in a symmetric co-flow channel.[3] Herein, we report the use of asymmetric microfluidic channels (*i.e.*, T-junction and cross junction) for the off-chip synthesis of shape-controlled polymeric microparticles, which is more efficient in the parallelization of microchannels for the industrial-scale production.[4]

Microchannels with rectangular cross section were fabricated on a synthetic silica substrate by the deep reactive ion etching (DRIE) technique. The microchannel consists of a upstream Y-shaped channel and a downstream T-junction/cross-junction. Around the T-junction/cross-junction, the channel widths were 100 μm with a uniform depth of 100 μm. An acrylate monomer (1, 6-hexanediol diacrylate, HDDA, 6.35 mPa s) was used as a polymerizable organic phase, and silicone oil (SO, 9.34 mPa s) or perfluorocarbon fluid (PFC, 1.6 mPa s) was used for the non-polymerizable organic phase. Aqueous solution of sodium dodecyl sulfate (SDS, 0.3 wt%) or polyvinyl alcohol (PVA, 2.0 wt%) was used as external aqueous phase. UV polymerization was continuously performed off the fluidic module.

A two-phase parallel stream was formed at the Y-junction when the two organic phases were infused separately into the two arms of the Y-junction. This two-phase stream then entered a shearing stream of the aqueous phase at the T-junction, resulting in the rapid formation of biphasic organic droplets by a shear-rupturing mechanism (~300 drops/s). The droplets were highly monodisperse in size, typically with a coefficient of variation (CV) around 2 %. When we used the combination of two Y-shaped channels and a cross-junction, biphasic droplets could be alternately formed at the two opposed inlet channels.

When SO was used as the non-polymerizable organic phase, the produced droplets formed Janus structure. The Janus droplets had a snowman-like appearance, with a curved internal interface separating the two organic phases. The satellite droplets (~20 μm) also had similar Janus structure. By varying the ratio of the flow rates of the two organic phases, it was possible to adjust the volume ratio of the two organic phases in the Janus droplets. We consider that the formation of these Janus geometries was driven only by the minimization of the interfacial free energies among the three liquid phases, because similar results were obtained when we exchanged the positions of the two organic streams in the microfluidic channels. By subsequent UV polymerization, we could synthesize monodisperse nonspherical particles. The particle shape could be tuned by varying the flow-rate ratio of the two organic phases between 0.1-10.0.

On the other hand, when PFC was used as the non-polymerizable organic phase and the PVA aqueous solution as the continuous aqueous phase, HDDA droplets engulfing PFC were reproducibly formed, because of the strong hydrophobicity of PFC. The UV polymerization of the external HDDA produced macro-porous shells with a small pore on their surfaces.

References

[1]D. Dendukuri et al., *Langmuir* **21**, 2113 (2005); S. Xu et al., *Angew. Chem. Int. Ed.* **44**, 724 (2005); R. F. Shepherd et al., *Langmuir* **22**, 8618 (2006); Z. Nie et al., *J. Am. Chem. Soc.* **127**, 8058 (2005).

[2]D. Dendukuri et al., *Nat. Mater.* **5**, 365 (2006); J.-H. Jang et al., *Angew. Chem. Int. Ed.* **46**, 9027 (2007).

[3]T. Nisisako et al., *IMRET9*, 190, Potsdam/Berlin, Germany (2006); T. Nisisako et al., *Adv. Mater.* **19**, 1489 (2007).

[4]T. Nisisako et al., *Lab Chip* **8**, 287 (2008).

Production of Hydrogen Peroxide via the Direct Combination of H_2 and O_2 in a Pilot Microreactor

Emmanuel Dada, **Dalbir Sethi** and **Dayi Deng**, FMC Corporation, US 1 & Plainsboro Road, Princeton, NJ 08543, USA

H_2O_2 is a chemical with diverse applications. Its current uses range from pulp and paper bleaching to health care; and to water purification where it is considered an environmentally friendly alternative to chlorine; and synthesis of various oxychemicals due to its high selectivity and effectiveness as an oxidizing agent. On-site production of H_2O_2 using a micro-channel reactor will reduce production energy and the costs associated with transportation, storage, and dilution of H_2O_2.

With funding from the Industrial Technologies Program, of DOE (Department of Energy), USA, FMC Corporation and the Stevens Institute of Technology designed a microchannel reactor and heat exchanger system for the on-site combination of H_2 and O_2 in an explosive concentration regime that ensures low-pressure, energy efficient, and safe reactor operation. Figure 1 shows a conceptual process flow diagram of the microchannel reactor system.

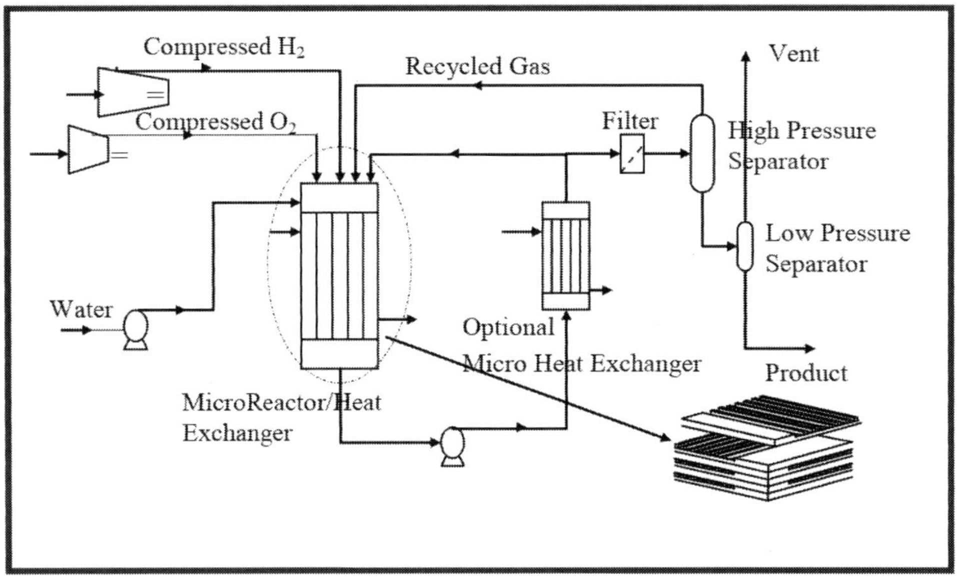

Figure 1: Conceptual Process Flow Diagram for the Microchannel Reactor System

A multichannel, multilayer microreactor pilot system comprising about 450 channels of dimension 800 microns by 800 microns was packed with a proprietary FMC catalyst and used to produce hydrogen peroxide by the direct combination of H_2 and O_2. The multi-channel high throughput micro-reactor pilot unit is shown in Figure 2. Using this propriety catalyst, we discovered operating conditions that enabled the production of over 2 wt% H_2O_2 in a single-channel reactor from a gas feed of 3% hydrogen in air and a solution feed of 1 wt % H_2SO_4 (with 10 ppm NaBr). Under the same operating conditions in the multi-channel pilot reactor, we achieved a concentration of 1 wt % H_2O_2 at a liquid (aqueous) flow rate of 1 kg/hr. The 1 wt % H2O2 is applicable for biocides and wastewater treatment.

Further optimization studies to improve gas-liquid feed distribution in the pilot reactor to achieve productivity equivalent to the single channel experiments are planned. Methods to uniformly pack catalyst into a multi-channel reactor and to remove the catalyst cleanly from every channel for replacement are discussed, as are the challenges associated with achieving uniform feed distribution to every channel in a gas-liquid system.

Figure 2: Multi-channel High Throughput Micro-reactor Pilot Unit

Microreactors with integrated porous silicon layer for reaction studies by matrix-free mass spectrometry

Wojciech. P. Bula, Vincent Verdoold and Han J.G.E. Gardeniers
Mesoscale Chemical Systems, MESA+ Institute for Nanotechnology, University of Twente, P.O. Box 217, 7500 AE, Enschede, The Netherlands

Porous silicon can be used as matrix material for the laser desorption/ionisation on silicon-mass spectrometry (DIOS-MS) [1] and porous silicon supports compatible with MALDI plates are commercially available. The DIOS-MS technique diminishes the utilization of a matrix – a chemical substance that absorbs the laser beam energy and facilitates vaporization and ionization. Furthermore, the utilization of porous silicon overcomes the problem of interference of a matrix with sample ions and therefore enables to identify low molecular weight compounds.

Here, we report on the integration of porous silicon spots in a silicon-glass microreactor to perform a chemical reaction in a continuous flow regime and an in-line investigation of a product formation by MS. The reaction products leave a microchannel network through an open outlet and an eluent is placed on a porous silicon spot on the front side of a chip (Figure 1). Two types of microreactors were designed and fabricated.

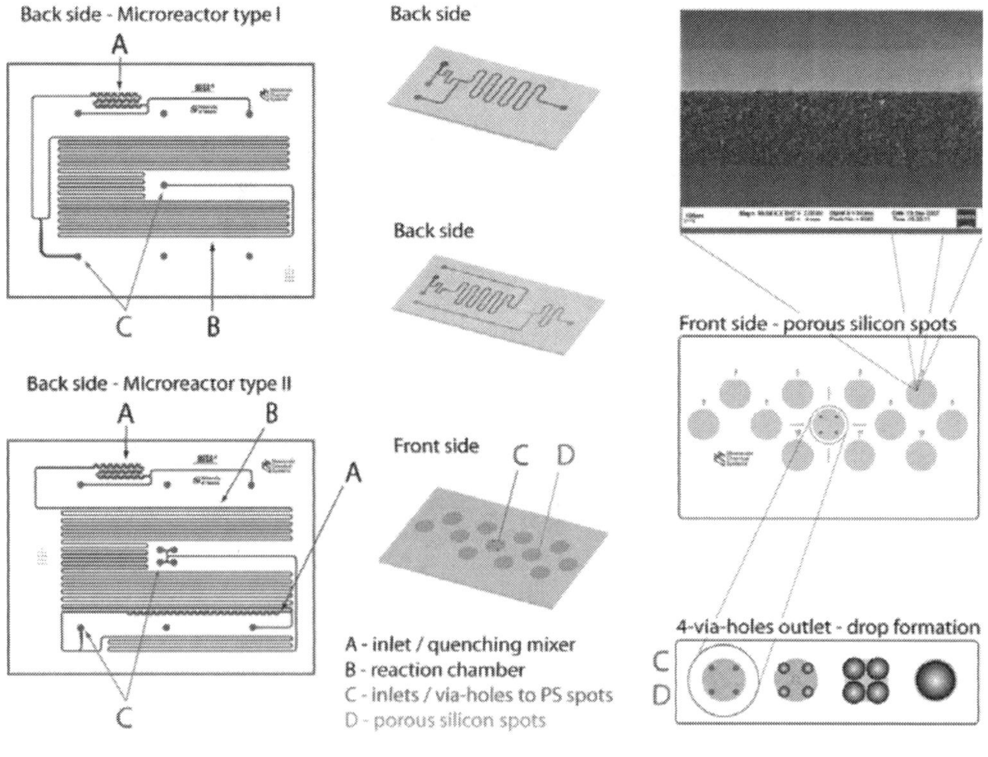

Figure 1 Schematic representation of the microreactor with integrated porous silicon spots (left).The spots are located on the front side of the chip.

The reactor of type I consists of an inlet mixer (A) where two reactants are mixed together and a chemical reaction is initiated. Subsequently, the active mixture passes into a reaction chamber. Both reactants are pumped into the chip with identical flow rates (constant volume pressure driven flow induced by a syringe pump). To avoid the mixture of undefined reagent ratio contaminate the reaction coil (B), an additional under-pressure line is introduced. By applying a small under-pressure all solutions are removed from the chip to the waste location. The residence time is defined as the period between shutting-off the waste line and the time the active mixture spends in the reaction coil. The reaction is terminated by solvent evaporation on the porous silicon spot.

When the droplet of an eluent is formed at the outlet of the channel, the flows are stopped and the droplet evaporates. Then, the residual liquids from the channels are removed from the chip by the waste line. Subsequently, the chip is removed from the holder, placed inside a chip-tailored steel MALDI plate, and put into a standard MALDI TOF MS device where an analysis of the product is conducted. After the experiment the chips are flushed with fresh solvents and cleaned in an ultrasonic bath (to remove the residuals deposited on the porous silicon spot). The details of the operation are presented in Figure 2.

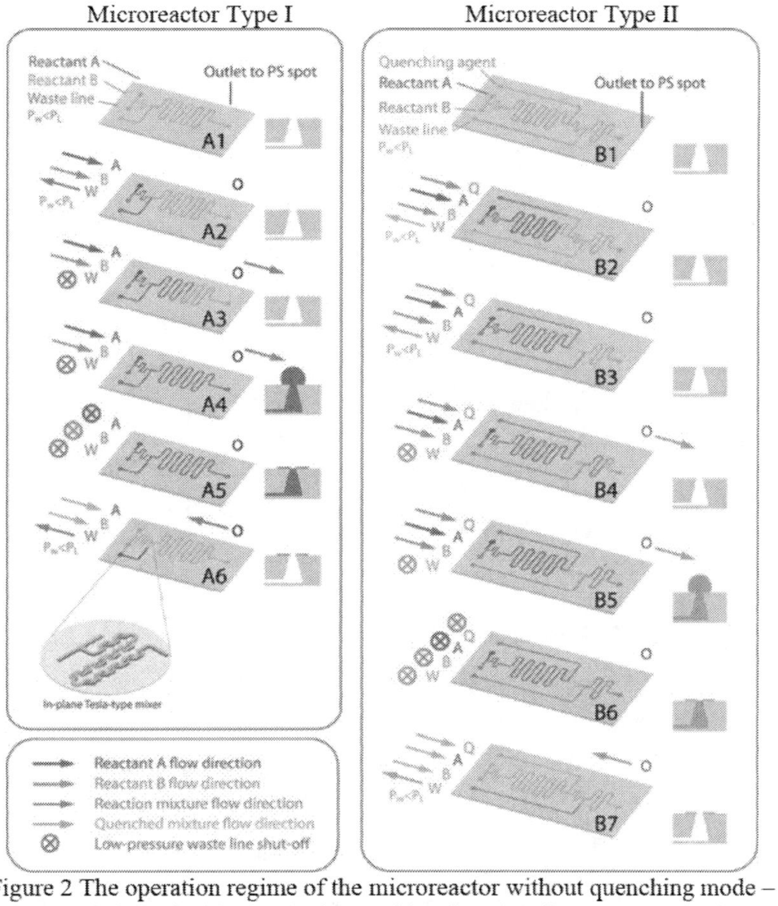

A1/B1 – Stand-by mode.

A2 – Pumping of reactants into the chip. Waste line OPEN - pressure of waste line lower than in the reactants lines – the active mixture passes into waste.

B2 – Pumping of reactants and quenching agent into the chip, chemical quenching of reaction in quenching mixer. Waste line OPEN, all liquids pass to waste.

A3/B4 – Waste line CLOSED – the reactants pass through the chip.

A4/B5 – Droplet formation on the porous silicon outlet.

A5/B6 – Reactants and waste lines CLOSED. Stopped flow, evaporation of solvent on PS spot.

A6/B7 – Waste line OPEN, removing of the liquids from microfluidic channels.

Figure 2 The operation regime of the microreactor without quenching mode – Type I (left) and with chemical quenching functionality – Type II (right).

In the microreactor type II an additional quenching line and a mixer are integrated (Figure 1-II). The reaction time is defined by the geometry – the time difference between the initiation of the reaction in the inlet mixer and the termination in the quenching mixer (where the active mixture is combined with a chemical inhibitor of the reaction). The waste line operates identically as in the design I. After the equilibration time, when all flows are stabilized, the waste line is closed and the inactive mixture runs through the outlet to be deposited on the porous silicon spot. The microreactor type II can be used to study the kinetics of chemical reactions. In contradiction to the type I, the evaporation of the solvent at the outlet does not increase the reaction rate by the increase of the reactants concentration due to the fact the reaction is already chemically quenched on the chip.

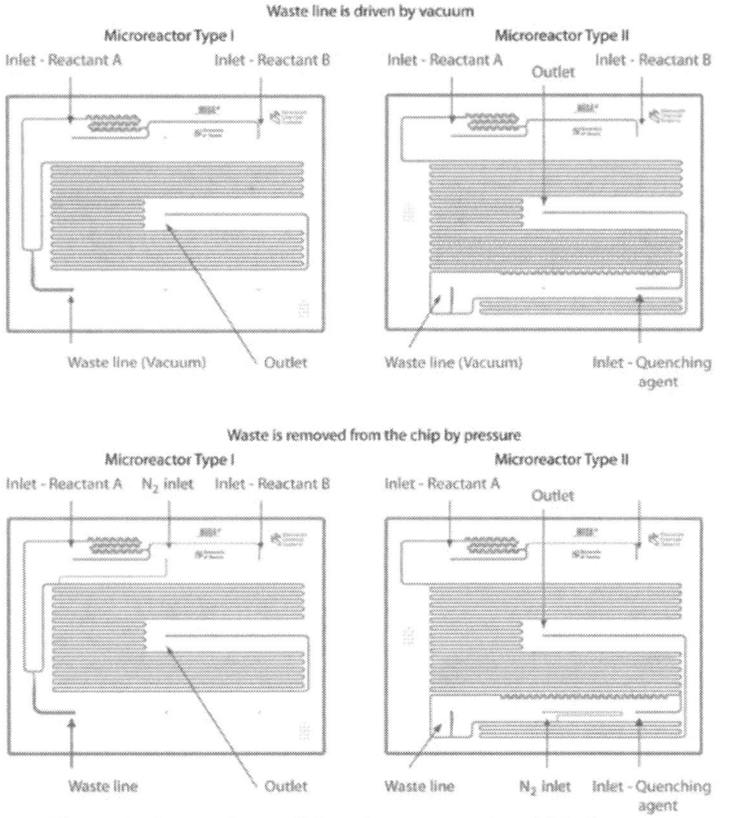

Figure 3 The topology of the microreactors in which the reactant mixture, during the equilibration step, is removed from the chip by the waste outlet connected to the vacuum line (top) and with additional pressure inlet to control the direction of flow (bottom).

The standard versions of both microreactors were designed to operate with the waste outlet connected to the vacuum line (underpressure) in order to remove the reactants from the chip during the equilibration step. The mixture of solutions does not flow into the main microreactor coil.

The requirement of having access to the vacuum line is overcome by an additional pressure line (Figure 3). The inherent gas such as nitrogen is pumped into the chip during the equilibration step. The hydraulic resistance of the reaction coil is much higher than the resistance of a short channel between the pressure line and the waste outlet. The pressure of the chemically inherent gas in the reaction coil will prevent the entrance of the reactants mixture into reaction coils.

The silicon-glass microreactors were fabricated in a standard microfabrication processes [2] (Figure 4).

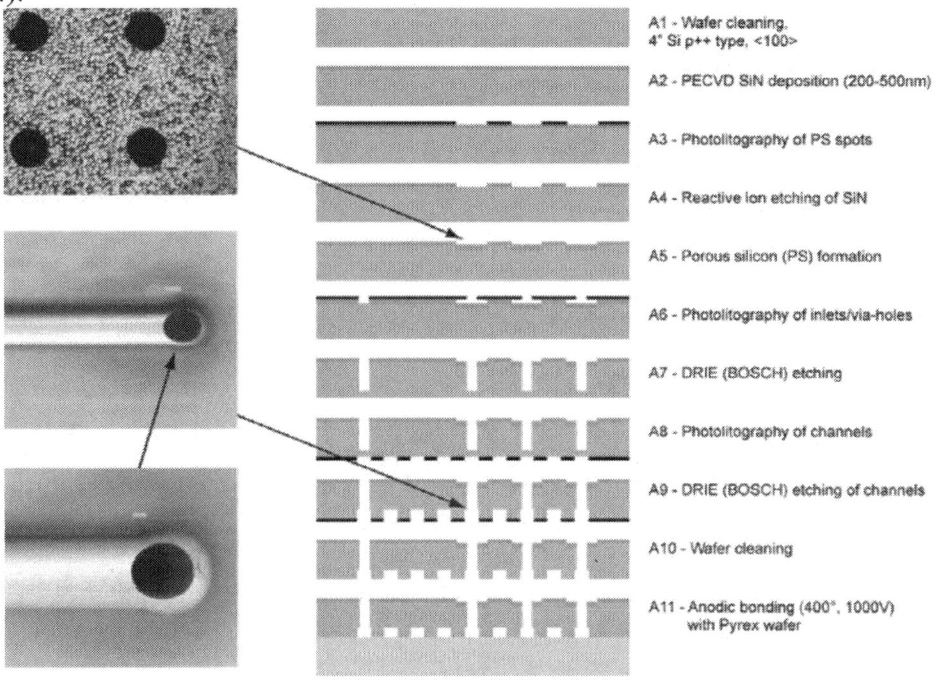

Figure 4 The photography of the via-holes and process scheme for the fabrication of the microreactor (not to scale).

The channels, 100μm wide and 100μm deep, were formed on the backside of a 4 inch highly p-doped (0.01 Ω·cm) (100) silicon wafer by deep reactive ion etching (BOSCH process). The porous silicon spots (thickness of ca. 0.9 μm, diameter of 2 mm) were created locally via hydrofluoric acid anodization process [3] using a PECVD silicon nitride mask. The inlets and via-holes were etched in silicon by BOSCH process. The diameters of the via-holes are 120 and ca. 90 microns on the topside and the backside of the silicon wafer respectively. The silicon wafer was anodically bonded to Pyrex glass and diced into separate chips.
Two types of outlet configurations were proposed. The conventional via-hole with the outlet opening in the center of the porous silicon spot and the 4-outlet located close to the perimeter of the porous silicon area (see Figure 1). In case of problems with the homogeneity of the product concentration in an evaporated eluent, four droplets growing on the outlet should reach the size when they will merge and cover the central part of the porous silicon spot.

The feasibility of the porous silicon spots was tested by spotting a solution of Angiotensin I human (M=1296.48) on a microreactor-DIOS plate and a conventional MALDI plate without a matrix. The correct mass spectrum was observed from porous silicon spots (Figure 5), while no signal was observed in a case of the MALDI plate without the utilization of a matrix.

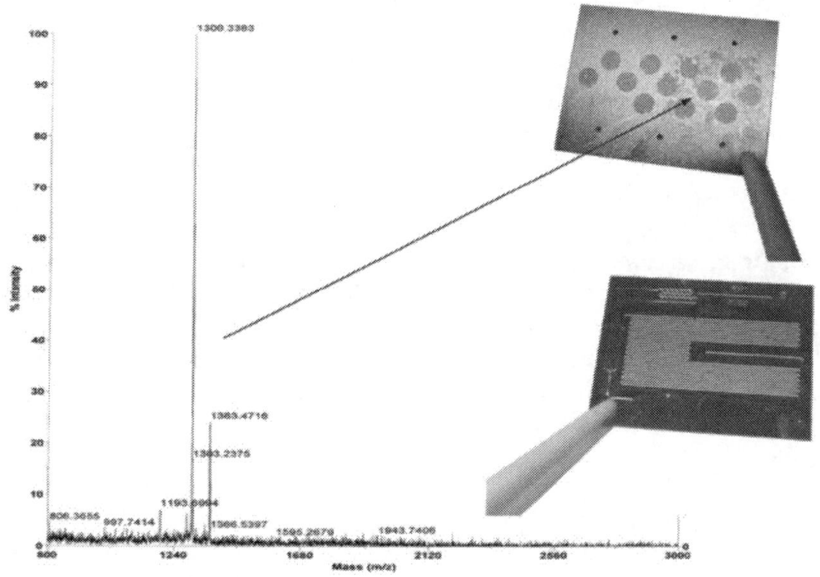

Figure 5 The spectrogram of Angiotensine I human from the porous silicon spot of the microreactor

The opportunity of the integration of porous silicon spots with multiline microreactors reduces the experimental time and effort. The concept can be easily extended to a higher number of reaction channels in parallel, as was reported recently [4]. In this case, a wide range of residence time points on a reaction kinetic curve can be covered just by a single flow rate.

References

[1] J. Wei, J.M. Buriak and G. Siuzdak, Desorption–ionization mass spectrometry on porous silicon, Nature 399, 1999, pp. 243–246.

[2] J. G.E. Gardeniers, R.E. Oosterbroek, A. van den Berg, Silicon and glass micromachining for μTAS, in: Lab-on-a-chip: Miniaturized systems for (bio)chemical analysis and synthesis, eds. R.E. Oosterbroek and A. van den Berg, Elsevier, Amsterdam, 2003, pp. 37-67.

[3] R.W. Tjerkstra, M.J. de Boer, J.W. Berenschot. J.G.E. Gardeniers, A. van den Berg and M.C. Elwenspoek, Electrochim. Acta, 1997, 42, pp. 3399-3406.

[4] W. P. Bula, W. Verboom, D. N. Reinhoudt and J.G.E. Gardeniers, Lab Chip, 2007, 7, pp. 1717-1722

209g New Micro Viscosity Sensor – a Novel Analytical Tool for Online Monitoring of Polymerization Reactions in a Micro Reaction Plant

Thomas R. Dietrich, mikroglas chemtech GmbH, Galileo-Galilei-Str. 28, Mainz, D-55129, Germany and Ulrich Schlecht, Research Institute caesar, Ludwig-Erhard-Allee 2, Bonn, 53175, Germany.

During the last 15 years, it has been shown that micro reactors can help to develop new and optimize conventional chemical reactions. Different micro mixers, micro heat exchangers, micro reactors for homogeneous or heterogeneous liquid-liquid reactions or for gas/liquid reactions have been developed and tested for development as well as for production. Even the handling of solids, e.g. manufacturing of particles or polymerization reactions could be shown. Very important for running micro reactors is the detection of the products formed during the process. This is necessary to optimize the reaction as well as to control the process during the production. Even though a variety of different reactions in micro reaction plants could be realized, still not many online analytical methods for micro reactors are known. Within this project it was necessary to develop a method to control the chain length of the produced polymer. During the polycondensation of the monomers the viscosity of the solution changes. The higher the chain length and the lower the concentration of monomers in the solvent, the higher becomes the viscosity of the solution. Therefore, it was decided to use this parameter as a signal to follow the reaction. With this new analytical method it is possible to monitor the reaction and control the chain length and molecular weight distribution. While the viscosity of small volumes can be measured with a classical rheometer, this analysis cannot be performed "online" during production. On the other hand, online analysis can be performed with conventional viscosity sensors. However, large flow-through volumes are required. Therefore, the project presented here combines online analysis with the handling of small volumes and flow rates as used within micro reaction plants. The sensor was developed at the Research Center caesar. The polymer solution has to flow through a stainless steel capillary (1/16") with an inner diameter of 0.75 mm, which is excited by an ac voltage through an electrical coil in a homogeneous magnetic field to perform shear waves (torsion) at a frequency of approx. 20 kHz (see fig 1a (drawing) and b (photo)). The resonant frequency of the capillary depends on the viscosity of the liquid going through it. Using the frequency and amplitude information from such a sensor, it is possible to measure the viscosity of the polymer solution. The respective calibration curve is shown in Figure 2. Higher viscosity leads to damping of the amplitude as well as to a shift in resonance frequency. Knowing the viscosity of a given concentration of a polymer in a solvent, it is possible to calculate the molecular weight of this polymer. The presentation will show the first results with this sensor used in a Suzuki reaction. The sensor allows the measurement of the viscosity of liquids over a wide range (1 – 350 mPa s). The signals depend not only on the viscosity of the solutions but also on the density and temperature. The authors developed within a project funded by the German government a micro reaction plant which allows to produce polymers out of up to 8 monomers by using the Suzuki coupling method. Micro reactors can have a positive influence on the quality of the polymer product. Especially the mixing of the monomers, the catalyst solution, and end capper solutions have a great influence on the molecular weight distribution. With micro reactors it is possible to have a very narrow distribution which is not achievable in typical batch processes. The authors would like to thank VDI-VDE-IT and the German Ministry for Education and Research for funding this work within the POKOMI project as well as Merck (Frankfurt), IMM (Mainz), hte (Heidelberg), Jumo (Fulda) for their help within this project. Fig 1a and b. Viscosity sensor, measuring the torsion of a stainless steel capillary. Fig 2: Calibration curves for the Viscosity Sensor

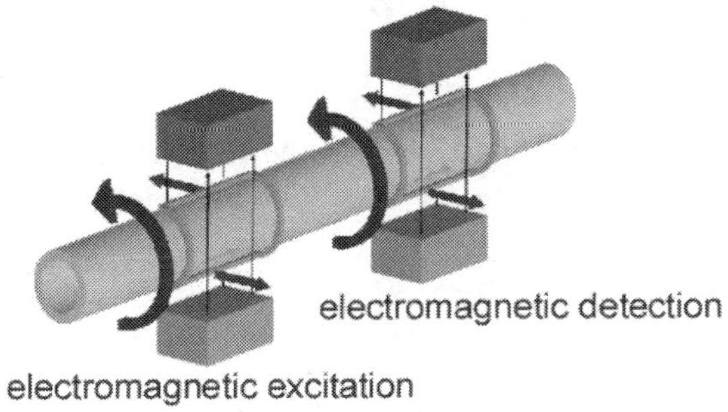

electromagnetic detection
electromagnetic excitation

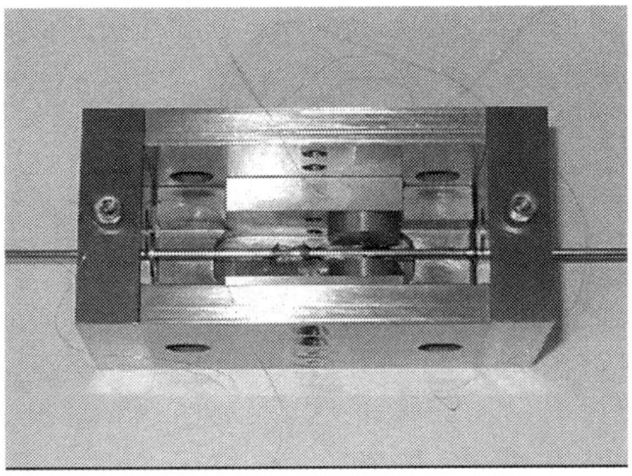

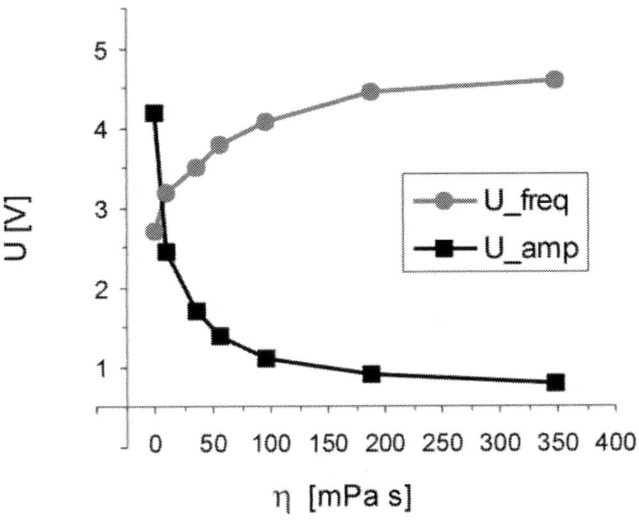

U [V]
η [mPa s]
U_freq
U_amp
0 50 100 150 200 250 300 350 400
1 2 3 4 5

128b In-Situ Cast Micro Reactor Catalyst Supports for Methanol Reforming

George J. Iwaszek, Travis Conant, and Abhaya Datye. Chemical and Nuclear Engineering, University of New Mexico, 1 The University of New Mexico, MSC 01 1120, Albuquerque, NM 87131-0001

Due to its high energy density and low reforming temperature, methanol is an ideal choice for fueling portable proton-exchange membrane (PEM) fuel cells. It has been shown that CuO/ZnO/Al2O3¬ catalysts are able to produce hydrogen while maintaining a high selectivity to CO2 at relatively low temperatures (230°C), keeping CO from poisoning the catalyst at the PEM anode. For a 20W fuel cell system operating at a 25% overall efficiency, about 3g of catalyst would be needed for the reformer operating isothermally at 230°C. Portability requires compact dimensions and reliable operation. A variety of approaches have been reported in the recent literature including catalyst-coated annular-flow micro reactors [1] and powdered catalyst packed micro reactors [2]. We have previously shown the feasibility of coating non-porous, micro-channel surfaces with a commercial catalyst and reported that there is an inherent limit on the one-pass coating thickness which is related to the physical properties of the coating fluid, the reactor geometry, and the coating procedure [3, 4]. While wall-coated reactors do operate isothermally with low-pressure drop, they suffer from low volumetric catalyst loadings. In-situ cast monoliths as catalyst supports in micro-channels are an attractive alternative, offering high loading potential and porosity. Monolithic supports have been reported for various materials, including porous alumina, silicon carbide and mesoporous carbon [5,6,7]. In this study, we report the fabrication and characterization of silica monolith catalyst supports cast in-situ in 250-530 micron diameter capillary reactors. Monolithic silica columns were first described for liquid chromatography applications [5], but we have successfully applied the methods to fabricate micro reactors suitable for methanol reforming. The relative ease of silica monolith fabrication in micro-channels and good control of parameters such as porosity and domain size (defined as the combined size of a through-pore and a skeleton) make this an attractive approach for studying monolith/catalyst performance. A typical silica monolith cross-section showing thee through-pore and skeleton is shown in Figure 1. The silica monoliths were prepared from tetramethyloxysilane (TMOS) using poly-ethylene glycol (PEG) as the template material. Porosity and domain size are controlled by the ratio of TMOS to PEG. A TMOS/PEG/urea gel is prepared and injected into fused-silica capillary tube with subsequent aging, mesopore formation and heat treatment to decompose the organic moieties to form the silica monolith. At this point, catalyst can be loaded by either washcoating or, as in our study, repeated injection and drying of a prepared solution of palladium nanoparticles. Our presentation will describe in detail the following: • Preparation of silica monoliths with controlled domain size and porosity in micro channels. • Critical micro reactor parameters: porosity, monolith uniformity, total pressure drop. • Catalyst loading by the impregnation method using prepared nanoparticle solutions. • Performance of the reforming micro-reactor. • Extensions of this method to monolith supports using other materials. Figure 1: In-situ cast silica monolith in a 530µ capillary at 2000x magnification showing the through-pore and skeleton detail for a particular preparation. This work has been funded by the U. S. Army Research Laboratory under the Collaborative Technology Alliance Program, Cooperative Agreement DAAD19-01-2-0010 References 1. Yu X, Tu ST, Wang Z, Qi Y Chem. Eng. J 116, (2006) 2. Yoshida K, Tanaka S, Hiraki H, Esahi M Jrn. Micromechanics MicroEng., 16 (2006) 3. Conant T, Karim A, Rogers S, Samms S, Randolph G, Datye A Chem. Eng. Sc. 61, 17 (2006). 4. Karim, A, Bravo J, Gorm D, Conant T, Datye A Catal. Today 110, 86 (2005). 5. Ganley JC, Reichmann KL, Seebauer EG, Masel RI J. Cat, 227 (2004) 6. Christian M, Kenis P, Lab Chip, 6 (2006) 7. Chang H, Joo SH, Pak C J. Mater. Chem., 17 (2007) 8. Hara T, Kobayashi H, Ikegami T, Nakanishi K, Tanaka N Anal. Chem. 78 (2006)

2.0kV ×2.00k SE
20.0um

103b Separation of Ethane-Ethylene in High Flux Microchannels

Ravi Arora[1], Anna Lee Tonkovich[1], Maddalena Fanelli[1], Laura Silva[1], Jennifer Marco[2], Daniel Weidert[1], Jeff Marco[3], Paul Neagle[4], and Thomas Hickey[1]. (1) Velocys, Inc., 7950 Corporate Boulevard, Plain City, OH 43064, (2) Velocys Inc., 7950 Corporate Boulevard, Plain City, OH 43064, (3) Velocys Inc, Plain City, OH 43064, (4) Technology Development, Velocys, Inc., 7950 Corporate Boulevard, Plain City, OH 43064

Ethylene is one of the largest volume petrochemical produced worldwide. This important intermediary is produced in extremely large plants that make 20 billion lb/year ethylene or more. These plants feature 20 or more fractionation columns, which are needed to separate and purify the key products (ethylene and propylene) from the gas effluent leaving the cracking furnaces. The most challenging separation is the splitting of ethylene from ethane in the 'C2 splitter.' The relatively small volatility ratio of ethylene in ethane results in conventional towers which are very large, up to 300 feet tall. This size is limited by manufacturing and transportation constraints, limiting the benefits from economies-of-scale. Microchannel distillation technology holds the potential to greatly improve the 'C2 splitter' by shrinking the size and cost of distillation hardware, and reducing energy consumption.

Microchannels with small characteristic dimension have low mass transfer resistance due to smaller diffusion distances. The increased mass transfer in microchannels enables a large reduction in the Height of an Equivalent Theoretical Plate (HETP). This permits separation hardware based on microchannel technology to have shorter height, smaller footprint, and more stages for efficient separation. Microchannel technology is scaled up by numbering up, adding identical channels or more modules. This modularity allows microchannel technology to benefit from the economy-of-mass-production.

A microchannel device was built and tested to evaluate the ethane-ethylene separation. A steady and consistent separation performance was achieved in the microchannel device. Evaluation of the data showed an achievement of HETP of 63 mm compared to more than 500 mm in a conventional distillation column. This presentation will discuss the experimental set-up and results, as well as the commercial application of this innovative technology.

83e Numerical Study of Droplet Formation Patterns in Flow Focusing Microfluidics

Peng He, Haejune Kim, Dawei Luo, and Zhengdong Cheng. Artie McFerrin Department of Chemical Engineering, Texas A&M University, College Station, TX 77843

The formation of droplets resulting from breakup of aqueous thread into immiscible liquid in microfluidics device with flow focusing geometry at various conditions was investigated. The full transient from startup to breakup into droplets was simulated numerically by solving the time-dependent Navier-Stokes equations of two phase flow using volume-of-fluid (VOF) method and level set method. The numerical simulation results were compared with experimental data at different flow rates, and the mechanisms of droplets formation were discussed.

Web Page: che.tamu.edu/orgs/groups/Cheng/

240d Controlled Deposition of Metallic Nanopartices for Enhanced in Situ Raman Characterization

Michael L. Ramirez, Orlando Ruiz, Marcos Barreto, and Samuel P. Hernandez-Rivera. Chemistry, University of Puerto Rico-Mayaguez, Center for Chemical Sensors Development, PO Box 9019, Mayaguez, PR 00681

The Raman characterization of trace level residues on surfaces was improved by up to ten times (10x) by the addition of metallic nanoparticles to contaminated areas. The enhancement in the Raman scattering when a molecule is adsorbed on a roughened metal surface has been reported since the late seventies. This enhancement comes as a result of an increase in the magnitude of the induced dipole therefore increasing the intensity if inelastic scattering. A related modality, Surface Enhanced Raman Resonance Spectroscopy (SERRS), consists on matching the incident light to the resonance frequency of an electronic transition of the analyte. Both approaches increase sensitivity, reduce fluorescence and lower detection limits of Raman overcoming traditional limitations.

In this study different deposition techniques were evaluated for the deposition of metal nanoparticles to a contaminated area for surface enhanced Raman spectroscopy. The intent is to combine SERS and a Raman standoff detection system for the remote assessment of surfaces at trace levels. The distribution and morphology of silver particles and nanogregates was compared for several deposition techniques differing in drop sizes from microliters to picoliters. Metallic silver and gold nanoparticles of sizes between 30 and 60 nm were deposited using techniques such as thermal inkjet cartridge, ultrasonic sprayer and an electrospray needle to deposit SERS active colloids on glass, silicon and stainless steel. The spectroscopic evaluation of surfaces was made by Raman microscopy. In all cases an enhancement of the spectroscopic signature was observed. These enhancements have applications for trace detection and surface interactions studies.

CONTROLLABLE SYNTHESIS OF PEPTIDE POLYMERS IN MICROFLUIDIC REACTOR

Masaya Miyazaki,[1,2] Jun Kaneno,[3] Hitonobu Shimada,[3] and Hideaki Maeda[1,2]*
[1]Nanotechnology Research Institute, National Institute of Advanced Industrial Science and Technology (AIST), Tosu, Saga, Japan
[2]Interdisciplinary Graduate School of Engineering Sciences, Kyushu University, Kasuga, Fukuoka, Japan
[3]NS Materials Co., Chikushino, Fukuoka, Japan

Biography
[1]Nanotechnology Research Institute, National Institute of Advanced Industrial Science and Technology (AIST), 807-1 Shuku, Tosu, Saga 841-0052, Japan; Tel: +81(942)81-3676; Fax: +81(942)81-3657
[2]Department of Molecular and Material Sciences, Interdisciplinary Graduate School of Engineering Sciences, Kyushu University, 6-1 Kasuga Koen, Kasuga, Fukuoka 816-8580, Japan; Tel & Fax: +81(92)583-8891
[3]NS Materials Co., 3-2-16 Kamikoga #101, Chikushino, Fukuoka 818-0041, Japan; Tel & Fax: +81(92)405-0290
**Corresponding authors: E-mail: m.miyazaki@aist.go.jp*

Microfluidic system possesses several advantages for polymer synthesis over the conventional reaction systems. Reactions using micromixing devices give better results than that of batchwise reactions. In the present study, we performed synthesis of peptide polymer such as collagen and elastin model peptides. We examined three microreaction systems for polypeptide synthesis. First we developed a continuous-flow circulation microreaction system to obtain higher molecular weight species by expanding reaction time. By using this system, we could obtain polymers at the same molecular weight as the batchwise reaction (Mn>10000) with nallower molecular weight distribution (less than Mw/Mn=1.2). However, we could not obtain higher molecular weight species by changing the dimension of microreactor. Thus we examined rapid microreaction by using chaotic micromixer. By this reactor, we could obtain very high molecular weight species (Mn = 200000 for collagen and 160000 for elastin) with very narrow molecular weight distribution (Mw/Mn = 1.05) by reaction at relatively high flow rates. These result indicates that microreactor with chaotic micromixer is suitable for precise synthesis of peptide polymers.

Introduction

Recently, many biopolymers with interesting functionalities have been developed and applied in various fields. In particular, synthetic biopolymers including polypeptides such as collagen and elastin are very important in fields such as biomedical and tissue engineering. Such functional polymers have been significantly developed owing to their specific properties such as various biological activities, biodegradability, and biocompatibility. Although several useful polymers were developed, it is usually effort and time consuming to obtain useful polymers which can be commercialized. One of the underlying problems deals with design and screening of polymer molecules. Although there have been demands for high-throughput and combinatorial synthesis system to rapidly and effectively obtain useful polymers, it is generally difficult to control polymer properties such as average molecular weight and molecular weight distribution, thus, a method which would enable control of polymerization reactions is desirable.

Microreaction systems are expected as an efficient reaction apparatus that enables high controllability of chemical reactions [1]. Such controllability results from efficient heat transfer, mass transport, and/or larger surface/interface area. Recent studies have shown the potential benefits of applying the various chemical reactions to microfluidic reactors. Several polymerization reactions have been performed using microreaction systems [2]. Controlling of rapid reactions such as living polymerization was successfully performed.

Our group has been interested in chemical reactions and analysis methods utilizing hydrodynamic behaviors generated characteristically in microfluids, including laminar flow, secondary flow, and shearing force [3,4]. We have shown that biopolymer chain can be expanded and oriented in microfluidic system using DNA as a model [5]. We also demonstrated that amino acid polymerization by NCA can be strictly controlled in microreactor [6]. These results clearly demonstrated that Microfluidic reactor is a excellent tool for biopolymer synthesis. In the present study, we expand our approach for polypeptide synthesis. We performed polymerization reaction of peptides in microreactor using collarge and elastin as models.

Collagen(polyPHG):	H-(Pro-Hyp-Gyl)$_n$-OH
Elastin(polyGVGVP):	H-(Gly-Val-Gly-Val-Pro)$_n$-OH

Scheme 1 Polypeptides synthesized in this study. (Hyp: Hydroxyproline)

Experimental

Materials

Amino acids were purchased from Watanabe Chemical Industries Ltd. (Hiroshima, Japan) or Peptide Institute (Osaka, Japan). Sigma-Aldrich Co. (St. Louis, MO) supplied most of chemicals used in this study. Poly(tetrafluoroethylene) (PTFE) microtube (250 μm of inner diameter) and PTFE T-shaped connector were obtained from Flon Chemical Inc. (Osaka, Japan). Glass plates and silicon plates were purchased from PMT inc. (Fukuoka, Japan).

The monomeric peptides, Pro-Hyp-Gly (PHG) for collagen and Gly-Val-Gly-Val-Pro (GVGVP) for elastin, were prepared by conventional solution-phase peptide synthesis with standard Boc-/Benzyl- protective groups [7]. After removal of N-and C-terminal protective groups, purity of the peptides was confirmed as >95% by RP-HPLC analysis.

Microchannel Reaction in Continuous-Flow Mode

Microreactor was assembled by connecting microcapillaries, gear and syringe pumps, with tee conectors as shown in Fig. 1. Reaction was started by gradual addition of reagents from syringe pumps, then resulting reaction mixture was circulated by the gear pumps.

Fabrication of Micromixer

For the bottom part of the mixer, silicon plates (70 mm x 30 mm x 1 mm) were covered by Cr sputtering followed by spin-coating of SU-8 photoresist. Microchannel patterns were projected using a Mask-less lithography system [8]. The plates were subjected to dry-etching, and then the size of fabricated microchannel was analyzed by Laser microscopy (Keyence Co., Tokyo,

Japan). For the top plate, a pyrex glass plate (70 mm x 30 mm x 1 mm) was subjected for micromachining using Robodrill equipped with diamond milling bar. The structure of fabricated microchannel was confirmed by laser microscope, as in silicon plate. These plates were bonded by anodic bonding to give micromixer chips.

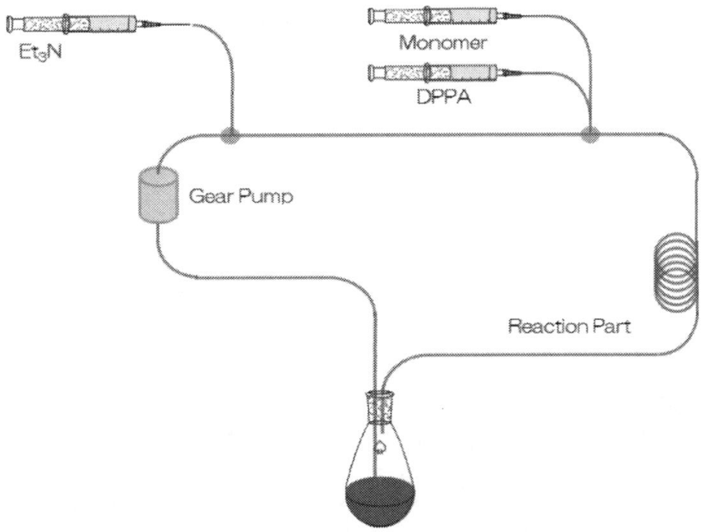

Figure 1. The continuous-flow microreactor

Microchannel Reaction Using Chip-Type Micromixer

The monomer and TEA solution and DPPA solution in DMSO, were separately charged into a 1 ml Hamilton Gastight Syringe (Hamilton Co., Reno, NV). The solutions were supplied to the microreactor using a KDS 230 syringe pump with a parallel syringe holder (KD Scientific Inc., New Hope, PA). The two syringes were set at identical pumping rates so that the final mixed solution would have a constant reagent concentration. The flow rate affects the reaction time, that is, the residence time in which the reaction mixture flowed through the microreactor.

Evaluation of polymer properties

The obtained product was immediately quenched by heating, and subsequently analyzed by GPC. The GPC analysis was performed by Hitachi LaChrome System with RI detection using COSMOSIL 5GPC-100 and 300 (Nakarai Tesque Co., Kyoto, Japan) and DMSO (Sigma, HPLC grade) as the eluent.

Results

Microreaction by continuous-flow mode

In batchwise reaction, polymerization of peptide is a kind of slow reaction and therefore needs longer reaction time [9]. Thus, we designed a continuous-flow microreaction system. By this system, activated monomer was gradually added to the reaction mixture, and therefore, total reaction time can be extendable. We examined this system with polymerization of collagen tripeptides. As shown on Table 1, although the reaction time was extended, not so high molecular weight species were obtained. Only slight improvement of the molecular weight distribution (Mw/Mn=1.2) was observed in this system. We also examined effect of dimension of

Table 1 Comparison of polymer properties obtained by continuous-flow reactor.

Entry	Polymer synthesized	Reactor &Condition*	Mn	Mw/Mn
1	poly(PHG)	ϕ 0.5mm x l 1000mm	14,300	1.22
2	poly(PHG)	ϕ 0.5mm x l 1500mm	16,600	1.38
3	poly(PHG)	ϕ 0.5mm x l 3000mm	15,000	1.25
4	poly(PHG)	ϕ 0.25mm x l 1000mm	14,200	1.17
5	poly(PHG)	ϕ 0.25mm x l 3000mm	15,000	1.25

*All reaction was performed at 20°C with constant pumping rate (gear pump 150 μl min^{-1}; syringe pump, 17μl min^{-1})

the capillary of the microreaction part. However, no difference was observed in the properties of the obtained polymers. This result means influence of reaction time for the reaction is not so effective. Therefore, we changed total strategy.

Microreactor with chaotic micromixer chip

For the most of polymerization reaction, it is important to control mixing of the reagent to obtain nearly monodisperse polymers. Next, we examined the effect of mixing on polymerization of peptide by performing reaction using chaotic micromixer. This type mixer has typical feature; chaotic advection at higher Reynold's number flow yields better mixing [10]. Thus, this system works at higher flow rates.

This micromixer was connected to a Teflon tube (500μm i.d., 3000mm l.). The polymerization was performed by simple loading of monomer and DPPA solutions into micromixer by syringe pumping. First, we performed reaction at lower flow rate (50ml min^{-1}). At this flow rate, the mixing mainly depends on the diffusion of each solution, and the longer reaction time was ensured. As shown on table 2, the resulting polymer has modelately high molecular weight (Mn=60968) but broader distribution (1.56). When raising flow rates, the molecular weight of resulting polymer become higher (>100kDa) with narrower molecular weight distribution (approx. Mw/Mn=1.2). This means chaotic mixing affect on the reaction, not only for molecular weight distribution, but also the molecular weight of the polymers. To confirm this result is not specific for collagen polymer synthesis, we also performed polymerization of GVGVP peptide. As

Table 2 Comparison of polymer properties obtained by chaotic micromixer system.

Entry	Polymer synthesized	Reactor &Condition*	Mn	Mw/Mn
1	poly(PHG)	Batchwise	17,700	2.56
2	poly(PHG)	Chaos mixer (50μl min^{-1})	61,000	1.56
3	poly(PHG)	Chaos mixer (100μl min^{-1})	79,800	1.42
4	poly(PHG)	Chaos mixer (300μl min^{-1})	203,100	1.14
5	poly(PHG)	Chaos mixer (400μl min^{-1})	204,900	1.20
6	poly(PHG)	Chaos mixer (500μl min^{-1})	207,300	1.18
7	poly(PHG)	Chaos mixer (600μl min^{-1})	196,900	1.26
8	poly(PHG)	Chaos mixer (700μl min^{-1})	199,800	1.23
9	poly(PHG)	Chaos mixer (800μl min^{-1})	198,700	1.25
10	poly(GVGVP)	Chaos mixer (400μl min^{-1})	147,900	1.06

*All reaction was performed at 20°C

shown on table 2 (Entry 10), similar to collagen polymer, we could obtain high molecular weight polymer at narrower molecular weight distribution (Mw/Mn=1.06). Perhaps this difference is result from the sequence of the peptide. Anyhow, these result means that chaotic mixer is suitable for pepetide polymer synthesis.

Discussion

Polymerization of peptide is considered as a kind of difficult to control polymerization because of slow reaction. Because of this feature, the reaction was performed for several days in batchwise reaction. In the present study, we examined the effect of microreaction system for this peptide polymer synthesis by two different approaches. One is the continuous-flow reaction system. This system covers short reaction time of microreactor (<hr) by circulating reaction mixture and gradual addition of the monomer and reagent. However, this system did not gave strong effect. Resulting polymer had almost same molecular weight as the batchwise reaction. Only molecular weight distribution was improved. This result mean the reaction was not so affected by the reaction time; rather mixing of two solutions seemed important. Thus, we examined micromixer system. The microreactor equipped with chaotic mixer yielded results with strong impact. The resulting polymer at higher flow rates was >200kDa for collagen and >150kDa for elastin polypeptides. Such high molecular weight species are difficult to obtain by batchwise reaction. Surprisingly, these high molecular weight polymers have narrow molecular weight distribution (Mw/Mn≦1.2). In batchwise reaction system, synthesis of such high molecular weight species usually yielded very broad molecular weight distributions. This result indicates that our microreaction system is strong tool for peptide polymer synthesis.

We have shown that polymer chain can be expanded and oriented in microfluidic system using DNA as a model [5]. In our previous studies of polyamino acid synthesis, this share stress does not affect on the reaction, because the resulting molecule is not so high molecular weight [6]. In this study, we could obtain much high molecular weight species by chaotic micromixing system. We examined the effect of flow rate for polymerization, but raising flow rates did not affect on the properties of obtained polymers. These result supports that conformational change did not affect on the stepwise polymerization reaction. Such conformational change induced by laminar flow system might be advantageous for substitution of polymers or graft polymerizations, which target of reaction locates middle of the polymer molecule.

Reactions involved in chemical polymerization employed for synthesis of biomaterials should be highly controlled because polymer properties such as average molecular weight and molecular weight distribution strongly affect the function of the polymer. For instance, polylysine functions are recognized in gene and drug delivery, cell adhesion, and cytotoxicity depending on the polymer size. For collagen and elastin, molecular weights strongly affect the properties of polymer. Small molecules are soluble for water and therefore used for cosmetics. Whereas the high molecular weight species are used for biomaterial, especially in the tissue-engineering field. When using biopolymers of less well known biological function, the molecular weight of biopolymer should be carefully selected, and the molecular weight distribution should be controlled as narrow as possible. It is expected that such requirements could be satisfied by using the present microfluidic system which has high controllability of molecular weight distribution and which enables simple control of molecular weight of amino acid-based biopolymers. Present-day investigations for biomaterials are directed toward the development of novel functional substances by using high-throughput and combinatorial synthesis method.

The device developed in our laboratory could be a basic model of a simple, controllable, and continuous combinatorial synthesis process with a high-throughput system for biomaterial libraries that has high potential with respect to rapid optimization of polymerization reaction.

Conclusion

In this study, we examined effect of microreaction system for peptide polymerization. Microreactor possesses multilamination micromixwer yielded only small molecular weight speces (≦10kDa). By using continuous-flow microreactor, we could obtain modelately high molecular weight species (≧20kDa) at narrower molecular weight distribution (Mw/Mn≦1.2). These results indicates that these systems are simple expansion of batchwise reactions, which require several days for obtaining modelately high molecular weight species at broader molecular weight distribution (Mw/Mn>1.5). These microreactors could slightly improved molecular weight distribution, by taking accout of small dimension. Striking difference was observed in microreactor which possesses chaotic micromixer chip. By this system, we could obtain higher molecular weight species (>100kDa) at very narrow molecular weight distribution (≦1.2). Such high molecular weight species could not obtain by batchwise reactions. These features might be useful for further development of polypeptide-based biomaterials.

Acknowledgment

We thank Dr. T. Honda, Dr. T. Tani, and Ms. A. Koga for technical supports. Part of this work was supported by Grants in Aid for basic scientific research (B 18310094) from JSPS.

Literature Cited

[1] W. Ehrfeld, V. Hessel, and H. Lowe, *Microreactors, New Technology for Modern Chemistry*, Wiley-VCH, Weinheim (2000).

[2] V. Hessel, C. Serra, H. Lowe, and G. Hadziioannou, *Chemie Ingenieur Technik*, **77**, 1693-1714 (2005); M. Miyazaki and H. Maeda, *Encyclopedia of Micro- and Nanofluidics* (D. Li ed.) Springer Heidelberg, in press (2008).

[3] M. Miyazaki, K. Yamashita, Y. Yamaguchi, T. Honda, H. Nakamura, M. Fujii, and H. Maeda, *New J. Chem.*, **28**, 1622-1627 (2004).

[4] K. Yamashita, Y. Yamaguchi, M. Miyazaki, H. Nakamura, H. Shimizu, and H Maeda, *Lab Chip*, **4**, 1-3 (2004).

[5] K. Yamashita,Y. Yamaguchi, M. Miyazaki, H. Nakamura, H. Shimizu and H. Maeda, *Anal. Biochem.*, **332**, 274-279 (2004).

[6] T. Honda, M. Miyazaki, H. Nakamura, and H. Maeda, *Lab Chip.*, **5**, 812-818 (2005).

[7] M. Bodanszky, *Principles of Peptide Synthesis*, Springer (1993).

[8] M. Miyazaki, T. Honda, H. Nakamura, and H. Maeda, *Chem. Eng. Technol.*, **30**, 300-304 (2007)

[9] T. Kishimoto, Y. Morihara, M. Osanai, S. Ogawa, M. Kamitakahara, C. Ohtsuki, and M. Tanihara, *Biopolymers*, **79**, 163-172 (2005).

[10] H. M. Xia, C. Shu, S. Y. M. Wan, and Y. T. Chu, *J. Micromech. Microeng.*, **16**, 53-61 (2006).

83g Development of An Active Rotary Reactor and Application to Emulsion Production

Ken-Ichiro Sotowa, Atsushi Yamamoto, Keizo Nakagawa, and Shigeru Sugiyama. Department of Chemical Science and Technology, The University of Tokushima, 2-1 Minami-Josanjimacho, Tokushima, 770-8506, Japan

This paper is related to production of emulsion in microstructured channels. There are many publications dealing with production of fine droplets in microchannels. Although they were successful in producing emulsions with very sharp size distrubtions, it is hard to control the si ze of droplets.

The most commonly used technique to control the droplet size is to manipulate the flow rate of the continuous phase. However, in this method, the concentration of the emulsion is also changed, and when fine droplets are produced, only sparse emulsion can be obtained.

In this study, we propose a new method to produce size-controlled emulsions. The details are to be presented at the conference.

PRESSURE DROP IN THE MICRO T-JUNCTIONS

A. Soleymani, W. Ratchananusorn, H. Yousefi, and I. Turunen

Department of Chemical Technology, Lappeenranta University of Technology, 53851 Lappeenranta, Finland

Abstract

Results are presented from a numerical study examining the pressure drop of the liquid phase inside micro T-junctions at moderate Reynolds numbers. The effects of variation in design and operating parameters, such as the aspect ratio of the mixing channel and the volume flow rate on the flow dynamics and pressure gradient were studied. The main aim was to propose an explicit equation to predict the friction factor inside T-junctions. Furthermore a model was proposed to determine the mixing zone in micro T-mixers.

Keywords: Laminar flow, Mixing, Simulation, Mathematical modeling, Microstructure, T-mixer

1. Introduction

The rapid progress in fabricating and utilizing the microstructure systems during the last past decade has not been matched by corresponding advances in the understanding of the unconventional physics involved in the manufacturing and operation of small devices. Providing such understanding is crucial to designing, optimizing, fabricating and utilizing improved microdevices. The results of experiments with microdevices can not always be explained via traditional flow modeling. The pressure drop in a long microduct is observed to be nonconstant. The Navier-Stokes equations and the no-slip boundary condition at a fluid-solid interface break down as the Knudsen number, defined as the ratio between the mean free path and the characteristic length, is above 0.1. The Mach number, a dynamic measure of fluid compressibility, increase downstream, limited only by chocked-flow condition. Gaseous flows even at Mach number smaller than 0.3 are often compressible in microdevices. The differences between the behavior of the fluid flows in microscale devices and that in macroscopic counterparts can be attributed to the fact that the small scale makes molecular effects such as wall slip more important, and it amplifies the magnitudes of certain ordinary continuum effects to extreme levels.

The pressure gradient becomes especially large in small cross-section channels. For fixed volume flux, the pressure gradient increases as l^{-4}, where l is the length scale. Despite the fundamental simplicity of laminar flow in straight ducts experimental studies of microscale flow have often failed to reveal the expected relationship between the friction factor and the Reynolds number. Microscale measurements of the friction factor by Mala and Li (1999) and Qu et al. (2000) revealed higher values for the friction factor than theoretical macroscale predictions. In contrast, the works of Pfahler et al (1990) and Yu et al (1995) are among the studies finding a decrease in friction factor on the microscale.

The purpose of this work was to study the flow dynamics of the liquid phase inside T-type micromixers by means of computational fluid dynamics. Over the past decade, computational methods have proven to be an effective tool to design and develop microprocesses. This opens up the possibility of ensuring the feasibility of processes in microreaction devices or to optimize these devices on a computer with a minimum expenditure of the time and money. Here the aim was to propose a correlation between the friction factor, Reynolds number and the geometrical parameters of T-junctions. To end this, the effects of variation in operating and design parameters, such as the volume flow rate, the aspect ratio of the mixing channel or the hydraulic diameters of the channels on the pressure drop were studied.

2. Simulation method

Computational Fluid Dynamics simulation was carried out with the commercial code FLUENT 6.3.26 (ANSYS Inc.). The model used considers the laminar mixing of the species in a T-type micromixer resulting from the contact between two identical fluids having the same viscosity ($\mu = 0.001 \text{ Pa s}$) and the same density ($\rho = 998.2 \text{ kg m}^{-3}$). The density and viscosity of the flow were assumed to be constant along the channels, despite the very large pressure gradients that characterize microscale flow. This can be attributed to the fact that the effects of internal heating due to viscous dissipation are much less significant in liquid flows, owing to the much higher heat capacity of liquids relative to gases.

The changes in mass and momentum are then described by

$$\nabla \cdot \mathbf{V} = 0, \tag{1}$$

$$\rho \frac{D}{Dt}\mathbf{V} = -\nabla P - \nabla \cdot \boldsymbol{\tau}, \tag{2}$$

where $\mathbf{V}$ is the velocity vector. P and $\boldsymbol{\tau}$ are pressure and shear stress, respectively.

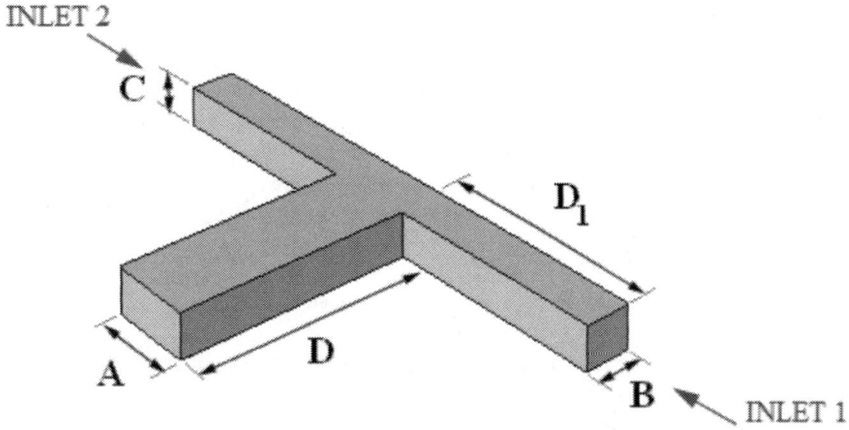

Fig. 1. Schematic picture of the T-mixer with dimensions $A \times B \times C$

The field of velocity in the three-dimensional T-mixers as schematically illustrated in Figure 1 was sought through discretization of the conservation equations (1) and (2) using the finite volume formulation (Patankar, 1980). A uniform velocity profile was assumed at the inlets and the pressure at the exit was assumed to be fixed to the local atmospheric pressure. No-slip boundary condition at the side walls was applied (Gad-el-Hak, 2002). The SIMPLE-Consistent algorithm (Vandoormaal, and Raithby, 1984) was implemented for pressure-velocity coupling. To achieve realistic simulations, the lengths of the side branches were chosen to be long enough ranging from 3000-4000 μm so as not to influence significantly on the simulation of the fluid behavior inside the mixer channel. The non-influence of the cell size on the results was verified for different calculations.

To obtain a correlation between the friction factor, the Reynolds number and the geometrical parameters of micro T-mixers a set of numerical type simulation was performed for 30 different mixers in different volume flow rates.

3. Control volume

Figure 2 demonstrates the variation in the cross-sectional pressure along the mixing channel length at moderate Reynolds numbers for a $600 \times 300 \times 300$ mixer. For a moderate Reynolds number the values of the pressure are sharply decreased at the entrance of the mixing channel due to the deflection of the inflowing streams.

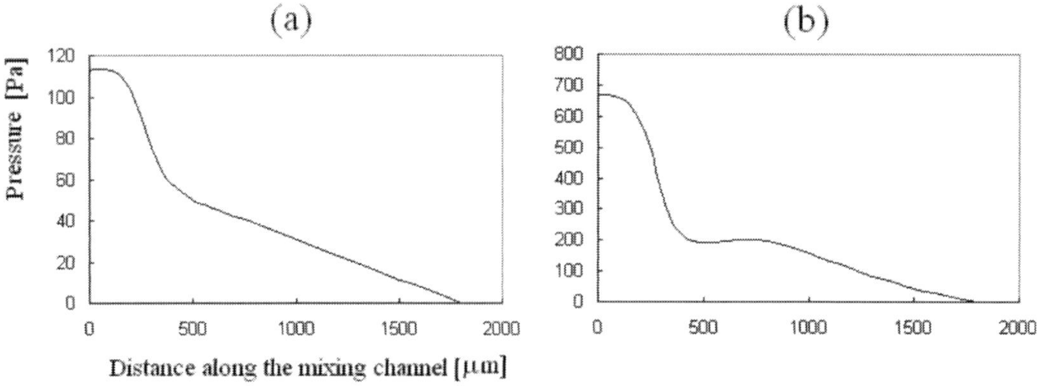

Fig. 2. Cross-sectional pressure along the mixing channel for a $600 \times 300 \times 300$ mixer at (a) $\mathrm{Re} = 80$, and (b) $\mathrm{Re} = 240$.

In a T-type micromixer the mixing zone is defined as the region where convection is the main mixing process owing to the very small diffusion constant of liquids in the orders of 10^{-9} m^2/s. Figure 3 describes the streamlines in the mixing zone for a $600 \times 300 \times 300$ mixer at $\mathrm{Re} = 240$. The swirling of the fluid flow results in dragging fluids from the middle to the top and bottom sides of the mixer.

After the mixing zone, a linear pressure dependence is observed as expected for fully developed duct flow. As depicted in figure 2, by increasing the Reynolds number the mixing zone becomes larger and the pressure drop significantly increases.

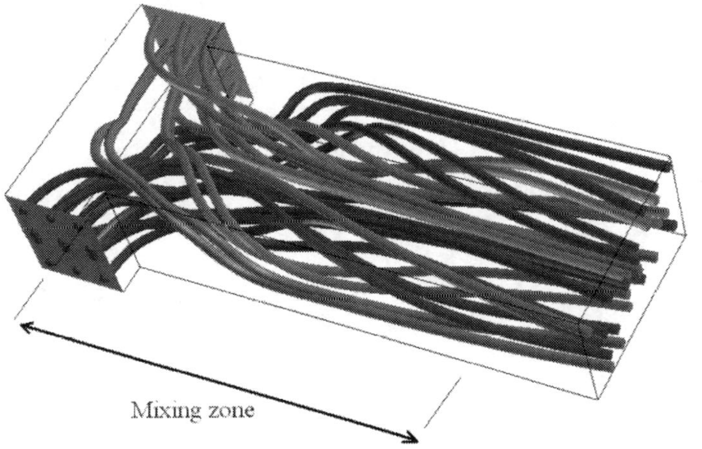

Fig 3. Streamlines at the entrance of the mixing channel for a $600 \times 300 \times 300$ mixer at Re=240. Colour is used to distinguish between the streams entering each of the inlet channels.

In the light of the above discussion special care has to be taken in choosing the control volume for the calculation of the pressure drop. As the concern in this study is the pressure drop resulted from the deflection of the inflowing streams, the control volume was set to be the mixing zone whose dimensions varies by both mass flow rate and the geometry of the T-mixer.

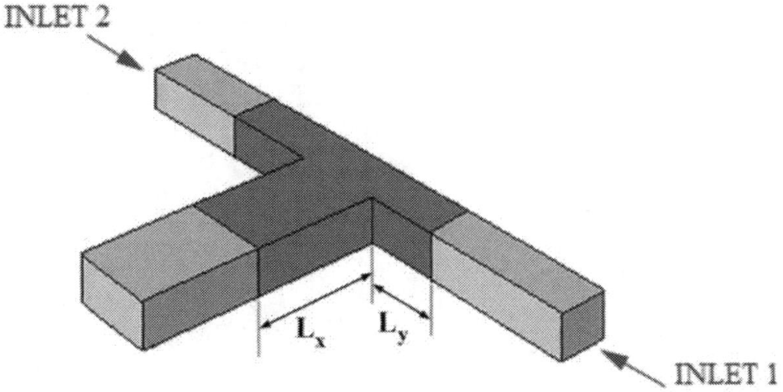

Fig 4. Mixing zone in a T-mixer. The mixing zone is set to be the control volume for the calculation of the pressure drop and friction factor in T-micro mixers

Figure 4 depicts the schematic of the control volume for the calculation of the pressure drop in a T-mixer. L_x and L_y defined as the lengths of the mixing channel and inlet channels where the flow is deflected by the T-junction. Here L_x is defined as the location away from the junction in the mixing channel where the longitudinal component of the velocity is 99% of the fully developed velocity magnitude. Likewise L_y is defined as the location away from the junction in the inlet channels where the longitudinal component of the velocity is 99% of the fully developed velocity magnitude.

3. Results and discussion

3. 1. Mixing zone

Analyzing the set of data obtained from numerical calculations by means of the differential evolution algorithm resulted in the following expression for the L_x defined as the length of the control volume located in the mixing channel

$$\frac{L_x}{D_h} = (0.00156\,\text{Re} + 0.0576)(4.2433\,\text{Dr}^2 - 0.1635\,\text{Dr} + 1) \tag{3}$$

where Re is the Reynolds number in the mixing channel based on the hydraulic diameter of the mixing channel, D_h. Here Dr is the ratio between the hydraulic diameter of the mixing channel and the hydraulic diameter of the inlet channels, D_h/D_{h_in}. Equation 3 implies that in moderate Reynolds number range, the dimensionless entrance length in the mixing channel L_x/D_h, depends not only on the volume flow rate and the dimensions of the mixing channel but also depend on the geometrical parameters in the upstream of the flow in the inlet channels. This dependency appears in the second parenthesis in the right side of equation (3).

The results of the proposed model for L_x/D_h were compared to that of from simulation results in Figure 5. The equation in the form of Y=AX+B in Figure 5, Y=0.9975X , is the equation of the adapted least regression line with the best state at Y= X. As the model results are exactly the same as the simulation ones, all the points fall exactly on a line at 45°. In addition to the adapted least regression line, the regression constant (R^2 =0.9643) in Figure 5 describes the excellent agreement between the simulation and the model data for L_x/D_h . In an ideal situation, when these parameters are exactly the same, $R^2 = 1$.

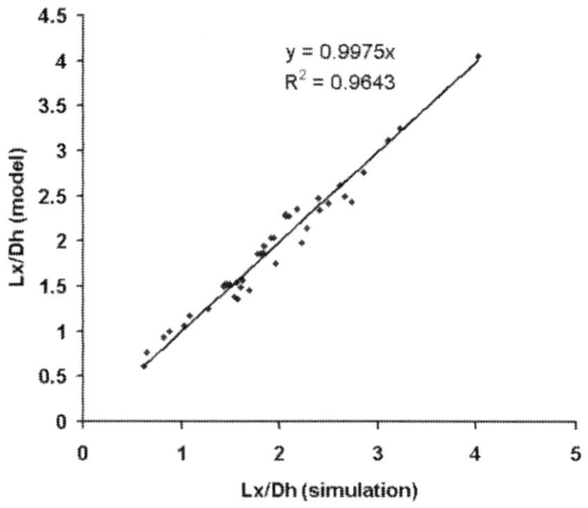

Fig. 5. The comparison of the simulation results for $\frac{L_x}{D_h}$ and that from Equation (3)

Analyzing the set of data obtained from numerical calculations resulted in the following expression for the L_y defined as the length of the inlet channels of the control volume

$$\frac{L_y}{D_{h\_in}} = \left(0.0038\,\mathrm{Re}_{in} + 0.6342\right)\left(-0.3531\,\mathrm{Dr}_{in}^{-1} + 1 + 0.131\,\mathrm{Dr}_{in}\right) \quad (4)$$

where Re_{in} is the Reynolds number in the inlet channels based on the hydraulic diameter of the inlet channels, D_{h_in}. Here Dr_{in} is defined as

$$\mathrm{Dr}_{in} = \mathrm{D}_{h\_in} / \mathrm{D}_{h\_A/2} \quad (5)$$

where $D_{h\_A/2}$ is the hydraulic diameter of a rectangular surface of dimensions $\frac{A}{2} \times C$. The surface is the fraction of the cross sectional face of the mixing channel in which the flow fed into one of the inlet channel passes into it. A is the width of mixing channel and C is the depth of the T-junction.

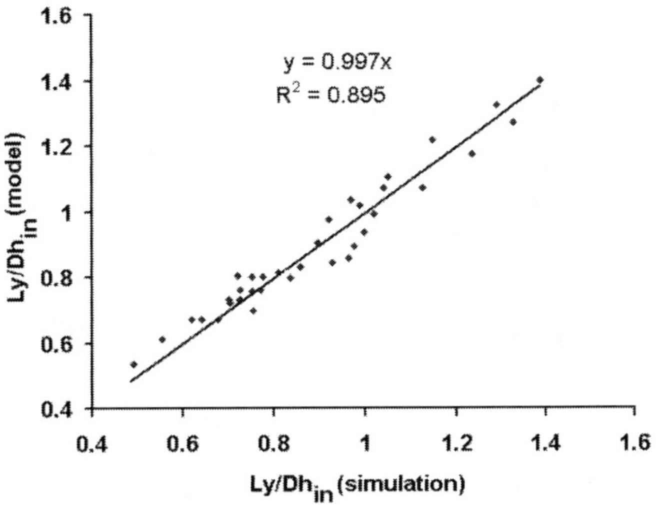

Fig 6. The comparison of the simulation results for $\frac{L_y}{D_{h\_in}}$ and that from equation (4)

Figure 6 describes the comparison of the simulation results for L_y/D_{h_in} and that from the proposed model (Equation 4). From the adapted least regression line, Y=0.997X, and the regression constant (R^2 =0.895) in Figure 6, it can be concluded that the equation 4 can predict L_y/D_h within a high degree of accuracy.

3. 2. Calculations of the friction factor

After the determination of the control volume for the specified mixer at a certain volume flow rate, characteristic length L and pressure loss ΔP were measured. Characteristic length L is defined as the ratio of the control volume V and the cross-sectional area of the mixing channel,

$$L = (L_{y,1} + L_{y,2})\frac{S_{in}}{S} + L_x + B \quad (6)$$

where. s_{in} and s are the cross sectional area of the inlet and outlet channels, respectively. The pressure loss inside the control volume, ΔP, is defined as the pressure drop resulted by bending of the streamlines and it is calculated as follow

$$\Delta P = \left[\frac{1}{2}(P_{in,1} + P_{in,2}) - P_{out}\right] - \left[\frac{1}{2}(\Delta P_{s,1} + \Delta P_{s,2}) + \Delta P_{s}\right] \tag{7}$$

where $P_{in,1}$ and $P_{in,2}$ are the measured pressure at the inlets of the control volume and P_{out} is the measured pressure at the outlet of the control volume. $\Delta P_{s,1}$ and $\Delta P_{s,2}$ are the pressure loss inside the inlet channels of the control volume if they were parts of a very long straight channel with the same cross section as the inlet channels. Likewise ΔP_{s} is defined for the outlet part of the control volume. The first bracket in the right side of equation 7 presents the total pressure loss inside the control volume. The second bracket of equation 7 describes the pressure loss resulted from the flowing of the fluid inside the straight channels of the same lengths in the fully developed form. The friction factor f is then calculated as

$$f = \frac{\left(\frac{\Delta P}{L}\right) D_h}{\rho U^2}. \tag{8}$$

The analysis of the data results in the following expression for the friction factor f in terms of the Reynolds number in the mixing channel, Re and the ratio between the hydraulic diameter of the mixing channel and inlet channels, Dr.

$$\begin{aligned} f(\mathrm{Re}, Dr) = {} & (1.0886Dr - 1.343)\times 10^{-5}\,\mathrm{Re}^2 \\ & -(0.0011Dr + 6.597\times 10^{-5})\,\mathrm{Re} \\ & -(1.3524Dr - 2.1598). \end{aligned} \tag{9}$$

The results of the proposed model for friction factor were compared to the simulation results for f in Figure 7. The equation of the adapted least regression line (y=0.9933x) and the regression constant (R^2 =0.9632) in Figure 7 describe the excellent agreement between the simulation and the model data for friction factor f.

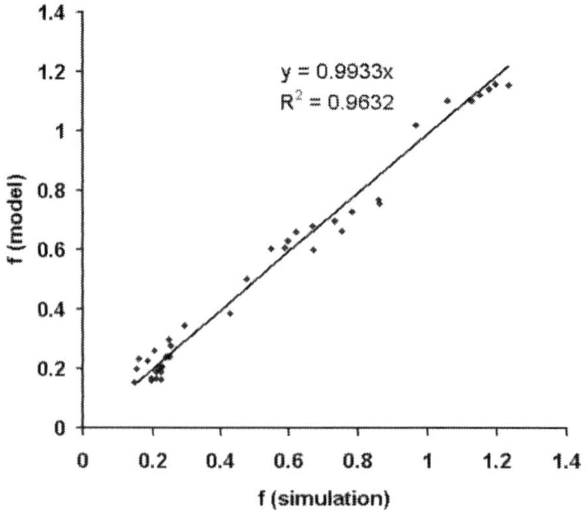

Fig 7. The comparison of the simulation results for friction factor f and that from Equation (8)

4. Conclusion

Flow dynamics and pressure drop of liquids in T-type micromixers were investigated. The aim was to develop a model to predict the friction factor and pressure drop in T-micromixers. To this end a new definition was set to define the control volume for calculation of the pressure drop. The control volume was defined as the mixing zone for the T-mixer. In moderate Reynolds number range, the dimensionless entrance length, models were proposed to estimate the mixing zone as a function of the volume flow rate and the mixer dimensions. Application of the differential evolution algorithm to the set of data obtained from the simulation resulted in a model for the friction factor in T-mixers. The model results for the friction factor were in excellent agreement with those from simulation. As a result the proposed equation can estimate the friction factor and pressure drop within a high degree of accuracy.

5. References

Bothe, D., Stemich, C., and Warnecke, H., 2006. Fluid mixing in a T-shaped micro-mixer, Chem. Eng. Sci. 61, 2950-2958.

Ehlers, St., Elgeti, K., Menzel T., and Wießmeier G., 2000. Mixing in the offstream of a microchannel system, Chem. Eng. Process 39, 291-298.

Engler, M., Kockmann, N., Kiefer, T., and Woias, P., 2004. Numerical and experimental investigations on liquid mixing in static micromixers, Chem. Eng. J., 101, 315-322.

Gad-el-Hak M., 2002. The MEMS handbook, CRC press.

Hardt S., Schönfeld F., 2003. Laminar mixing in different interdigital micromixers: II. Numerical simulations, AIChE Journal 49 (3), 578-584.

Mala, G. M., and Li, D., 1999. Flow characteristics of water in microtubes, Int. J. Heat Fluid Flow, 20, 142-148.

Pfahler, J. N., Harley, J., Bau, H., and Zemel, J., 1990. Liquid Transport in Micron and Submicron channels, Sensors and Actuators, 21-A23, 431-434.

Price, K. V., Storn, R. M., Lampinen, J. A., 2005. Differential Evolution: A Practical Approach to Global Optimization. Berlin. Springer.

Qu, W., Mala G.M., and Li, D., 2000. Heat transfer for water flow in trapezoidal silicon microchannels, Int. J. Heat Mass Transfer 43, 3925–3936.

Soleymani A., Kolehmainen E., and Turunen I., 2008. Numerical and experimental investigations of liquid mixing in T-Type micromixers, Chem. Eng. J., 135S, 219-228.

Vandoormaal, J. P., Raithby, G. D., 1984. Enhancements of the SIMPLE Method for Predicting Incompressible Fluid Flows. Numer. Heat Transfer 7, 147-163.

Yu, D., Warrington, R., Barron, R. and Ameel, T, 1995. An Experimental and Theoretical Investigation of Fluid Flow and Heat Transfer in Microtubes, Proc. ASME/JSME Thermal Engineering Joint Conf. Maui, Hi. 523-530.

83h Segmented Gas-Liquid-Liquid Flow and Use for Catalytic Reactions

Claude De Bellefon and **Gauthier Averlant**. LGPC ESCPE Lyon, CNRS, 43, bd 11 Novembre 1918, Villeurbanne, 69616, France

Three phase gas-liquid-liquid segmented flow in small pipes has seldom been reported. Such flow has been used for material synthesis, i.e. crystallization, the gas segments behaving as liquid slugs separators, or so called "coalescence inhibitors".[1] However, only one report focuses on the use of G/L/L segmented flow for reaction but without analysis of the flow pattern.[2] We would like to report some work on the use of segmented flow with a reacting gas, and organic and aqueous layers for performing hydrogenations catalysed with transition metal complexes and organic synthesis with a higly soluble gas. The impact of gas consumption on the flow will be analysed. Image analysis of the segmented 3-phase flow (see picture) also leads to propose a mapping with respect to superficial velocities of the fluids. [1] Chen D.L., Li L., Reyes S., Adamson D.N., Ismagilov R.F. Langmuir 2007, 23, 2255 [2] Onal Y., Lucas M., Claus P. Chem. Eng. Tech. 2005, 28, 972.

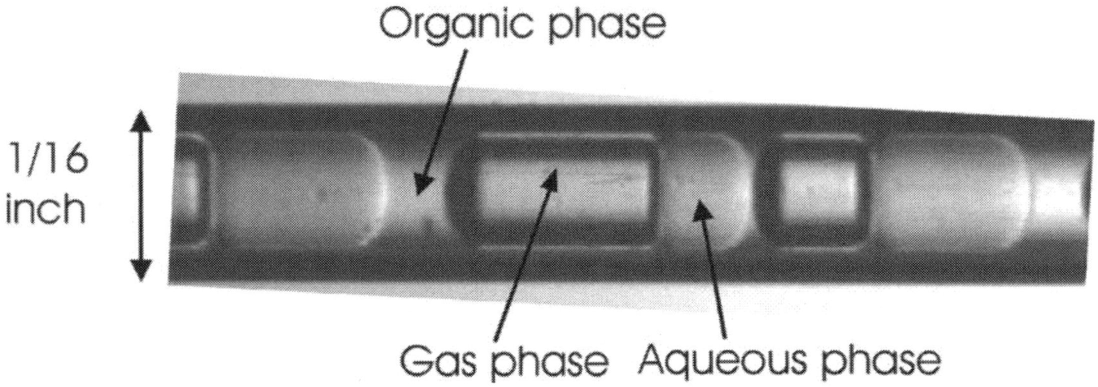

129b Mixing Performance of T-Shaped and Zigzag-Shaped Microreactors

Michael Schlüter, Marko Hoffmann, and Norbert Räbiger. University of Bremen, Institute of Environmental Process Engineering, Leobener Str., Bremen, 28359, Germany

Micro fluidic devices are successfully in use for several applications in chemical engineering and biotechnology. Nevertheless, there is still no breakthrough for micro process engineering because of a lack in understanding the mechanisms for local hydrodynamics and mass transfer on micro scales. Micro Particle Image Velocimetry (μ-PIV) in conjunction with Confocal Laser Scanning Microscopy (CLSM) enables the measurement of three-dimensional flow and concentration fields in micro devices for common stationary cases.

The visualization of the concentration field is carried out for the mixing of a passive tracer (Rhodamine B) as well as the reactive mixing (Calcium ions with an appropriate fluorescence tracer, i.e. "Fluo-4"). By using a confocal laser scanning microscopy (CLSM) it is possible to record image slices of a few microns thickness and a confocal pinholes removes out-of-focus emitted light and thus improves the lateral and axial spatial resolution. By taking image slices at different depths and assembling these slices together, a three-dimensional concentration field inside the mixing channel can be rendered. With the capability to visualize the tracer distribution at different cross sections a quantitative analysis of the mixing performance of different micromixer, i.e. T-shaped micromixers (typical dimensions: mixing channel width 400 micron, height 200 micron) and zigzag-shaped micromixers (typical dimensions: mixing channel width and height: 200 micron) is possible. The mixing quality can then be quantified using Danckwerts' intensity of segregation and the intensity of mixing respectively. Since these measures are insensitive to the length scales on which segregation occurs, the potential for diffusive mixing is used. This potential is a measure for the driving force for diffusive fluxes (the reciprocal value is a measure for the scale of segregation). The experimental analysis of the velocity field (2 dimensions + 2 components), i.e. the velocity components u and v, on a number of planes, e.g. 50 planes for a microchannel with a height of 200 micron) are basis for the calculation of the out-of-plane velocity w by using the continuity equation for incompressible fluids. With the knowledge of all three velocity components the 3D streamlines can be visualized. This allows a detailed characterisation of mixing performance for T-shaped and zigzag-shaped micromixers.

Web Page: www.iuv.uni-bremen.de

Micro particle image velocimetry (μ-PIV) in porous media using refractive index matching

Michael Heiland, Michael Schlüter, Norbert Räbiger
University of Bremen, Institute of Environmental Process Engineering, Bremen, Germany

Abstract

For the description of porous media or filtration, often integral methods are used where the properties of the fluid and the porous medium are combined in empirical factors like e.g. in Darcy's Law. These models are often based on averaged (integral) measurements and assumptions made about the actual local structures. For understanding the mechanisms inside the porous medium it is however necessary to study the local hydrodynamic and deposition effects. In this investigation a method is used where the measurement of the flow velocity field inside a simple porous medium is carried out with micro particle image velocimetry (μ-PIV). The optical access for the measurement is obtained through the method of refractive index matching (RIM). In the presented experiments, the porous sample consists of a transparent glass structure (borosilicate 3.3) with a polished glass sphere of d=1.3mm sintered into a glass capillary D=1.7mm. The experimental results show that it is possible to measure the velocity field behind the sphere with the μ-PIV system at a resolution of 25 microns vector-to-vector spacing and thus introduce a method which allows the measurement of the local velocity distribution inside a porous medium at micro-scale experimentally. This investigations will help to get a deeper insight into transport phenomena on micro scales within porous media to understand the mechanisms of momentum and mass transfer in filtration and adsorption processes.

Introduction

In the process of deep bed filtration, dissolved substances or solids in dilute suspensions are deposited when flowing through porous structures. There are many mechanisms responsible for the deposition, like adsorption on the surface, straining or hydrodynamic capture [1]. The process naturally occurs in the ground, where it is e.g. used for drinking water recovery through river bank filtration. There are also many industrial applications for deep bed filtration and research in this field has a long tradition (for an overview see Tien [1] or McDowell-Boyer [2]).

When modelling deep bed filtration, many investigations are based on integral experiments, measuring breakthrough curves of particles [3] or the permeability of the filter [4]. The modelling is then often based on empirical factors since the complex processes inside the porous medium are very difficult to describe. But the knowledge of the local phenomena is essential for the understanding of the macroscopic process. There are many recent works which investigate such local effects concentrating on the description of local deposition mechanisms [5-7] or flow inside a porous structures [8, 9]. For the local experiments some investigation use a model system [6] which represents actual porous structures or the access is provided through X-Ray-tomographic methods [10] or by optical access through refractive index matching (RIM) [7-9, 11]. Budwig [12] gives an overview of different RIM methods.

Due to progress in digital imaging the viewpoint onto the local phenomena is strongly enhanced, being able to study the processes on the time and length scales were the transport processes actually take place. This trend can also be found in the fast growing field of micro process engineering which makes use of the potential to directly control the governing transport processes [13].

In this report, a method will be introduced, which allows the investigation of the velocity field inside porous structures at micro scale. A micro particle image velocimetry system (µ-PIV) is used with a transparent glass model of a porous structure. The optical access is established with the method of refractive index matching (RIM). Building on the presented method, the correlation between local hydrodynamic and particle transport will be investigated to describe macroscopic effects like fluid dispersion, deposition or break through of particles which are of interest for the user of filters. The investigation aims at the advanced understanding of the phenomenology of deep bed filtration for the optimization of filtration processes.

Experimental setup

The method for experimental studies of the flow field in porous media is developed by means of a transparent model of a porous structure made of glass (borosilicate 3.3, DURAN [14]). The sample consists of a glass capillary with an inner diameter of D=1.7mm and a polished glass bead (d=1.3mm) that is sintered to the wall. The fabrication process is described in detail by Barg [15], here it should be stated that special care has to be taken concerning sintering temperature and time in order to avoid crystallization of the glass or heavy deformation due to viscous flow. The transparent structure used in the experiments is shown in Figure 1.

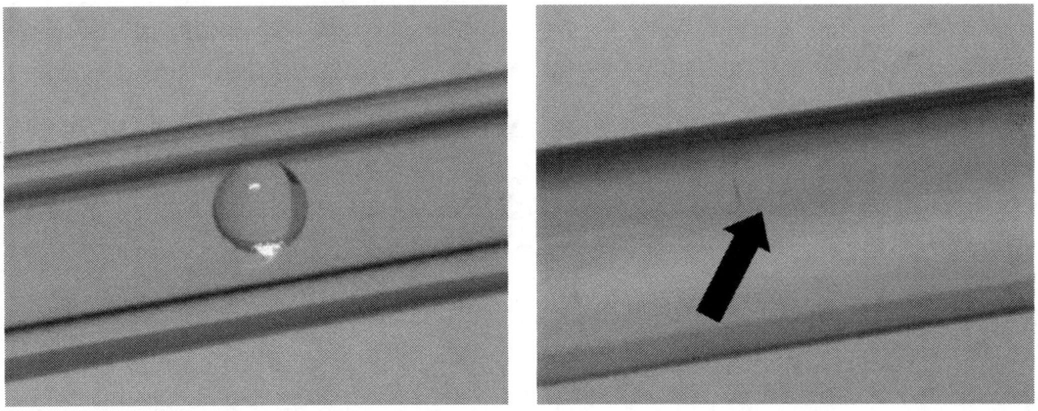

Figure 1. Image of the glass bead sintered into the capillary. Left: with air surrounding the bead; right: with refractive index matched fluid

The experiments are conducted with µ-PIV as an optical method which requires an optical access. This is established by matching the refractive indices of the glass and the surrounding fluid, thus avoiding light refraction at the interfaces [12]. In Figure 1 the used model structure is shown with and without refractive index matching (RIM). As material for the model, borosilicate 3.3 is chosen because this type of glass offers a high chemical and thermal resistance and is used as a standard material in many laboratory applications and access to many half finished and custom made products is good. The relatively low refractive index of n=1.473 facilitates the finding of a corresponding fluid and as the material has a standardized composition [14], a constant refractive index is ensured.

Table 1. Overview of different fluids for refractive index matching

material system	refractive index $n_{D,20}$	exposure risk	price	comment	literature
acetone- toluene	1.35-1.49	highly volatile	low	ventilation dissipation of comonents	[16]
water - DMSO	1.333-1.48	low	low	good cleaning as water based	[11]
water - glycerine	1.333-1.478	low	low	very high viscosity	[9, 12]
DC556-DC550 silicone oil	1.46-1.4935	medium	medium	high viscosity cleaning difficult	[8]
immersion oil	1.473	low	high	high viscosity, expensive	
DC200-DC500 silicone oil	1.375-1.4935	low	medium	DC200 is available in different viscosities, cleaning difficult	[9]

To match the fluid with the solid, two soluble liquids are mixed with one liquid having a lower and the other one having a higher refractive index than the solid. The refractive index of the solution is adjusted by varying the concentration of the components, but also the wavelength of the light and temperature have an influence on the refractive index. The solid has a much lower temperature-dependence on the refractive index than the fluid, which allows the final matching through temperature adjustment. In the Literature, different mixtures of materials are used for RIM which are listed in Table 1. For the practical use, matters of safety, price and handling are of special relevance.

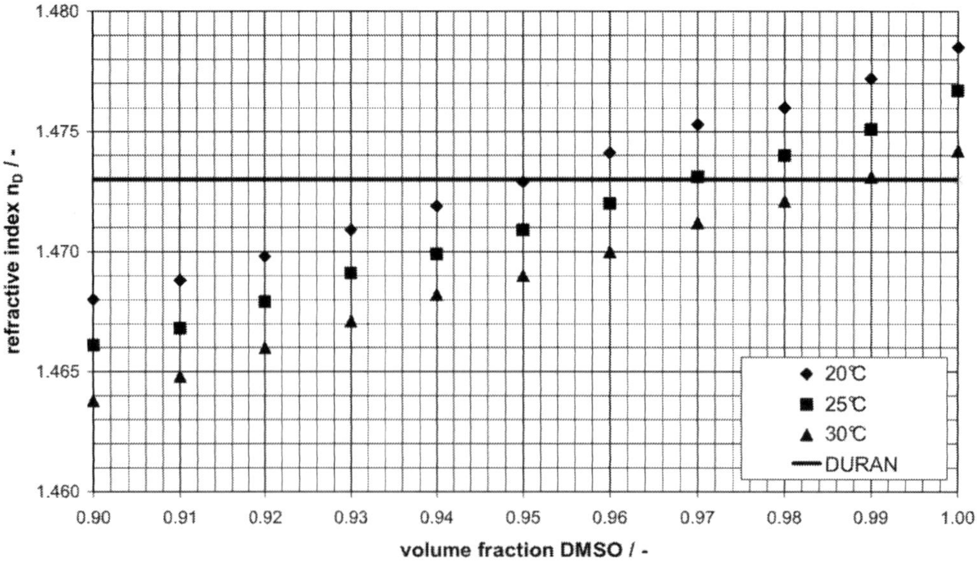

Figure 2. Refractive indices of the DMSO-water solution at different compositions and temperatures

For the experiments a solution of dimethyl-sulfoxide (DMSO) and water is chosen as both fluids are cheap and handling and cleaning is easy because the system is water-based. Figure 2 shows the results of refractive index measurements of DMSO-water with an abbe refractometer at a variation of temperature and composition. With the information from this diagram the composition of 95%vol DMSO and 5%vol water is selected to match at about 20°C. In the experiment the fine tuning of the refractive index can be achieved through temperature adjustment of the fluid.

The experimental investigation of this work is performed using the setup in Figure 3. For μ-PIV a conventional epi-fluorescence microscope (Olympus BX51W1) is used. A pressurized container delivers the refractive index matched liquid without pulsation and the liquid flow is measured by a mass flow meter (Bronkhorst High Tech BV, NL). A heat exchanger allows the control of the liquid temperature inside of the sample.

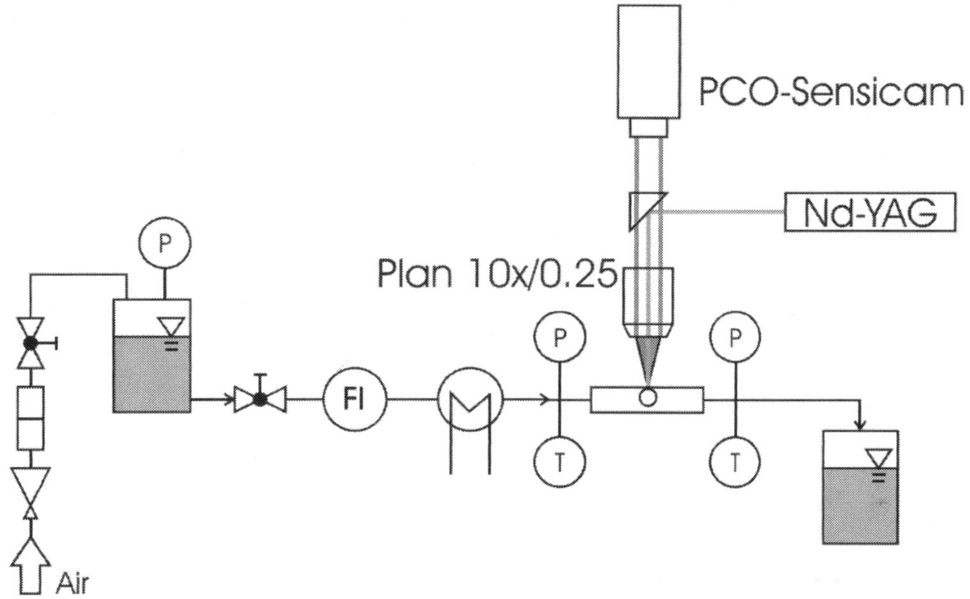

Figure 3. Experimental setup for the investigations of velocity fields in microstructures with μ-PIV

For the μ-PIV experiments a pulsed Laser (Nd-YAG 532nm, pulse width 5ns, New wave research, Inc.) is coupled to the microscope. As objective a Plan 10x/0.25 is used. The fluid is marked with fluorescent polystyrene tracer particles coated with Rhodamine B (diameter 500nm, micro particles GmbH, Berlin) which follow the liquid stream without inertial effects. A CCD camera (PCO Sensicam QE) is used for acquisition of the images of particle displacement during the two consecutive pulses at a defined pulse distance. For the analysis of the displacement, a standard cross correlation scheme is used to determine local velocity and direction. See Hoffman [17] for a more detailed description of the μ-PIV system. Figure 4 shows a cross section of the used model. For the reduction of refraction on the outer shell, a specially produced immersion oil (n=1.473, Dr. Tillwich GmbH Werner Stehr, Germany) is inserted into the gap between the capillary and the coverslip.

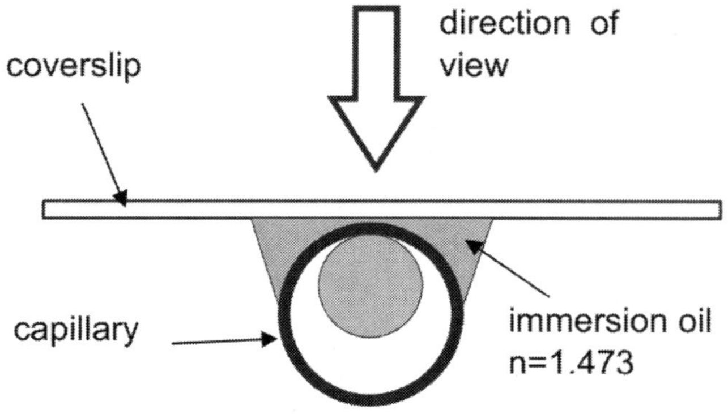

Figure 4. Cross section of the model structure. Immersion oil is introduced into the gap to minimize refraction

Results and discussion

Velocity measurements were conducted at different positions inside the sample at laminar flow conditions with a flow rate of 70 g/h which corresponds to a Reynolds number of Re=7 based on the free section of the capillary. As can be seen in Figure 5 the velocity field could be determined behind the sphere at a resolution of 25 microns vector-to-vector spacing.

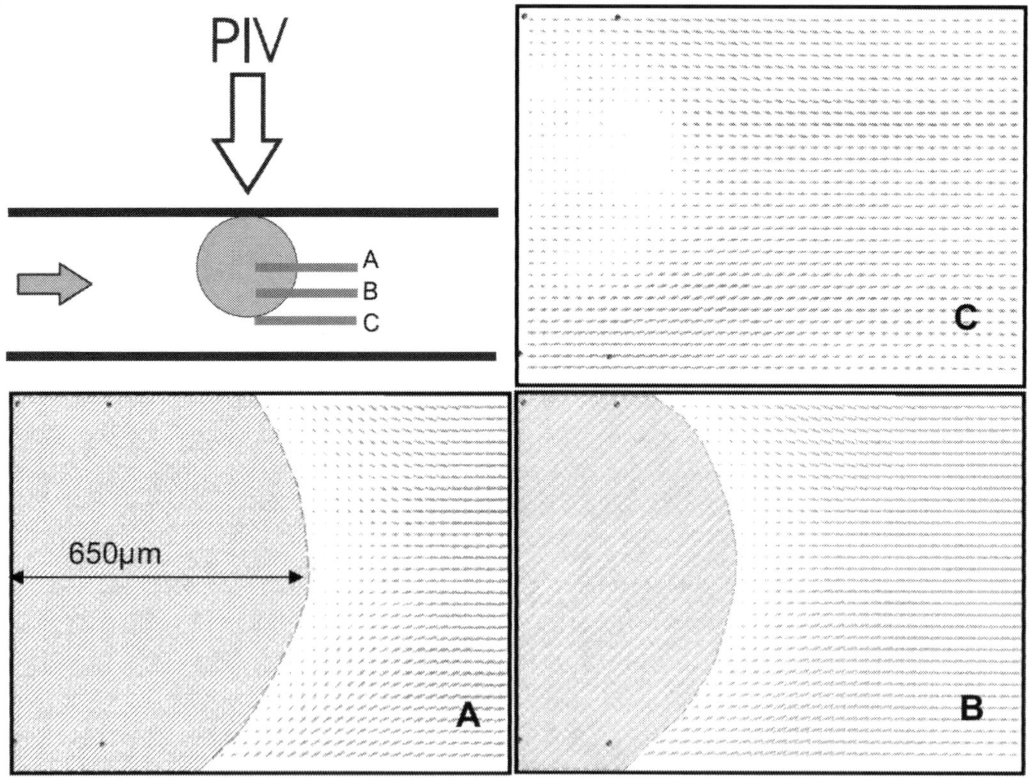

Figure 5. µ-PIV measurement of velocity field at three different section (A-equatorial; B-3/4; C-behind the sphere) with a resolution of 25 microns vector-to-vector spacing

Conducting the experiments, the deposition of the fine fluorescing particles on the surfaces leads to the loss of contrast in the images and thus has to be reduced in future experiments. Also the long term stability of the particles in DMSO is relatively low, so that an alternative liquid/particle system has to be considered.

Conclusions

An experimental method has been introduced, which allows the fabrication of porous structures and the measurement of local velocity fields inside at a resolution of 25 microns vector-to-vector spacing. The measurements are performed by µ-PIV and the optical access is obtained through refractive index matching.

Further investigation will concentrate on the extraction of the relevant data from the PIV-measurements and the model structure will be extended to a more complex structure with relevant d/D ratio. In parallel, macroscopic filtration experiments will be conducted to reveal relevant integral effects concerning particle transport. These results will be correlated to the local hydrodynamic investigations in order to enhance the modelling of deep bed filtration.

Acknowledgements

This work is part of the research program "Non-metallic Porous Structures for Physical-Chemical Functions" funded by the German Research Foundation (DFG), grant 'Graduiertenkolleg GRK 1375/1'. The authors gratefully acknowledge the financial support.

References

1. Tien, C. and A.C. Payatakes, *Advances in deep bed filtration.* AIChE Journal, 1979. **25**(5): p. 737-759.
2. McDowell Boyer, L.M., J.R. Hunt, and N. Sitar, *Particle transport through porous media.* Water Resources Research, 1986. **22**(13): p. 1901-1921.
3. Bradford, S.A., et al., *Physical factors affecting the transport and fate of colloids in saturated porous media.* Water Resources Research, 2002. **38**(12): p. 12.
4. Ergun, S., *Fluid flow through packed columns.* Chemical Engineering Progress, 1952. **48**(2): p. 89-94.
5. Yoon, J.S., J.T. Germaine, and P.J. Culligan, *Visualization of particle behavior within a porous medium: Mechanisms for particle filtration and retardation during downward transport.* Water Resources Research, 2006. **42**(6): p. 16.
6. Sirivithayapakorn, S. and A.A. Keller, *Transport of colloids in unsaturated porous media: Explaining large-scale behavior based on pore-scale mechanisms.* Water Resources Research, 2004. **40**(12): p. 8.
7. Ghidaglia, C., et al., *Hydrodynamic interactions in deep bed filtration.* Physics of Fluids, 1996. **8**(1): p. 6-14.
8. Northrup, M.A., et al., *Direct measurement of inerstitial velocity field variations in a porous medium using fluorescent particle image velocimetry.* Chemical Engineering Science, 1993. **48**(1): p. 13-21.
9. Stöhr, M., K. Roth, and B. Jähne, *Measurement of 3D pore-scale flow in index-matched porous media.* Experiments in Fluids, 2003. **35**(2): p. 159-166.

10. Al-Abduwani, F.A.H., et al., *Filtration of micron-sized particles in granular media revealed by x-ray computed tomography.* Review of Scientific Instruments, 2005. **76**(10).
11. Tallarek, U., et al., *Quantitative study of electrokinetic transport in porous media by confocal laser scanning microscopy.* Langmuir, 2003. **19**(11): p. 4527-4531.
12. Budwig, R., *Refractive index matching methods for liquid flow investigations.* Experiments in Fluids, 1994. **17**(5): p. 350-355.
13. Hessel, V., S. Hardt, and H. Löwe, *Chemical Micro Process Engineering - Fundamentals, modelling and reactions.* 2004, Weinheim: WILEY-VCH Verlag GmbH. Signatur Fb 011.
14. Schott, *Schott Technische Gläser.* 1999, Mainz.
15. Barg, S., et al., *Modeling of glass sintering applied for the fabrication of porous glass bodies.* Journal of Materials Science, 2008. **43**(2): p. 483-488.
16. Mier, P., *Experimentelle Untersuchung und numerische Simulation der Flüssigkeits-Feststoff-Strömung im Strahlschlaufenapparat.* 2006, Technische Universität Berlin: Berlin.
17. Hoffmann, M., M. Schluter, and N. Rabiger, *Experimental investigation of liquid-liquid mixing in T-shaped micro-mixers using μ-LIF and μ-PIV.* Chemical Engineering Science, 2006. **61**(9): p. 2968-2976.

Development of Microstructured Reactors and a Following Continuous Work-up System for Liquid/Liquid Reactions for Use in Polycondensation

Sonja Rothstock, Bernd Werner, Patrick Löb, Volker Hessel
Institut für Mikrotechnik Mainz GmbH, Mixing and Fine Chemistry Department, Mainz, Germany

In the German public funded BMBF project POKOMI (Förderkennzeichen 16SV1981) the synthesis of OLED materials by the Suzuki-Polycondensation using microreactor technology is investigated. Within the scope of this project IMM focus on one hand on the development, realization, characterization and application of suited microstructured reactors for biphasic liquid/liquid reactions and on the other hand also on the establishment of a pilot scaled following continuous work-up system.

The Suzuki polycondensation was developed by transferring the Suzuki cross coupling to polymer synthesis [1]. The Suzuki coupling is a homogeneous palladium catalysed C-C cross coupling reaction with great industrial importance for example for the manufacturing of flavours, fine chemicals and pharmaceuticals (API).The Suzuki reactions are conventionally carried out in a liquid-liquid biphasic solvent system as a batch process. In the scope the POKOMI project a continuous process for liquid-liquid biphasic reactions was established.

In order to ensure mass transfer in biphasic reactions a sufficient interfacial area is required. In the considered case the two phase system is not stable and therefore coalescence comes into play. In order to overcome this issue two different reactor concepts were developed. The first concept based on the repeated breakage of the droplets. The so called redispersion microreactor [2] consists of exchangeable redispersion units which are separated by multichannels and an inspection window which allows the investigation of the flow patterns.

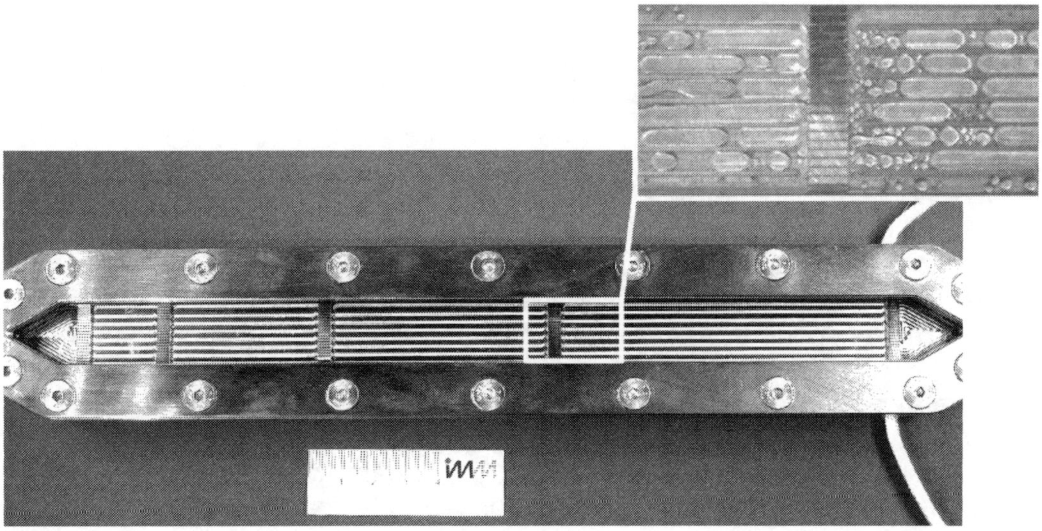

Figure 1. Redispersion microreactor with an inspection window

Before the liquid enters the redispersion unit big slugs are observable and behind the unit small droplets appear which directly form slugs again. Two different redispersion units were investigated.

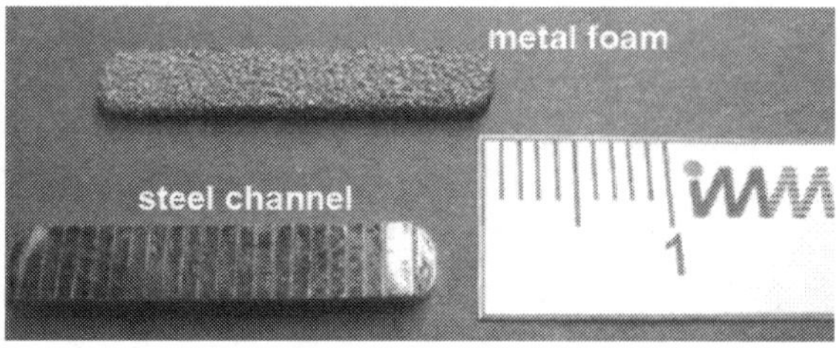

Figure 2. Investigated redispersion units

The so called steel channels contains 20 small channels (1 x 2 x 4 mm). The metal foam redispersion units consist of nickel-chromium foam with a average pore diameter of 0,4 mm and a surface area of 5400 m^2/m^3 and a porosity of 95%, which is produced by the Recemat Company.

The real case ternary biphasic solvent consisting of toluene, 1,4-dioxane and water has been used for the determination of the reactor performance. The 1,4-dioxane is soluble in water and toluene and lowers the interfacial tension between the both immiscible phases. Various flow patterns were observed inside the reactor, which were captured with a high speed camera. The different flow regimes are shown in the following figure.

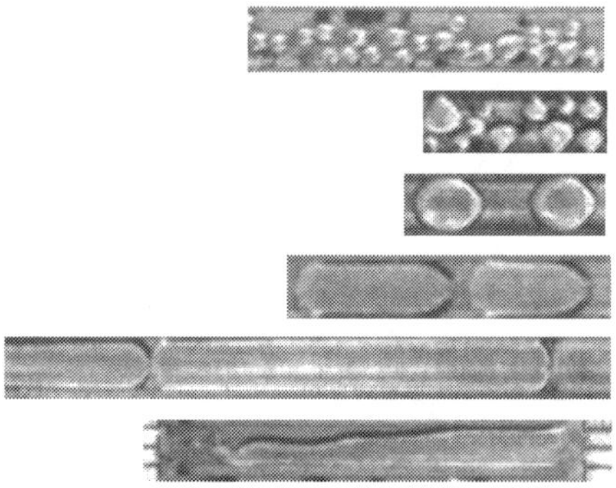

Figure 3. Observed flow regimes inside the channels of the redispersion microreactor

In the fine dispersed and dispersed flow regime the drop diameter is much smaller than the channel width. The drop flow regime is characterised by the formation of isolated drops with a diameter as large as the channel width. In the slug flow regime the length of the drops are larger than the channel width. With an increasing slug length the regime passes into the elongated slug flow. Here the slugs are no longer isolated. The deformed interface flow is defined by the parallel flow of both phases with a wavy interface. The flow regimes 2-5 are occur at the same time and their appearance is mostly independent from the flow rate. The flow regimes 1 and 6 are observable for higher flow rates only. The temperature increase shows hardly any influence on the flow regime variation. The pictures were also used to estimate the specific interfacial area. Depending on the flow rate a specific interfacial area between 1000-1600 m^2/m^3 was formed using both kinds of redispersion units. For the measured flow rate range it was not possible to establish a drop flow all over the reactor. Even for higher flow rates the drop flow appeared just in a few channel sections. In the case of the metal foam redispersion units the pressure drop is much lower (Table 1).

As a second concept a metal foam filled tube reactor was developed. The metal foam is used as a form of static mixer-reactor.

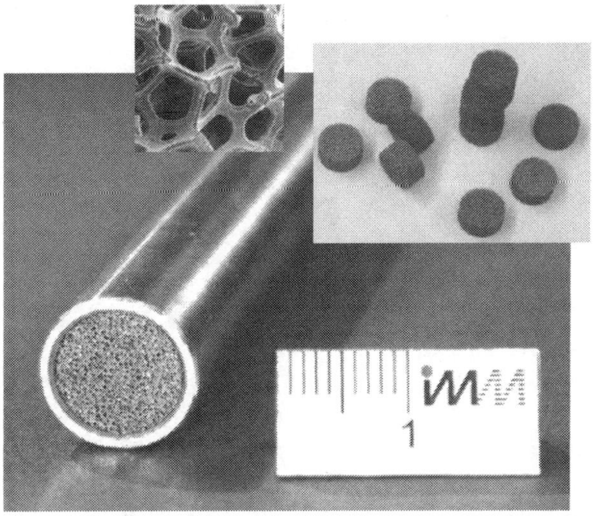

Figure 4. Metal foam filled tube reactor for liquid-liquid biphasic reactions

Due to the special combination of chemical, mechanical and physical properties metal foams are already used for a wide range of applications, for example heat exchangers [3]. The microstructure of the used open-cell metal foam provides a very high porosity (60-98 %) which leads to low pressure drop. The pressure drop over the reactor was measured for different velocities. The biphasic solvent system contains toluene, dioxane and water.

Table 1. Pressure drop over the Redisperison dispersion microreactor (length 0.03 m, for both kinds of redispersion units) and the tube reactor (length 1 m).

	redispersion microreactor			tube reactor	
flow rate (L/h)	velocity/10^{-3} (m/s)	metal foam Δp (bar)	steel channel Δp (bar)	velocity/10^{-3} (m/s)	Δp (bar)
0.36	6.3	< 0.1	0.3	1.2	0
0.63	10.9	0.1	0.8	2.2	< 0.1
0.90	15.6	0.1	1.1	3.1	0.5
1.17	20.3	0.2	1.5	4.1	1.2

In the present case the same Nickel-Chromium metal foam was used as for the redispersion units in the redispersion microreactor. Given that the reactor consists of tube sections of different length the reactor length and consequently the residence time can be varied.

As a test reaction the palladium catalyzed Suzuki cross coupling of phenylboronic acid and 4-bromotoluene in the presence of a palladium-tri-(o-tolyl)-phosphine catalyst was carried out.

The reaction is given in the following figure.

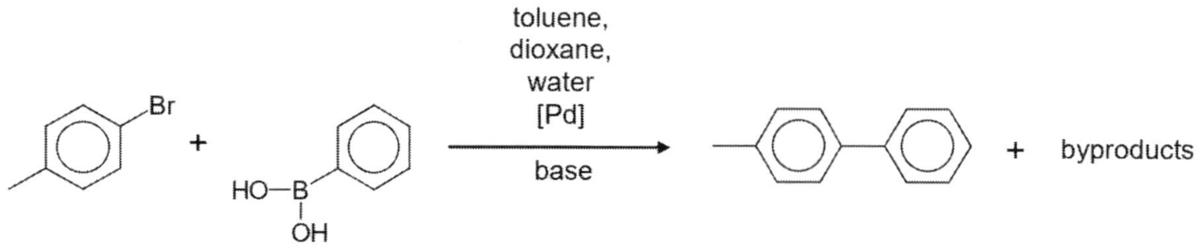

Figure 5. reaction equation of the test reaction

The substrates and the catalyst are dissolved in the organic phase. The aqueous phase contains the base.
To characterize the tube reactor and to optimize the reaction conditions for a start the reactions temperature was varied for a residence time of 10.5 min.

The conversion increases with increasing reaction temperature. Conversions of 90% are achieved which competes with the best literature know micro processing of Suzuki couplings [4]. The best conversion obtained by the redispersion reactor was no more than 30%. The superior performance of the metal foam filled tube reactor is explained by increased residence time, better and more uniform dispersion, and ease of achieving high-temperature operation. Thus, for the medium slow Suzuki coupling milli process technology, rather than micro process technology, is the right approach for efficient and intensified processing.

It should be noted that the reaction temperature is over the boiling point of the solvent system, following the ideas proposed by the *Novel Process Windows* concept to actively adjust reaction time to achievable residence time in the new tools [5]. Therefore a system pressure has to be applied to ensure that the solution is liquid inside the reactor. For the given reaction temperature a pressure between 7-14 bars was applied.

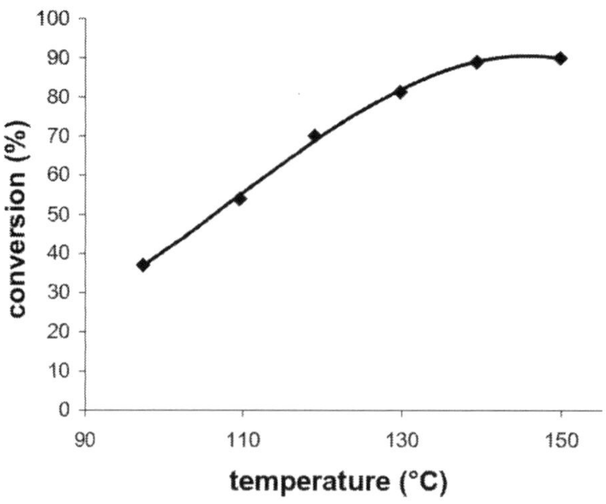

Figure 6. Variation of the reaction temperature

As a side reaction the homocoupling of the boronic acid derivatives takes place (< 1%) but this can be minimized by inert conditions. So the conversion is equivalent to the yield. With a conversion of 90 % and the given flow rate the space-time-yield amounts to 125 kg/m^3 h. A TON of 424 and a TOF of 121 h^{-1} was achieved for the Suzuki coupling. Additionally the variation of the flow rates and the residence time the reaction conditions are optimized. The optimized parameters shall be transferred to the polycondensation reaction.

In the scope of the POKOMi project a continuous process shall be established to come to a complete continuous process it was necessary to develop a continuous working work-up system following the microstructured reactor.

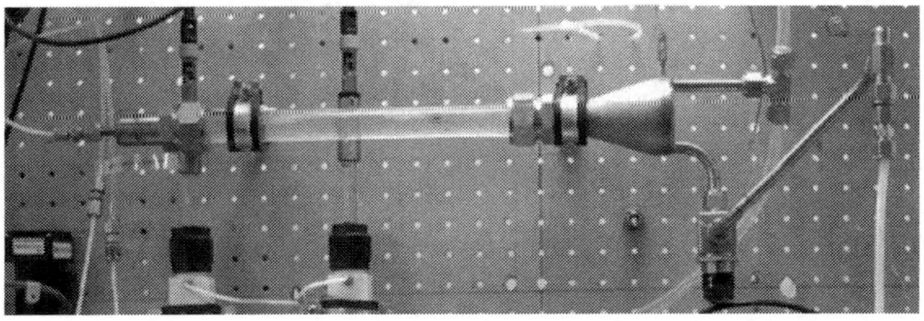

Figure 7. Picture of the developed mini-settler

The realized set-up consists of a phase separator to remove the aqueous phase and two wash steps including again phase separation. In order to mix the washing water with the organic phase caterpillar micromixers are used. Phase separation thereby is realized by gravity driven mini-settlers (Figure 7) with a volume of 246 mL for each settler.

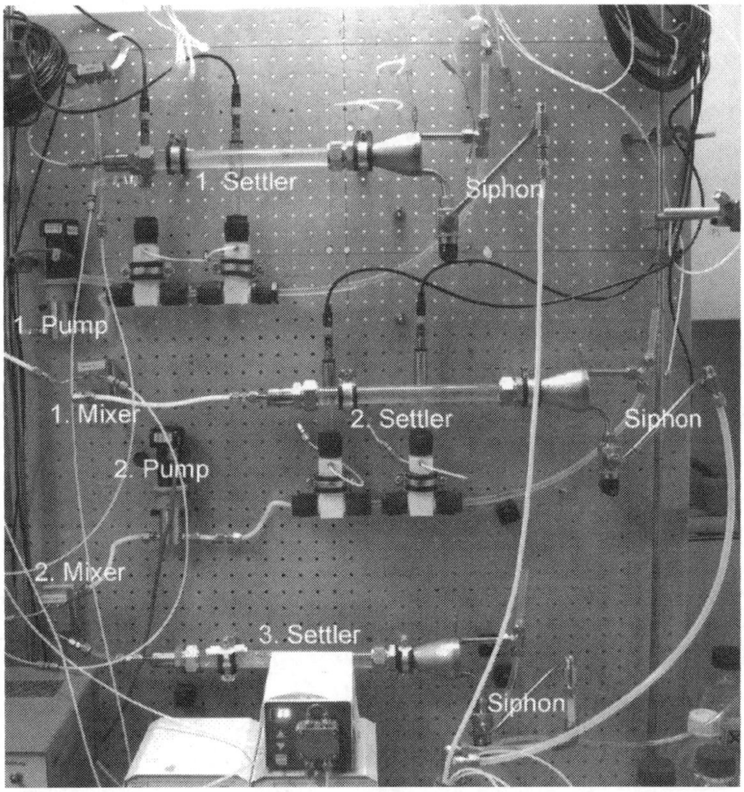

Figure 8. Picture of the complete mixer-settler unit

The reaction mixture enters the first settler and is separated. The aqueous phase (waste) leaves the settler at the lower outlet. The organic phase leaves the settler at the upper outlet and is pumped into a caterpillar micromixer. Here the organic phase is mixed with fresh water in order to extract the by products. The resulting mixture is separated in the second settler and then a second washing stage follows.

The developed washing unit is suited for total flow rates between 1.8 and 6 L/h. The testing of the mixer-settler system was so far restricted to practical issues regarding to the continous coupling of the single purification steps so that no negative interference establishes. It could be in particular proven that a stable and time-independent filling and discharging of the settler occurs without any flooding or draining effects. Also, the dynamic operation during system start-up has been investigated, especially concerning the filling of the siphon. Challenges, difficulties and first experiences with this work-up system will be reported.

[1] Schlüter, A. D., J. Polym. Sci. Part A: Polym. Chem. 39 (2001), 1533-1556

[2] Hessel, V., Rothstock, S., Agar, D.W., Jadhavrao, P., Löb, P., Werner, B., in: Conference proceedings, 2007, AIChE spring national meeting: April 22-27, 2007, Houston. – New York: AIChE, 2007, 74d.

[3] Boomsma, K., Poulikakos, D., Zwick, F., Mechanics of Materials 35 (2003), 1161-1176.

[4] Haswell, S. J., Middleton, R. J., O´Sullivan, Skelton, V., Watts, P., Styring, P., Chem. Commun. (2001), 391-398.

[5] Hessel, V., Löb, P., Löwe, H., Curr. Org. Chem. 9 (8) (2005), 765-787.

45d "Sandwich Microreactor" for Heterogeneous Catalytic Processes:N2O Decomposition as a Case Study

Lioubov Kiwi-Minsker[1], Bryan Bromley[2], Volker Hessel[3], and Albert Renken[2]. (1) Institute of Chemical Sciences and Engineering, Ecole Polytechnique Fédérale de Lausanne, EPFL-ISIC-GGRC, Station 6, Lausanne, 1015, Switzerland, (2) Isic-Ggrc, Ecole Polytechnique Fédérale de Lausanne, EPFL-ISIC-GGRC, Station 6, Lausanne, 1015, Switzerland, (3) Institut für Mikrotechnik Mainz GmbH, Carl-Zeiss-Straße 18-20, Mainz, D-55129, Germany

The manufacture of chemicals in catalytic micro-structured reactors (MSR) has become recently a new branch of chemical reaction engineering focusing on process intensification and safety. MSR have high specific surface area in the range of 10000 to 50000 m2/m3 which allows an effective mass and heat transfer if compared to traditional chemical reactors. The important feature of MSR is an integrated heat exchange, which makes the key difference between MSR and other structured reactors, like honeycombs. MSR are operated under laminar flow with the heat transfer coefficient for liquids about 10 kW/(m2K). This is one order of magnitude higher than in the traditional heat exchangers avoiding hot-spots formation, allowing higher reaction temperatures and reducing reaction volumes. This in turn improves the energy efficiency and the operational cost.

One of the main problems in using MSR for heterogeneously catalyzed gas-phase reactions is the introduction of the catalyst in the reaction zone. The straight forward way is to fill microchannels by catalyst powder, but this leads to high pressure drop. In addition, each channel must be packed identically to avoid maldistribution, which is known to broaden residence time distribution diminishing reactor performance.

Another approach is MSR with catalytically active walls. The specific surface area is increased by chemical treatment of the channel walls or by their coating with a porous layer. The porous layer can serve as a catalyst or a support for a catalytic phase. The main limitation for catalytic wall MSR is the thickness of the porous layer. Since the majority of MSR are used for fast highly exothermic reactions, the layer should be < 1-2 micro-m in order to avoid mass/heat transfer limitations. Therefore, the total mass of the catalyst is too small for achieving process intensification referred for a unit of the reactor volume.

In this paper we present a new design of MSR: the "Sandwich Microreactor". It was constructed by IMM and consisted of thin (~300micro-m) porous plates of sintered metal fibers (SMF) serving as structured catalytic beds. SMF plates are sandwiched between regular metallic plates, which provide integrated heat exchange. Each SMF plate possesses a 3-dimensional regular microstructure, giving low pressure drop during the passage of reacting gases. The pressure drop in the laminar flow regime shows linear dependence on the flow rate.

The residence time distribution was measured indicating on flow pattern close to ideal plug-flow reactor.

The SMF was coated by catalytically active Fe-ZSM5 thin film (< 2fÝm) and the Sandwich MSR was tested in the N2O decomposition as a model highly exothermic reaction. The temperature was measured at the outlet confirming the isothermal reactor operation. The results will be presented in detail.

Nematic Liquid Crystal Phases Observed in Ionic Liquid Crystal Polyurethane-ureas

S. N. Jaisankar,[†,§] *Donna J. Nelson,*[†,*] *Christopher Brammer,*[†] *and Heather Rhoads*[†]

[†]*Department of Chemistry and Biochemistry, University of Oklahoma, Norman, OK 73019*

[§]*Polymer Division, Central Leather Research Institute, Adyar, Chennai - 600 020, India*

Liquid crystal (LC) characteristics of polyurethane (PU) produced from 2,4-toluenediisocyanate (24TDI) and 4,4'-bis(6-hydroxyhexoxy)biphenyl (BHHBP) are reported to be due to structural features, including the asymmetric methyl in 24TDI, rigid biphenyl units, flexible methylene spacers of BHHBP, but less from intermolecular hydrogen bonding of urethane linkages.[1a] Polyurethane ureas (PUUs) might be more likely liquid crystals due to a potential for different hydrogen bonding, consisting of more but weaker bonds. Hydrogen bonding in PUs was reported to be "only of secondary importance" in mesophase formation and morphology,[1a] but important to dimensional stability.[1b] In contrast, hydrogen bonding in PUUs is expected to contribute differently, not only because NH bonds in urea linkages will provide additional protons for donation, but also because replacing O by N will allow weaker hydrogen bonding. Indeed, reports have indicated that hydrogen bonding is important in PUUs.[2a,b] However, although many PUUs have excellent thermoplastic characteristics,[2] their design as liquid crystals has not been adequately explored.[2d]

The effects of PU and PUU precursors on product characteristics guide designing a PUU likely to have LC properties. Many PUU syntheses reported using a diol, a diisocyanate, and a diamine chain extender.[2] We collected from different polymeric systems, characteristics which yielded the desired properties: (1) using an aromatic diamine as the chain extender,[2d] (2) using BHHBP as the mesomeric diol, which yielded a LC in the PU system,[1a] (3) designing unbranched hard segments, which reportedly increase packing and hydrogen bonding in PUUs,[2a] (4) using an ionic hard segment, which gives better mechanical properties in multiple systems,[1c,3] (5) selecting sulfonate as the ionic group because it was reported to produce stronger intermolecular interactions in PUs.[1c] Although the ionic nature of the product might be expected to reduce its LC properties,[4] this is also expected to increase its usefulness in polymer blends.[5]

We report herein results and product characteristics from reacting BHHBP with four representative diisocyanates (Scheme 1), followed by treatment with the ionic chain extender sodium 2,5-diaminobenzenesulfonate (DBS), by iterating 4X through a modified two-step procedure[2a] to give the corresponding PUU (eq 1, R = diisocyanate).

FTIR was used for structural characterization of

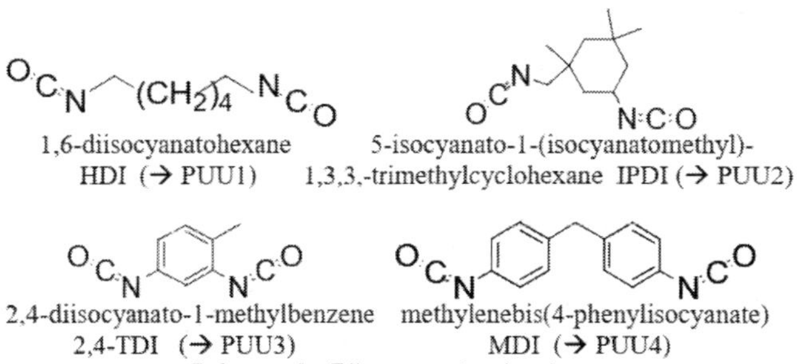

Scheme 1. Diisocyanate structures.

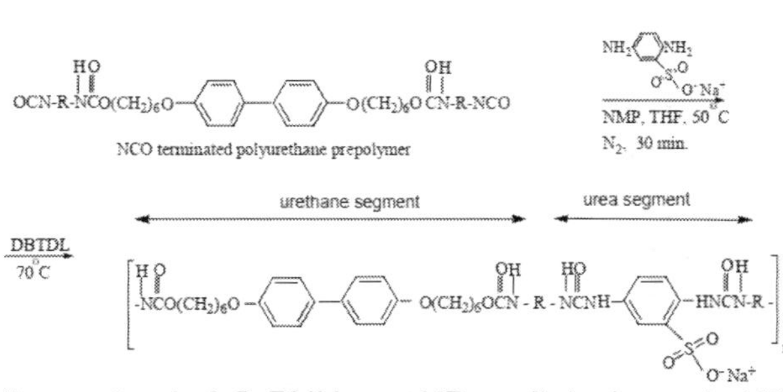

(1)

thermotropic LC PUUs as KBr pellets (sample:KBr = 1:80). FTIR spectra (Figure 1) confirmed each PUU formation by disappearance of the very broad peak at 3350-3535 cm^{-1} (O-H stretch) and appearance of a new sharper peak at 3330 cm^{-1} (urethane N-H stretch). Absence of an isocyanate antisymmetric stretching band at 2260-2280 cm^{-1} also evidences complete reaction between hydroxyl and diisocyanate. Similarly, IRs of PUUs1-4 each show a band at 1610 cm^{-1} (urethane C=O stretch) with a shoulder at 1540-1545 cm^{-1} (urethane N-H bend), and a second band at 1225-1258 cm^{-1} (N-C stretch of N-COO); these indicate formation of functionalities critical to the PUUs. The spectra for PUUs1-4 have the usual overlapping free and hydrogen bonded N-H stretch bands at about 3300-3500 cm^{-1}. This agrees with known changes in polyurethane IR data produced by the addition of polyamides.[6]

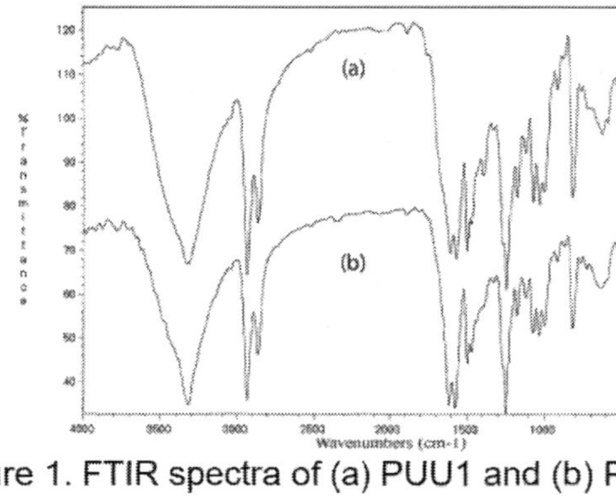

Figure 1. FTIR spectra of (a) PUU1 and (b) PUU2.

Na^+ is counterion to SO_3^- in PUUs; its complexation to nitrogen lone pair electrons reduces the N-H bond length and produces the observed shift of –NH peaks from 3500 to 3330 cm^{-1} similarly to polyester PUUs.[7]

Ionic LC-PUUs1-4 were also characterized by using cross polarization/magic angle spinning (CP/MAS) ^{13}C NMR. The ^{13}C NMR of PUU1 shows peaks (in ppm) at 29.5 (CH_2), 45.2 (CH_2), 65.3 (CH_2-

OPh), 69.6 (CH_2-OC(=O)NH), 119.0 (C=C), 131.8 (C=C), 160.1 (urea C=O), and 162.1 (sm shoulder, urethane C=O). PUUs1-4 show number average mol wts (M_n) in the range 25,000 - 35,000, as determined by VPO (Vapor-Phase Osmometry).

PUU thermal properties (Table 1) were determined by differential scanning calorimetry (DSC, Seiko SSC 5200H, 5°C/min) and thermogravimetric analysis (TG/DTA 220 module, 15°C/min). Each shows a two-stage decomposition, T_{d1} and T_{d2}.

Table 1. PUU heating temperatures (°C) of glass transition (T_g), melting (T_m), first (T_{d1}), and second decomposition (T_{d2}).

Codes	T_g	T_m		T_{d1}	T_{d2}
	DSC	DSC	POM	TGA	TGA
PUU1	-	185.2	189.2	343	399
PUU2	123.9	179.1	164.9	359	445
PUU3	-	255.4	248.0	315	467
PUU4	-	269.8	262.5	340	496

PUU2 DSC shows two endothermic transition temperatures,[1b] glass (T_g) and melting (T_m) at 123.9 and 179.1°C respectively (Figure 2a).

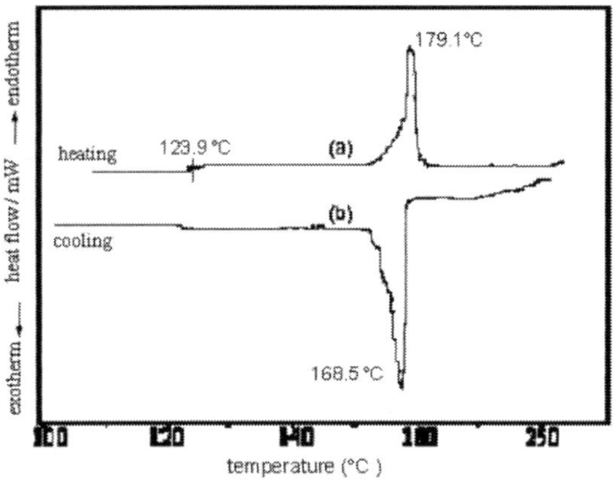

Figure 2. DSC traces of PUU2 (a) heating (b) cooling.

Lliquid crystal PUU textures were observed by using a cross polarized optical microscope (POM) with a hot stage (OLYMPUS BX50, with LINKAM THMS 600). PUU1-4 POM temperatures (Table 1) agree well with melting points determined by DSC. Ionic PUU2 began melting at 164.9°C and reached its isotropic state at 290°C; upon cooling, the LC starts to appear at 168.5°C. This mesophase can probably be attributed to its inherent viscosity.[1a]

A PUU2 micrograph (Figure 3) taken after cooling from its isotropic state shows a nematic phase with schlieren LC double axial texture, characterized by point singularities from which two or four dark brushes radiate.[8] The optical textures of the other ionomers also appear to be typical[8] nematic LC phases: schlerien (PUU1) and grain (PUU3 and PUU4). These are obtained by slowly cooling the PUU sample from its isotropic melt to room temperature; the nematic phase often starts to form in birefringent (double reflection) droplets.[8] The nematic texture indicates each is a monotropic LC mesophase.[1a]

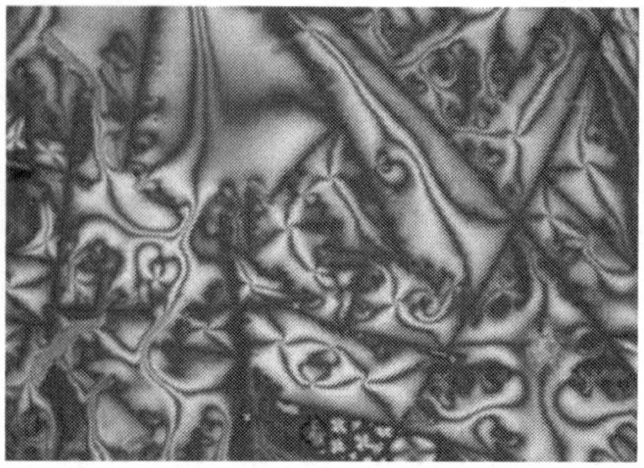

Figure 3. PUU2 photomicrograph at T=126.8°C (cooling, nematic texture 100X).

We are aware of no other report of a PUU exhibiting this nematic mesomorphic property. Our PUUs each exhibit a nematic liquid crystal mesophase upon cooling, which is rare in ionic PUUs; thus they could be useful for compatibilizers and processing aids in polymer blends.[5] Both structure and properties of each PUU were dependent upon the isocyanate used; although PUUs1-4 had different isocyanate precursors, all were monotropic LCs.

Acknowledgment

We acknowledge the National Science Foundation and Ford Foundation for support. SNJ appreciates financial support from the Science and Engineering Research Council, Department of Science and Technology (SERC-DST) India.

References

1. (a) Papadimitrakopoulos, F.; Hsu, S. L.; MacKnight, W. J. *Macromolecules* **1992**, *25*, 4671, and references therein. (b) Sheth, J. P.; Aneja, A.; Wilkes, G. L. *Polymer* **2004**, *45*, 5979. (c) Santerre, J.; Brash, J. *Ind. Eng. Chem. Res.* **1997**, *36*, 1352.
2. (a) Sheth, J.; Wilkes, G.; Fornof, A.; Long, T.; Yilgor, I. *Macromolecules* **2005**, *38*, 5681. (b) Curgul, S.; Yilgor, I.; Yilgor, E.; Erman, B.; Cakmak, M. *ibid.* **2004**, *37*, 8676. (c) Sung, C. S. P.; Hu, C. *ibid.* **1981**, *14*, 212. (d) Wolinska-Grabczyk, A.; Muszynski, J.; Penczek, P. *Polimery* **1997**, *42*, 55.
3. (a) Xue, Y.; Hara, M. *Macromolecules* **1997**, *30*, 3803. (b) Xue, Y.; Hara, M. *Polymer* **2005**, 46, 7293.
4. Xue, Y.; Hara, M.; Yoon, H. *Macromolecules* **1998**, *31*, 7806.
5. (a) He, J.; Liu, J. *J. Appl. Polym. Sci.* **1998**, *67*, 2141. (b) Zhang, A.; Zhang, B.; Feng, Z. *J. Appl. Polym. Sci.* **2002**, *85*, 1110.
6. Mattia, J.; Painter, P. *Macromolecules* **2007**, *40,* 1546.
7. Wang, H-L.; Kao, H. M.; Digar, M.; Wen, T. C. *Ibid.* **2001**, *34*, 529.
8. Wang, X-J.; Zhou, Q-F. *Liquid crystalline polymers*; World Scientific Pub: New Jersey, 2004; pp 209-217.

Selective hydrogenation of phenylacetylene to styrene on a Pd/TiO_2 coating in a microreactor

Evgeny V. Rebrov[1], Angel Berenguer-Murcia[2], Helen E. Skelton[2], Brian F.G. Johnson[2], Jaap C. Schouten[1]

[1]*Department of Chemical Engineering and Chemistry, Eindhoven University of Technology, P.O. Box 513, 5600 MB Eindhoven, the Netherlands*
[2] *Department of Chemistry, Cambridge University, Lensfield Road, CB2 1EW Cambridge, the UK*

Abstract

The selective hydrogenation of phenylacetylene to styrene by hydrogen was studied in the 25-50 °C temperature range in a continuous flow microchannel of 250 μm i.d. with a Pd supported on mesoporous TiO_2 catalyst. The inner wall of the microreactor was dip-coated with a prehydrolyzed titania precursor solution containing an $EO_xPO_yEO_x$ amphiphilic triblock copolymer (EO = ethylene oxide, PO = propylene oxide, x = 106, y = 70) and suspended palladium nanoparticles with an average diameter of 4 nm. Solvent was withdrawn from the channel at a rate of 1 cm/s yielding a 120 nm continuous titania layer with a cubic mesostructure after drying and calcination. The organic copolymer was removed by calcination at 300°C under a residual pressure of 15 mbar. The coating porosity was 0.4, resulting in overall coating mass of 1.32 mg with a 1 wt.% Pd content. The liquid flow was varied between 1 and 10 μL/min and the hydrogen flow between 250 and 550 μL/min (STP). This corresponds to superficial velocities of 0.034-0.34 cm/s and 8-18 cm/s for liquid and gas, respectively. The maximum selectivity to styrene of 95 % was observed at a phenylacetylene conversion of 88%. The reaction rate in terms of TOF was found to be up to 2 s^{-1}.

Keywords: hydrogenation of phenylacetylene; microchannel; Pd nanoparticles

1. Introduction

The synthesis of a large number of fine chemicals, particularly in the field of flavor and fragrance chemistry and pharmaceuticals, involves the selective hydrogenation of organic compounds as a critical step. Those multiphase reactions can be performed in microchannels with a catalytic coating deposited on the channel wall. Usually, the geometric surface of the microchannels is not sufficient for performing catalytic reactions. Small surface enhancement factors can be sufficient for fast reactions [1]. The efficient use of microstructured catalytic reactors requires a shaping of the catalyst by deposition of thin catalytic coatings at the walls of the reactor channels to increase the surface area. So far, we applied different methods for coating microreactors such as anodic oxidation of aluminum alloys [2], cathodic reduction in ionic melts [1,3], hydrothermal synthesis [4,5], atomic layer deposition (ALD) [6], and sol-gel methods [7]. However, those methods are limited to coating before microreactor bonding/assembling. Furthermore, they can also be limited by the size (channel length) and the composition of substrate materials which should fit to the deposition equipment and be compatible with the synthesis solutions. The deposited coatings require that bonding of the individual plates is performed at low temperatures to avoid catalyst sintering or the reactors to be assembled using mechanical connections between components inside a holder (e.g. by stacking together all the plates in the housing), which limits their applications to low-pressure processes to avoid leaking. Thus, coating of assembled microchannels opens new perspectives in the development of single phase and multiphase catalytic microreactors. The main problem is, however, how to obtain a uniform coating thickness along the channel, and if supported catalysts are used, how to obtain a uniform distribution of an active component throughout the coating thickness.

Several methods are reported for coating of assembled microchannels. Pfeifer et al. [8] used an aqueous dispersion of 1 wt.% hydroxyethyl cellulose containing 20 wt.% of ZnO particles to fill in the microchannels with a cross section of 100 μm ×100 μm. A homogeneous ZnO layer was produced by drying at room temperature. The thickness of the resulting layer was 20 μm. The dried layer was impregnated dropwise with a saturated solution of palladium acetate in toluene, which was fully evaporated after addition of each droplet. The authors observed that the coated foils showed extremely dark regions at the ends of the channels, and the middle of the foils was nearly white, due to the absence of Pd. Reuse et al. coated the 200 μm × 100 μm microchannels with a suspension of a micro-milled commercial copper based catalyst (G-66MR, Südchemie) [9]. About 10 mg of catalyst were deposited on each plate resulting in a catalytic layer of ca. 5 μm. The catalytic activity of the coating in methanol steam reforming was found to exceed the values of the pelletized G-66MR catalyst; an effect often observed for thin catalytic layers. However, no long-term deactivation data were reported.

Bravo et al. coated capillaries of quartz and fused silica with a commercial BASF $CuO/ZnO/Al_2O_3$ catalyst [10]. Prior to coating, the inner surface of capillaries with 530 μm in diameter was cleaned with a hot sulfuric acid/NOCHROMIX® (from GODAX Laboratories) mixture for 30 min at 100–120°C. The surfaces were rinsed with water followed by ethanol and dried in an oven at 80°C. This step was performed to remove organic material from the surface and to increase the surface hydrophilicity. Then, a slurry of catalyst with a boehmite binder was filled in the capillary and removed by a stream of inert gas to produce coatings ranging from 1 to 25 μm in thickness depending on the slurry viscosity. The coating was adherent and did not peel off during catalytic tests in methanol steam reforming for at least 10 days. The activity of the coated catalyst was superior to that of the same catalyst in a packed bed. It should be noted that a bulk catalyst was coated onto the microchannels in all studies mentioned above. Although there have also been several reports concerning the immobilization of metals on a glass wall, lowered reactivity and leaching of the metals during reactions remain serious problems [11-13]. Greenway et al. used a 300 μm x 115 μm microchannel for a Suzuki coupling reaction (synthesis of 4-cyanobiphenyl from 4-bromobenzonitrile and phenylboronic acid) on a silica supported 1.8 wt.% Pd catalyst [11, 13]. In those studies, a silica slurry containing a desired amount of catalyst precursor was physically positioned in the channel and dried at 100°C for 1 h to produce the catalytic coating. Palladium contamination in the product was found indicating Pd leaching from the catalytic coating. Recently, Kobayashi et al. proposed an efficient way of stabilization of Pd catalyst on the channel walls [14]. At first, amine groups were introduced onto the surface of a glass microchannel with a cross section of 200 μm x 100 μm, which was followed by cross-linking of a Pd containing precursor during the heating. The reactor was tested in hydrogenation of benzalacetone, 1,2-diphenylacetilene and several other substrates. The authors observed no leaching of the Pd from the wall coated catalyst. However, even following this approach, the tuning of the size and composition of the catalyst particles remains unsolved. Despite the large diversity of methodologies [15], reproducible and cost-effective preparation of the catalytic coating remains a crucial step in fabrication of multiphase micro fluidic devices.

In this work we propose a new method for stabilization of an active component in supported metal catalysts. This method is based on incorporation of metal colloidal nanoparticles into inorganic mesoporous thin films with large surface areas and narrow pore size distributions [16]. The preparation of bimetallic colloids has been attracting an increasing attention during the last years [17-23]. The addition of a second metal generally improves some properties of the catalyst, such as activity, selectivity and/or stability in certain reactions [24]. Recent studies have shown that colloidal nanoparticles can be excellent hydrogenation catalysts when they are activated on mesoporous supports [19]. In this way, the preparation of bimetallic nanoparticles that can be later deposited on a selected support is seen as a suitable alternative to other methods since both composition and particle size can be controlled [25, 26]. Thus, it is possible to prepare bimetallic alloy nanoparticles with a perfectly defined composition and structure [25], even though the synthesis parameters must be carefully controlled.

Hydrogenations are important industrially relevant reactions. Phenylacetylene removal by selective hydrogenation is a process of current industrial importance [27, 28] since phenylacetylene is a poisoning impurity in styrene feedstocks which causes deactivation of the styrene polymerization catalyst. This makes the decrease of phenylacetylene concentration to values close to 10 ppm mandatory. The hydrogenation of phenylacetylene proceeds at 25-50 °C and the desired product is the intermediate (styrene), which makes it very convenient for the evaluation of process design [28, 29] and for the testing of hydrogenation catalysts [30-33]. In this sense, we can find a substantial amount of literature in which both homogeneous [34-37] and heterogeneous [38-43] catalysts have been tested in batch reactors trying to reach highly selective processes towards the semi-hydrogenation of alkynes. As hydrogenations are highly exothermic reactions, an efficient heat removal is crucial for efficient performance. A microstructured reactor has a large specific interfacial area per unit of volume of ca. 5000 m^2/m^3, as opposed to only 100 m^2/m^3 for batch reactors.

Having said this, the objective of the present study is to apply a novel methodology for the development of a continuous multiphase capillary reactor that allows selective conversion of phenylacetylene to styrene. The microreactor should be operated either in slug-flow or ring flow regimes [44], which guarantees a high mass transfer rate from the gas bubble to the catalyst.

2. Experimental

Colloidal palladium nanoparticles of 4 nm in diameter were synthesized following the polyol reduction method, as described in our previous work [45]. The only substantial difference between the approach described here and that available in the literature is that in this case, the nanoparticles were dispersed in absolute ethanol instead of dry methanol. Nanoparticle-doped mesoporous titania thin films were prepared using a precursor solution with the following composition: 1 TiO_2 (Titanium tetrabutoxide (TTB)): 0.005 Pluronic F127: 40 EtOH: 1.3 H_2O: 0.13 HNO_3. The incorporation of the Pd nanoparticles was done by using the ethanol-suspended nanoparticles solution instead of absolute ethanol to obtain a final 1 wt.% Pd loading in mesoporous titania. The solution was then prepared by adding the appropriate amounts of F127, water, and TEOS, in that respective order. The bottle was then capped and the mixture stirred at room temperature for 2 hours. After aging, the solution was used to dip-coat the internal surface of a fused silica capillary (Varian Inc.) with an internal diameter of 250 μL and a length of 9 m.

Prior to dip-coating, the internal surface of the capillary was activated in a flow of 1 M NaOH aqueous solution (0.03 mL/min) for 15 minutes in order to enhance the adhesion of the coating to the channel wall. Then, the capillary was flushed for 15 minutes (0.03 mL/min) with demineralized water and the titania sol, respectively. Finally, the titania sol was withdrawn from the capillary at a rate of 1 cm/s. The capillary was dried and calcined in an oven at a residual pressure of 15 mbar. The heating rate from 25 to 300°C was 1 °C/min with a 1-hour dwelling time every 25°C below 200°C, a 2-hour dwelling time at 225, 250 and 275°C, and a 4-hour dwelling time at 300°C. A similar coating was deposited on 1 cm x 1cm flat silicon substrates to perform characterization. Low-angle X-ray diffraction spectra were recorded on a Rigaku Geigerflex device with Cu Kα radiation (40kV, 40 mA), 0.01-0.02° step size, with a counting time of 4.0 s. High-resolution Transmission Electron Microscopy studies were performed in a JEOL JEM-3011 electron microscope operating at 300 kV with a structural resolution of 0.16 nm. The samples were prepared by scratching off the silicon wafer-deposited films using a clean scalpel blade using a droplet of demineralised water as suspending fluid. A drop of the suspension was deposited on a carbon-coated copper grid, followed by drying at ambient conditions. The porous structure of the films was determined by ellipsometric porosimetry (EP) using adsorption-desorption of ethanol at 14°C. In these experiments both the effective refractive index and the film thickness were measured. The film thickness was determined from the change in the effective refractive index in the range of 300-1200 nm by spectroscopic ellipsometry.

Hydrogenation of phenylacetylene was studied in the 25-50°C temperature range in the microchannel with the Pd/TiO_2 coating. A mixture of 10 vol.% phenylacetylene in methanol and hydrogen (99.9 vol.%) were mixed in a T-mixer prior to the microchannel. The liquid flow was varied between 1 and 10 μL/min and the hydrogen flow between 250 and 550 μL/min (STP). This corresponds to superficial velocities of 0.034-0.34 cm/s and 8-18 cm/s for liquid and gas, respectively. Samples were collected at the outlet of the capillary, diluted to a 1:10 fraction in methanol and analyzed on a Varian CP-3800 gas chromatograph equipped with a CP-Sil 5CB capillary column operating at 85°C and a FID detector.

Results and discussion

Considering the small size of the nanoparticles that were prepared for this study, it is reasonable to assume that they will have a strong inclination towards agglomeration. If we then take into further account that the nanoconfined colloids were submitted to a temperature of 300°C under a residual pressure of 15 mbar, significant nanoparticle aggregation could be expected. Nevertheless, our observations indicated that due to confinement within the mesopores of the inorganic matrix, the particles remain stable. Figure 1 shows the TEM images obtained from the

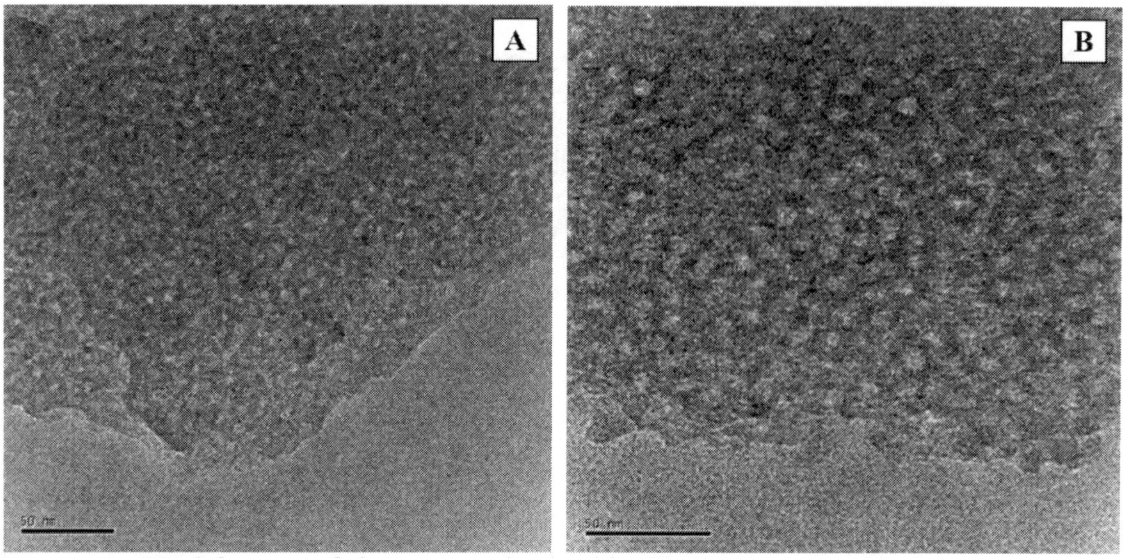

Figure 1. TEM images of the porous titania thin film. (a) Image showing some of the nanoparticles (dark spots) embedded in the inorganic matrix. (b) Image showing the porous network of the deposited thin film.

thin films collected from the flat silicon substrates, in which the porous structure and the nanoparticles are visible.

It is remarkable from the TEM images that the nanoparticles could retain their small sizes throughout the thermal treatment, which clearly indicates that this synthetic protocol is a suitable one for the incorporation of colloidal nanoparticles into porous inorganic matrixes. It should also be noted that, upon incorporation of the palladium nanocatalysts, the formation of the final mesostructure film might be compromised due to the presence of these species.

In order to confirm the formation of the desired mesostructure (which is apparent from Fig. 1B), an ethanol adsorption-desorption isotherm was obtained at 14°C using one of the non-doped thin films deposited on a silicon wafer. According to the applied models, the coating thickness was estimated to be 120 nm, and the coating porosity was 0.4, resulting in total coating mass of 1.32 mg with 1 wt.% Pd content. As Fig. 2 shows, the obtained isotherm is a typical IV isotherm, characteristic of mesoporous solids. The hysteresis loop present in the isotherm is normally indicative of large pores with slightly smaller openings, which results in the relative pressure at which desorption takes place being significantly lower than that at which

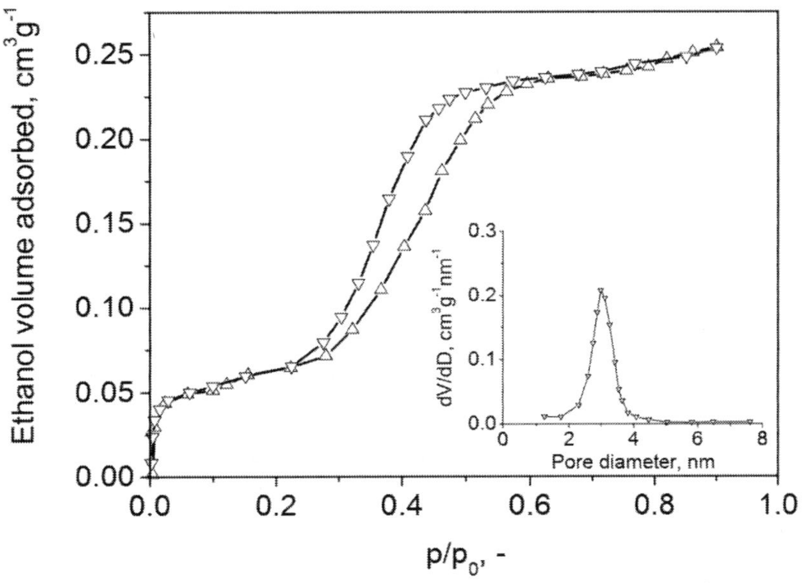

Figure 2. Ethanol adsorption-desorption isotherms at 14 °C on TiO_2 thin films and corresponding pore size distribution.

filling of the mesopores takes place. In the case of mesoporous materials, this behavior might be associated with cubic mesoporous networks [46]. This result shows that although the mesostructure is not completely developed, the sharp ethanol uptake at $P/P_0 \approx 0.4$ clearly indicates the presence of pores with a narrow size distribution within the mesopore region. Thus, the methodology that has been described here can produce thin films with the desired mesostructure, which can in turn host the small nanoparticles inside its matrix. These systems may be readily used for a variety of reactions, thus showing the great versatility of the prepared composites.

The microreactor was then tested in the hydrogenation of phenylacetylene under different conditions of liquid flow and temperature in order to assess which conditions would be the optimal ones for the successful conversion of phenylacetylene to styrene in a continuous reactor. Figure 3 shows the results obtained under different liquid flow rates of 3-11 μL/min, using a constant hydrogen flow of 550 μL/min corresponding to a Taylor flow regime. It can be seen from Fig. 3 that as the residence time is decreased, conversion of phenylacetylene is significantly reduced. This is especially notorious in the case of the reaction being performed at 30°C, where the conversion drops from almost 100% to 20% as the liquid flow rate is increased from 3 to 6 μL/min. Another important observation is that selectivity increases with the liquid flow rate.

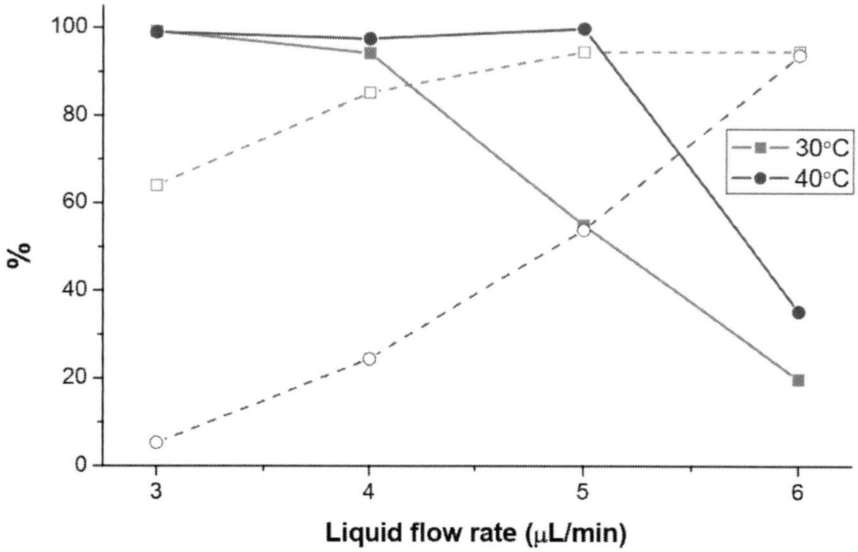

Figure 3. Conversion (solid line and full symbols) and selectivity (dashed line and hollow symbols) plot versus flow at different temperatures for the capillary reactor. The hydrogen flow rate was kept constant at 550 μL/min (STP). The liquid consists of a mixture of 10 vol.% of phenylacetylene in methanol.

As previous results have shown, the Pd-catalyzed hydrogenation of phenylacetylene proceeds until almost complete depletion of the phenylacetylene from the reaction vessel, even when styrene is present from the beginning of the reaction [47]. However, in a microreactor, in the absence of mass transfer limitations, hydrogenation keeps occurring, leading to the formation of the fully hydrogenated product, ethylbenzene. This is also evidenced as we increase the temperature in the capillary (see Fig. 3) which causes the conversion to increase, but also yielding lower selectivity values at low liquid flow rates. Full conversion of phenylacetylene was observed at liquid flow rates below 3 μL/min in the studied range of temperatures. Table 1 shows the results obtained using the capillary reactor. Table 1 clearly shows that it is possible to control both the conversion and selectivity of the process by careful control of the flow and temperature conditions on the microreactor. By doing this, it is possible to obtain phenylacetylene conversion values close to 94% with selectivities towards styrene over 85%.

Table 1. Performance of the capillary microreactor in hydrogenation of phenylacetylene under different conditions.

Liquid flow rate (μL/min)	Gas flow rate (μL/min)	Temperature (°C)	Conversion (%)	Selectivity to styrene (%)
3	550	30	99.2	63.9
3	550	40	98.9	5.3
3	550	50	99.8	1.6
4	550	30	94.2	85.2
4	550	40	99.8	53.8
4	550	50	100	44.8
5	550	30	40.3	94.4
5	550	40	54.9	94.5
5	550	50	97.5	24.4
6	550	30	19.8	94.6
6	550	40	35.2	93.8
7	550	50	73.2	88.3
11	900	50	64.6	93.3

These performance values are very promising and quite close to those found for the same reaction performed with Pd nanoparticles under homogeneous conditions [47]. It must be noted that under fast flow conditions (11 μL/min liquid flow rate, 900 μL/min hydrogen flow rate), it was possible to obtain conversion values over 60%, with very high selectivity. It is remarkable that at these fast flow conditions it is possible to obtain TOF values of over 2 s^{-1}, which are significantly higher than those reported in the literature, even for homogeneous reaction conditions. Thus, these novel systems clearly show the numerous practical possibilities that may arise from the efficient integration of thin film technology with colloidal chemistry to produce mesoporous matrixes that can host a variety of mono- and bimetallic nanoparticles. Considering the range of nanoparticle compositions and metallic ratios that can be derived from the polyol reduction method, the versatility of such composites should be regarded as highly promising in the development of state-of-the-art microreactors.

Conclusions

A novel approach for catalyst deposition onto the walls of microchannels is proposed in which air-stable metal nanoparticles are directly incorporated into the mesoporous inorganic matrix of a thin film delivered on the internal surface of a microreactor. The nanoparticles retain their small particle size even after calcination, and are accessible for reaction thereafter. Hydrogenation of phenylacetylene in the microreactor showed that the reaction occurs under kinetically controlled conditions. As the liquid flow rate increases, the conversion decreases with simultaneous increase of selectivity towards styrene. The maximum selectivity to styrene of almost 95% was observed at a high liquid flow rate of 6 μL/min. The reaction rate in terms of TOF was found to be up to 2 s^{-1}.

Acknowledgements

The financial support of the Netherlands Organization for Scientific Research (NWO) and British Council (British Council-NWO Partnership Programme in Science, projects PPS 888 and 894) is gratefully acknowledged.

References

[1] E.V. Rebrov, S.A. Kuznetsov, M.H.J.M. de Croon, J.C. Schouten, Catal. Today 125 (2007) 88.

[2] I.Z. Ismagilov, R.P. Ekatpure, L.T. Tsykoza, E.V. Matus, E.V. Rebrov, M.H.J.M. de Croon, M.A. Kerzhentsev, J.C. Schouten, Catal. Today 105(3-4) (2005) 516.

[3] S.A. Kuznetsov, A.R. Dubrovskiy, E.V. Rebrov, J.C. Schouten, Z. Naturforsch. 62a (2007) 647.

[4] E.V. Rebrov, G.B.F. Seijger, H.P.A. Calis, M.H.J.M. de Croon, C.M. van den Bleek, J.C. Schouten, Appl. Catal. A 206 (2001)125.

[5] M.J.M. Mies, E.V. Rebrov, J.C. Jansen, M.H.J.M. de Croon, J.C. Schouten, J. Catal., 247(2) (2007) 328.

[6] C. Dezelah, L. Niinistö, E.V. Rebrov, M.J.F. Warnier, M.H.J.M. de Croon, J.C. Schouten, Synthesis and characterization of ALD-Pd catalyst for application in multiphase microstructured reactors, in Proc. EuropaCat VIII, Turku, Finland, P2-81 (2007).

[7] O. Muraza, E.V. Rebrov, T. Khimyak, B.F.G. Johnson, P.J. Kooyman, U. Lafont, M.H.J.M. de Croon, J.C. Schouten, Chem. Eng. J. 135 (2008) S99.

[8] P. Pfeifer, K. Schubert, M. A. Liauw, G. Emig, Appl. Catal. A. 270 (2004) 165.

[9] P. Reuse, A. Renken, K. Haas-Santo, O. Görke, K. Schubert, Chem. Eng. J. 101 (2004) 133.

[10] J. Bravo, A. Karim, T. Conant, G.P. Lopez and A. Datye, Chem. Eng. J. 101 (2004) 113.

[11] G. M. Greenway, S. J. Haswell, D. O. Morgan, V. Skelton, P. Styring, Sens. Actuators B 63, (2000) 153.

[12] X. Ouyang, R. S. Besser, Catal. Today 84 (2003) 33.

[13] P.D. Christensen, S.W.P. Johnson, T. McCreedy, V. Skelton, N.G. Wilson,. Anal. Commun. 35 (1998) 341.

[14] J. Kobayashi, Y. Mori, K. Okamoto, R. Akiyama, M. Ueno, T. Kitamori, S. Kobayashi, Science 304 (2004) 1305.

[15] J. Haber, J.H. Block, B. Delmon, Pure&Appl. Chem. 67 (1995) 1257.

[16] T.S. Glazneva, E.V. Rebrov, J.C. Schouten, E.A. Paukshtis, Z.R. Ismagilov, Thin Solid Films 515 (2007) 6391.

[17] J.H. Sinfelt, "Bimetallic Catalysts, An Exxon Monograph", Wiley, New York, 1983.

[18] H. Kuroda, Y. Iwasawa, Intl. Rev. Phys. Chem. 8 (1989) 207.

[19] (a) B.F.G. Johnson, Topics in Catalysis, 24(1-4) (2003) 147; (b) Robert Raja, Vladimir B. Golovko, John M. Thomas, Ángel Berenguer-Murcia, Songhai Xie, Wuzong Zhou, Brian F.G. Johnson, Chem. Comm. 15 (2005) 2026.

[20] R.H. Baughman, A.A. Zakhidov, W.A. de Heer, Science 297 (2002) 787.

[21] Y.S. Choi, Y.S. Cho, J.H. Kang, Y.J. Kim, I.H. Kim, S.H. Park, H.W. Lee, S.Y. Hwang, S.J. Lee, C.G. Lee, T.S. Oh, J.S. Choi, S.K. Kang, J.M. Kim, Appl. Phys. Lett. 82 (2003) 3565.

[22] C. Niu, E.K. Sichel, R. Hoch, D. Moy, H. Tennent, Appl. Phys. Lett. 70 (1997) 1480.

[23] B. Kleinsorge, V.B. Golovko, S. Hofmann, J. Geng, D. Jefferson, J. Robertson, B.F.G. Johnson, Chem. Comm. (2004) 1416.

[24] M.C. Román-Martínez, D. Cazorla-Amorós, S. de Miguel, O. Scelza, J. Japan Petroleum Institute 47(3) (2004) 164.

[25] R W. Wunder, J. Philips, J. Phys. Chem. 100 (1996) 14430.

[26] H. Bönnemann, R.M. Richards, Eur. J. Inorg. Chem. (2001) 2455.

[27] B.R. Maurer, M. Galobardes, U.S. Patent 4,822,936 (1989).

[28] T. Vergunst, F. Kapteijn, J.A. Moulijn, Ind. Eng. Chem. Res. 40 (2001) 2801.

[29] B.A. Wilhite, R. Wu, X. Huang, M.J. McCready, A. Varma, AIChe J., 47 (2001) 2548.

[30] D. Duca, L.F. Liotta, G. Deganello, Catal. Today 24 (1995) 15.
[31] M.A. Aramendia, V. Borau, C. Jimenez, J.M. Marinas, M.E. Sempere, F.J. Urbano, Appl. Catal. 63 (1990) 375.
[32] G. Carturan, G. Cocco, G. Facchin, G. Navazio, J. Mol. Catal. 26 (1984) 375
[33] X. Huang, B. Wilhite, M.J. McCready, A. Varma, Chem. Eng. Sci.58 (2003) 3465
[34] M.M. Dell´Anna, M. Gagliardi, P. Mastrorilli, G.P. Suranna, C.F. Nobile, J. Mol. Catal. A: Chemical 158 (2000) 515.
[35] M. W. van Laren, C. J. Elsevier, Angew. Chem. Int. Ed. 38 (1999) 3715.
[36] M. Costa, P. Pelagatti, C. Pelizzi, D. Rogolino, J. Mol. Catal. A: Chemical 178 (2002), 21.
[37] L.B. Belykh, T.V. Goremyka, L.N. Belonogova, F.K. Schmidt, J. Mol. Catal. A: Chemical 231 (2005) 53.
[38] S. Mandal, D. Roy, R.V. Chaudhari, M. Sastry, Chem. Mater. 16(19) (2004) 3714.
[39] L. Guczi, Z. Schay, G. Stefler, L.F. Liotta, G. Deganello, A.M. Venezia, J. Catal. 182(2) (1999) 456.
[40] A. Mastalir, Z. Király, J. Catal. 220(2) (2003) 372.
[41] S.U. Son, K.H. Park, Y.K. Chung, Org. Lett. 4(22) (2002) 3983.
[42] J.-L. Pellegatta, C. Blandy, V. Colliere, R. Choukroun, B. Chaudret, P. Cheng, K. Philippot, J. Mol. Catal. A: Chemical 178(1-2) (2002) 55.
[43] D.S. Shephard, T. Maschmeyer, G. Sankar, J.M. Thomas, D. Ozkaya, B.F.G. Johnson, R. Raja, R.D. Oldroyd, R.G. Bell, Chem.-A European Journal 4(7) (1998) 1214.
[44] M.J.F. Warnier, E.V. Rebrov, M.H.J.M. de Croon, V. Hessel, J.C. Schouten, Chem. Eng. J., 135 (2008) S153.
[45] S.Dominguez-Dominguez, A. Berenguer-Murcia, D. Cazorla-Amoros, A. Linares-Solano, J. Catal. 243 (2006) 74.
[46] K. Wang, M.A. Morris, J.D. Holmes, Chem. Mater. 17 (2005) 1269.
[47] S. Domínguez-Domínguez, A. Berenguer-Murcia, B.K. Pradhan, D. Cazorla-Amorós, Á. Linares-Solano, J. Phys. Chem. C, accepted.

AUTHOR INDEX

AUTHOR INDEX

AUTHOR INDEX

AUTHOR INDEX

AUTHOR INDEX

AUTHOR INDEX

AUTHOR INDEX